NORTH AMERICA'S
AMAZING MAMMALS

Outdoorsman, wildlife enthusiast, and nature photographer and videographer Lochlainn Seabrook is a popular award winning writer and historian, and the author and editor of nearly 100 books on topics ranging from science and nature to history and religion. He does not pen books for fame and fortune, but for the love of writing and sharing his knowledge with others.

NORTH AMERICA'S AMAZING MAMMALS

An Encyclopedia for the Whole Family

BY AWARD-WINNING WRITER-NATURALIST AND HISTORIAN

LOCHLAINN SEABROOK

Diligently Researched and Generously
Illustrated for the Elucidation of the Reader

2020

Sea Raven Press, Park County, Wyoming, USA

NORTH AMERICA'S AMAZING MAMMALS

Published by
Sea Raven Press, Cassidy Ravensdale, President
Park County, Wyoming, USA
SeaRavenPress.com

SEA RAVEN PRESS
SCIENCE BOOKS FOR NATURE LOVERS!

1st SRP paperback edition, 1st printing, April 2020; 2nd printing, June 2023 • ISBN: 978-1-943737-77-2
1st SRP hardcover edition, 1st printing, April 2020; 2nd printing, June 2023 • ISBN: 978-1-943737-78-9

ISBN: 978-1-943737-77-2 (paperback)
Library of Congress Control Number: 2020933443

North America's Amazing Mammals: An Encyclopedia for the Whole Family, by Lochlainn Seabrook. Includes illustrations, an index, appendices, and bibliographical references.

Front and back cover design and art, book design, layout, and interior art by Lochlainn Seabrook
All images, graphic design, graphic art, and illustrations copyright © Lochlainn Seabrook
All images selected, placed, manipulated, and/or created by Lochlainn Seabrook
Cover art: "Grizzly Bear in the Rocky Mountains," by Larry Jacobsen

WRITTEN, PUBLISHED, PRINTED AND MANUFACTURED IN THE UNITED STATES OF AMERICA

SEA RAVEN PRESS

Dedication

To the preservation of North America's remaining wild spaces and the 458 currently known mammal species that inhabit them.

Epigraph

"WE AND THE BEASTS ARE KIN."

Ernest Thompson Seton, 1898

CONTENTS

Warning: SEA RAVEN PRESS BOOKS WILL EXPAND YOUR ★ MIND!

NOTES TO THE READER

SCIENTIFIC CLASSIFICATION
☛ In most cases I have adhered to the common names and scientific names of my subjects based on those that were accepted at the time of publishing (early 2020). These in turn are based on the morphological and genetic studies that were then current. Sometimes, however, I have chosen to use what I consider the most logical taxonomy, and so my classifications may not always coincide with those of other authorities.

As taxonomy (based primarily on comparative anatomy, genetics, physiology, and serology) is an always expanding, ever changing science—with much that is still unstable, confusing, complex, and even unknown—classification is often more theory than fact, with best guesses, educated inferences, and generalized assumptions sometimes necessarily taking the place of hard science. It is not surprising to realize then that as more is learned about our mammals, many of the cladistic tables and taxonomic charts we take for granted today will be thrown out and redrawn at some point in the future. Indeed, many are being revised by systematists as you read this, and—just as mine differs from past mammal guides because of this—future mammal guides will differ from mine.

American beaver.

While taxonomical debates are ongoing, however we choose to classify the creatures themselves, the enthralling beauty and astonishing traits possessed by our amazing North American mammals will not change.

CATEGORIZATION
☛ Many natural history guidebooks and references categorize their chapters by species, genus, or family. Due to the nature of this work, however, I have eliminated chapters completely and instead list my mammalian subjects alphabetically, in the traditional style of encyclopediae.

WEIGHTS, MEASUREMENTS, & TEMPERATURE
☛ All measurements in the entries are for adult males and for the upper extreme of the scale, unless otherwise indicated—as, for example, in those rare instances when females are larger than males.

The total body length given always includes the tail—unless otherwise indicated. Since we cannot know if we have obtained either the smallest or the largest specimen of a given mammal species, in many cases the measurements provided here must be regarded as general—the results of often limited study.

While this is indeed a zoological work, it is directed at the general public. Thus I have done away with the scientific custom of using the metric system in favor of the simpler U.S. Customary System of Weights and Measurements. For the same reason I also use the Fahrenheit Scale rather than Centigrade Scale.

NEW WORDS
☛ In some instances I have found it necessary to coin new words and terms in order to describe my subjects properly. If you discover a word you are unfamiliar with, and it is not found in the dictionary, you have probably come across one of my linguistic inventions. For convenience I have listed these in the Glossary whenever possible.

DEFINING DIET TYPES
☛ For simplicity's sake I define an herbivore as an organism that eats *only* plants; I define an omnivore as an organism that eats *both* plants and animals; I define a carnivore

as an organism that eats *only* animals. These differ from some sources that, for instance, define an herbivore as an organism that eats *mostly* plants and occasionally meat, and a carnivore as one that eats *mostly* meat and occasionally plants. Under my more strict rule both of these animals would be considered omnivores. On the other hand, my definitions for subdiets, like frugivory (fruit-eating), granivory (seed-eating), and lignivory (wood-eating), are more flexible. Thus I define an avivore (bird-eater) as an organism whose diet is comprised partly, primarily, or solely of birds.

PREDATORS
☛ I do not generally include humans under "predators" since it is taken for granted that we intentionally hunt and trap most mammal species in North America. For the purpose of clarity I therefore usually list only *natural* predators; that is, undomesticated carnivores that a mammal encounters in the wild.

SUBSPECIES
☛ In keeping with tradition, I have simplified the text by sticking solely to accepted individual recognized species; that is, I have left out what I consider to be subspecies. If I had included them, this book would have been much larger and more complex, and, as a result, much less accessible to the public.

NUMBERS
☛ In the mammal entry section I adhere to one of the encyclopedist's great space-saving devices by writing numbers as numerals instead of words.

EASE OF USE
☛ To make this book as family-friendly, compact, easy to read, and enjoyable as possible for the layperson, for the most part I have excluded such standard details as footprint and track descriptions, dental formulae, skeletal measurements, scat characteristics, chromosome specifications, taxonomic discussions, and graphic mating behaviors, and in many instances, economic importance and conservation statuses, as well. For those seeking more information on these particular subjects please see my bibliography.

Bison skull.

THE ALL-IMPORTANT GLOSSARY
☛ Note that in spite of the family-oriented manner in which this work is presented, in many entries I intentionally use "big words" (scientific language) to encourage my readers (of all ages) to utilize the Glossary. Here, he or she will discover a fascinating, indispensable, and highly informative introduction to the world of natural history, one that will further one's appreciation for our North American mammals.

TO THOSE WHO CAME BEFORE
☛ I am indebted to those stalwart, pre-20th-Century naturalists who came before me; who bravely explored the uncharted wilds of North America; who carefully and scientifically cataloged the myriad of new mammals they discovered; and who arduously authored hundreds of volumes—many of whose findings and illustrations I was able to bring back to life in *North America's Amazing Mammals*.

FOR CURRENT INFORMATION REGARDING MAMMALS
☛ As a nonaffiliated general reference intended to both educate and entertain, this book should not be considered as representing any specific group, organization, or government. For up-to-date information regarding North American mammals (such as scientific data, hunting requirements, wildlife statutes, conservation rules, state laws, etc.), the reader is advised to contact the proper authorities governing his or her area.

MAMMAL CLASSIFICATION
How Mammals Are Scientifically Grouped & Arranged

TAXONOMIC HIERARCHY OF MAMMALS
KINGDOM: Animalia ("animals")
SUBKINGDOM: Bilateria ("bilateral" [symmetry])
INFRAKINGDOM: Deuterostomia ("second -mouth")
PHYLUM: Chordata ("chordates")
SUBPHYLUM: Vertebrata ("vertebrates")
INFRAPHYLUM: Gnathostomata ("jaw-mouth")
SUPERCLASS: Tetrapoda ("four-footed")
CLASS: Mammalia ("mammals")
SUBCLASS: (group)
INFRACLASS: (group)
ORDER: (group)
FAMILY: (group)
SUBFAMILY: (group)
GENUS: (group)
SPECIES: (individual)
SUBSPECIES: (individual)

EXAMPLE: THE AMERICAN BISON
KINGDOM: Animalia
SUBKINGDOM: Bilateria
INFRAKINGDOM: Deuterostomia
PHYLUM: Chordata
SUBPHYLUM: Vertebrata
INFRAPHYLUM: Gnathostomata
SUPERCLASS: Tetrapoda
CLASS: Mammalia
SUBCLASS: Theria (placental and marsupial mammals)
INFRACLASS: Eutheria (placental mammals)
ORDER: Artiodactyla (even-toed ungulates)
FAMILY: Bovidae (bovids: cattle, antelopes, goats, sheep)
SUBFAMILY: Bovinae (bison, cattle, oxen, buffalo, kudu)
GENUS: *Bison* (from the Latin word *bison*, "wild ox")
SPECIES: *bison* (American bison)
SUBSPECIES: plains bison (*Bison bison bison*), wood bison (*Bison bison athabascae*)

THE TREE OF LIFE

Where Mammals Fit In

The position of mammals on the evolutionary Tree of Life.

INTRODUCTION
The Importance of North America's Mammals

THERE ARE FEW THINGS MORE thrilling than seeing a wild mammal in its natural habitat. What is not often considered is that having some basic knowledge of our mammalian cousins adds interest and excitement to the experience. This brings us to the function of this book.

A long-eared bat species.

It is one thing to see a porcupine waddling through the woods, but it is quite another to know that it cannot "throw" any of its 30,000 quills at you. It is one thing to spot a pronghorn antelope grazing lazily on the prairie, but it is quite something else to know that it is related to the giraffe and is the fastest land animal in the Western Hemisphere. It is one thing to observe the magnificent sperm whale swimming gracefully out in the open ocean, but it is quite another to know that it can dive nearly two miles down and hold its breath for as long as 1½ hrs.

Facts like these are certainly fun. But they are much more than answers to board game trivia questions. Having a basic understanding of a particular animal brings meaning and enjoyment to our interaction with it. Knowing a creature's length, weight, diet, predators, range, and special traits, as just a few examples, are actually vital if we are to gain a fuller understanding of a species. And it is this knowledge, combined with the exhilaration of the visual experience, that is so pleasurable to both the novice animal-watcher and the professional field biologist.

OUR AMAZING MAMMALIAN COUSINS
Currently there are 5,416 different species of mammals that have been cataloged worldwide (this list grows yearly). Here in North America we share our land and waters with an estimated 458 of these. All, from the tiny pygmy shrew (which weighs less than a dime) to the gigantic blue whale (which is one-third the length of a football field), are amazing in their own individual ways, hence the title of this work.

Though this book can certainly be used in the field (while bearing in mind that identification must be achieved using written descriptions rather than color photos and illustrations), it is not meant to be a *complete scientific* field guide to

American beaver lodge and dam.

Northern American mammals, nor does it chronicle every one of our 458 species. Rather, based on a lifetime of working with and around mammals, it is a general reference to 378 of what I consider our most important, interesting, and well-known mammals.

MAMMALS PROVIDE NUMEROUS BENEFITS
Now a word on mammals themselves. *All* mammals are, in one way or another, priceless and should be protected and preserved to the best of our ability through wise conservation. Though some may be rightly regarded as "destructive pests" (such as our invasive wild boar), it is debatable if even these animals merit extinction. What they *do*

Ringtail.

deserve is judicious management. As humans this is our solemn responsibility, for due to our intelligence, widespread population, worldwide influence, and powerful technologies, we are the natural stewards of both the earth and its many inhabitants—including our fellow mammalian relations.

While I am quite aware of the negative impact many mammals can have on the environment as well as on human society (this damage is well-known to farmers, ranchers, hunters, biologists, zoologists, and wildlife conservationists), I believe this fact should be balanced with a knowledge of how they profit not only us, but the global ecosystem as well, for these benefits are numerous and invaluable, but little discussed, little understood, and vastly underappreciated:

• Predatory mammals (secondary consumers) help keep prey species strong and healthy by culling the weak, sick, and aged.
• Prey mammals (primary consumers) help keep predatory species strong and healthy by supplying them with nourishment.
• Mammals eat and control the population of creatures we consider pests—such as insects, mice, and rats—which spread disease, consume crops, and destroy property. Fishers, for instance, prey on porcupines, a mammal that debarks and kills trees, adversely affecting timber production.
• As ecosystem engineers, their rooting, digging, and burrowing helps stimulate plant growth by mixing the soil and softening the ground, which allows roots, water, and oxygen to more easily penetrate the surface. Thus, by positively influencing microtopography, mammals both increase plant diversity and encourage habitat heterogeneity.

Gray wolf.

• Mammals help maintain the equilibrium of the world ecosystem by checking the populations of creatures that would otherwise overfeed on terrestrial and aquatic plants—which consume carbon dioxide (damaging to human life) while producing oxygen (necessary for human life).
• Mammals disperse plant seeds and fungi spores through their dung, assisting in the distribution of vegetation while aiding in the health and succession of various ecosystems, such as savannas, taiga, and chaparral. In turn, the new vegetation they help produce enhances and enriches soil quality, boasting the overall ecological well-being of their habitats. In this way they play a significant role in forest nutrient cycles, such as facilitating the establishment of tree seedlings and reforestation after a habitat has been disturbed from, for example, fire, a volcano, or an avalanche. Squirrels and deer are just two of our better known seed dispersers.
• Mammal dung feeds trillions of microbial decomposers that help expedite the delivery of water and various elements to plant life (thereby *conserving* ecosystems), that aid in regulating the climate (thereby *stabilizing* ecosystems), and that assist in transforming vital nutrients, such as oxygen, phosphorous, sulfur, carbon, and nitrogen, into more readily available forms for consumption by higher animals, including humans (thereby *enhancing* ecosystems).
• By aiding in the breakdown of plants and animal carcasses, mammals play a large and important role in the biodegradation process—essential to the health and balance of the world ecosystem.
• Mammal dung from bats and seals, as just two examples, contains copious amounts of

vital plant nutrients that feed and sustain vegetation, which in turn feed and sustain insects, fish, amphibians, reptiles, birds, and, of course, other mammals.

- Food-gathering mammals leave behind tons of unconsumed and unused biomass every year. Pikas, for instance, form small haystacks of vegetation and dung, which, when abandoned, act as both fertilizer for the soil and food for other animals (from microbes to large mammals). Thus food-gathering mammals increase the biodiversity of a habitat.

- By consuming invertebrates considered detrimental to the environment, mammals help reduce levels of atmospheric carbon dioxide, or CO_2. The sea otter, for example, feeds on sea urchins, whose natural predator, the starfish, is disappearing along the Pacific Coast (due to disease). The resulting overpopulating sea urchin, in turn, has begun overgrazing on sea kelp, an all-important plant in coastal communities—for it captures and converts carbon dioxide to oxygen during photosynthesis, while providing habitat and food for abalone and other human-valued marine life. By checking the burgeoning urchin population these playful mustelids contribute to the biological balance of kelp ecosystems.

Jaguar being attacked by a boa constrictor.

- The burrows and tunnels of mammals not only loosen and aerate the soil (helping prevent anoxic and hypoxic ground conditions), they also recycle soil nutrients while providing gravitational water channels, all which facilitates drainage in soils with poor water absorption properties. By shaping both macrotopographical and microtopographical features, these biopedturbational activities help filtrate water, augment soil fertility (by conveying nutrient-rich material down to the subsoil), push subsoil up to the surface (after which it is turned into topsoil, forming new soil), lower soil compaction, stimulate the growth of new vegetation, and provide microhabitat housing for thousands of other types of animals, from insects and reptiles to birds and other mammals. Additionally, the dung mammals leave behind in subterranean chambers helps fertilize the soil. Even digging and rooting (by mammals such as wild boar and hog-nosed skunks) can have mutualistic species benefits, for this activity creates microhabitats for various types of invertebrates. The same holds true for marine mammals. While bottom-feeding, for instance, the gray whale excavates large holes in the mud that provide convenient homes for a myriad of benthic organisms. Additionally, the silt it churns up brings a wide variety of invertebrates to the surface, supplying food for both fish and sea birds. In these ways, and many others, mammals are vital in helping maintain the functionality and biodiversity of ecosystems.

- Mammals, such as pronghorn, eat plants that are considered harmful to the environment. This includes invasive vegetation that competes with native plants.

- Predatory mammals keep prey mammals, like deer, from remaining in one place for too long, preventing the damaging effects of overgrazing: loss of habitat, loss of wild and domestic animals, erosion, desertification, and pollution of soil and water.

Moose (male).

- Conversely, the grazing of mammals can also be helpful. For example, rabbit grazing—which is known to actually promote the growth and spread of certain types of grasses—is indispensable in protecting a number of ecosystems, such as

threatened calcareous grassland, which is home to many insect species.
• The living bodies of mammals provide homes and food for countless parasites, both internal (endoparasites) and external (ectoparasites).
• When mammals die, their decomposing bodies add important nutrients back into the soil, where it is utilized by both flora and fauna.
• Mammals aid humans by providing us with revenue and recreation (for example, ecotourism, recreation, sports, birdwatching, hunting, fishing, trapping), and by supplying us with food (from their meat), assorted medical and utilitarian products (from their bones and organs), and clothing (from their skins), all which make our lives easier and more enjoyable while strengthening the economy.

Common mole.

These, along with their numerous other contributions, make mammals an integral link in both the food chain and the world ecosystem. More to the point, our planet would be much different without them: we ourselves would not exist.

As we are mammals too, we of course have many of the same body parts, traits, and habits as our wild and furrier cousins. Thus studying nonhuman mammals can teach us much about ourselves as well. Take the bat, for instance. One of its outstanding characteristics, and one by which it can often be identified, is its tragus, an auditory prominence that comes in many shapes and sizes, some quite extraordinary. This backward-facing flap of skin is located in front of the external opening of each ear. A zealous insectivore, the bat uses its two tragi to detect the nearly inaudible sounds of insects that may be lurking behind it—one of Nature's ingenious adaptations to a hunting style in which 360° hearing is all-important. How interesting to realize that we humans also have tragi (one on each ear), and that we use them for the same purpose as bats (that is, to pick up sounds coming from behind us).

These are just a few examples of how mammals not only aid in maintaining a balanced and healthy ecosphere, but just as importantly, help enrich the human experience through knowledge of their ecologies and life histories. They are, in short, an intrinsic part of our world, each performing a vital role in the Circle of Life.

A PRECIOUS LIVING HERITAGE

North America would not be North America without her nonhuman mammals, and it is for them, both massive and small, imposing and inconspicuous, beautiful and homely, appealing and plain, fearsome and non-threatening, aggressive and docile, that I wrote this book. It is my hope that in perusing and studying these pages, my readers will become better informed and sensitized to the unique mammalian world in which each species lives; to their stupendous powers, adaptability, and ruggedness; to the difficulties and complications they face each day—many of these created not only by Mother Nature, but by we ourselves.

Common skunk.

If you are not a mammalophile before you read this book, you will most likely be one by the time you finish it! In fact, those who peruse the following pages with an open mind and heart will develop a new level of awe, fascination, respect, and reverence for them. In that spirit, let us all, each and every one, do our part, in whatever way we can, to conserve and preserve this precious living heritage: *North America's Amazing Mammals*.

LOCHLAINN SEABROOK
Jackson Hole, Wyoming, USA
April 2020

MAMMALIAN HABITATS

Common Biomes & Ecosystems Where Mammals Are Found

Mixed forest community.

Ericaceous-muskeg bog community.

Black spruce-tamarack community.

Pine community.

Open disturbed community.

Cattail-sedge marsh community.

Woodland pond community.

Urban farm community.

Large lake community.

Alder-willow community.

Mature deciduous stand.

Young deciduous community.

Young coniferous community.

White cedar community.

How To Use This Book

THIS IS NOT A FIELD guide in the traditional sense, and it is not meant to be. It is an encyclopedia, with every mammal listed alphabetically by its common name, and, when possible, accompanied by a black and white illustration. Thus, unlike a standard field guide (which, for example, aids the reader with silhouettes, distribution maps, individual color illustrations, and photos of tracks, scat, and subspecies), when using *North America's Amazing Mammals*, one must first know the common name of the mammal he or she is interested in studying. Thus, if you have the common name, you can begin right away by looking up your subject alphabetically.

If you do not have the common name of the subject you are seeking, but only its scientific name, or merely one of its colloquial or regional names, simply turn to the Index, where I have listed the accepted common name and scientific name, as well as the alternate names and nicknames, of each mammal.

To assist the reader further, below are the 13 categories of the 378 mammals I have included, along with an alphabetized list of the individual mammal species in each category—by custom, common name first, scientific name second:

1. ARTIODACTYLA (even-toed hoofed mammals): American bison (*Bison bison*), Barbary sheep (*Ammotragus lervia*), bighorn sheep (*Ovis canadensis*), caribou (*Rangifer tarandus*), collared peccary (*Pecari tajacu*), Dall's sheep (*Ovis dalli*), elk (*Cervus elaphus*), fallow deer (*Dama dama*), moose (*Alces americanus*), mountain goat (*Oreamnos americanus*), mule deer (*Odocoileus hemionus*), muskox (*Ovibos moschatus*), pronghorn (*Antilocapra americana*), white-tailed deer (*Odocoileus virginianus*), wild boar (*Sus scrofa*).

Elk.

2. CARNIVORA (carnivores): American badger (*Taxidea taxus*), American black bear (*Ursus americanus*), American hog-nosed skunk (*Conepatus leuconotus*), American marten (*Martes americana*), American mink (*Neovison vison*), arctic fox (*Vulpes lagopus*), black-footed ferret (*Mustela nigripes*), bobcat (*Lynx rufus*), brown bear (*Ursus arctos*), Canada lynx (*Lynx canadensis*), coyote (*Canis latrans*), eastern spotted skunk (*Spilogale putorius*), ermine (*Mustela erminea*), fisher (*Pekania pennanti*), gray fox (*Urocyon cinereoargenteus*), gray wolf (*Canis lupus*), hooded skunk (*Mephitis macroura*), jaguar (*Panthera onca*), jaguarundi (*Puma yagouaroundi*), kit fox (*Vulpes macrotis*), least weasel (*Mustela nivalis*), long-tailed weasel (*Mustela frenata*), margay (*Leopardus wiedii*), mountain lion (*Puma concolor*), North American river otter (*Lontra canadensis*), ocelot (*Leopardus pardalis*), polar bear (*Ursus maritimus*), raccoon (*Procyon lotor*), red fox (*Vulpes vulpes*), red wolf (*Canis rufus*), ringtail (*Bassariscus astutus*), sea otter (*Enhydra lutris*), striped skunk (*Mephitis mephitis*), swift fox (*Vulpes velox*), western spotted skunk (*Spilogale gracilis*), white-nosed coati (*Nasua narica*), wolverine (*Gulo gulo*).

Wolves.

3. CETACEA (dolphins, porpoises, whales): Atlantic spotted dolphin (*Stenella frontalis*), Atlantic white-sided dolphin (*Lagenorhynchus acutus*), beluga (*Delphinapterus leucas*), Blainville's beaked whale (*Mesoplodon densirostris*), blue whale (*Balaenoptera musculus*), bowhead whale (*Balaena mysticetus*), common bottlenose dolphin (*Tursiops truncatus*), common minke whale (*Balaenoptera acutorostrata*), Cuvier's beaked whale (*Ziphius cavirostris*), dwarf sperm whale (*Kogia sima*), false killer whale (*Pseudorca crassidens*), fin whale (*Balaenoptera physalus*), Gervais' beaked whale (*Mesoplodon europaeus*), gray whale (*Eschrichtius robustus*), harbor porpoise (*Phocoena phocoena*), humpback whale (*Megaptera novaeangliae*), long-finned pilot whale (*Globicephala melas*), narwhal (*Monodon monoceros*), Northern bottlenose whale (*Hyperoodon ampullatus*), North Atlantic right whale (*Eubalaena glacialis*), orca (*Orcinus*

Bowhead whale.

orca), pygmy sperm whale (*Kogia breviceps*), Risso's dolphin (*Grampus griseus*), rough-toothed dolphin (*Steno bredanensis*), sei whale (*Balaenoptera borealis*), short-beaked common dolphin (*Delphinus delphis*), short-finned pilot whale (*Globicephala macrorhynchus*), sperm whale (*Physeter macrocephalus*), striped dolphin (*Stenella coeruleoalba*), True's beaked whale (*Mesoplodon mirus*), white-beaked dolphin (*Lagenorhynchus albirostris*).

4. CHIROPTERA (bats): Allen's big-eared bat (*Idionycteris phyllotis*), big brown bat (*Eptesicus fuscus*), big free-tailed bat (*Nyctinomops macrotis*), brazilian free-tailed bat (*Tadarida brasiliensis*), California leaf-nosed bat (*Macrotus californicus*), California myotis (*Myotis californicus*), cave myotis (*Myotis velifer*), eastern pipistrelle (*Pipistrellus subflavus*), eastern small-footed myotis (*Myotis leibii*), evening bat (*Nycticeius humeralis*), fringed myotis (*Myotis thysanodes*), ghost-faced bat (*Mormoops megalophylla*), gray myotis (*Myotis grisescens*), hairy-legged vampire (*Diphylla ecaudata*), hoary bat (*Lasiurus cinereus*), Indiana myotis (*Myotis sodalis*), Keen's myotis (*Myotis keenii*), lesser long-nosed bat (*Leptonycteris yerbabuenae*), little brown bat (*Myotis lucifugus*), long-eared myotis (*Myotis evotis*), long-legged myotis (*Myotis volans*), Mexican long-nosed bat (*Leptonycteris nivalis*), Mexican long-tongued bat (*Choeronycteris mexicana*), northern yellow bat (*Lasiurus intermedius*), pallid bat (*Antrozous pallidus*), pocketed free-tailed bat (*Nyctinomops femorosaccus*), Rafinesque's big-eared bat (*Corynorhinus rafinesquii*), red bat (*Lasiurus borealis*), Seminole bat (*Lasiurus seminolus*), silver-haired bat (*Lasionycteris noctivagans*), southern yellow bat (*Lasiurus ega*), southeastern myotis (*Myotis austroriparius*), southwestern myotis (*Myotis auriculus*), spotted bat (*Euderma maculatum*), Townsend's big-eared bat (*Corynorhinus townsendii*), Underwood's mastiff bat (*Eumops underwoodi*), Wagner's bonneted bat (*Eumops glaucinus*), western bonneted bat (*Eumops perotis*), western pipistrelle (*Pipistrellus hesperus*), Yuma myotis (*Myotis yumanensis*).

Hoary bat.

5. CINGULATA (armadillos): nine-banded armadillo (*Dasypus novemcinctus*).

Armadillo.

6. EULIPOTYPHLA (moles, shrews, hedgehogs, gymnures, solenodons, desmans): American shrew mole (*Neurotrichus gibbsii*), arctic shrew (*Sorex arcticus*), Arizona shrew (*Sorex arizonae*), broad-footed mole (*Scapanus latimanus*), cinereus shrew (*Sorex cinereus*), coast mole (*Scapanus orarius*), desert shrew (*Notiosorex crawfordi*), dwarf shrew (*Sorex nanus*), eastern mole (*Scalopus aquaticus*), hairy-tailed mole (*Parascalops breweri*), least shrew (*Cryptotis parva*), long-tailed shrew (*Sorex dispar*), marsh shrew (*Sorex bendirii*), Merriam's shrew (*Sorex merriami*), montane shrew (*Sorex monticolus*), ornate shrew (*Sorex ornatus*), Pacific shrew (*Sorex pacificus*), pygmy shrew (*Sorex hoyi*), short-tailed shrew (*Blarina brevicauda*), smoky shrew (*Sorex fumeus*), southeastern shrew (*Sorex longirostris*), southern short-tailed shrew (*Blarina carolinensis*), star-nosed mole (*Condylura cristata*), Townsend's mole (*Scapanus townsendii*), Trowbridge's shrew (*Sorex trowbridgii*), vagrant shrew (*Sorex vagrans*), water shrew (*Sorex palustris*).

Star-nosed mole.

7. LAGOMORPHA (hares, rabbits, pikas): American pika (*Ochotona princeps*), antelope jackrabbit (*Lepus alleni*), arctic hare (*Lepus arcticus*), black-tailed jackrabbit (*Lepus californicus*), brush rabbit (*Sylvilagus bachmani*), collared pika (*Ochotona collaris*), desert cottontail (*Sylvilagus audubonii*), eastern cottontail (*Sylvilagus floridanus*), European hare (*Lepus europaeus*), European rabbit (*Oryctolagus cuniculus*), marsh rabbit (*Sylvilagus palustris*), mountain cottontail (*Sylvilagus nuttallii*), mountain hare (*Lepus timidus*), New England cottontail (*Sylvilagus transitionalis*), pygmy rabbit (*Brachylagus idahoensis*), snowshoe hare (*Lepus americanus*), swamp rabbit (*Sylvilagus aquaticus*), white-sided jackrabbit (*Lepus callotis*), white-tailed jackrabbit (*Lepus townsendii*).

Hares.

8. MARSUPIALIA (marsupials): Virginia opossum (*Didelphis virginiana*).

9. PERISSODACTYLA (odd-toed hoofed mammals): wild horse (*Equus ferus*).

Virginia opossum.

Wild horse.

10. PINNIPEDIA (seals, walruses): bearded seal: (*Erignathus barbatus*), California sea lion (*Zalophus californianus*), gray seal (*Halichoerus grypus*), Guadalupe fur seal (*Arctocephalus townsendi*), harbor seal (*Phoca vitulina*), harp seal (*Pagophilus groenlandicus*), hooded seal (*Cystophora cristata*), northern elephant seal (*Mirounga angustirostris*), northern fur seal (*Callorhinus ursinus*), ribbon seal (*Histriophoca fasciata*), ringed seal (*Pusa hispida*), Steller sea lion (*Eumetopias jubatus*), walrus (*Odobenus rosmarus*), West Indian monk seal (*Monachus tropicalis*).

Walrus.

Sasquatch.

11. HOMINIDAE (great apes and humans): Sasquatch (*Homo sapiens cognatus*).

12. RODENTIA (rodents): Abert's squirrel (*Sciurus aberti*), agile kangaroo rat (*Dipodomys agilis*), alpine chipmunk (*Neotamias alpinus*), American beaver (*Castor canadensis*), arctic ground squirrel (*Urocitellus parryii*), arctic lemming (*Dicrostonyx torquatus*), Arizona cotton rat (*Sigmodon arizonae*), Arizona gray squirrel (*Sciurus arizonensis*), Arizona woodrat (*Neotoma devia*), Bailey's pocket mouse (*Chaetodipus baileyi*), banner-tailed kangaroo rat (*Dipodomys spectabilis*), Belding's ground squirrel (*Urocitellus beldingi*), black rat (*Rattus rattus*), black-tailed prairie dog (*Cynomys ludovicianus*), Botta's pocket gopher (*Thomomys bottae*), brown lemming (*Lemmus trimucronatus*), brush mouse (*Peromyscus boylii*), bushy-tailed woodrat (*Neotoma cinerea*), cactus mouse (*Peromyscus eremicus*), California chipmunk (*Neotamias obscurus*), California ground squirrel (*Otospermophilus beecheyi*), California kangaroo rat (*Dipodomys californicus*), California mouse (*Peromyscus californicus*), California pocket mouse (*Chaetodipus californicus*), California vole (*Microtus californicus*), Camas pocket gopher (*Thomomys bulbivorus*), canyon mouse (*Peromyscus crinitus*), Chihuahuan grasshopper mouse (*Onychomys arenicola*), chisel-toothed kangaroo rat (*Dipodomys microps*), cliff chipmunk (*Neotamias dorsalis*), Colorado chipmunk (*Neotamias quadrivittatus*), Columbian ground squirrel (*Urocitellus columbianus*), Columbia Plateau pocket mouse (*Perognathus parvus*), cotton mouse (*Peromyscus gossypinus*), creeping vole (*Microtus oregoni*), dark kangaroo mouse (*Microdipodops megacephalus*), deer mouse (*Peromyscus maniculatus*), desert pocket gopher (*Geomys arenarius*), desert kangaroo rat (*Dipodomys deserti*), desert pocket mouse (*Chaetodipus penicillatus*), desert woodrat (*Neotoma lepida*), Douglas squirrel (*Tamiasciurus douglasii*), Dulzura kangaroo rat (*Dipodomys simulans*), dusky-footed woodrat (*Neotoma fuscipes*), eastern chipmunk (*Tamias striatus*), eastern fox squirrel (*Sciurus niger*), eastern gray squirrel (*Sciurus carolinensis*), eastern harvest mouse (*Reithrodontomys humulis*), eastern woodrat (*Neotoma floridana*), Florida mouse (*Podomys floridanus*), Franklin's ground squirrel (*Poliocitellus franklinii*), fulvous harvest mouse (*Reithrodontomys fulvescens*), giant kangaroo rat (*Dipodomys ingens*), golden-mantled ground squirrel (*Callospermophilus lateralis*), golden mouse (*Ochrotomys nuttalli*), gray-collared chipmunk (*Neotamias cinereicollis*), gray-footed chipmunk (*Neotamias canipes*), Great Basin pocket mouse (*Perognathus mollipilosus*), Gunnison's prairie dog (*Cynomys gunnisoni*), Harris' antelope squirrel (*Ammospermophilus harrisii*), heather vole (*Phenacomys intermedius*), Heermann's kangaroo rat (*Dipodomys heermanni*), hispid cotton rat (*Sigmodon hispidus*), hispid pocket mouse (*Chaetodipus hispidus*), hoary marmot (*Marmota caligata*), Hopi chipmunk (*Neotamias rufus*), house mouse (*Mus musculus*), Idaho pocket gopher (*Thomomys idahoensis*), insular vole (*Microtus abbreviatus*), least chipmunk (*Neotamias minimus*), Labrador collared lemming (*Dicrostonyx hudsonius*), little pocket mouse (*Perognathus longimembris*), lodgepole chipmunk (*Neotamias speciosus*), long-eared chipmunk (*Neotamias quadrimaculatus*), long-tailed pocket mouse (*Chaetodipus formosus*), long-tailed vole (*Microtus longicaudus*), marsh rice rat (*Oryzomys palustris*), meadow jumping mouse (*Zapus hudsonius*), meadow vole (*Microtus pennsylvanicus*), Merriam's chipmunk (*Neotamias merriami*), Merriam's ground squirrel (*Urocitellus canus*), Merriam's kangaroo rat (*Dipodomys merriami*), Merriam's mouse (*Peromyscus merriami*), Merriam's pocket mouse (*Perognathus merriami*),

Porcupine.

Mexican ground squirrel (*Ictidomys mexicanus*), Mexican spiny pocket mouse (*Liomys irroratus*), Mexican vole (*Microtus mexicanus*), Mexican woodrat (*Neotoma mexicana*), Mojave ground squirrel (*Xerospermophilus mohavensis*), montane vole (*Microtus montanus*), mountain beaver (*Aplodontia rufa*), mountain pocket gopher (*Thomomys monticola*), muskrat (*Ondatra zibethicus*), narrow-faced kangaroo rat (*Dipodomys venustus*), narrow-headed vole (*Microtus gregalis*), Nelson's antelope squirrel (*Ammospermophilus nelsoni*), Nelson's pocket mouse (*Chaetodipus nelsoni*), North American porcupine (*Erethizon dorsatum*), northern bog lemming (*Synaptomys borealis*), northern flying squirrel (*Glaucomys sabrinus*), northern grasshopper mouse (*Onychomys leucogaster*), northern Idaho ground squirrel (*Urocitellus brunneus*), northern pocket gopher (*Thomomys talpoides*), northern pygmy mouse (*Baiomys taylori*), northern red-backed vole (*Myodes rutilus*), northern rock deer mouse (*Peromyscus nasutus*), Norway rat (*Rattus norvegicus*), nutria (*Myocastor coypus*), oldfield mouse (*Peromyscus polionotus*), olive-backed pocket mouse (*Perognathus fasciatus*), Ord's kangaroo rat (*Dipodomys ordii*), Pacific jumping mouse (*Zapus trinotatus*), pale kangaroo mouse (*Microdipodops pallidus*), Palmer's chipmunk (*Neotamias palmeri*), Panamint chipmunk (*Neotamias panamintinus*), Panamint kangaroo rat (*Dipodomys panamintinus*), piñon mouse (*Peromyscus truei*), Piute ground squirrel (*Urocitellus mollis*), plains harvest mouse (*Reithrodontomys montanus*), plains pocket gopher (*Geomys bursarius*), plains pocket mouse (*Perognathus flavescens*), prairie vole (*Microtus ochrogaster*), red squirrel (*Tamiasciurus hudsonicus*), red-tailed chipmunk (*Neotamias ruficaudus*), red tree vole (*Arborimus longicaudus*), Richardson's ground squirrel (*Urocitellus richardsonii*), rock pocket mouse (*Chaetodipus intermedius*), rock squirrel (*Otospermophilus variegatus*), rock vole (*Microtus chrotorrhinus*), round-tailed ground squirrel (*Xerospermophilus tereticaudus*), round-tailed muskrat (*Neofiber alleni*), sagebrush vole (*Lemmiscus curtatus*), salt-marsh harvest mouse (*Reithrodontomys raviventris*), San Diego pocket mouse (*Chaetodipus fallax*), San Joaquin pocket mouse (*Perognathus inornatus*), shadow chipmunk (*Neotamias senex*), silky pocket mouse (*Perognathus flavus*), singing vole (*Microtus miurus*), Siskiyou chipmunk (*Neotamias siskiyou*), Sonoma chipmunk (*Neotamias sonomae*), southeastern pocket gopher (*Geomys pinetis*), southern bog lemming (*Synaptomys cooperi*), southern flying squirrel (*Glaucomys volans*), southern grasshopper mouse (*Onychomys torridus*), southern plains woodrat (*Neotoma micropus*), southern pocket gopher (*Thomomys umbrinus*), southern red-backed vole (*Myodes gapperi*), spiny pocket mouse (*Chaetodipus spinatus*), spotted ground squirrel (*Xerospermophilus spilosoma*), Stephens' kangaroo rat (*Dipodomys stephensi*), Stephens' woodrat (*Neotoma stephensi*), taiga vole (*Microtus xanthognathus*), tawny-bellied cotton rat (*Sigmodon fulviventer*), Texas antelope squirrel (*Ammospermophilus interpres*), Texas kangaroo rat (*Dipodomys elator*), Texas mouse (*Peromyscus attwateri*), Texas pocket gopher (*Geomys personatus*), thirteen-lined ground squirrel (*Ictidomys tridecemlineatus*), Townsend's chipmunk (*Neotamias townsendii*), Townsend's ground squirrel (*Urocitellus townsendii*), Townsend's pocket gopher (*Thomomys townsendii*), Townsend's vole (*Microtus townsendii*), tundra vole (*Microtus oeconomus*), Uinta chipmunk (*Neotamias umbrinus*), Uinta ground squirrel (*Urocitellus armatus*), Utah prairie dog (*Cynomys parvidens*), Washington ground squirrel (*Urocitellus washingtoni*), water vole (*Microtus richardsoni*), western gray squirrel (*Sciurus griseus*), western harvest mouse (*Reithrodontomys megalotis*), western jumping mouse (*Zapus princeps*), western pocket gopher (*Thomomys mazama*), white-ankled mouse (*Peromyscus pectoralis*), white-eared pocket mouse (*Perognathus alticola*), white-footed mouse (*Peromyscus leucopus*), white-footed vole (*Arborimus albipes*), white-tailed antelope squirrel (*Ammospermophilus leucurus*), white-tailed prairie dog (*Cynomys leucurus*), white-throated woodrat (*Neotoma albigula*), woodchuck (*Marmota monax*), woodland jumping mouse (*Napaeozapus insignis*), woodland vole (*Microtus pinetorum*), Wyoming ground squirrel (*Urocitellus elegans*), Wyoming pocket gopher (*Thomomys clusius*), yellow-bellied marmot (*Marmota flaviventris*), yellow-cheeked chipmunk: (*Neotamias ochrogenys*), yellow-faced pocket gopher (*Cratogeomys castanops*), yellow-nosed cotton rat (*Sigmodon ochrognathus*), yellow-pine chipmunk (*Neotamias amoenus*).

Manatee.

13. SIRENIA (manatees, dugongs): West Indian manatee (*Trichechus manatus*).

American bison, Jackson, Wyoming, with the Teton Range (part of the Rocky Mountains) in the background. The official national mammal of the United States, the American bison is the premier symbol of our region's wild mammals.

ENTRIES

WHAT IS A MAMMAL?
Definitions & Statistics

The first clue comes from the word itself: the English word mammal is from the Latin word *mamma*, meaning "mother," an indirect reference to the female mammary gland or "breast."

Biologically speaking then, a mammal is an animal that belongs to the Mammalia class: a group of warm-blooded, higher vertebrates comprising humans and all other animals that nourish their young with milk secreted by mammary glands, who bear live young, and whose skin is more or less covered in fur or hair. This makes us human mammals.

We now come to a group of related words. Mammalogy is a branch of zoology concerned with the scientific study of mammals. A mammalogist is one who studies mammals. A mammalophile is one who loves mammals.

Of the 6,399 currently recognized living species of mammals in the world, approximately 458 mammal species, or 29 percent, live in our region: North America, a 9,365,000 sq., mi. area located in the Western Hemisphere.

To put it another way, nearly ⅓ of all the species of mammals that inhabit our globe live in North America, which is defined as Canada, the United States (but not Hawaii), Mexico, Central America, the Caribbean Islands, and Greenland. This book, *North America's Amazing Mammals*, covers 378, or 83 percent, of these 458 species.

Mother whale nursing her calves.

ABERT'S SQUIRREL (*Sciurus aberti*): It grows to a total length of 23", its tail is about 10" long, and it can weigh up to 2 lb. The upper body is grizzled gray, the back is reddish, the lower body is white, the tail is gray above, whitish below. Most noticeable are its 1¾" ears, each with a ¾" tuft extending from the eartip, a characteristic from which it derives its other common names: the "tassel-eared squirrel" and the "tuft-eared squirrel." Its furry tussocks disappear in the summer and regrow in the fall.

Abert's squirrel.

Found in coniferous forests in the southwestern U.S. (Utah, Arizona, Colorado, New Mexico) and south to northern and central Mexico. This large diurnal sciurid is an herbivore with a lignivorous, folivorous, granivorous appetite. Besides wood, leaves, and seeds, it consumes fungus and dirt, hence it is also a mycophage and a geophage. Also known to eat mistletoe, though pine nuts are its preferred food; is active all year; does not seem to hibernate.

Mates once a year between late winter and early spring. This tree squirrel is polygynandrous. The female gestates for about 6 wks., giving birth to 3-4 altricial young who are weaned 2½ mos. later. Constructs either a bolus nest or a broom nest near the tops of trees, ponderosa pines in particular, lining them with grass, bark, moss, leaves, and feathers.

Mainly solitary, it communicates and perceives via normal Sciuridae channels: sight, hearing, smell, and touch. Vocalizations include barking, clucking, and screeching. Its natural enemies are birds of prey, wild felids, and wild canids. Abert's squirrel probably lives 3-4 yrs. in the wild.

AGILE KANGAROO RAT (*Dipodomys agilis*): Like all kangaroo "rats," it is not a true rat, but a relative of the pocket gopher. This medium-sized rodent grows to a total length of 13", its pes or hind foot measures nearly 2", and it can attain a weight of up to 3 oz. Its dorsum is gray-brown or rusty-brown to dark brown, its ventrum is white, tan, or light gray. The 8" tail is long, thin, dichromatic, and whip-like, with a whitish fur stripe above, a dark fur stripe below; the proximal half of the tail is covered in short hairs, while a longish tuft of hair grows over the distal half of the tail; there is a white lateral stripe curving across and down over each thigh.

The body is small and ovoid in shape; the head is large with a rounded convex profile; the eyes are large, dark, and ringed in tan (sometimes the rings do not entirely circumscribe the eyes); the ears are large, erect, round, and located high on the head; the nose is short and tapered with a black nose ring (a thin vertical band running from one side to the other); long sensitive nasal vibrissae sprout from the end of the muzzle; the front feet are small and retrogressed; the 5-toed pedes or hind feet are long and powerful. Occasionally walks on all fours, but when speed is necessary it prefers hopping on its large back feet, lending it its common name.

This heteromyid inhabits arid scrubland, brushland, sageland, rangeland, and chaparral with soft, loose, or gravelly soil. Its range is small, encompassing only coastal southern California and northern Baja California. Transports food in its large cheek pouches. Can leap 8'-9' when fleeing danger; acute hearing allows it to detect predators from long distances.

Agile kangaroo rat.

This species has not yet been studied in detail. However, it seems to share much of the same ecology and life history as Heermann's kangaroo rat, and in appearance it is almost identical to Stephens' kangaroo rat and the Dulzura kangaroo rat—whose ranges all overlap. The life span of the agile kangaroo rat is unknown.

ALLEN'S BIG-EARED BAT (*Idionycteris phyllotis*): This chiropterid grows to a total length of nearly 5", its forearm is close to 2" long, its wingspan is over 13½", and it weighs ½ oz. The hair is tawny-gray on top but black at the base, giving it a somewhat grizzled appearance. Its underside is pale. It has a tall tragus, postauricular whitish patches, and hairless wings. Most prominent are its gigantic ears, which grow to almost 2" (nearly half its body length) and their 2 lappets—which attach at the base of the ears and project over the forehead (the only bat with such appendages). While awake or hunting it keeps its ears erect; while at rest it relaxes them in a folded-back position.

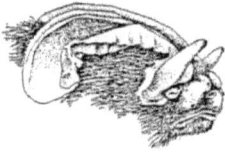

This fast-flying vespertilionid shelters in caves, cliffs, mines, and rocky outcroppings in montane forests with reliable water sources. A secondary consumer, it ranges through the American Southwest, south into Mexico. Females give birth to a single pup in the nursery roost, which they form in the summer. During this period the males live separately in bachelor colonies. A nocturnal insectivore, it leaves its roost after dark, hunting down its prey (typically over water) using echolocation. Allen's big-eared bat lives to about 3 yrs. in the wild.

Long-eared bat with ears folded back.

ALPINE CHIPMUNK (*Neotamias alpinus*): North America's smallest chipmunk, it grows to a total length of about 7½" and a weight of around 1½ oz. Its overall pelage coloration is silvery gray with a pale yellowish-orange or rusty wash over the shoulders and sometimes the sides. The belly is whitish-gray or tan. Dorsal and lateral stripes run alternately between white, blackish-brown, and brownish-gray. The face mask is made up of several sharply defined malar stripes that flow horizontally front to back; the dark eyes are ringed in white; the 3" tail is a grizzled orangish-black with an indistinct blackish tip. Its genus name derives from the Greek *neo* ("new") and *tamias* ("dispenser"), the latter word a reference to its role as a seed disperser.

Both diurnal and nocturnal, this energetic, solitary, petraphilic sciurid inhabits rocky locations, including rock-strewn cliffs, talus slopes, meadows, and stony edgeland around bodies of water (ponds, lakes) throughout, as its common name indicates, subalpine fields and forests. Its range is quite narrow: the Sierra Mountains of central California. Builds its house in rocky settings, and, being fossorial, excavates burrows below ground for safety and warmth.

Alpine chipmunk.

Though terrestrial, it is partially scansorial, and readily climbs trees when

necessary. Builds up reserves of fat, then hibernates during cold weather—though this might be more accurately described as a periodic state of torpor since it experiences frequent arousals, waking up often to feed from its storehouse of edibles. An omnivore, its diet centers around seeds, grasses, nuts, fruit, and fungi, but it also consumes insects, amphibians, reptiles, birds' eggs, and chicks. Transports its food in large cheek pouches. Its breeding season occurs in spring. The doe gestates for about 1 mo., giving birth to 4-6 altricial pups in a nest constructed deep within a rock crevice. She weans them 5-6 wks. later, after which they become independent. The buck does not contribute to the rearing of his offspring.

Communicates and perceives via normal Sciuridae channels: sight, hearing, smell, and touch. Vocalizations include chucking and chipping (the sound from which all chipmunks derive their name), but it is best known for its sweet, high pitched, bird-like call. Its natural enemies are reptiles, birds of prey, mustelids, procyonids, wild felids, and wild canids, as well as other members of its own family (such as the red squirrel). The alpine chipmunk probably lives 1-2 yrs. in the wild; perhaps twice that long in captivity.

AMERICAN BADGER (*Taxidea taxus*): The official state animal of Wisconsin, this somewhat scaled down version of its much larger cousin, the wolverine, grows to 35" in total length and can weigh up to 25 lb. It has a wide flattish head and body, thick neck, stocky legs, small ears, short tail, a pointed black-and-white stripped face, an upturned nose (for rooting), large front feet and shovel-like claws (for digging), and loose skin (which helps it maneuver through its burrows). Its long shaggy fur

American badger.

is grizzled gray in color, and is typically mixed with brown, rust, and white tones. It has a buff-colored chin and throat. Some individuals have whitish lateral and dorsal stripes along the sides and back as well. Gender dimorphism is present, with the male being larger than the female.

A fossorial mustelid, its habitats include grassland, woodland edges, plains, fields, meadows, deserts, parkland, marshland, pastures, sagebrush, and farmland. Its range extends over most of the U.S. (except the southeast), through southern Canada, and south into Mexico. Though not territorial, it establishes a home range of up to 3 sq. mi.

Like all members of the weasel family, it is ideally suited to a carnivorous lifestyle, and possesses keen senses and the ability to climb, swim, dive, and run (up to 20 mph). A solitary hunter and an opportunistic carnivore with an omnivorous side, it feeds on insects, amphibians, birds, small mammals (prairie dogs, mice, rats, voles, gophers, marmots, squirrels), snakes, lizards, and carrion. When necessary it also consumes vegetation; hoards and caches surplus food. Both a mutualist and a symbiotic species, it will sometimes hunt cooperatively with coyotes, with which it shares much the same carnivorous diet: the coyote chases a prey item into the badger's burrow; if the prey sees the badger in time, it jumps back out and the coyote snatches it; if it does not see the badger in time, the badger grabs it (thus, one or the other, often both, are rewarded with a meal).

Though mainly nocturnal, the badger is also active during the day. Contrary to popular belief, it does not hibernate; instead it employs torpor, going into a deep sleep during frigid months, with frequent arousals, waking up every month or so. If food is scarce it sustains itself on its body fat, stored up from the previous summer.

A skilled and rapid digger, it builds massive, multi-chambered underground burrows up to 30' long and 10' deep, and can excavate a tunnel on the

American badger.

spot to evade predators—disappearing from view within seconds. Will take over the unoccupied burrows of other animals when necessary. Like many fish, reptiles, and birds (but unlike most mammals), it has a nictating membrane (known as a "third eyelid"), that helps protects its eyes from dirt, sand, and grit while digging or traveling underground.

Its breeding season runs from summer to early autumn. Sows employ delayed implantation and gestate for about 7 mos., giving birth to 1-5 altricial young in the spring in a grass-lined nest situated underground. The cubs are weaned at around 2 mos., becoming independent by 6 mos. of age. The boar plays no role in the rearing of his offspring.

Possessed of extraordinary physical senses, in the typical manner of most members of the Mephitidae family, it uses sight, hearing, touch, and smell to communicate with conspecifics. A naturally cantankerous mustelid that is constructed like a miniature tank, this secondary consumer is a ferocious fighter with only a few natural enemies: birds of prey (such as eagles), wild felids (such as mountain lions), and wild canids (such as coyotes). Though it is wary and retiring rather than aggressive, if provoked it will hiss, bite, growl, snarl, and squeal, finally driving off its opponent by excreting a foul-smelling oil. Beneficial as a soil aerator and recycler, a commensal homebuilder, a pest controller, and as food for carnivores; detrimental as a farm and ranch pest (whose burrow entrances can injure the limbs of livestock). The American badger lives up to 10 yrs. in the wild; twice that long in captivity.

American beaver.

AMERICAN BEAVER (*Castor canadensis*): The official state animal of New York and Oregon, this is our region's largest rodent. It grows to a total length of nearly 47" and can attain a weight of up to 100 lb. (though 60-70 lb. is more usual). Its dense, soft, oily, waterproof pelage runs from yellow-brown to burnt reddish-brown to dark brownish-black; the cheeks may be lighter in color; the eyes are small and beady; the ears are small, erect, and rounded; the large upper and lower incisors are orangish or golden-brown; the long black 18" tail, which is used in an up-and-down motion, is broad, horizontally flat, scaly, nearly hairless, and paddle-shaped; the front legs are short; the front paws are hand-like and heavily clawed; being natatorial, the large webbed back feet are used to propel it through the water. The body is stout; the head is massive with a convex profile; the muzzle is robust and short; the nose is large, black, and snubbed; the upper lip is large. As adult males and females are not readily distinguishable, it is not gender dimorphic.

This large, sedentary, aquaphilic castorid is a nearctic native that inhabits a wide variety of temperate streams, rivers, ponds, wetlands, reservoirs, marshes, lakes, swamps, ditches, and islands, particularly in and around muskeg-rich forests (the source of its building materials and much of its food). One of the most widely distributed of all North American mammals, it is found across our entire region, from Alaska and Canada in the north to Arizona, Texas, Florida, and northern Mexico in the south; from Washington, Oregon, and California in the west to Maine, Virginia, and Georgia in the east. Being riparian, it avoids extremely dry areas such as deserts. Indeed, it is found in nearly every part of North America except arid locations. Does not hibernate and is active all year, but will spend extremely cold days inside its home, eating cached food and living off fat it stores in its tail. Territorial, the beaver viciously defends its home and local turf from conspecifics.

When necessary it will create a watertight dam out of small tree trunks, logs, reeds, leaves, rocks, grasses, aquatic vegetation, reeds, saplings, mud, branches, and sticks to weaken and decelerate a river's current or to raise the water level. The average dam is about 15'-20' in length and 5'-10' deep. A beaver can cut down a 6" thick tree in only a few minutes; transports trimmed tree parts to the dam site in its mouth. The ultimate purpose of the dam is to create a pond (at least 4' deep) and a surrounding wetland, for it is here that it builds its home and lives out its life.

Beaver lodge.

The beaver's house is either a simple burrow dug into the side of a river bank, or, more likely, a "lodge," a massive dome-shaped structure that can reach 8'-10' in height and span some 40' in diameter. Made of woven branches, sticks, grass, bark, and aquatic detritus, and covered over in mud (which acts as caulking and insulation), the lodge can be located in the middle of a body of water or on the water's edge (a river bank or lake shore). It has a large central chamber at the top (always above the water line) with a wood-chip floor, an air vent in the wall, and several portals that open up underwater. This isolated "island home," with its underwater entrances and exits, provides protection against most predators. A wide variety of building designs may be used as each situation is different, requiring different types of construction. Both its dam and its watery abode necessitate continual maintenance, and so grow larger with the passage of time as repairs are made and new material is added. Thus, a few old dams have been discovered that are many hundreds, even thousands, of feet long—the result of numerous beaver communities living in the same area over many generations.

A nocturnal and crepuscular herbivore, its diet is focused mainly on the cambium and bark of trees (preferred types: poplar, cottonwood, alder, cherry, birch, aspen, willow, beech, and maple). However, it also consumes roots, tubers, stems, twigs, leaves, herbs, buds, and aquatic vegetation (cattails, water lilies, pond weeds). A modified digestive system helps it break down and assimilate its woody fare. As noted, it stores surplus edibles in larder chambers. Highly social, it lives in groups called "colonies," comprised of up to a dozen (but usually half that many) related individuals: typically a mated father and mother and their offspring from the past 2-3 yrs. (2-3 generations).

Its breeding season begins in winter and may run to early spring, depending on geography. Males and females are monogamous and form lifelong pair-bonds. The female produces 1 litter a year and gestates for about 90 days. She gives birth to as many as 8 fully furred, open-eyed, precocial kits. Weighing roughly 1 lb., they can swim within 24 hrs. and explore outside the lodge within a few days. The mother cuddles and suckles them, carrying them about on her back or tail as she goes about her daily duties. Both parents provide protection and socialization; when the young are old enough to consume solid food, the father helps collect it for them. While weaning takes place between a few weeks and several months after birth, the kits remain with their parents until the age of 2. At this time they are exiled

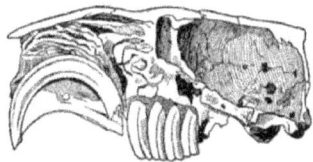

Beaver skull, showing curved incisor and molars.

and begin their lives of independence, usually dispersing to start their own colonies.

An important keystone species, the beaver communicates and perceives via typical Castoridae channels: sight, hearing, touch, and smell. The latter function is especially important to the beaver, whose castoreum or castor glands (large secretory organs located on the rump) release a pungent oil (the source of its genus name) that is used both to waterproof its fur and to mark the scent mounds it constructs around its home. Made out of grass, sticks, and mud, these malodorous piles of debris serve as territorial markers, and may measure up to 1' in height and 3' in width. It is well-known for another form of communication: tail-slapping the water surface when alarmed.

Extremely wary on land because it is semiaquatic, it is ideally suited to life in and around water: its eyes have protective membranes that act like a pair of diving goggles (enhancing vision underwater); both its ears and its nose are valved and can close when submerged; 2 mouth flaps (located between the incisors and rear molars) seal out water while leaving the front teeth accessible—which allows it to eat and carry wood underwater; a thick layer of fat helps it maintain a stable body temperature in cold water; can swim 6 mph and hold its breath for up to 15 mins. underwater; uses its sturdy tail as a rudder

American beaver.

while swimming and as a balancing stand while on its hind legs. Its sharp teeth, which grow continually throughout its life, are specifically designed for cutting down trees. This it accomplishes by gnawing circles around the trunk, removing large chips, until the weakened tree falls over of its own weight.

Its natural enemies are birds of prey (owls, eagles), other mustelids (fishers, river otters, wolverines), wild felids (bobcats, lynxes, mountain lions), wild canids (foxes, coyotes, wolves), and ursids (black bears, brown bears). A highly effective allogenic engineer, it benefits us by modifying old biomes and creating new ones, such as wetlands: a large natural sponge that protects and improves water quality, houses floodwater, moderates the climate (by storing carbon), maintains water flow, controls erosion, and provides habitat, shelter, and food for flora and fauna, as well as food and recreation for humans. If and when a beaver colony evacuates its dam and wetland, the area eventually gives birth to a new ecosystem: a meadow, the biological foundation of a forest. Thus, this animal plays a vital role in ecosystem succession.

A vital mammalian resource throughout much of North America, humans have long relied on its pelt and meat for income and sustenance. Can be detrimental if dammed water begins flooding the surrounding area, destroying crops, timber, and homes. Once nearly hunted to extinction in many regions, protective laws have allowed it to fully recover and reestablish itself across North America. Today it is so common in some places that it is considered a pest. The American beaver lives about 10 yrs. in the wild; twice that long in captivity.

AMERICAN BISON (*Bison bison*): This iconic all-American animal—the official national mammal of the United States, the official state animal of Oklahoma, and the official state mammal of Wyoming—grows to nearly 12' in total length, stands 6' at the shoulder, and can weigh 2,200 lb. (a little over 1 ton), all on a vegetarian diet. The thick coat is brown, varying from dark brown (almost black) at the head, to cinnamon- or light-brown at the rump. Easily identified by its massive wooly head, long shaggy frontal fur, dark "beard," large shoulder hump, and tuft-tipped tail, it is the largest land animal in North America.

Both genders have permanent gray-black horns, the male's growing up to 26" long and reaching 2½' from tip to tip. Used for defense, generally speaking, the

American bison (male).

adult male's horns point straight up, the female's curve inward (these traits, however, may vary depending on gender, age, and the individual). Though bulls and cows can indeed sometimes be difficult to distinguish, the mature male has a wider more triangular head, the mature female has a narrower more gracile head. With the male being larger than the female, gender dimorphism is present.

This terricolous cursorial ungulate is a nearctic native that inhabits open plains, prairies, grassland, river valleys, scrubland, meadows, burnovers, mixed woodland, wetland, and savanna. Its present-day range extends over the northwestern tier of our region, encompassing portions of the Yukon, Northwest Territory, British Columbia, Alberta, Saskatchewan, Montana, Wyoming, South Dakota, and Utah. A number of controlled herds exist elsewhere, such as in New Mexico.

Typically traveling in a line, this huge diurnal herbivore is also a folivore, grazing on grasses, sedges, lichen, sagebrush, roots, forbs, berries, and tree parts (leaves, bark, wood, stems, twigs). Highly gregarious, it forms various types of groups, including bachelor herds (made up of mature bulls and young adult males) and matriarchal herds (made up of mature cows and juvenile females and males).

Bison foraging.

Its mating season runs from early summer to early fall. Males shove and head-butt one another during the rut. The cow mates once a year and gestates for about 9 mos., giving birth to a single, reddish-colored, precocial calf in the spring in (when possible) an isolated area away from the main herd. Nicknamed "red dogs" due to their cinnamon-colored coat, baby bison can stand and walk within a few hours and are fully independent at about 1 year of age. The polygynous bull plays no role in caring for his offspring.

Feisty and majestic, this massive bovid once roamed over most of the U.S. (except the northeast and Atlantic Coast) from Alaska to Mexico, with an estimated total population of 60 million. Nearly pushed to extinction in the 19th Century, its numbers are returning under careful management, with about 500,000 now living on protected lands in the western U.S. and Canada.

Complacent unless disturbed, its size, power, speed, and cantankerous personality make it a truly perilous foe. Nonetheless, this primary consumer has a number of natural enemies: wolves, bears, and mountain lions. Besides its dangerous pointed horns (with which it can stab, tear, gore, lift, and toss its tormentor), it has other means of defense and evasion: it is an excellent swimmer, it can jump 5' into the air (easily leaping over tall fences), and can run up to 45 mph for 5 mi. Communicates and perceives along standard Bovidae channels: sight, hearing, touch, and smell. Vocalizations include grunting, bellowing, snorting, bawling, and bleating. Engages in dirt-wallowing, such as mud-bathing and dust-bathing, which protects the skin from sun and insects.

With an evolutionary history that is complex and murky, bison taxonomy remains a highly debated topic. Two subspecies of this fleecy artiodactyl are currently recognized by most authorities: the American plains bison and the wood bison. Its scientific family, Bovidae, traces back to the Miocene (from 5 million to 23 million yrs. ago). Beneficial as a soil enricher, a zoo ambassador, a focus of ecotourism, a cross-breeder with cattle (creating the "beefalo"), and as food for carnivores (including humans); detrimental as carrier of dangerous diseases that can affect livestock (though livestock diseases can be deadly to bison as well). The average life span of the American bison is 15-20 yrs. in the wild; twice that long in captivity.

Bison herd.

AMERICAN BLACK BEAR (*Ursus americanus*): The official state mammal of Alabama and the official state animal of New Mexico and West Virginia, it is also known as the "black bear" and the "cinnamon bear." It stands 3½' at the shoulder, grows to a total length of 7', and can attain a weight of up to 600 lb. The smallest of our three bear species, as with many other mammals its common name can be misleading: though it is often black in eastern regions, it is just as likely to be brown, cinnamon, or blond in western regions. It can be white or bluish in various parts of Alaska, and a white-blue phase, known as a "glacier bear," appears only along the coast of British Columbia. Its muzzle is often tan and the chest may have a white patch. Sometimes confused with the grizzly bear (particularly when both have brown coats), the two can be easily distinguished by the following marked traits: the grizzly has a concave nose and a large shoulder hump; the black bear has a convex nose and a small shoulder hump.

American black bear.

This terricolous solitary ursid is a nearctic native that inhabits temperate, mixed coniferous-deciduous forests, fields, brushland, meadows, swamps, ridges, tundra, rangeland, chaparral, tidelands, bays, parks, campgrounds, roadside pullouts, clearcuts, garbage dumps, avalanche chutes, and mountainous regions. Prefers riparian environments with abundant food sources. Normally it does not migrate above 7,000'. Found only in North America, its range is widespread, covering Alaska, nearly all of Canada, and much of the U.S. (it is found in some 40 states, and is the only bear in the eastern states), south to northern Mexico.

An adaptable and opportunistic carnivore, its diet consists mainly of roots, grasses, forbs, pine needles, honey, grains, cambium, twigs, seeds, leaves, fruit, tubers, vegetables, nuts, grubs, beetles, crickets, ants, agricultural crops, and, as a synanthrope, human refuse. Digs up the ground, tears apart decaying wood, and turns over logs, stumps, and rocks in search of food. It does not actively hunt living animals, usually only taking meat (such as small mammals, fish, carrion, the young of deer and moose, and sometimes livestock) when the opportunity presents itself.

Normally wary of humans, this secondary cavity user is also highly intelligent and curious, which can bring it into unwanted contact with us. A crepuscular ursid, it may also be diurnal or nocturnal, depending on various conditions (region, weather, degree of hunger). It is a shy solitary creature, but will sometimes congregate in temporary groups where food is plentiful. It excels at swimming, fishing, and climbing trees (which it can scale with remarkable speed and agility). Sleeps in shallow depressions that it scrapes out of the forest floor. Both males and females form territories, scent-marking the boundaries to warn off conspecifics.

Despite its ungainly appearance, it is quick on its feet, and can attain a speed of 30 mph for short distances. Its home range may extend out to 15 sq. mi. An identifying sign of this species are "bear trees": trees that have been torn, ripped, bitten, slashed, and rubbed. Its 9" long by 5" wide footprints can appear humanlike, and are therefore often mistaken for Sasquatch prints, and vice versa. But close inspection reveals many obvious differences. First, bears have claws—which leave marks in soil or snow, while Sasquatch have toenails—which do not leave marks. Second, a bear's toes are arranged

American black bear.

in the opposite order of Sasquatch (and humans), with the smallest toe on the inside of the foot, the big toe on the outside. At just 1', the bear's walking stride length is also quite different when compared with that of Sasquatch (which has a stride of up to 5') and

humans (whose average stride is around 2½').

Depending on the region, the black bear's breeding season generally runs from May to August. A slow breeder, the sow typically mates every other year, employs delayed implantation, gestates for about 7 mos., then, while in a torpid state (black bears do not hibernate), gives birth to 1-5 altricial cubs in midwinter inside a cave, rock crevice, or hollow log. Weighing as little as ½ lb., they are the smallest babies in relation to the adult's size of any terrestrial placental mammal. Cubs are weaned at around 7 mos. old, becoming independent at 1½ yrs. of age. Boars do not contribute to the raising of their offspring.

Communicates and perceives using channels standard to the Ursidae family: sight, body language, hearing, touch, and smell. A secondary consumer, its natural enemies are grizzly bears, mountain lions, and wolves. Cannibalistic, it sometimes preys upon its own kind. Though it usually flees from people, like other large carnivores it is unpredictable, and thus can be extremely dangerous and even deadly when encountered. Beneficial as a seed disperser, a pest controller, and a favored

Bear skull.

species of human hunters; detrimental as farm, ranch, hiker, and camp pest. The American black bear lives an average of 15-20 yrs. in the wild; up to twice that long under optimum conditions.

AMERICAN HOG-NOSED SKUNK (*Conepatus leuconotus*): One of the largest of the North American skunks, it grows to 3' in total length, its hind foot measures nearly 4", and it can attain a weight of up to 10 lb. Its coat is long with coarse fur; its dark eyes and rounded ears are small. It is distinguished by its heavy body, thickset legs, plantigrade paws, and single broad dorsal stripe that runs from the top of the head to the tip of its all-white bushy tail (hence its species name *leuconotus*, "white back"); its ventrum is generally all-black (or sometimes brownish-black). Its long, wide, pig-like snout and bare nose pad, designed for churning up soil, give it its common name (as well as one of its nicknames: the "rooter skunk"). Its powerful front feet have longer claws than the hind feet, ideal fossorial adaptations for its speciality: excavation. This solitary badger-like skunk, which has its origins in the Miocene Epoch over 5 million yrs. ago, is gender dimorphic, with the male being larger than the female.

A riparian mephitid, it inhabits a wide variety of biomes and ecosystems, including brushland, pasture, open woodland, canyons, coastal plains, savanna, cornfields, rocky locations, stream banks, foothills, sageland, thorny understory, forest, chaparral, ranchland, shrubland, seepage slopes, marshland, desert, gulches, cactus-rich areas, mesas, and grassland, always near water. Also known as the "white-backed hog-nosed skunk," its range covers the southwestern U.S. (Texas, Colorado, Oklahoma, New Mexico, Arizona), south into Mexico and beyond. Forms grass-lined dens in rock crevices, caves, brush, and mines, as well as the abandoned burrows of other animals.

An omnivorous and opportunistic carnivore that is primarily insectivorous (beetles, larvae, bees), its diet also consists of fruit, seeds, nuts, worms, gastropods, reptiles, birds, small mammals (bats, rats, rabbits), and carrion. Besides being a skilled digger, it also has robust jaws, sharp teeth, and a keen sense of smell; it is an expert climber and can run up to 10 mph, abilities well-suited to both foraging and defense.

Its breeding season runs from February to March. The doe gestates for around 55 days, giving birth to 1-5 altricial kits in mid to late spring. They become independent in late summer. The buck does not contribute to the rearing of his offspring.

Mainly nocturnal and crepuscular, during the day it can often be found in an underground den avoiding the heat. A secondary consumer, its natural enemies are birds of prey (hawks, eagles, owls), badgers, foxes, bobcats, lynxes, mountain lions, and coyotes. Its primary defensive weapon is its stark aposematic black and white coloration, an anti-predator adaptation that serves as an obvious warning sign to would-be attackers. If this fails, it will stand on its hind legs, hissing and growling. It has one last line of defense: it sprays a foul-smelling, yellowish musk oil from two nipple-like glands under its tail, which it can aim with some precision. This deters most attackers. The American hog-nosed skunk lives 2-4 yrs. in the wild; 7-9 yrs. in captivity.

Note: The western hog-nosed skunk (*Conepatus mesoleucus*), also known as the "common hog-nosed skunk," was once considered a separate species from the American hog-nosed skunk. Recent scientific studies, however, have shown them to be conspecific, and thus they have been combined under the binomial *Conepatus leuconotus*.

AMERICAN MARTEN (*Martes americana*): About the size of a house cat, it grows to 26" in total length, its hind foot measures 3½", and it can attain a weight of up to 4 lb.

Its rich shiny dorsal pelage is a light reddish-brown with blackish tips; the legs and feet are generally dark brown to black; the head, face, and ears are grayish to brown; the tail is black-tipped. Has a distinct cream, cinnamon, or orange throat patch or "bib." Also known as the "pine marten" and the "American sable," this shy slender member of the weasel family has large alert eyes, pert triangular ears, and a sharply pointed snout, resulting in a somewhat miniature fox-like face. Its curved claws provide extra traction; its long, brushy 10" tail acts as a counterbalance while climbing and leaping; feline-like ears increase hearing perception; sharp teeth easily penetrate, hold, tear, and crush the flesh and bone of its prey. A gender dimorphic species, the male is larger than the female.

American marten.

It inhabits mature coniferous, dense deciduous, and old-growth mixed hardwood forests, scrubland, taiga, and rocky areas (with ample snag and deadfall) along the northern tier of our region. A nearctic species, its range stretches over 1 million sq. mi., from Alaska to Nova Scotia, from California and the Rocky Mountain states south through the Midwest and on into the Northeast. Territorial, sedentary, and largely asocial, it aggressively guards its 15 sq. mi. home range against conspecific intruders. Uses hollow logs, empty squirrel nests, tree cavities, tree crowns, snow cavities, and rock piles for denning and resting. During rare moments of inactivity, it may be found on or around the middens of other mammals.

This solitary arboreal creature is an agile climber and a swift sprinter, due, in part, to its semi-retractable claws. It can also swim, both above and below the surface. It does not hibernate, is at ease in cold conditions, and readily tracks its quarry in snow tunnels—particularly mice and other small subnivean prey. Its large furry paws allow it to walk over the surface of the snow. Like most mustelids, it emits a noxious oil from special scent glands, which, in its case, it rubs on branches (known as "scent posts") to mark its territory.

American marten.

Inquisitive, fearless, and omnivorous, the American marten is a ferocious hunter that can easily chase down its favorite game, the red squirrel, either on the ground or in the treetops. This opportunistic forager is a crepuscular and nocturnal feeder that also consumes fruit, seeds, nuts, honey, insects, rabbits, carrion, and other rodents. Robs nests when in search of birds and eggs; resorts to cannibalism when food is scarce. Kills with a rapid and piercing bite to the neck. Stores surplus food underground.

Its breeding season takes place over the summer months. However, the female employs delayed implantation, her fertilized eggs attaching to the wall of her uterus about 6 mos. later (in February). She gestates for a total of 8-9 mos., giving birth to 1-5 altricial kits in midspring. Weaned at around 6 wks., the young grow rapidly, nursed, fed, groomed, and protected in the nesting den by their attentive mother until they can fend for themselves. Reaching adult size quickly under her care, they disperse in the fall to establish their own home ranges. Although the polygynous male sometimes temporarily pairs up with the female, it is not known if he contributes to the rearing of his offspring.

American marten.

Like other members of the Mustelidae family, it communicates and perceives via standard mammalian channels: sight, hearing, touch, and smell. Its vocalizations include yowls, screams, huffs, and chuckles. A secondary consumer, its natural enemies are birds of prey, bobcats, foxes, coyotes, and other American martens. It lives an average of 4-6 yrs. in the wild; up to 15 yrs. in captivity.

AMERICAN MINK (*Neovison vison*): It grows to nearly 30" in total length, has a 3" hind foot, and weighs up to 3½ lb. The natural color of its sleek coat is a uniform chocolate brown, with lighter mottling on its throat, chest, and belly—a cryptic coloration that is ideal for the generally dark and shadowy environments it occupies. There is often a small white chin patch as well. Due to escapees from mink farms, however, its fur color can vary widely over its range, from white to black. It possesses the physical morphology typical of its family: a slightly flattened head, dark protruding eyes, a pointed snout, rounded ears, a long supple body, short legs, and a 10" tail (that makes up ⅓ of its total body length). Gender dimorphism is present, with the male being larger than the female.

This petraphile inhabits rocky, muskeg-rich, forested areas, and watery environments, haunting thick hiding cover along the shores of pools, creeks, streams, ponds, rivers, estuaries, marshes, swamps, lakes, wetlands, and coastlines. Though it is most at home in littoral zones, lacustrine environments, and riverine habitats, it is also found in taiga, grassland, and savanna. This North American native ranges across most of Alaska, Canada, and the U.S. (excluding the southwestern desert regions). Dens in protected riverbanks, streambanks, lake banks, under tree roots, and in hollow trees,

American mink.

brushpiles, or abandoned rabbit and muskrat burrows, moving from lair to lair frequently. Aggressively defends its 2 sq. mi. home range against conspecific males, scent-marking it with a foul-smelling oil from its rump glands.

A member of the order of carnivores, its sharp teeth are set in powerful crushing jaws, excellent weaponry for hunting, killing, and defense. Depending on the season, its diet consists of worms, snails, crayfish, crabs, fish, snakes, frogs, turtles, water fowl, shrews, voles, mice, rats, chipmunks, rabbits, hares, and muskrats (the latter, is its favorite prey). Will take poultry, neonatal lambs, and other farm animals when available. Like its cousin the weasel, it delivers a quick kill-bite to the neck. Stockpiles and caches surplus food.

Solitary, fearless, secretive, and nocturnal, it is an excellent climber, swimmer, and diver. Its silky, luxurious, highly prized coat is waterproof and its feet are webbed, making it well-suited to its semiaquatic life as a riparian mammal. This agile, natatorial mustelid can swim up to 100' underwater and dive down to a depth of 15'.

Its breeding season takes place in late winter to early spring under a polygynandrous mating system. The sow gestates for around 2 mos. or so, giving birth

to 2-10 altricial cubs in late spring in a grass- and fur-
lined nesting den. Weaning takes place at 6 wks., with
the young dispersing at around 6 mos. of age. The
equally indiscriminate non-paternal boar does not form
a pair-bond with the female, nor does he assist in the
rearing of his offspring.

Communication and perception occur via the
customary channels of most members of the Mustelidae
family: sight, hearing, smell, and touch. Though
normally quiet, its vocalizations include hissing, snarling,
purring, screaming, and screeching. A secondary

American mink.

consumer, its natural enemies include coyotes, foxes,
bobcats, snakes, and birds of prey (in particular owls). Douses predators with a noxious
liquid (though it cannot aim its spray like its cousin the skunk). The American mink lives
5-6 yrs. in the wild; up to 10 yrs. in captivity.

AMERICAN PIKA (*Ochotona princeps*): Also known as the "whistling hare" or "piping
hare," this asocial solitary lagomorph grows to a total length of 8½", its hind foot
measures 1⅜", and it weighs up to 4½ oz. Its silky dorsal coat is cinnamon-brown to
grayish brown, its ventral coat is lighter, usually buff to
whitish. It has the typical appearance and traits of the
genus *Ochotona*: dense fur; a stubby guinea pig-like
body; a mouse-like face; small rounded ears; short
limbs; and a concealed (though, in its case, quite long)
tail. The front feet have 5 toes, the back feet have 4
toes; all 4 paws are heavily furred, except the ends of
the toes, which have naked black pads. The ears have
dark hair on the inside and are encircled with a white
band of fur around the outer edge. It is gender
dimorphic, with the male being slightly larger than the
female.

American pika.

One of the smaller members of the rabbit family, this diurnal petraphilic herbivore
does not hibernate, nor does it make burrows. It colonizes isolated rocky embankments
and steep talus slides that are situated near open fields. Prefers cool environments and
high altitudes across its range: montane regions of the North American West, from
British Columbia to New Mexico.

This herbivorous ochotonid feeds on a wide variety of green vegetation, from
grasses, bark, sedges, and lichens, to shrubs, clover, thistles, and pine needles,
sometimes consuming its food immediately, other times caching it for future use.
During the summer it engages in "haying": pruning then drying its food on rocks in the
sun. Once cured, it gathers the cuttings into haystacks (up to the equivalent of 8-9 dry
gallons), moving them into storage during winter as food reserves.

The male establishes territories, which it marks
off and defends using both vocalizations and various
bodily secretions, including an oil from a scent gland
located in its cheeks. Thrusting its body up and
forward while calling enables it to throw its voice
some distance, a defensive strategy that confuses and
distracts its enemies. This makes it one of the few
animals in the world to employ the art of
ventriloquism.

The female gestates for about 1 mo., giving
birth to 1-6 altricial pups in late spring. After about

American pika.

1 mo. of nursing and care the young are weaned and
become independent. Extremely sensitive to climate changes and availability of food.
Near the bottom of the food chain, this member of the "crying hares" group is a primary

consumer that has numerous predators, including carnivorous birds (ravens, hawks, eagles) and small to medium-sized mammals (weasels, foxes, bobcats). The American pika lives for an average of about 3 yrs. in the wild; 7 yrs. in captivity.

AMERICAN SHREW MOLE (*Neurotrichus gibbsii*): It grows to a total length of 5", its tail is about 1½" long, its hind foot measures ¾", and it weighs up to ⅜ oz. This makes it the smallest mole in our region. Its pelage is uniformly gray above and below, with some grizzling on the hair tips. Though due to its dentition and skull size it is a true mole, its body size, forefeet, and overall appearance are shrew-like, thus its common name. It has a cone-shaped head, long whiskered muzzle, pink nose, small (nearly unuseable) eyes, no visible external ears, a hairy tail, and slightly enlarged, non-webbed front paws designed for excavation.

This North American eulipotyphlid is a riparian that inhabits wetlands and damp shrubland, grassland, thickets, chaparral, and coniferous and deciduous rainforests with soft deep soil, mature understory, abundant leaf litter, moist vegetation, and rotting logs. Prefers forest edges along streams. It ranges along the northwest Pacific Coast, extending from southern British Columbia south to central California.

Common shrew mole.

Known for being a carnivorous talpid and a rapacious vermivore, it is actually an omnivore that consumes fungi, seeds, lichens, nuts, grains, insect larvae, gastropods, and chilopods. Diurnal, nocturnal, and fossorial, its hunting methods include digging, running, swimming, and climbing, abilities in which it excels. Detects prey with its tactile snout. With its high energy needs it must feed nearly constantly, consuming at least its own body weight (or more) everyday in order to survive. To accomplish this it hunts for several minutes then naps for several minutes—round the clock.

Its polygynandrous breeding season runs from midwinter to midsummer. The female probably gestates for about 4-5wks., giving birth to 1-3 altricial young in a nest that, unusually, is constructed on the forest floor. Common, social, and extremely active both above and below ground, the natural enemies of this secondary consumer are reptiles, birds of prey, procyonids, mustelids, wild felids, and wild canids. Does not hibernate; may move about in small groups of a dozen or so. As an insectivore it is of great benefit to us. The American shrew mole probably lives 3-5 yrs. in the wild.

ANTELOPE JACKRABBIT (*Lepus alleni*): One of the largest hares in our region, it grows to a total length of 26", its hind foot measures nearly 6", and it weighs up to 10½ lb. Its upper coat is a grizzled cinnamon-brown with gray mottling. Its lower half is sandy or buff colored with hints of white and gray. The face, throat, and chest are cinnamon stained; its ears are a buffish color from top to bottom; its tail is white; its flanks have white hairs with black tips; its feet are furred.

It inhabits desert, grassland, foothills, chaparral, rangeland, arid shrubland, and savanna throughout its range: southern Arizona south into coastal Mexico. Does not make burrows; spends the daylight hours in various types of concealed shelter forms or shaded rock crevices in an attempt to avoid the heat.

Its large lean frame and long lanky limbs are not only designed for speed, they also help it regulate heat: the larger the surface area of an animal, the easier it is to rid the body of excess warmth. Its long 8" ears also play an important role in maintaining proper body temperature: in the summer months, when the blood vessels in its ears dilate, it holds the appendages erect, which helps dissipate heat. In winter months, when the vessels constrict, the ears are held close to the body, which conserves heat. Such adaptations make it well-suited to the weather-extreme biomes in which it lives.

Solitary, crepuscular, and nocturnal, this herbivore comes out between dusk and dawn to feed. Using its large eyes it is able to locate its favorite foods in near total darkness: coarse grass, cacti, leaves, twigs, buds, stems, and bark. As a number of hare species dine on animal flesh, it is possible that it also eats meat, perhaps mice and other types of rodents. Does not need to drink water as it metabolizes what it needs from the solid food it consumes. A graminivore and folivore, it is also both a coprophage and a geophage—though concerning the latter practice, it is not known if the dirt it consumes is intentional (that is, it is actively seeking minerals from the soil) or accidental (that is, soil is swallowed unintentionally along with the foods it consumes).

Though considered nonterritorial, there is often rivalry between males, which is usually resolved by boxing: the pair stand on their hind legs and punch at one another with their front paws. These bouts seldom lead to death, yet some can become quite violent, leaving one or both individuals with serious even life-threatening injuries.

Males and females may also behave agonistically toward one another, particularly during the breeding season, which takes place from early winter to early fall. The polyestrous female, who bears up to 4 litters a year, gestates for about 6 wks., giving birth to as many as 4 precocial leverets in a maternity nest lined with her own fur. As with many other hares species, the doe does not place all her young in one location. She spreads them about in different carefully concealed nests to protect them from predators (better that 1 neonate perishes rather than the entire litter). She nurses them only once every 24 hrs., always at night, covering over the entrance when done—more methods that help avoid drawing unwanted attention. The buck plays no role in infant care.

Antelope jackrabbit.

This lagomorph communicates via standard Leporidae channels: mainly through scent (it possesses special glands on its rump that emit a powerful musk), though it also squeals, cries, grunts, growls, and foot-stomps. The natural predators of this primary consumer are birds of prey, wild felids, and wild canids. It possesses all of the evasion tactics that typify the hare family, including simple concealment, flattening out its body, freezing in place, hopping on its hind legs, and huge antelope-like leaps, from which it derives its common name.

It is best known, however, for its ability to run up to 45 mph (it is probably the fastest of our hares), while both zigzagging and flashing its white flanks and white rump (the latter which also contributes to its common name). Such maneuvers can bewilder its pursuer, giving it enough time to escape. The life span of the antelope jackrabbit is unknown, though it is probably from 1-5 yrs., with the average being 1-2 yrs.

Arctic fox.

ARCTIC FOX (*Vulpes lagopus*): Also known as the "snow fox," it stands about 1' tall at the shoulder, grows to 3½' in total length, and can attain a weight of up to 21 lb. Its summer coat ranges in color from tawny, tan and brown, to gray, black, and grizzled, as well as mixtures of all these; its winter coat, however, is creamy or pure white. In some areas its thick wooly fur takes on a bluish sheen, prized by trappers and fur traders. This small carnivore is well adapted to its subzero habitat: its reduced surface area (compact body, small rounded, heavily furry ears, short legs, and snubbed muzzle) minimizes exposure to cold; its deep rich fur coat and bushy tail decrease heat loss; and its heavily furred paws (from which it derives its species name: *lagopus* means "hare foot"), provide warmth and traction in snow and on ice.

A chionophile, it inhabits tundra, ice floes, forest edges, and coastal areas throughout its range: mainly Alaska and northern Canada, but also sea ice as far north as the North Pole. This nomadic circumpolar canid travels further than any other mammal, sometimes journeying as far as 3,000 mi. during the winter months. It does not hibernate. In summer it dens up on hill slopes, inclines, cliffs, and banks; in winter it dens up in snow banks.

An opportunistic hunter and often solitary, at times it forms small hunting packs, feeding on lemmings, birds and their eggs, insects, fruit, squirrels, hares, fish, seal pups, carrion, and crustaceans, the latter meal which turns its white coat slightly pink. Follows polar bears in order to scavenge their kills; caches surplus food. Front-facing ears combined with sensitive hearing aid in hunting, which it does by listening for prey that burrows beneath the snow. When it locates its prey, the arctic fox jumps high into the air. The impact of its fall breaks through the top layer of snow, its body landing unerringly upon its victim.

Arctic foxes.

Its breeding season begins in midwinter and ends in midspring. Vixens (adult females) and dogs (adult males) form temporary pair-bonds and establish territories, which they scent-mark with urine and defend against conspecifics. The polyestrous female may bear 2 litters a year. She gestates for 6-7 wks., giving birth to 5-25 pups between April and June in large, often well-used den complexes that may extend to 300 sq. ft. and have over 100 entrances. The male contributes to the rearing of the young by guarding the family and bringing solid food to the den while the mother nurses. Pups become independent at around 2 mos. of age.

Communication and perception take place via standard Canidae channels: touch and smell, and to a lesser degree sight and hearing. A secondary consumer, its natural enemies are birds of prey, wolverines, and other foxes. The arctic fox lives around 5 yrs. in the wild; it lives for up to 15 yrs. in captivity.

ARCTIC GROUND SQUIRREL (*Urocitellus parryii*): The largest of the North American ground squirrels, it grows to a total length of 16", its pes is nearly 3" long, and it weighs up to 28 oz. Its pelage is golden to cinnamon-brown around the head and shoulders. Its back is grizzled brownish-gray to reddish, sometimes marked with small white speckles; its abdomen and legs are light brown to orangish-brown. The body is cylindrical; the head is prairie dog-like; the eyes are large; the ears are small; the legs are short but powerful; the manus are clawed and built for digging; the 6" tail is thinnish. Gender dimorphism is present: the male is larger than the female.

Holarctic, it inhabits alpine and subalpine meadows, taiga, tundra, riverbanks, lakeshores, brushland, forests, and mountains throughout Alaska, west to Siberia and east through northwestern and north-central Canada. There is no other mammal like it in this region, making it the world's most northern-dwelling ground squirrel. Extremely philopatric, this boreal creature lives in extensive, shallow, underground tunnels less than 3' deep, often above the timberline. Burrows have multiple levels and portals and serve a myriad of purposes: they provide protection from predators, insulation from the cold, nesting sites, and hibernaculum. It avoids permafrost areas. Males mark off their territories with scent glands.

A generalist feeder, the diet of this opportunistic omnivore includes leaves, fruits, sedges, grasses, roots, flowers, woody plants, stalks, foliage, mushrooms, and seeds (its genus name, *Spermophilus*, means "seed-lover"). As a carnivore it eats invertebrates, birds' eggs, birds, carrion, and other mammals, including its own kind. Caches surplus food. Stores up fat reserves for its 7-mo. hibernation period, which begins in the fall.

A colonial sciurid with a polygynandrous mating system, its breeding season takes place in early spring, shortly after emerging from hibernation. Males violently compete over estrus females, who gestate for about 1 mo., giving birth to 5-10 altricial pups in

May. At this time blood-related mothers form matricentric groups for the purpose of aiding each other in caring for neonates, and as added protection from both predators and males—the latter who will often engage in infanticide to prevent the spread of genes in offspring they have not sired. By about 1 mo. the pups are weaned; by 2-3 mos. of age they become independent and disperse.

It communicates and perceives via normal Sciuridae channels: sight, sound, smell, and touch. Extremely loquacious, its vocalizations include chattering, clicking, whistling, and various "cheek" and "chick" sounds and alarm calls. Its natural enemies are birds of prey (owls, hawks, eagles, falcons), mustelids, wild felids, wild canids, ursids, and conspecifics. Eskimos use its pelt to line their coats, hence its nickname, the "parka squirrel." The arctic ground squirrel probably lives about 5-7 yrs. in the wild.

ARCTIC HARE (*Lepus arcticus*): The largest hare in North America, it grows to a total length of 26", its hind foot measures nearly 7", and it can attain a weight of up to 15 lb. In our region, under the influence of photoperiodism, during the winter months its dorsal coat molts to white; during the summer months it molts to grayish-brown; the ventral coat remains whitish all year; the tips of the ears remain black all year.

As both its common name and its species name indicate, this lagomorph is a polar native who hails from the northernmost sections of our region, ranging across the upper tier of Canada, east to Newfoundland and Labrador. Its habitat is mainly open tundra, montane plateaus, and rocky plains. As an adaptation to its harsh environment, it has densely furred paws (insulation against the cold) and long sharp claws (for traction on snow and ice—but which it also uses for digging, rivalry, and self-defense). Will sometimes hop bipedally on its hind feet, leaving odd tracks in mud or snow. Primarily solitary and nocturnal, it can sometimes be found in super droves (groups) of 1,000-2,000 individuals. It spends the daylight hours in its burrow (either underground or in snow) napping and grooming.

Arctic hare.

An omnivore with a folivorous appetite, this primary consumer feeds mainly on willow tree parts; but it also eats grass, moss, lichen, roots, and seaweed; during cold months it will consume tree leaves, twigs, and bark. Like some of its lagomorph cousins, it will occasionally eat meat (mainly carrion) and consume snow as its water source. It breeds between early spring and late summer, the doe bearing 1 or 2 litters per year. She gestates for around 50 days, giving birth to as many as 8 leverets. The buck does not assist in infant care.

Communication with conspecifics occurs through boxing, licking, and scratching. Along with its keen hearing, its broad-field vision is augmented by lateral eye placement, which gives it the ability to see 360 degrees. The natural enemies of this leporid are birds of prey, ermine, foxes, lynxes, and wolves. It evades predators through stealth, concealment, camouflage, leaping, freezing in place, and running (it can attain a top speed of 40 mph). The average life span of the arctic hare in the wild is unknown, but it is probably 1-2 yrs.; in rare cases it may live as long as 3-4 yrs.

ARCTIC LEMMING (*Dicrostonyx torquatus*): Also known as the "banded lemming," "varying lemming," and "pied lemming," it grows to a total length of nearly 7", its hind foot measures ⅞", and it can weigh up to 4 oz. The pelage on its dorsum is yellowish-gray, the ventrum is rusty-gray to whitish; there is a thin black dorsal stripe. A whitish fur ruff encircles the neck, lending it another one of its many common names: the "collared lemming." It is the only member of the Rodentia order in our region whose coat turns white in winter. The body is stocky; the eyes are medium-sized; the ears are small and concealed under fur; the short ¾" tail is furred; its paws are specially designed for excavating in snow.

This cricetid is a palearctic native whose main habitation is tundra, and whose range, as its common name implies, extends across the northernmost arctic regions of Alaska and Canada. Shelters in rock crevices and snowdrifts; excavates tunnel systems in thawed soil or beneath the snow; subsidiary burrows contain sleeping, waste, and nesting chambers; constructs bolus-style natal nests of grass about 7"-8" in diameter. The summer diet of this herbivore consists of grasses, fruit, and sedges; in winter it feeds on buds, as well as on woody material such as twigs and tree bark. Its breeding season runs from early spring to late summer. The iteroparous polyestrous female bears at least 2 litters annually; she gestates for 2-3 wks., giving birth to about a half dozen pups. The male assists in the rearing of his offspring, suggesting that this species may be at least serially monogamous.

Communication and perception occur via standard Cricetidae channels: smell, touch, hearing, and sight. Its natural enemies are gulls, birds of prey, mustelids, wild felids, and wild canids. Shares much of the ecology and life history of its close cousin, the Labrador collared lemming (*Dicrostonyx hudsonius*). The arctic lemming probably lives an average of 1-2 yrs. in the wild.

ARCTIC SHREW (*Sorex arcticus*): This medium-sized soricid grows to 5" in total length, its hind foot measures ½", and it weighs up to ⅜ oz. Its long, nearly 2" tail is almost half its body length. Uniquely, it has tricolored pelage: a dark brown to blackish dorsum, light brown flanks, and a gray-brown ventrum. This coloring gives it its alternate common names: the "black-backed shrew" and the "saddle-backed shrew." Has a tapered muzzle, tubular body, small eyes, and short ears and fur.

A solitary riparian mammal, it haunts bogs, meadows, taiga, swamps, forests, mixed grass fields, hummocks, marshes, ditches, and wetlands throughout its range: Alaska and the Arctic Circle, south through Canada and into the northern Midwest states. Both diurnal and nocturnal, this voracious, highly active shrew has an elevated metabolism. Carnivorous and insectivorous, its diet is comprised primarily of insect larvae, sawflies, caterpillars, grasshoppers, centipedes, worms, and beetles; but it will also eat carrion and other small mammals (such as voles).

The breeding season of this North American native runs from spring through summer. The female gestates for nearly 3 wks., giving birth to an average of 6 offspring, weaning them by around 1 month of age. Communicates and perceives using standard Soricidae channels: smell, hearing, sight, and touch. A secondary consumer, its main natural predators are owls. The arctic shrew may live up to 1½ yrs. in the wild.

ARIZONA COTTON RAT (*Sigmodon arizonae*): The ecology and life history of this large vole-like cricetid is nearly identical to that of its close cousin the hispid cotton rat. The main difference, besides having less chromosomes than the hisbid, is that—as its common name suggests—the Arizona's range lies further south, extending from the southern U.S., southward all the way to South America.

ARIZONA GRAY SQUIRREL (*Sciurus arizonensis*): This tree squirrel grows to a total length of about 23" and a weight of around 1½ lb. The upper pelage is a grizzled steel gray, the lower pelage is white or off-white. Some individuals may have a mottling of dull rust along the back, the chest, the ears, and on the side of the face. The 12" tail is grayish-black above, yellowish below, and is edged with a thin white stripe, separating the upper and lower halves. Has a typical squirrel-shaped body, head, and limbs. Lacks ear tufts.

Gray squirrel, black variety.

This riparian sciurid is found in canyon woodlands and thick deciduous forests near water. Prefers broadleaf lowlands, especially walnut, cottonwood, and sycamore tree habitat. Its range runs from northern Mexico into New Mexico and, as its scientific name indicates,

Arizona. Shy, wary, quiet, and uncommon, its omnivorous diet varies with the season and population, but in general it feeds on seeds, nuts, bark, fungi, flowers, roots, stalks, fruit, and insects.

Its breeding season takes place in spring, with females gestating 60 days or so, then giving birth to around 3 pups. Dreys (nests) are constructed with leaves in the craws of trees. It communicates via sound, touch, and smell, and to a lesser degree, sight. Vocalizations include barking, chucking, and clicking sounds. Its primary natural enemies are snakes, mephitids, and wild felids. The Arizona gray squirrel may live 3-5 yrs. in the wild.

ARIZONA SHREW (*Sorex arizonae*): One of the smallest of the world's shrews, it grows to a total length of about 4" (nearly half of this is its 1¾" tail), has a ½" hind foot, and weighs less than ⅛ oz. Countershaded, its dorsum is gray to brown, its ventrum is light. This soricid possesses a tapered nose, small eyes, tiny ears, and short legs.

A secondary consumer, it inhabits mesic canyons and montane forests with thick hiding cover, from Arizona and New Mexico south into Mexico. Its breeding season occurs from late summer to midfall. The Arizona shrew probably lives less than 2 yrs. in the wild. More study is needed.

ARIZONA WOODRAT (*Neotoma devia*): Very similar if not almost identical to the desert woodrat, this cricetid occurs, as its name indicates, only in western Arizona, sharing much of the same ecology and life history as its close cousin.

ATLANTIC SPOTTED DOLPHIN (*Stenella frontalis*): Also known as the "spotted porpoise," this energetic muscular little delphinid grows to 7½' in total length and weighs up to 315 lb. Its upper body is grayish; the lower body is whitish. Irregular light spots, dashes, and swashes cover the entire body from nose to tail. It is the only truly spotted dolphin. Despite its common name, however, not all individuals are spotted. Spotting increases with age in those who have them.

Atlantic spotted dolphin.

Has up to 42 pairs of peg-like teeth in the upper and lower jaws; a pronounced melon (a rounded echolocation organ located in the forehead area); a long white-tipped beak; long curved flippers; and a recurved dorsal fin situated near the center of the body. Gregarious, it gathers in loosely formed pods of 6 to several hundred individuals, joining and leaving the group at will. Inhabits temperate and tropical waters of the Atlantic Ocean, ranging from Massachusetts to South America. Feeds on cephalopods and schooling fish; hunts the ocean floor for invertebrates; occasionally hunts in groups. Known to associate with its cousin the common bottlenose dolphin.

The female has one unspotted calf every 3 yrs. (on average), which she nurses for up to 5 yrs. Social and intelligent, like many other dolphin species, this fast-swimming, carnivorous marine mammal delights in playful activities, from surfing bow and stern waves to jumping high into the air. Can dive down to 200' and hold its breath for up to 10 min. A secondary consumer, its natural predators are orcas, sharks, and other toothed whales. The life span of the Atlantic spotted dolphin in the wild is unknown; possibly 30-50 yrs.

ATLANTIC WHITE-SIDED DOLPHIN (*Lagenorhynchus acutus*): Placed by some authorities in the genus *Leucopleurus*, this close cousin of the white-beaked dolphin (with which it is sometimes confused) grows to a total length of 9' and can attain a weight of up to 500 lb. As in all animals, there is individual variation in coloration. However, its dorsum is normally black, the ventrum is white, and the sides are gray and contain tan, yellow, or light gray lateral markings that swash across the skin. The body is robust; the beak is quite short; the pectoral flippers are long, curved, and narrow; the tail is stout; and the hooked dorsal fin is located in the middle of the back.

Each jaw is well equipped with nearly 2 dozen pairs of teeth. As is true of all beaked dolphins and whales, this delphinid's beak is not a nose, for the nostrils are located on top of the head—an adaptation to life as a marine mammal. In keeping with its dolphin profile, it is playful and enjoys riding the bow and stern waves of boats and large whales. Uses echolocation to track down its prey: crustaceans, cephalopods, and fish.

After a gestation period of about 10 mos., the cow bears one 50 lb. calf every 2 yrs. or so. This is

Atlantic white-sided dolphin.

a pelagic, cold water-loving species; thus, in our region, it prefers the deep cool waters of the North Atlantic Ocean (the source of its common name). It is abundant off the coast of Newfoundland, south to Massachusetts. Highly social, it travels in groups of all sizes, including superpods of 1,000 individuals or more; known to strand. A secondary consumer, its natural predators are sharks and orcas. The Atlantic white-sided dolphin may live to 30 yrs. in the wild. Little else is known.

Pronghorn.

B

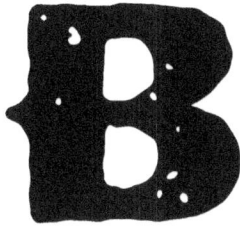

BAILEY'S POCKET MOUSE (*Chaetodipus baileyi*): One of the larger species of pocket mice, it grows to a total length of about 9 ½", its hind foot measures around 1", and it can attain a weight of up to 1⅜ oz. The pelage on the upper body is a mix of gray and yellowish hairs; the lower body is whitish; the rump is blackish; the long 5½" tail is bicolored with a terminal tuft. It is gender dimorphic: the male is larger than the female.

This coarse-haired heteromyid is a nearctic native that inhabits foothills, dunes, and slopes in or near flat, rocky desert locations with stony ground and thin vegetation. Its range is limited to the extreme southwestern area of the U.S. (small portions of southeastern California, southern Arizona, southwestern New Mexico), Baja California, and northwestern Mexico. Lives in subterranean burrows; establishes a home range of several thousand square feet.

Solitary and nocturnal, it is an omnivorous generalist with a diet centered around seeds, grains, leaves, nuts, and insects. Caches surplus food in a winter larder, transporting it in its cheek pouches or "pockets," from which it derives its common name. Its breeding season occurs in spring and summer. The doe probably gestates for about 3 wks., producing a litter of 2-4 altricial pups in the natal nest. She weans them around 1 mo. later. It is unlikely that the buck participates in the rearing of his offspring.

Communication and perception run along standard Heteromyidae channels: sight, smell, hearing, and touch. This terricolous rodent does not hibernate and is active all year long. Its natural enemies are reptiles, birds of prey, mustelids, wild felids, and wild canids. Beneficial as a soil aerator, seed disperser, and as a prey base for carnivores. Bailey's pocket mouse probably lives an average of 1 yr. or so in the wild; twice that long in captivity.

BANNER-TAILED KANGAROO RAT (*Dipodomys spectabilis*): The largest and heaviest of the kangaroo rat family, it grows to a total length of nearly 15" and a weight of almost 5 oz. Its upper body is a fulvous gray; the sides are golden brown; the underside is white. The front and back limbs are white; a white thigh stripe curves over and down the upper leg; each eye has a small supraocular white spot and each ear has a postauricular white spot. The 8½" tail is longer than the body, whip-like, and bicolored, with a partial white ring at the proximal end, and white and black lateral stripes meeting at the medial point of the tail, forming a subterminal black fur tuft succeeded by a white fur tuft at the distal point or tip. These striking flag-like tail markings give it its common name. The body is large and ovoid; the head is large with a convex profile; the eyes are large and dark; the nose is short with indistinct single white and black horizontal stripes running over the bridge onto the lower cheeks.

As befitting a rodent named after the kangaroo, it is well designed for a saltatorial lifestyle, with small, retrogressed front legs and large, powerful, 4-toed back legs. A gland on its back spreads a viscous liquid over the pelage, probably serving to coat the fur, providing insulation against heat, cold, and water. This species is one of the most highly gender dimorphic of the Heteromyidae family, with the male typically being much larger and heavier than the female.

This relatively big heteromyid inhabits arid regions in the southwestern U.S., such as desert, slopes, chaparral, creosote brushland, grassland, rangeland, shrubland, sand dunes, hills, and savanna, with gravelly desert pavement, dense soil, and sparsely scattered vegetation. Its range covers parts of Arizona, New Mexico, Texas, Utah, and northern Mexico. Each individual excavates and constructs its own massive burrow system, leaving conspicuous ejecta mounds and refuse heaps that can measure up to 4' high and 15' wide. These contain multiple side passages, galleries, auxiliary holes, and exits and entrances, the latter with radiating runways. It avoids soft soil in favor of hard soil, which provides better support for its many complex tunnels, portals, and escape shafts.

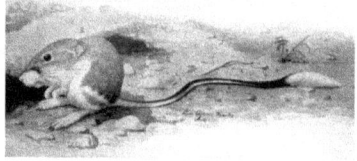
Banner-tailed kangaroo rat.

An herbivore with a strong granivorous inclination, its diet is built primarily on and around seeds. This is supplemented with grass tufts, mesquite, various green plants, and a myriad of other types of vegetation. Nocturnal and active all year, it forages at night, avoiding inclement weather by resting in its den. Appears to acquire most of its water from its food. Hoards and stores surplus edibles, transferring them to its storage den in its large cheek pouches.

This rat has no set breeding season and thus can mate at any time of the year. Bucks fight for rights to estrus does, who gestate for about 3 wks., giving birth to several altricial gray-furred pups in grass-lined natal nest chambers. Weaning takes place at around 30 days, though the offspring may choose to stay in their mother's burrow system for many months afterward. Being natally philopatric, most of the young will eventually return as adults to breed in the same area.

Communication and perception take place via the normal Heteromyidae suite of channels: sight, smell, touch, and hearing. Vocalizations include squealing, grumbling, shrieking, and squeaking. Well-known for its foot-thumping alarm signal. Its natural enemies are reptiles, birds of prey, mustelids, wild felids, and wild canids. Extremely fast and agile, it quickly disappears underground when trying to avoid predators. Considered pestilent due to destruction of grassland and the spreading of flea-borne bacteria. As a species it is on the decline. The banner-tailed kangaroo rat probably lives 4-5 yrs. in the wild.

BARBARY SHEEP (*Ammotragus lervia*): This large social artiodactyl stands 3½' at the shoulder, grows to about 6' in total length, and can weigh up to 300 lb. Similar to other North American sheep in appearance and behavior (but probably more closely related to goats), its overall coat is light brown; however, its long "beard" and chest fur are whitish. (This flowing fringe of hair, which grows from its throat down to it forelegs, is one of its main identifying traits.) Its horns grow up to 33" long, and curve up and backward over the neck and shoulders. It is gender dimorphic: the male is larger than the female.

Barbary sheep.

A terricolous diurnal native of northern Africa, it was introduced to the American southwest in 1950 as a game animal and is the only species in its genus. This agile climber and jumper lives in canyons, mountains, desert, and generally dry, arid, rocky regions. Prized for its meat, it is known as the "aoudad" by the Berbers of North Africa. A crepuscular grazer, this nonnative herbivorous bovid feeds on grasses, flowers, forbs, twigs, lichen, leaves, and shrubs, and is able to subsist for long periods without water—which it derives from the vegetation it consumes.

Breeding occurs during the fall months: the ewe mates once a year, gestates for around 5 mos., and gives birth to 1-2 kids in the spring. They can walk and climb almost immediately, and are weaned at 4 mos. of age. During the rut, rams head-butt and clash their horns together as they fight over dominance and females. Communication and perception occur via standard Bovidae channels: sight, hearing, smell, and touch. A primary consumer, it probably has the same natural predators as native North American ovids and caprids: wild felids, wild canids, and ursids. Beneficial as food for carnivores and as a game animal for hunters. The Barbary sheep lives for about 10 yrs. in the wild; 20 yrs. in captivity.

BEARDED SEAL: (*Erignathus barbatus*): The largest of the arctic seals, it attains a total length of 12' and can weigh up to 1,000 lb. Its fur is countershaded, with the upper body running from yellowish-gray to dark brown, the lower body yellow-gray to light silver. The head and flippers are often a reddish or rusty orange. Its blubber-laden, fusiform body is stout but long; the head is squarish and smallish; both the front and back flippers are well-clawed; the eyes are large and dark. The pelage lacks obvious markings. Its most distinguishing characteristic is its "mustache": dense tufts of bristly but sensitive whiskers that spring from each side of its snout. Despite its common name, it does not have a "beard" (though it may appear to have one from a distance), and would have been more appropriately called the "mustached seal." Its large squarish foreflippers have bestowed upon it the nickname the "square flipper seal." Gender dimorphism is present, with the female being larger than the male.

Bearded seals.

This littoral phocid is circumpolar. It inhabits the cold marine shoals of arctic and subarctic regions, where it hauls out on drifting pack ice to bask and rest, riding floes over the shallow feeding grounds that it prefers. It is sometimes found in open water. Two subspecies are recognized: *Erignathus barbatus nauticus* is found from Alaska east to Asia; *Erignathus barbatus barbatus* is found from eastern Canada to Scandinavia. Establishes a home range of about 5 sq. mi., which it defends against unwanted conspecifics.

This lethargic carnivore is also a molluscivore and a piscivore, with a benthic diet consisting primarily of bottom dwelling fish, crustaceans, and mollusks (which includes cephalopods). Occasionally pelagic fish, like cod, are consumed. Normally dives to a depth of about 300' in search of its prey, but is known to go down to over 1,000' when necessary.

Solitary and diurnal, this natatorial mammal is polygynandrous, with a breeding season that begins in early spring and ends in early summer. The female gestates for about 11 mos., employing delayed implantation so that her single 5', 70 lb. pup will be born at an optimum time of the year (namely spring or summer). Birth takes place on top of an ice floe, the mother staying with her infant until it is weaned 3-4 wks. later, after which she abandons it to its fate. During this period she does not eat, but instead devotes herself solely to nursing and protecting her pup, which quickly gains 100 lb. during its first week of life. The male plays no role in the rearing of his young.

One of the most vociferous of all oceanic creatures, the slow-growing bearded seal communicates via traditional Phocidae channels: touch, sight, smell, and sound. The male's remarkable vocalizations border on the strange and eerie, and are unique to each individual. Its mating song can be heard up to 10 mi. away. Its vocal repertoire includes trills, beeps, moans, bird-like tweets, pulses, sweeps, and siren-like whistles, some which can sound more like a UFO from a low-budget 1950s movie than a marine mammal.

Its main enemies are polar bears, orcas, and walruses (the latter which feeds on the young). It is hunted by native peoples for its meat, hide, and oil. In order to protect itself from predators, the bearded seal rides on small ice floes with its head facing the water. To avoid bears specifically, when it is on shorefast ice it remains close to ice holes. Both of these defensive strategies permit a quick and unimpeded escape when in danger. It sleeps "standing" upright in the water, with its head at the surface.

Because it relies on ice for resting, pupping, and feeding, it is known as an "ice seal" (one of 4 species), and is thus particularly susceptible to climatic changes that affect the arctic ecosystem. The bearded seal lives 25-30 yrs. in the wild.

BELDING'S GROUND SQUIRREL (*Urocitellus beldingi*): This small ground squirrel grows to a total length of around 11" and a weight of about 12 oz. Pelage coloration varies with the population. Generally the upper body is grayish and the belly is a grizzled gray, or the upper body may be brownish and the belly fulvous. The tail is grayish or brownish on the dorsum, rusty on the ventrum, with a darkish tip. Sometimes the chest is pinkish and there is a cinnamon-brown wash over the back. The head is small; the eyes are medium-sized and ringed in white; the ears are small; the limbs are shortish; the tail is flattish and, at 2½" in length, quite short.

An open area dweller, it inhabits short-grass environments such as hayfields, pastures, roadside brush, and meadows in the alpine and subalpine montane regions of the western U.S., a range that extends from Oregon and California east to Idaho, Nevada, and Utah.

This little sciurid is omnivorous with an herbivorous diet that includes nuts, leaves, weeds, bulbs, stalks, flowers, grains, roots, fungi, agricultural crops, and, as its scientific name *Spermophilus* ("seed-lover") indicates, seeds. Consumes animal matter as well, mainly insects, but also birds' eggs and carrion. Cannibalism is sometimes practiced by immature males, who will eat the young of unrelated females if given the opportunity.

Belding's ground squirrel.

Its promiscuous breeding season runs from late spring to early summer, though a female is only receptive for a few hours. Males compete violently, and even kill one another, for access to estrus females, instigating both battles and mating chases. Females gestate for 3-4 wks., giving birth to 2-10 altricial young (the number depends on the age of the mother) in an underground nest lined with grass. Mothers nurse their babies at 5 pairs of mammary glands for about 1 mo., after which they are weaned—the male young dispersing, the female young staying with the matriarchal group at or around the main burrow site. Mothers who have lost neonates to predation or inclement weather (quite common) sometimes travel to a new colony, where they may practice infanticide, killing the offspring of one of the new colony's mothers in order to take over her nesting burrows. The father plays no role in infant caretaking.

Hibernation begins in late summer and lasts for 8-9 mos., one of the longest known hibernation periods of any mammal. This means that each year an adult must perform nearly all of its life functions (such as searching for food, excavating burrows, finding a mate, and breeding) in the span of only 3 or 4 mos. This diurnal semicolonial rodent often stands up on its hind legs in order to survey its territory. Engages in altruistic behavior, such as warning its relatives of danger.

Communicates and perceives via traditional Sciuridae channels: sight, touch, smell, and hearing. Vocalizations include chirps, chips, whistles, trills, and various warning cries. Its natural enemies are birds of prey, mustelids, wild canids, and ursids. Belding's ground squirrel lives from 3-6 yrs. in the wild.

BELUGA (*Delphinapterus leucas*): Also known as the "beluga whale" and the "white whale," this medium-sized toothed cetacean grows up to 20' in total length and can attain a weight of 4,000 lb. (2 tons). The only all-white whale in the world (ideal camouflage for its snowy, icy home), unlike some other odontocetes, it has teeth in both the upper and lower jaws. Calves are gray, becoming lighter and finally white as they mature. Flippers are short, squarish, and paddle-shaped. The adult has a small head for its body, which

Beluga.

is stout and cylindrical. Despite this, its malleable forehead is quite massive—extra space needed to house its melon, a complex biological mechanism that it uses to pick up returning echolocation sounds.

A close cousin of the narwhal (with which it sometimes interbreeds), the beluga's neck is unusual for a cetacid: being jointed rather than fused (as in most other whale species), it is highly flexible and can turn sharply to each side—the better to hunt and evade predators. Its genus name, *Delphinapterus* (Greek for "wingless dolphin"), refers to the fact that it lacks a dorsal fin (possibly an evolutionary adaptation to a life spent swimming below ice sheets). Its species name, *leucas*, is Greek for "white," the color of its skin. Its 6" thick layer of blubber acts as both a nutrient reserve source and as insulation against the frigid seas in which it ranges: mainly Arctic and sub-Arctic waters, as well as around open polar ice packs. Migrates north in the spring; south in the fall.

Though it has no vocal cords, this loquacious marine mammal can "speak" through its nose (blowhole), which is surrounded by several sound-producing air sacs. It was once known as the "sea canary" or "canary whale" due to its ability to create a wide repertoire of bird-like—as well as a host of other—sounds, from grunts, trills, chirps, moos, clicks, groans, and whistles, to squeals, cries, bleats, clucks, clangs, moans, and shrieks. Wild belugas have even been known to imitate human voices.

Extremely sociable and tactile, this monodontid travels in pods (sometimes in gender-specific pods) and inhabits rivers, canals, shallow bays, coves, and estuaries, where thousands may gather to molt (scrape off old yellowish skin on the sandy or gravel bottom). A suction-feeder (with exceptionally flexile lips), it uses echolocation to hunt down its prey, which includes squid, octopus, aquatic crustaceans (shrimp, crab), sea snails, worms, and fish (cod, flounder, salmon, herring, smelt). Does not chew its food, but uses its teeth to grasp and swallow its prey whole.

Will hunt in small packs in order to drive its quarry into shallow water. Though

Beluga.

it spends much of its time at the water's surface, when necessary it can dive to a depth of a little over ½ mi. and remain underwater for nearly 20 min. Its highly maneuverable flippers makes it extremely agile in the water, even enabling it to swim in reverse. Breaks open ice with its head to create air holes. The cow gives birth to 1 calf every 2-3 yrs. A secondary consumer, its main enemies are polar bears and orcas. The gentle, slow-swimming, playful, highly intelligent beluga may live up to 80 yrs. or more in the wild; however, half that many yrs. is probably more common.

BIG BROWN BAT (*Eptesicus fuscus*): Its common name is well-earned: it grows to 5" in total length, its forearm is 2" long, it has a wingspan of nearly 13", and it weighs ⅝ oz. Its sizeable head holds 32 large teeth, a pair of large eyes, and a large nose. Its long fur is a grizzled, cinnamon, or reddish-brown on the dorsum; the ventrum is lighter in color, usually buff to gray-tan. The ears, nose, wings, and interfemoral membrane are generally darkish to black; the calcar is keeled; the tragus is wide; the wings are broad and short; the ears are rounded; the tail extends just beyond the uropatagium.

An international chiropterid, it inhabits mainly forests, but is also found in urban areas, residing in any building into which it can gain entrance. It ranges from Canada southward, covering most of the lower 48 U.S. states, continuing south into Central America. This brown widespread vespertilionid has a preference for sheltering in human-made structures, particularly houses; hence its scientific name: *Eptesicus* ("house flyer") *fuscus* ("brown"); that is, the "brown house flyer." When necessary this petraphile will also roost in caves, mines, trees, rock piles, and rocky cliffs. It hibernates during cold months.

Big brown bat.

Though she mates in autumn, the female delays fertilization until spring, gestates for about 60 days, then gives birth to 1 or 2 pups between late spring and early summer. Infant-rearing takes place in a female-only maternity colony, where the mother nurses, grooms, and protects her young until they are weaned. When a pup falls to the ground from its perch (a common occurrence), its loud desperate squeaks help the mother find and identify it among its fellow neonates.

Bat skeleton.

A highly beneficial insectivore, its diet consists primarily of agricultural pests, such as beetles (which it crushes with its powerful jaws and sharp teeth); it also feeds on wasps, ants, flies, dragonflies, and moths. It tracks down its prey, over both land and water, using echolocation. Since it cannot eat in winter, it must build up enough fat reserves to carry it through hibernation, which typically begins in the late fall. This nocturnal hunter is the fastest bat on record, with a top flying speed of 40 mph (as fast as a wild horse). A secondary consumer, its enemies include birds of prey (such as falcons), reptiles (such as snakes), and small mammals (such as raccoons). The big brown bat lives 15-20 yrs. in the wild.

BIG FREE-TAILED BAT (*Nyctinomops macrotis*): It grows to a total length of about 6", its forearm is around 2½" long, its wingspan measures nearly 17", and it can attain a weight of up to 1 oz. Its smooth fur is a rich cinnamon-brown to black on the dorsum, buff on the ventrum. Its ears face forward and connect at the root; the upper lip has vertical grooving. Its wings are long, pointed, and narrow, and are generally darker than the pelage. This molossid's large size, and the fact that its tail extends beyond its tail membrane, give it its common name. Gender dimorphic, the male is larger than the female.

A petraphilic secondary consumer, it shelters in rocky crevices, outcroppings, and cliffs, as well as shrubs, trees, and buildings throughout the temperate and tropical regions of its range: British Columbia through America's southwestern states, and south into Mexico and beyond.

A powerful flyer and a strong climber, this nocturnal hunter takes to the air after sundown, using echolocation to search out its prey, primarily moths, but also terrestrial insects—making it our benefactor. It mates in late winter. After forming a nursery colony, the females gestate for about 10 wks., giving birth to a single pup in late spring or early summer. They nurse, groom, and provide general care for their offspring for about 12 wks., at which time the pups begin to fly and feed on their own.

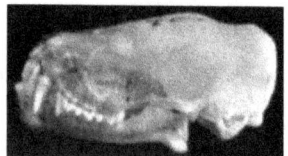

Bat skull, *Nyctinomops* genus.

Migrates each year. Highly vocal, while hunting its echolocation call resembles a loud chatter-like sound. During America's War for Southern Independence (1861-1865), the Confederate army used this chiropterid's nitrogen-rich dung, known as "guano," to produce gunpowder. Though widely distributed, this species is rare and in decline, and thus is not often seen. The life span of the big free-tailed bat is unknown.

BIGFOOT: See Sasquatch.

BIGHORN SHEEP (*Ovis canadensis*): The official state animal of Colorado, it is also known as the "Rocky Mountain bighorn sheep." It is similar to the mountain goat in size: it stands about 3½' at the shoulder, grows to about 6' in total length, and weighs up to 300 lb. Its summer pelage is brown; its winter pelage is grayish. The belly is sometimes buff-colored. The ram's horns, which form massive C-shaped curls around the sides of the head, can spread to 33" and weigh as much as 30 lb. If its horns grow too large and begin to block its vision, a ram will "broom" them by breaking off the tips or by sanding them down on rocks. It is gender dimorphic: males are larger than females.

Bighorn sheep.

Its powerfully built frame and well-designed hooves make it a nimble petraphilic climber, perfectly suited to its habitat: high rocky terrain, alpine meadows, craggy bluffs, desert, valleys, rangeland, chaparral, and steep mountain slopes and cliffs. A nearctic native, it ranges from southern Canada south into Mexico, covering numerous Rocky Mountain states and their neighbors, including California, Colorado, Nevada, and Texas.

Mildly migratory and quite social, it congregates in herds of 10-100 individuals—though mature rams (along with young males) and ewes (along with young females) usually form separate groups, the former known as bachelor herds, the latter best described as matriarchal herds. Unlike many other members of the Artiodactyla order, it will often use the same bed patch (a 4' wide depression) for many years rather than create a new one each night.

This diurnal herbivore feeds mainly on grasses, forbs, shrubs, and leaves, procuring much of its water needs from the vegetation it consumes. It communicates through various typical sheep vocalizations, as well as scent. The mating season begins in midsummer and ends at the start of winter. The ewe breeds once a year, gestates for about 6 mos., and bears 1-2 precocial kids in the spring. They are weaned and independent by 6 mos. of age. During the rut, males engage in head-butting tournaments, charging one another at 20 mph, creating a sharp "crack" that can be heard over a mile away. These violent contests can last 24 hrs. Their double-layered skulls, extra thick tendons, reinforced bones, and heavy neck and spine musculature help cushion the tremendous blows they experience during combat.

A primary consumer, this ovid has numerous predators, including the lynx, bobcat, wolf, mountain lion, bear, and coyote; golden eagles pursue lambs, sometimes purposefully knocking them off cliff faces. The bighorn has evolved numerous evasion tactics to avoid its enemies: not only can it swim, but on level ground it can bound 15' and run up to 30 mph; it can ascend a cliff face at 15 mph on as little as 2" of ledging; while descending a steep slope it can bound up to 30'. These feats are made possible, in great part, by its hard-rimmed hooves which contain soft absorbent centers (traits it shares with the mountain goat), ideal for rock-

Bighorn sheep.

climbing and cliff-jumping. Escape from predation is also aided by its extraordinary eyesight, with which it can see the slightest movement up to 18 football fields away. Communicates and perceives along standard Bovidae channels: vision, hearing, smell, and touch. The bighorn sheep lives an average of about 5-6 yrs. in the wild, though under ideal conditions it may live as long as 20 yrs.

BLACK-FOOTED FERRET (*Mustela nigripes*): The only ferret native to our region, it stands 6" at the shoulder, is about 24" in total length, and can attain a weight of up to 2½ lb. Sports a black eye mask highlighted by white surrounding fur. Has a tan coat, a yellowish throat patch, a black nose, and a black-tipped tail. Its common name derives from its distinctive black legs and feet, which help camouflage it in the prairie environment it occupies. Has a lithe tubular body, a tapered face, small rounded triangular ears, a long neck, sharp teeth, powerful jaws, short legs, and large clawed front feet (for digging and defense). Gender dimorphic, the male is somewhat larger than the female.

This motile endothermic mammal is a nearctic native that inhabits grass prairies, arid areas, hilly terrain, savanna, and prairie dog towns (the larger the better). Its present range is small, covering only portions of Wyoming, Montana, and South Dakota. Its home range covers about 150 acres.

Slender, alert, and agile, this obligate carnivore is physically well-adapted to its primary dietary preoccupation: stalking its favorite underground prey, the prairie dog—which makes up 90 percent of its diet. When necessary it augments this fare with reptiles, birds, eggs, mice, voles, rabbits, gophers, and squirrels. It can consume 100 prairie dogs a year (a benefit to humans since prairie dogs can carry bubonic plague).

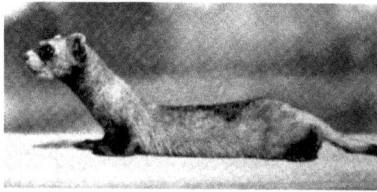

Black-footed ferret.

After surprising and killing its sleeping prey in its burrow, the ferret takes over its victim's domicile, often turning it into its home. Nocturnal, it spends nearly 90 percent of its life underground (making it difficult to locate, observe, and study). It does not hibernate. May travel up to 5 mi. a night in search of food.

A largely solitary mustelid, its breeding season occurs during the spring. The jill (female) employs delayed implantation and gestates for 5-6 wks., then gives birth to 1-7 altricial kits in early summer—usually in an old prairie dog burrow. The young leave the natal nest after about 6 wks. and become fully independent a few months later in autumn. The hob (male) does not participate in the rearing of his offspring.

Playful, curious, energetic, and acrobatic, the black-footed ferret has excellent senses and communicates and perceives using a variety of cues and methods that are standard to most members of the Mustelidae family. These include: vocal (chattering, whimpering, chortling, hissing), chemical (scent-marking using glands under its tail), and tactile (touching, rubbing). A secondary consumer, its natural predators are reptiles (snakes), birds of prey (hawks, owls, eagles), badgers, foxes, coyotes, and bobcats.

Once thought to be extinct, its distribution today is a mere fraction of its original historic range, which ran from Alaska to Mexico (centering mainly on America's Great Plains). Due to disease, habitat destruction, and most importantly, the mass elimination of prairie dogs, however, during the 20th Century its range shrank dramatically, and the black-footed ferret became (and still is) one of the rarest of all North American mammals. Careful conservation has brought it back from the brink of extermination. Reintroduced to some of it former areas (using offspring from the remaining original group of less than a dozen individuals), it is making a comeback in parts of Canada, Mexico, and the U.S. The highly endangered black-footed ferret lives up to 4 yrs. in the wild; 10 yrs. in captivity.

BLACK RAT (*Rattus rattus*): More commonly known as the "house rat," this medium-sized exotic rodent grows to a total length of 18" and a weight of up to 13 oz. Despite its common name, it is not generally black, but rather gray to brown in color on the dorsum, creamy light gray on the ventrum (subspecies come in a variety of colors), cryptic coloration that helps camouflage it in its shadowy nighttime environments. The body is slender and cylindrical; the fur is dense and coarse; the head is large with a convex profile; the ears are large, pinkish, and erect; the eyes are dark and beady; the nose is robust with a pinkish tip; the limbs are short; the feet are pinkish; the toes are clawed; the long, monochromatic, 10" tail is scaly, sparsely furred, and either pinkish, brown, or gray in color.

Black rat.

Since they can appear similar and both can have black or brown fur, it is commonly confused with its cousin the Norway rat. Yet the two display overt differences: the body of the black is generally much lighter, slimmer, and lighter; its head is shorter, more gracile, and straighter; its eyes and ears are larger; its tail is longer in comparison to body length; and its nose is pointed while the Norway's is blunt. The black rat is gender dimorphic, with the male being larger than the female.

This cosmopolitan nonnative is an invasive Asian species that came to North America with the English settlers at Jamestown, Virginia, in 1607. From there it has radiated out across the entire coastal region of lower North America, from British Columbia, Canada, to California, from Mexico to Georgia, and from Florida to New England, northward to Ontario and Nova Scotia. Though unlike its larger, more aggressive cousin the Norway rat, it does not like to swim, it is a synanthropic species that lives chiefly in and around ships and oceanside buildings—lending it one of its many nicknames, the "ship rat." Away from its urban and suburban haunts, it inhabits a myriad of biomes and ecosystems, from northern temperate scrubland, chaparral, rangeland, and woodland, to tropical grassland, savanna and forest. It is found as far inland as Nevada in the west, and Kentucky in the east. Outside North America it has spread to every continent and is one of the world's most ubiquitous mammals.

Social, non-hibernating, riparian, and nocturnal, it lives in large colonies with individuals of both genders and all ages. Alpha males and alpha females dominate the group, which operates according to strict hierarchical social rules.

Rattus rattus is both a highly adaptable terricolous creature and an arboreal one, taking up residence where convenient, from sewers, basements, tunnels, caves, and the abandoned underground burrow systems of other mammals, to trees, hollow logs, and snag, as well as the rafters, walls, attics, and ceilings of houses, barns, factories, warehouses, and other human-made structures—hence another nickname: the "roof rat." Has a particular fondness for islands, seaports, docks, shipyards, and the hulls and holds of ships, on which it regularly travels as an unwanted stowaway. Will often burrow under buildings to den, where it constructs stick and grass natal nests. It is especially common along waterways and in large Southern and Pacific coastal cities. Establishes a home range of about 500 sq. ft.

This omnivorous, opportunistic murid is a dedicated granivore whose preferred diet is comprised of wild seeds and cultivated grains and cereal crops—which, to the detriment of the farmer, it destroys by both consumption and contamination from bodily waste. A gastronomic generalist, these plant foods

Black rat.

are supplemented with woody edibles, such as leaves, stalks, bark, and wood, and also nuts and fruits. When available, it will also consume gastropods, arachnids, insects, birds' eggs, and young birds.

Its polygynous breeding season runs all year, with the maximum matings occurring during the late summer. The most aggressive bucks accrue the majority of breeding rights, passing their genes onto the next generation. The polyestrous doe of this species is one of the world's most prolific mammalian reproducers: gestating for about 3 wks., she can bear as many as 5 litters (of up to 20 altricial pups each) in a single year. The young are weaned and independent at 1 mo. of age, and are themselves ready to breed just 2-3 mos. later. The father does not assist in the rearing of his offspring.

Communication and perception occur via standard Muridae channels: hearing, smell, and touch (its sense of sight is probably poor). Vocalizations include hissing, chittering, squeaking, chattering, and squealing. Scent-marks its territory as a defense against outsiders. As this skillful climber runs and leaps about, it uses its lengthy naked tail—which is longer than its head and body combined—to help it balance.

It has long been considered a dangerous pest and a notorious health menace due both to the massive damage it inflicts on crops, and for hosting the flea-borne bacteria that carries "black death" (bubonic plague)—among many other dangerous diseases that are deadly to both humans and other mammals. Its impact is so destructive that it is, in fact, responsible for the catastrophic decline and extinction of untold numbers of plant and animal species worldwide. Benefits the global ecosystem, however, by providing food for carnivores and for hosting a number of endoparasites and ectoparasites. It is useful to us directly as a subject of scientific research.

Its natural enemies are reptiles, birds of prey, wild felids, and wild canids. The black rat lives an average of 1-2 yrs. in the wild; twice that long in captivity.

BLACK-TAILED JACKRABBIT (*Lepus californicus*): Despite its name, it is not a rabbit, it is a hare; which is why it is larger and thinner, with longer ears and legs, than a true rabbit. This lagomorph grows to a total length of 25", its tail can reach 4½", and it weighs up to 8 lb. The coloration on its upper body is a grizzled brown-gray; its lower body is white to buff-brown. Distinguishing features are its large ringed eyes; its long

Black-tailed jackrabbit.

mule-like ears (inspiring one of its original names, "jackass rabbit"—later abbreviated to "jackrabbit"), which grow to 5" and are black-tipped and trimmed with light fur; its massive 6" hind feet; and its tail, which has a black dorsal stripe (from which it gets its common name) that extends up onto the hindquarters (and sometimes up the back). Gender dimorphism is present, with the female being larger than the male.

This leporid is most often found in arid biomes and ecosystems, including desert, prairie, scrubland, rangeland, rough grassland, and dunes. Also inhabits cultivated fields and agricultural areas, such as farms, cropland, and pastures. Its range covers much of the western U.S., from Washington and California to Colorado and Missouri, running south to Baja California. Mainly nocturnal, it shelters during the day in forms (rather than burrows), where it naps and grooms itself.

The most populous and widely distributed of all the North American jackrabbits, this endothermic herbivore feeds mainly on grasses, tubers, sagebrush, cacti, and fruit during the warm months, and woody foods (twigs, leaves, stems, bark) during the cold ones. It seldom needs to drink due to the amount of water found in the plants it consumes.

Its breeding season runs year-round. The doe mates 3-4 times a year, gestates for about 5 wks., and bears as many as 8 precocial leverets in a nest lined with fur plucked from her belly. She nurses her young for only a few days—and then only at night, so as not to draw the attention of predators. The babies that survive the rigors of infancy become independent within 3 wks. The buck does not contribute to infant care.

The natural enemies of this primary consumer are snakes, owls, hawks, weasels, badgers, bobcats, foxes and coyotes, which it evades by freezing in place, swimming (dog-paddling), leaping (up to 20'), and running in a zigzag pattern (at speeds of up to 35 mph). It uses its camouflaging fur to great advantage, and altruistically alerts other jacks to danger by squealing, thumping its hind feet, and flashing the white underside of its tail. The black-tailed jackrabbit lives 1-2 yrs. in the wild; up to 6 yrs. in captivity.

BLACK-TAILED PRAIRIE DOG (*Cynomys ludovicianus*): Also known as the "plains prairie dog," this medium-sized cynomyid grows to a total length of around 16" and can weigh as much as 3 lb. On the dorsum its pelage runs from buffish-brown to auburn-brown; on the ventrum it is yellowish-white; the 4" tail is thin, rusty-brown, and tipped with black, the source of its common name. Gender dimorphism is present, with the male being larger and heavier than the female.

This robust sciurid is a nearctic native that inhabits temperate arid prairies rich in short grass, cacti, and sagebrush. Also found in and around cropland, ranches, mixed grass savanna, steppe, rivers, hedgerow, and disturbed areas, nearly all with silty, loamy or pebbly soil. Its range stretches in a thin strip from southern Canada (Alberta and Saskatchewan) to northern Mexico, roughly following the arch of the Great Plains (and Rocky Mountains) from north to south. States covered: Montana, North Dakota, Wyoming, South Dakota, Nebraska, Colorado, Kansas, Arizona, New Mexico, Oklahoma, and Texas.

Black-tailed prairie dog.

Fossorial and terricoulous, it digs large burrow systems, some as deep as 12' and over 100' long. Ejecta mounds of excavated earth, which mark the entrances and exits, serve as surveillance perches from which it keeps watch for predators. Chambers within its burrows are used for a myriad of purposes: eating, resting, sleeping, shelter, protection, waste, mating, nesting, and raising offspring. Cuts down and eats much of the vegetation around its burrow in order to eliminate places where carnivores might hide. The most gregarious of the sciurids, and one of the most social of all mammal species, the black-tailed prairie dog forms massive colonies or "towns" comprised of thousands of individuals. This huge community, in turn, is made up of small family groups known as "coteries," each one whose members include an adult male, several adult females (the male's harem), and 2-3 generations of their offspring. The coterie covers a large area, upwards of 45,000 sq. ft. or more.

A diurnal, crepuscular, sedentary, opportunistic omnivore, its diet surrounds green vegetation, which includes numerous types of grasses (bluegrass, purple needle grass, wheatgrass, blue grama, burro grass, etc.). It also consumes cacti, thistle, leaves, seeds, forbs, flowers, sagebrush, stalks, nuts, wood, saltbush, grains, roots, bison dung, an occasional insect, and more rarely small mammals. It is able to meet its daily water requirements from the succulent foods it eats.

Its polygynandrous breeding season runs from late winter to early spring (depending on geography). The female gestates for about 5 wks., giving birth to an average of 4 altricial pups in a subterranean grass-lined maternal nest. Weaning occurs by about 6 wks. or so, though independence does not arrive for another 12-18 mos. At this time the young males strike off on their own, while the young females remain permanently within their mother's coterie. The father does not participate in the rearing of his offspring, though his natural territoriality prevents outside males from intruding onto the group's turf.

Communication and perception take place via standard Sciuridae channels: sight, touch, smell, and hearing. Its many vocalizations include chirps, yips, squeals, snarls, whistles, and barks—the latter sound from which it derives both its common name and its genus name (*Cynomys* is Greek for "dog mouse"). This keystone species estivates during extreme heat; does not hibernate, but remains active all year, storing up extra body fat for the lean cold months; engages in allogrooming.

Beneficial as a parasite host; soil aerator, fertilizer, and recycler; commensal shelter supplier (for other animals); and food for carnivores. Its role as an ecosystem engineer has both positive and negative effects on the land. Though it would seem that many of the disagreeable aspects commonly ascribed to this mammal have been overstated, nonetheless many regard it as a pestilent species simply for being a carrier of deadly bacteria (that produce several serious diseases, such as bubonic plague, sylvatic plague, pneumonic plague, and spotted fever), a destroyer of agricultural crops, an aggressive consumer of grasses (that is often needed for livestock), and the excavator of ground holes that pose a perpetual danger to the fragile legs of farm and ranch animals.

Its natural enemies are reptiles, birds of prey, mustelids, wild felids, and wild canids. The male black-tailed prairie dog lives an average of 4 yrs. or so in the wild; females may live a few years longer.

Black-tailed prairie dog.

BLAINVILLE'S BEAKED WHALE (*Mesoplodon densirostris*): Also known as the "dense-beaked whale" (for its extremely dense jawbones), it gets it common name from the French zoologist Henri de Blainville, who first documented it in 1817. One of nearly

Blainville's beaked whale.

two dozen species of beaked whales, it attains a total length of 15½' and weighs up to 2,300 lb. (a little over a ton). Possessing a flat forehead, a small forward-projecting blowhole, smallish flippers, a triangular hooked dorsal fin, and an arched lower jaw, its most distinguishing characteristics are two large teeth or tusks that grow upward out of the lower jaw from bumps located on the upper face. Coloring is dark gray-blue on the dorsum, light gray-blue along the ventrum, with irregular tan-gray spots occurring randomly over the skin—good camouflage for its ocean-going lifestyle.

While it prefers deep water, it sometimes roams the shallow waters around islands. A nearctic, palearctic, and neotropical ziphiid, it inhabits all three major oceans, from tropical to temperate regions. Feeds on fish and squid, sucking in its prey with seawater rather than biting down on it. It can dive to a depth of 5,000' and stay underwater for nearly 1 hr. It has been known to strand; its vocalizations include whistling while diving. A secondary consumer, its natural predators are sharks and orcas. Little else is known of its ecology and life history.

BLUE WHALE (*Balaenoptera musculus*): At 110' long, it is the exact same length as a Boeing 737-700 jetliner; this is over ⅓ the length of a football field or as long as 3 school buses parked end to end. At 420,000 lb. (or 210 tons), it weighs as much as 1 train locomotive, 17 school buses, 42 elephants, or 105 cars. Thus it weighs about 2 tons a foot. Its heart tips the scale at 2 tons, the weight of a rhinoceros; its tongue alone weighs 3 tons, the same as a hippopotamus.

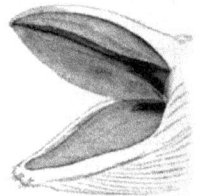

This balaenopterid has the distinction of being the biggest marine creature, the biggest whale, and the biggest *mammal* that has ever lived. It is also the biggest animal in the world today, and it is one of the largest animals that has ever lived. It has the thickest blubber relative to its size of any rorqual. It can

Open mouth of a balaenopterid showing its baleen.

also claim the title of being the heaviest known beast that has ever existed, though it is not the longest. (This record goes to a group of long-necked dinosaurs known as titanosaurs, who went extinct some 66 million yrs. ago: the largest species grew to over 130' in length.)

Despite its massive size, the blue whale is a baleen whale, and thus eats some of the smallest animals that have ever lived: zooplankton, infinitesimally tiny crustaceans (mainly krill), and fish larvae. It strains this food through its 40" long baleen plates, 300-400 tough bristly fibers that hang down like ash-colored curtains from the inside of its rostrum or upper jaw. Some 100 throat pleats expand to allow in the seawater that is strained through the baleen, then expelled once the plankton has been filtered out and trapped. In this way it can consume up to 4 tons or more of food a day.

The dorsum is a slate-blue and often mottled; the ventrum is whitish—but the latter is sometimes tinted yellow due to tiny sea organisms, mainly algae, that attach themselves to the whale's belly, lending it the early nickname, "sulphur-bottom whale." It has a nearly u-shaped rostrum, from which a ridge runs back to its double blowholes; its small dorsal fin is located near the back third of the body. It can spout as high as 30' into the air, the highest of all the whales. May travel alone or in small- to medium-sized pods. Its powerful slender body is hydrodynamically designed for speed, and it can swim up to 30 mph (2 mph quicker than the fastest recorded human can run).

Blue whale.

As a planktivore it does not need to dive deeply, for its food, zooplankton (microscopic aquatic animals), feeds on phytoplankton (microscopic aquatic plants), which needs to live near the surface of the water to capture sunlight. Thus the blue whale seldom dives down deeper than 300-400'. Inhabits deep waters around the world and ranges across every ocean, from the poles to the equator; migrates to cool waters in the spring, to warmer waters in the fall.

It is one of the great communicators of the animal world: it is capable of making the lowest frequency, as well as some of the loudest, sounds of all the whales—intense low frequency clicks and pulses that can be heard by conspecifics thousands of miles away. Females gestate for about 1 yr., giving birth to a 4-ton, 26' calf. Due to its immense size, this secondary consumer has few natural predators, except the orca; occasionally the young may be preyed upon by sharks. It is believed that the blue whale lives to about 110 yrs. in the wild.

BOBCAT (*Lynx rufus*): The official state wildcat of New Hampshire, it is the most common and widespread spotted cat in North America. It stands up to 21" high at the shoulder, grows to around 4' in total length, and can weigh as much as 70 lb. Its short thick dorsal coat, which ranges from tawny to light gray, from grizzled brown to red-brown, is covered with indefinite or "blurry" spots, dapples, lines, streaks, and bars. The ventral pelage is usually lighter in color. Earbacks are black with a white spot. Has a powerful stocky body, buff-colored chin and chest, long powerful legs, and large flared cheek ruffs. This medium-sized felid may be thought of as a smaller version of its close cousin the lynx. The bobcat, however, is more conspicuously spotted, has shorter fur, smaller paws, shorter legs, shorter ear tufts, and a longer banded tail—and, unlike the lynx, its dichromatic tail is black above and white below. Gender dimorphism is present, with males being larger than females.

Highly adaptable, this motile solitary mammal is a nearctic native that inhabits a myriad of biomes and ecosystems, including scrubland, brushland, swamp, arid desert, forest, rocky areas, broken terrain, farmland,

Bobcat.

intermontane regions, and mountains, particularly areas containing good hiding cover. Its range covers southern Canada south into Mexico. In the U.S. it can be found in all 48 of the lower states, from California to Virginia, from New England to Florida, but it is scarce in the central Midwest and most common in the Western states.

The petraphilic bobcat—named, not for a person, but for its short "bobbed" tail—is considered nocturnal. It is also somewhat crepuscular, however, and, in some regions, even diurnal, so it can be seen at any time day or night. Its diet is comprised mainly of rabbits and hares, but as an opportunistic carnivore it will eat nearly anything it can scavenge or catch, including reptiles, birds, mice, squirrels, porcupines, bats, beavers, peccaries, hogs, deer, and farm animals, such as poultry, pigs, and sheep (when available). This highly intelligent cat is a skillful hunter, climber, and swimmer, and uses its camouflaged coat, stealth, strength, agility, and speed (it can run 30 mph) to ambush and chase down prey—which can be much larger than itself. Hoards and caches leftover food.

Its main den is set up in a variety of microhabitats, such as rock crevices, caves, hollow trees, and stumps, or beneath logs and ledges. A secondary den, known as a "shelter den," is sometimes built in piles of vegetation or under stumps and rocky shelves. The territory of the male extends up to 30 sq. mi; the territory of the female is around 5 sq. mi. These areas are scent-marked using urine, dung, and a musky oil emitted from a gland located under the tail.

Normally solitary, males and females do not associate except during the breeding season, which typically occurs in late winter. The queen gestates for about 60 days, and gives birth to 1-8 altricial kittens in the spring. The kittens become independent at 8-10 mos. of age. Toms do not contribute to the rearing of their offspring.

Bobcat.

Vocalizations are similar to those of the domestic house cat, to which may be added the yowls, barks, hisses, and screams more typical of large wild felines. The natural enemies of bobcat kittens are birds of prey (hawks, owls) and wild canids (coyotes, wolves). Secondary consumers, due to their fierce nature (fearlessness), weaponry (teeth, claws), furtiveness, agility, and speed, adults are usually safe from predation. The secretive bobcat lives up to 10 yrs. in the wild; in captivity it may live as long as 30 yrs.

BOTTA'S POCKET GOPHER (*Thomomys bottae*): It grows to a total length of almost 11", its hind foot measures around 1", and it can attain a weight of up to almost 9 oz. Its pelage occurs in more colors than any other mammal in North America. Indeed, its coat can vary from white to black and nearly every color and combination in between (including varying tinges of pink, orange, and purple). Generally speaking, however, because fur color is linked to topographic color, the upper body is usually grayish-, rusty-, or yellowish-brown and the lower body is grayish-tan. The body is stout and cylindrical; the head is robust; the dark eyes and rounded ears are small; the muzzle is thick; the nose is snubbed and covered in sensitive vibrissae; the chestnut-colored incisors are exposed; the limbs are short; the paws are pink and clawed; the nearly 4" tail is mostly hairless and tannish-gray. Since the male is much larger and heavier than the female, this species is gender dimorphic.

A fossorial geomyid and a nearctic native, it inhabits a truly wide variety of biomes and ecosystems, from valleys to mountains, from deserts to forests, from sandy dunes to open meadows, from streambanks to rocky hillsides, from prairie to scrubland, from hedgerows to chaparral, from rangeland to ranchland, from savanna to farmland, from shrubland to orchards. Prefers semisoft friable soil that facilitates digging and tunneling, such as loam and clay. It is a purely western species, with a range that extends over parts of Oregon, California, Nevada, Utah, Colorado, Arizona, New Mexico, and Texas, south to Baja California and northern Mexico. Does not hibernate; active all year. Excavates large burrow systems, some over 100' long. Will readily defend its home against conspecifics.

Pocket gopher.

A solitary herbivore, its diet consists chiefly of subterranean plant parts, such as roots, tubers, and bulbs. It also eats corms, forbs, grass shoots, and other types of green vegetation; forages above ground and caches surplus food in its burrow larders. Its breeding season probably runs year-round in the southern end of its range. The female gestates for about 3 wks., giving birth to an average of 3-4 altricial pups in an underground nest. Weaning takes place within 5-6 wks.; independence occurs at about 2 mos. of age.

Some believe that this rodent is conspecific with the southern pocket gopher and Townsend's pocket gopher, while others maintain that all 3 are separate species. One fact is uncontested: Botta's is smaller than Townsend's and larger than the southern.

Communication and perception run along typical Geomyidae channels: while sight is not of great importance to pocket gophers, smell and touch are vital to an animal that spends most of its life in the dark. Botta's pocket gopher lives an average of 1-2 yrs. in the wild; perhaps slightly longer under optimal conditions.

Bowhead whale.

BOWHEAD WHALE (*Balaena mysticetus*): The official state marine mammal of Alaska, it is also known as the "arctic whale" and the "Greenland right whale." It is the second largest whale in the world (after the blue whale), growing to a total length of 70' and a weight of up to 100 tons. A filter feeder, at 14' in length its black baleen is the longest of any whale, giving it the species name *mysticetus*: from the Greek *mystax*, meaning "mustache" (an allusion to its baleen), and the Latin *cetus*, meaning "whale."

The common name bowhead comes from its huge bow-shaped mouth, which wraps around a giant head that makes up at least ⅓ of its body length—the only whale with such enormous head proportions. Its powerful compact body, encased in a 2' thick layer of blubber, is black with irregular white patches on the chin and sometimes the throat. It lacks a dorsal fin, while its paddle-shaped pectoral flippers are small for its physical dimensions. Its paired blowholes produce 2 columned, 20' spouts that form a "v" shape in the air. Will use its massive head to break through ice if trapped.

This massive balaenopterid inhabits the cold waters (including bays, inlets, estuaries, and straits) of the Arctic Ocean, with a range that extends south into the northern Pacific and Atlantic Oceans. It haunts these waters year-round, the only whale to do so. A planktivore, its diet consists of typical baleen whale fare: zooplankton, which includes copepods, amphipods, and various benthic organisms. It can consume nearly 2 tons of these microscopic animals a day. Every 3 or 4 yrs. the female gestates for around 14 mos., giving birth to a 17', 2,000 lb. blue-gray calf that grows about a ½' a day. It takes 15-20 yrs. before the infant bowhead is mature enough to have offspring of its own.

This migratory species is not as social as other cetacids, and is often found alone or in small pods of 5-10 members. It communicates with

Bowhead being attacked by orcas.

conspecifics through distinct low frequency songs that can be heard over long distances underwater. A slow-swimming whale (maximum speed 5-6 mph) and a shallow diver (goes down to a depth of only 400'-500'), it can hold its breath for up to 45 min. Almost hunted to extinction, it is now—like a number of other whales—rare, endangered, and protected.

A secondary consumer, its only natural predator is the orca, and many individuals retain scars from killer whale attacks. Though it was long believed that the life span of this highly disease-resistant whale is only around 75 yrs. in the wild, recent studies have shown that it lives up to 200 yrs. or more, making it the longest living mammal in the world.

BRAZILIAN FREE-TAILED BAT (*Tadarida brasiliensis*): The official state flying mammal of Oklahoma and Texas, it is one of the smallest of the world's 110 species of free-tailed bats, and the smallest in our region. Also known as the "Mexican free-tailed bat," it grows to a total length of nearly 4½", its forearm is about 2" long, its tail is 1¾", it has an 11" wingspan, and it weighs ½ oz. Half of its tail extends out past the uropatagium, hence its common name: "free-tailed bat." Its ears are large and rounded. Its coat is cinnamon to dark gray-brown on the upper body and is sometimes grizzled or frosted; its underside is usually lighter in color. It is one of the few bats with vertical wrinkles on its upper lip.

Brazilian free-tailed bat (museum speciman).

One of the most common chiropterids in our region (and thus one of the most numerous mammals in the U.S.), this tiny molossid shelters in caves, mines, bridges, hollow trees, wells, and buildings, from Oregon and California across the bottom half of the U.S., to Virginia and Florida, extending south into Mexico (and beyond to Central and South America). Highly social, it forms cave colonies of up to 20 million individuals, with over 200 bats compressed into a single square foot. (They are well-known for their spectacular emergence and reentry at one of their major roosts, Carlsbad Cavern, New Mexico, creating a sensational sight in the evening and morning skies.)

This nocturnal insectivore will fly 50 mi. in search of food. Moving at speeds up to 25 mph, it detects its prey (moths, bees, ants, wasps, leafhoppers, dragonflies) using echolocation, then uses the trawling foraging method to "net" insects in its interfemoral membrane. Its narrow pointed wings permit high-speed, long-distance flight, an ability it uses to great advantage while soaring at altitudes of 2 mi. or more (the highest on record for any bat species) over its vast 150 sq. mi. home range. It does not hibernate, but may employ torpor during cold weather.

The Brazilian free-tailed bat's tail extends beyond the uropatagium, giving it its common name.

It breeds in early spring. The female does not delay fertilization (as in *Myotis*), but goes immediately into gestation, which lasts for about 85 days. At her nursery colony she gives birth to one precocial pup (rarely twins) in the late spring or early summer. She nurses her neonate with milk containing the highest fat content of any bat species, caring for it until it is weaned and independent, about 5-6 wks. later. A secondary consumer, its natural enemies are reptiles (such as snakes), birds of prey (such as owls), and small mammals (such as skunks, opossums, and raccoons). The Brazilian free-tailed bat lives to 5-10 yrs. in the wild; a few yrs. longer in captivity.

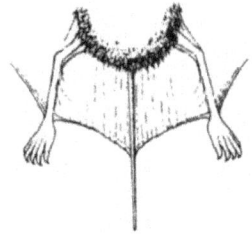

BROAD-FOOTED MOLE (*Scapanus latimanus*): It grows to a total length of 7", its tail is 1¾" long, and its hind foot measures 1". Its pelage is dark brown to silver gray. Has a stubby tubular body, cone-shaped skull, pointed muzzle, small eyes, short legs, short tail, and wide front paws for digging—the source of its common name. Its scientific name derives from the Greek *skapanetes* ("digger"), and the Latin *lat*

Broad-footed mole.

("wide") and the Latin *manu* ("hand"). Its velvety fur is designed to allow frictionless forward and backward movement as it moves through its underground burrow system. This eulipotyphlid is gender dimorphic: the male is slightly larger than the female.

Inhabits humid grassland, valleys, wet meadows, orchards, farmland, and open forest throughout its range: from southern Oregon through nearly all of California (except the state's arid, hot, and dry regions), to western Nevada, and south into Mexico. To make burrowing and foraging easier, it prefers friable moist soil near streams (though it avoids flooded areas). Territorial, it agonistically defends its tunnels from conspecifics and other intruders.

A fossorial talpid, it can be active at any time day or night. A solitary subterranean omnivore, it eats mainly earthworms, snails, slugs, and insects, but will also consume seeds, crustaceans, and small mammals. Its breeding season runs from midwinter to midspring. The female gives birth to 2-5 offspring once a year in a grass-lined nest, weaning them at about 1 month of age. A secondary consumer, its primary natural enemies are martens, hawks, and owls. The broad-footed mole lives for about 3 yrs. in the wild.

BROWN BEAR (*Ursus arctos*): The most prototypical of all bear species, its scientific name was intentional: *Ursus* is Latin for "bear," and *arctos* is Greek for "bear." More popularly known as the "grizzly bear," this awe-inspiring North American icon stands 4' high at the shoulder, reaches a total length of over 9', and can weigh up to 2,500 lb. (over 1 ton)—though 1,200-1,500 lb. is more typical. Its front claws are 4" in length, longer than the average human finger (from first knuckle to tip). Its massive shoulder hump and concave muzzle distinguish it from the less robust black bear, which has a smaller shoulder hump and a convex muzzle.

Brown bear.

The brown bear reaches its greatest dimensions in the northwest coastal region, where it was once wrongly regarded as a separate species, the "Alaskan brown bear." Standing as tall as 8' on its hind legs, it holds the world record for being the heaviest terrestrial carnivore—outweighing even the gigantic polar bear by 300 lb. or more. Its common name "grizzly" derives from its shaggy grizzled (white-tipped) fur, though its overall pelage may vary between blond, cinnamon, brown, and black. In the coastal regions of Canada and Alaska it is still correctly called the "brown bear," while in the lower 48 contiguous states, inland forms are known as "grizzlies." The aforementioned Alaskan brown bear is actually a subspecies of the brown, one more commonly known as the "Kodiak brown bear" (*Ursus arctos middendorffi*), which, at over 10' in height, is the largest bear in the world. The brown bear is gender dimorphic, with males being larger than females.

It lives in a wide variety of habitats, from ice fields and deserts to alpine forests and mountains. However, it prefers semi-open to open terrain, such as meadows, coastal areas, and tundra. The male's home range may

Brown bear preying on bison.

Brown bear (Kodiak subspecies).

extend up to 700 sq. mi. Its population once numbered in the many tens of thousands, at which time it roamed over much of the North American continent, from the U.S. West Coast to the Midwest, and from Canada to Mexico. Today a mere fraction of that number thrive in our region, concentrated mainly in the northwest, from Wyoming and Montana to Alaska and British Columbia.

This solitary ursine is an opportunistic omnivore, feeding on everything from roots, seeds, tubers, grasses, plants, bulbs, mosses, fungi, fruits, and nuts, to insects, fish, small mammals, mountain goats, mountain sheep, deer, elk, moose, carrion, and livestock. Will kill and eat its cousin the black bear when possible. Though a loner, it congregates with conspecifics when food is abundant; for example, along rivers during the salmon run. Roots up the ground (leaving large pits) and overturns logs and rocks in search of prey. Caches its food under branch piles and dirt. Another sign of its presence are "bear trees," which have been ripped, marked, rubbed, girdled, scratched, de-barked, and slashed.

Both crepuscular and nocturnal, it is most active in the morning and evening. Though it may also be seen during the day, generally this is when it rests, usually in a concealed area. Its matted bed is a 3' by 4' depression in the ground, filled with tree boughs and leaves. Not a true hibernator, after gaining as much as 400 lb., individuals in northern climates den up in winter under rocks or tree roots, or in caves and crevices, often on the side of a protected slope or hill. It then goes into a temporary state of torpor (generally similar to a deep sleep, but one from which it can be awakened), living off its stored fat reserves. It becomes active again in spring.

The Alaskan forms deeply imprinted 16" long by 11" wide hind tracks are sometimes mistaken for Sasquatch prints, which indeed share a resemblance. But bears leave claw marks, while bipedal primates, who have toenails rather than claws, do not. Also identifying is that a bear's big toe (the hallux) is outermost, while its smallest toe (the quintus) is innermost—the opposite of Sasquatch (and humans).

Though it has a cumbersome walk, this territorial nomadic plantigrade can easily run 40 mph (the speed of many deer and horses) and reach a stride length of up to 9' during a bounding run. Will stand upright when on the alert. Fast, powerful, impulsive, unpredictable, cantankerous, erratic, and an excellent swimmer and (contrary to popular opinion) an expert tree climber, it has fair eyesight, excellent night vision, and a hypersensitive nose (it is able to detect an unopened can of food in a locked car from miles away). Such traits and abilities make it the most potentially lethal of all North American bear species.

Its breeding season runs from midspring to midsummer, at which time males begin competing for females. As with other ursids, after mating, the sow delays implantation (which prevents her eggs from attaching to the walls of her uterus) for about 5 mos., allowing her to genetically coordinate her pregnancy with advantageous conditions, such as warm weather and an abundant food supply. After implantation, she gestates for approximately another 4 mos., making her total gestation period about 7-9 mos. long. She gives birth to 1-4 tiny altricial cubs (weighing only 1 lb. each) between late winter and early spring. Typically the sow breeds only once every other year. The young are weaned at about 2-3 yrs. of age, after which they become independent. The polygynandrous boar does not assist with the rearing of his offspring.

Brown bears.

It communicates and perceives via standard Ursidae channels: body language (yawning, tree-climbing, rubbing, bluff-charging, ear-flattening, sniffing, jaw-opening), chemical odors (scent-marking with urine and dung), and various sounds and vocalizations (whining, teeth-clacking, grunting, bawling, blowing air, screaming, huffing, tongue-clicking, jaw-chomping, mouth-popping, lip-smacking, humming, moaning). A secondary consumer and umbrella species that lives at the top of the food chain, its natural enemies are few, but cubs are sometimes preyed upon by mountain lions, wolves, and other bears. The brown bear or grizzly bear lives to an average age of 25 yrs. in the wild; longer under optimum conditions; up to 50 yrs. in captivity.

BROWN LEMMING (*Lemmus trimucronatus*): Also known as the "common lemming," it grows to a total length of nearly 7", its hind foot measures almost 1", and it weighs as much as 4 oz. The upper body is cinnamon-brown with a grayish wash; the lower body is yellowish-gray; the 1" tail is short and brownish. The body is thick and cylindrical; the fur is dense and rough; the eyes and ears are small; the limbs are short. It is gender dimorphic, with the male being larger than the female.

This solitary cricetid is a palearctic and nearctic native that inhabits wet tundra, flat mossy lowland, grassy meadows, and moist fields in both alpine and subalpine regions. Its range covers much of the treeless regions of northern Alaska and northwestern Canada. Outside North America its range extends westward into the arctic tundra of

Russia, where it is sometimes confused with its close cousin the Siberian brown lemming (*Lemmus sibiricus*). In winter it creates ball-shaped nests beneath the snow; in summer it excavates a subterranean shelter that contains multiple chambers designed to serve different functions (resting, protection, storage, nesting, etc.). Establishes a small home range of 50 sq. ft. or less.

The warm weather diet of this large herbivore consists of grasses, leafy vegetation, mosses, fruit, and sedges; its cold weather diet is made up of lichens, shoots, mosses, roots, and woody material from trees. Its breeding season occurs in summer.

Brown lemming.

Females can begin reproducing at less than 1 mo. of age. They gestate for about 20-22 days, giving birth to an average litter of 6-8 pups in a round, grass- or fur-lined underground nest. The male does not seem to participate in the raising of his offspring.

Communication and perception run along standard fossorial mammalian channels: touch and smell, and to a lesser degree, sight and hearing. When its population explodes every 3 yrs. or so, a natural culling process takes place, with the majority of individuals dying off from starvation, predators, and disease. Its enemies are birds of prey (owls), mustelids (weasels), wild canids (foxes), and ursids (brown bears). The brown lemming lives an average of 1-2 yrs. in the wild.

BRUSH MOUSE (*Peromyscus boylii*): Also known as the "brush deer mouse," it grows to a total length of nearly 10", its hind foot measures around 1", and it can attain a weight of up to 1 oz. or so. The upper body pelage is grayish, brownish, or grayish-brownish; the lower body is white; the sides are buff; the long 4" tail is furred and dichromatic with a tufted tip; its ankles are grayish. The ears are ¾" long. Diagnostic marking: a rusty-orange body stripe or lateral line that runs horizontally from face to rump.

This medium-sized riparian cricetid is a nearctic native that inhabits temperate, semiarid woodland, rocky scrubland, scree, loggy chaparral, steep talus slopes, canyons, and forested mountains, with abundant brush, blowdown, snag, deadfall,

Brush mouse.

and hiding cover. Its range extends over portions of California, Nevada, Utah, Colorado, Arizona, New Mexico, Texas, and Oklahoma, south into Baja California and central Mexico. Does not hibernate; active all year. A petraphile, it shelters and nests in shrubbery, rocky ledges, caves, duff piles, tree craws, boulder fields, and rock cracks, and also beneath stumps and fallen timber.

A nocturnal sedentary omnivore, its diet consists of seeds, grains, stalks, fruit, bark, nuts, leaves, fungi, cacti, and insects; it is particularly fond of acorns and conifer seeds. Its polygynandrous breeding season runs year long with a peak in spring and a decline in winter. The female gestates for around 3 wks., giving birth to 3-4 altricial young that she weans 1 mo. later. The male does not seem to participate in the rearing of his offspring.

Communication and perception occur via standard Cricetidae channels: sight, smell, touch, and hearing. Scansorial, it is an adept climber and can often be seen scurrying about in trees. Its natural enemies include reptiles, birds of prey, mustelids, wild felids, and wild canids. Beneficial as a prey base for carnivores; detrimental as a carrier of deadly diseases that can harm humans (such as the hantavirus). The brush mouse probably lives an average of 10-12 mos. in the wild.

BRUSH RABBIT (*Sylvilagus bachmani*): This small lagomorph grows to a total length of around 15", its hind foot is 3½" long, and it weighs up to 2 lbs. Its coat is gray and black with reddish, orange, or cinnamon-colored mottling. Its winter pelage is lighter. It has short ears, a small tail, and short limbs. It is gender dimorphic, with the female sometimes being slightly larger than the male.

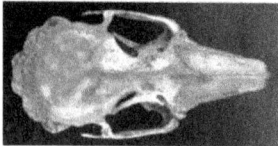

Brush rabbit skull.

Its common name comes from its favored habitat: thick brushland. It is also found, however, in mixed forest, grassland, and dense chaparral vegetation. Its range covers the coastal western region of North America, from Oregon south to Baja California. Does not burrow; instead it appropriates the burrows and tunnels of other animals.

This nocturnal herbivorous leporid feeds mainly on grasses and clover, but it also consumes berries, forbs, plantain, and leaves. It is sometimes seen in daylight during clement weather. When alarmed it may cry out and stamp its hind feet. If pursued it can run 25 mph; when this is not possible it prefers climbing bushes or trees rather than darting underground.

The viviparous polyestrous doe has 8 mammary glands and bears 3-5 litters a year. She gestates for about 1 mo., giving birth to 1-6 altricial kits, which she weans in 2-3 wks. The natural predators of this primary consumer are reptiles, birds of prey, weasels, raccoons, foxes, lynxes, bobcats, and coyotes. The life span of the brush rabbit is unknown; it probably lives at least 5 yrs. in the wild.

BUSHY-TAILED WOODRAT (*Neotoma cinerea*): This big rodent, the largest of the woodrats, grows to a total length of 19" and a weight of about 16 oz. Its countershaded pelage is generally yellowish-black on the upper body, white on the lower body (coloration varies among its 13 subspecies). Sometimes categorized as a "furry-tailed" woodrat, its long bicolored 9" tail is bushy and squirrely, giving it its common name. Darkish above and light below, the tail is used for balance when climbing and for warmth when the temperature drops. Its fur is long and thick; the body is compact and slightly ovoid; the head is small with a convex profile; the eyes are beady and dark; the nose is robust, snubbed, and covered in long sensitive nasal vibrissae; the ears are large, rounded, erect, and set well up on the head; the limbs are shortish; the digits are sharply clawed. It is highly gender dimorphic, with the male sometimes weighing nearly twice as much as the female.

This cold-weather loving, riparian cricetid is a nearctic native that inhabits a wide variety of biomes and ecosystems. These seem to include everything from rocky cliffs, scree, chaparral, shrubland, boreal steppe, rangeland, sageland, woody covering, coniferous forests, hollow logs, and mountainous regions, to prairie, grassland, taiga, caves, moist woodlands, alpine meadows, mine tunnels, tree snag, sand dunes, creeksides, and open desert. Its range covers western North America, from the Canadian arctic (British Columbia, Saskatchewan, Alberta, Yukon) south into Arizona and New Mexico. Establishes a home range of about 1,500 sq. ft., which it defends with scent-marking and agonistic territorial displays.

Bushy-tailed woodrat.

Omnivorous, its diet consists primarily of green plant matter, as well as nuts, shoots, needles, forbs, seeds, twigs, fruit, cacti, and fungi; in addition, it has an appetite for arthropods (spiders, scorpions, ticks, centipedes, mites, pillbugs). It is also coprophagous, which permits it to live through extreme conditions, such as droughts, or in regions where food is not regularly available. Does not need or use free water as it metabolizes what it needs from its diet.

Its breeding season runs from midspring to midsummer. Bucks compete violently for mating rights. The polyestrous doe may bear 3-4 litters a year, gestates for 4-5 wks., and gives birth to an average of 4 altricial pups. Independence and dispersal seem to arrive at around 2-3 mos. of age. As in most mammalian species, the father does not exhibit paternal behaviors.

This solitary, crepuscular, arboreal, nocturnal mammal is the archetypal "packrat," and will build up large collections of items that attract it, mainly stones and sticks. It has a particularly strong affinity for shiny objects, however, such as coins, bullets, necklaces, silverware, rings, and tinfoil wrappers, as many a camper has discovered to his or her chagrin.

Another identifying trait of this rodent is its midden-building ability, by which it constructs great refuse heaps carelessly thrown together using leaves, bark, bones, branches, rocks, dung, grass, feathers, and human-made objects. Located within rock crevices, on talus slopes, or in caves, mines, trees, woodpecker holes, or buildings, the woodrat midden usually contains both its den (where it eats, rests, and sleeps) and its grass-lined natal nest (where the young are born and reared). These slipshod stick structures contain numerous side passages and chambers, the latter which are used to cache surplus edibles. Due to its philopatric inclinations, a bushy-tailed woodrat midden is built up over time by succeeding generations, and can thus be thousands of years old.

Communicates and perceives via smell, touch, and hearing, and to a lesser degree, sight. Engages in hind foot-thumping as an alarm signal. Benefits the world ecosystem by serving as a bodily host to various parasites and by providing structural homes for a myriad of other animals, from reptiles and amphibians, to other small mammals. Its natural enemies are reptiles, birds of prey, mustelids, wild felids, wild canids, and ursids. The average life span of the bushy-tailed woodrat is about 3 yrs. in the wild.

Gray fox.

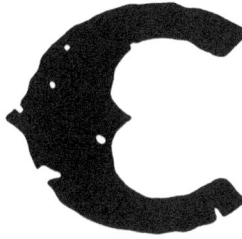

CACTUS MOUSE (*Peromyscus eremicus*): Also known as the "cactus deer mouse," it grows to a total length of nearly 9", its hind foot measures around 7⁄8", and it can attain a weight of 1 oz. or more. The fur is soft, satiny, and long; the lateral line is tannish. Has highly variable pelage coloration, from buff and brown to gray and black on the dorsum, white to pale gray on the ventrum. Its long 5" tail is bicolored, sparsely furred, ringed, and lacks a tuft.

Gentle and nonaggressive, this timid nocturnal cricetid is a nearctic native that inhabits cactus-rich deserts, the source of its common name. It is also found in and around badlands, semiarid steppe, chaparral, mesquite grass, rangeland, sage scrub, malpais lava fields, yucca stands, creosote shrubland, sandy plains, and loamy hillsides with abundant rock and brush. It is a southwestern species whose range covers portions of California, Nevada, Arizona, New Mexico, Texas, Baja California, and north central Mexico. Builds bolus-like nests in rock crevices, cactus clusters, or beneath shrubbery, lining them with grass, leaves, twigs, and feathers; will appropriate the abandoned burrows of other rodents; enters a state of torpidity under extreme conditions (heat, cold, drought, hunger, etc.).

Badlands: cactus mouse habitat.

A sedentary, terricolous omnivore, it consumes mainly seeds, grains, nuts, flowers, leaves, fruit, green vegetation in general, and also insects. Drinks little as most of its water needs are met by the foods it eats. Its breeding season runs year-round. The female gestates for approximately 3 wks. or so, giving birth to an average of 2 altricial pups in the natal nest. Weaning probably takes place within 2-4 wks. The male does not play a role in the raising of his offspring.

Communication and perception occur via standard Cricetidae channels: sight, smell, touch, and hearing. Engages in foot-drumming. Its natural enemies are reptiles, birds of prey, wild felids, and wild canids. Beneficial as a seed disperser, an insect predator, and as food for carnivores. On average the cactus mouse lives for less than 1 yr. in the wild.

CALIFORNIA CHIPMUNK (*Neotamias obscurus*): Once thought to be conspecific with, or a subspecies of, the similar Merriam's chipmunk, based on recent genetic studies it is now recognized as a separate species. Morphologically, however, it is nearly identical to Merriam's, with a homogeneous ecology and life history. They also share overlapping ranges—though the California thrives as far south as the Baja Peninsula in Mexico, while Merriam's is not generally found outside California. A few of the primary physical differences (which are difficult to observe in the wild) are that the California is about 1" shorter, around 1 oz. lighter, its head is more gently curved, its coloration is paler and less distinct, and its hind foot is larger than Merriam's.

CALIFORNIA GROUND SQUIRREL (*Otospermophilus beecheyi*): Also known as the "Beechey ground squirrel," it grows to a total length of 20", its pes is about 2" long, and it can weigh up to 30 oz. It is generally a mottled brownish or grayish on the dorsum, with small white or silver speckles and spots peppered over the coat. May be gray, tan, or yellowish on the ventrum. It has a lithe cylindrical body; small head; white eye rings; small, pointy, erect ears (without tufts); snubbed nose; and a brown-gray, bushy, 8" tail with white edging. May have a large, distinctive mantle or collar: a light-colored "V" that runs from the ears down to its shoulders, with the bottom (or point) of the "V" located on its head (facing forward).

California ground squirrel.

This fossorial sciurid prefers open grassy, serially occupied habitat, such as temperate meadows, pastures, chaparral, roadsides, farmland, and fields with loose soil; it can also be found in rocky settings, mountains, hills, valleys, and woodlands. Widely distributed over both wilderness and urban areas, its geographical range extends throughout the northwestern part of the U.S., covering the states of Washington, Oregon, California, and Nevada.

Though not particularly social, this ecosystem engineer lives in loose colonies in large burrow systems. Its 6" wide tunnels may be up to 200' long, with separate entrances for each tenant. Burrows are used repeatedly by succeeding generations and are often constructed under rocks and trees. Establishes a somewhat circular 150' home range around its burrow. If necessary it will estivate during exceptionally hot conditions; hibernates during extreme cold, particularly between November and February.

An omnivore with granivorous and insectivorous proclivities, its diet includes seeds, nuts, grains, flowers, fungi, tubers, stalks, bulbs, fruit, roots, and leaves; also consumes grasshoppers, caterpillars, beetles, crickets, birds' eggs, and small vertebrates. Hoards and caches surplus food.

California ground squirrels.

The breeding season of this common ground squirrel takes place after hibernation ends in early spring. The female gestates for around 30 days, giving birth to 5-15 altricial pups in a nesting chamber below ground. She weans her young by 2 mos. of age. As the father does not assist in the rearing of his offspring, the nuclear family of this species includes only the mother and her young.

Communicates and perceives via normal Sciuridae channels: sight, hearing, smell, and touch. Can carry bubonic plague. Its natural enemies are birds of prey, reptiles, mustelids, wild felids, and wild canids. The California ground squirrel lives about 5 yrs. in the wild; 8-10 yrs. in captivity.

CALIFORNIA LEAF-NOSED BAT (*Macrotus californicus*): This endothermic phyllostomid grows to a total length of nearly 4", its forearm is a little over 2", its wingspan is 15", and it weighs roughly ½ oz. It has large ears, big eyes, short broad wings, and a large snout. Its most distinguishing trait, however, is the triangular leaf-like appendage on its nose, the source of its common name. Its body fur is gray to brown in color, with lighter undersides. It is the only chiropterid north of Mexico with all of these characteristics.

Petraphilic, it inhabits caves, rock shelters, and abandoned mines in the desert scrublands of southern California, southern Nevada, and southern Arizona, south into northern Mexico. It roosts during the day in small colonies of up to 500 individuals, often hanging from the ceiling by one foot.

A nocturnal insectivore, it drops from its perch after sunset to go in search of its prey: ground-living, largely flightless or wingless insects, such as crickets, grasshoppers, beetles, katydids, caterpillars, and sometimes moths. While using both its eyesight and echolocation to hunt, it also has the ability to hover in midair. Also a frugivore, it eats the fruits of various cactus plants, a rare behavior among bats.

California leaf-nosed bat.

Does not hibernate or migrate. The female gestates for 8 mos., giving birth to one or two pups in June. A secondary consumer, its main natural enemies are reptiles (snakes), birds of prey, procyonids, wild felids, and wild canids (mainly coyotes). The California leaf-nosed bat can live up to 30 yrs. in the wild—though 10 yrs. is more typical.

CALIFORNIA KANGAROO RAT (*Dipodomys californicus*): Shares much of the same ecology and life history as its close cousin, Heermann's kangaroo rat. The primary difference is the range of the California, which extends from central Oregon south to central California.

CALIFORNIA MOUSE (*Peromyscus californicus*): The largest deer mouse in North America, it grows to a total length of nearly 9", its hind foot measures about ⅞", and it can attain a weight of almost 1½ oz. As with many other mammals, its body is cryptically colored and countershaded. In its case the dorsum is rusty-yellow with a gray wash, the ventrum is white. The fur is thick, long, and soft; the lateral line is buff-orange; the ears are large, thin, and erect; the feet are whitish; the long 6" tail is furred and grayish-yellow. These colors and markings provide ideal protective camouflaging in the specific biomes and ecosystems it inhabits. Gender dimorphism is present: the male is larger than the female.

This riparian cricetid is a nearctic native that occupies temperate chaparral, brushland, sageland, foothills, ravines, rangeland, coastal scrubland, ranchland, and oak and redwood forests, preferably with logs, good hiding cover, and plenty of forest duff. As its common and scientific names indicate, its range centers on California (in its case, mainly across the southern part of state). However, it is also found further south into Baja California, and perhaps as far east as Arizona and New Mexico.

This rodent is both terricolous and arboreal with excellent scansorial abilities. Appropriates the burrows of other animals, such as the woodrat; builds massive bolus-style nests out of forest debris, locating them in crevices and under stumps and fallen trees. Does not hibernate; active all year. Extremely territorial and aggressive, both the male and the female energetically defend the maternal nest chamber from conspecifics.

Also known as the "California deer mouse," the diet of this normally easy-going, sedentary, nocturnal omnivore is based around typical rodent fare: seeds, grasses, nuts, grains, fruit, flowers, fungi, leaves, forbs, and insects. Its water needs are met primarily through the food it consumes. This mouse will readily climb bushes, thickets, and trees in order to forage.

Its monogamous breeding season runs year-round, with bucks and does forming stable pair-bonds around their offspring. The polyestrous female, who can produce several litters each year, employs brief delayed implantation in order to time parturition with the most advantageous period for her neonates. After gestating for about 1 mo., she gives birth to a pair of altricial pups which she weans within a month or so. Unlike most other male mammals, the male of this species is quite paternal and helps raise his young, principally by protecting the nest site and collecting food for them.

Communication and perception run along standard Cricetidae channels: sight smell, touch, and hearing. Its natural enemies are reptiles, birds of prey, mustelids, wild felids, and wild canids. Beneficial as a seed disperser and as a prey base for carnivores. Potentially harmful as a carrier of viruses that are deadly to humans, such as the hantavirus. The California mouse lives around 1 yr. in the wild; at least twice that long in captivity.

CALIFORNIA MYOTIS (*Myotis californicus*): One of the few mammals with matching common and scientific names, this small vespertilionid grows to a total length of a little less than 3½", its hind foot is only ¼" long, its forearm is just over 3". Its calcar is keeled and its ears and wing membranes are dark brown to black. Its long dorsal pelage ranges from flat tan to cinnamon to dark brown, while the lower body is usually lighter in color. Its scientific name, *Myotis*, means "mouse-eared," an appropriate description of this entire genus, which tends to sport smallish, rounded mouse-like ears.

A secondary consumer, this chiropterid shelters alone or in small colonies in caves, bushes, canyons, snag, rocky cliffs and hillsides, buildings, mines, and under tree bark and bridges, inhabiting a wide variety of terrain, from arid deserts to humid forests, from coastal regions to mountainous areas. Prefers to be near a clean water source. Its range covers the western section of our region, from lower Alaska south to Mexico, and includes the U.S. states of Washington and Montana in the north to New Mexico and Texas in the south.

Wing of a vespertilionid bat.

A slow and erratic but highly nimble flyer, this tiny nocturnal echolocater emerges from its roost after dark, voraciously feeding on insects (mainly moths and flies) until sunrise—making it of great benefit to us. It mates in autumn; the female, however, delays fertilization until spring, giving birth to a single pup in early summer. *Myotis californicus*, the "mouse-eared Californian," lives to about 15 yrs. in the wild.

CALIFORNIA POCKET MOUSE (*Chaetodipus californicus*): The most common pocket mouse in our region, this large rodent grows to a total length of nearly 10", its large hind foot measures about 1", and it can attain a weight of almost 1 oz. The dorsum color is brownish-gray, the ventrum color is white; the long, nearly 6" tail is bicolored (brown on top, buffy-white underneath) and heavily tufted at the tip; the ½" ears are large; the lateral line, which divides the upper pelage from the lower, is brown. Diagnostic marking: a scattering of stiff bristly white spines on the posterior.

This solitary, nocturnal heteromyid is a nearctic native that inhabits primarily temperate arid grassland and chaparral slopes; also found in forests, deserts, sageland, mixed woodlands, shrubland, coastal scrubland, and mountainous locations. As its common and scientific names imply, it is indigenous to California, with a small range extending over portions of the southern half of the state. Also found in the northern section of Baja California. Terrestrial and fossorial, it spends much of its time in its underground burrow system; scent-marks its rather large home range. Semi-scansorial, it is an adept climber and can sometimes be found in bushes and trees. May enter a brief period of torpor under extreme conditions.

A sedentary terricolous omnivore, its diet consists of seeds, leaves, nuts, grains, forbs, and insects. Hoards and caches surplus food in burrow larders, transporting it in its external cheek pouches or "pockets" (the source of its common name). Its

California desert: California pocket mouse habitat.

water requirements are scant as it receives the liquid it needs from the green plant matter it consumes.

Its breeding season takes place between spring and summer. The polyestrous doe produces 1-2 litters a year; gestates for about 1 mo., giving birth to an average of 3-4 altricial pups in a subterranean natal nest. Weaning occurs at around 21 days. It is not known if the buck displays paternal behavior.

Communication and perception are normal for this member of the Heteromyidae family: sight, smell, touch, and hearing. Engages in dust-bathing and teeth-chattering. Its natural enemies are reptiles, birds of prey, mustelids, wild felids, and wild canids. Beneficial as a seed disperser and a soil aerator and recycler, and also as food for carnivores. The California pocket mouse probably only lives 1-2 yrs. in the wild; perhaps twice that long in captivity.

CALIFORNIA SEA LION (*Zalophus californianus*): It can reach a total length of nearly 8' and a weight of up to 850 lb. Its pelage is countershaded, with the dorsal fur being dark brown or gray, the ventral fur being tan to yellowish-gray. The flippers are usually darker than the rest of the body. When wet the coat may appear black. It has a dog-like face, a muscular, torpedo-shaped, fusiform body, and long wing-like flippers. The muzzle is covered in whiskers; the tail has been reduced to a stub; the external ears are small and aquadynamic. It is gender dimorphic, with the female being smaller than the male, who is often darker in color, more robustly built, and sports a distinctive sagittal crest (appearing as an enlarged pale-colored forehead).

This boisterous, social otariid lives in temperate, coastal marine habitats, preferring sandy and rocky beaches on islands that are protected by cliffs. It is also fond of rivers, and in particular, edificarian objects and environments, such as piers, jetties, buoys, and oil platforms. Its range extends from British Columbia to California, and south to Mexico.

California sea lion.

A carnivore, piscivore, and molluscivore, its diet consists mainly of various fish species (such as hake, salmon, and herring), but it also readily consumes octopus, squid, and abalone. Will hunt alone, in small groups, or in large groups, depending on the availability of food. It is known to engage in interspecies hunting, cooperating with other creatures, such as dolphins, whales, and seabirds, to locate and capture prey. Rests on beaches during the day; usually hunts at night. Uses sonar when foraging, and will dive to a depth of nearly 1,000' and remain submerged for as long as 20 mins. It can swim up to 25 mph, making it one of the fastest of the aquatic meat-eaters.

Breeding takes place in summer, typically on protected beaches. The female gestates for about 11 mos., giving birth to a single 35 lb. pup the following summer. The neonate is weaned by 1 yr. of age. The polygynous male is not paternal and does not assist in the raising of his young.

One of the most vocal mammals in the world, its dog-like "bark" is familiar to many. Other vocalizations include growling, bleating, and wailing. Its natural enemies are sharks and orcas. This naturally playful pinniped is one of the most easily trained mammals in North America, which is why it is so often seen performing at marine parks, circuses, and zoos. The California sea lion lives about 15 yrs. in the wild; may live to about 30 yrs. in captivity.

CALIFORNIA VOLE (*Microtus californicus*): This medium-sized vole grows to a total length of nearly 9", its hind foot measures 1", and it can attain a weight of up to 3 oz. The upper body coloration varies from yellowish-brown to gray-brown, from reddish-black to grizzled black; the lower body is generally grayish with white tipped hairs; at

California vole.

2½" the dichromatic tail is long for a vole. The body is stocky; the ears are rounded, short, and small, lending it its genus name: *Microtus* means "little ear." It is gender dimorphic, with the male being substantially larger and heavier than the female.

This riparian cricetid is a nearctic native that inhabits temperate grassland, meadows, and savanna, from low to high elevations, and from freshwater to saltwater coastal areas. A highly adaptable rodent, it is also found in and around sand dunes, wetlands, forests, moist foothills, deserts, chaparral, shrubland, and farms. As its common and scientific names indicate, this is a West Coast species whose range extends from Oregon through California, south into Baja California, Mexico. Semi-fossorial, it spends the majority of its time in subterranean tunnels; cuts down grass around the entrance, stacking it in little piles; small grass runways mark the surrounding area. Establishes a home range of about 1,000 sq. ft., and though normally social, it will defend its turf against unwanted intruders.

A crepuscular sedentary herbivore, its diet consists mainly of seeds, roots, grains, forbs, grasses, fruit, sedges, nuts, leaves, and various other green matter. It is also a coprophage, which enables it to procure life-sustaining nutrients during lean times.

Its monogamous-oriented breeding season is geographic-dependent and so may occur at almost anytime during the year. The polyestrous female, who is able to reproduce at just 3 wks. of age, can bear at least 4 litters annually. She gestates for about 3 wks., giving birth to as many as 10 altricial pups in a concealed grass-lined nest. Weaning and independence come quickly—only 2 wks. later. In vole communities where monogamy holds sway, the father aids in the rearing of his offspring; in those where polygyny dominates, he is less likely to participate. As in some other mammal species, the male may kill and eat neonates that he did not sire.

Communication and perception run along standard Cricetidae channels: smell, sight, touch, and hearing. Does not hibernate; active all year. The coat is loose, allowing extra freedom of movement when navigating in the confines of its underground home. The natural enemies of this keystone species are reptiles, birds of prey, ardeids, mustelids, wild felids, and wild canids. The California vole lives quickly: its average life span in the wild is about 6 mos.

CAMAS POCKET GOPHER (*Thomomys bulbivorus*): This hefty gopher, the largest in the genus *Thomomys*, grows to a total length of about 13", its hind foot measures nearly 2", and it can attain a weight of almost 1½ lb. Though, depending on numerous factors (population, geography, gender, calendar), its pelage coloration can be highly variable, the dorsum is generally dark grayish-brown, the ventrum is sooty-gray; the nose and ears are dark gray to black; the irregularly-shaped throat marking is whitish; may

A pocket gopher.

have an indistinct blackish dorsal stripe; the sparsely furred tail grows to about 3".

The body is stout and muscular; the paws are well-clawed and built for digging; its large yellowish incisors are exposed, yet—similar to the American beaver—are designed to allow the mouth to close, preventing dirt from entering the oral cavity while it is burrowing. As befitting a fossorial animal, its eyes and ears are small, its feet are oversized, and its muzzle is covered with sensitive nasal vibrissae. Gender dimorphism is present: the male is much larger than the female.

This solitary sedentary geomyid is a nearctic native that inhabits temperate, well-drained grassland, savanna, orchards, hedgerows, meadows, farmland, clearings, fields, and disturbed locations, preferably with moist clay, or loamy or sandy soil. It has one of the most restricted ranges of any mammal in our region, one that extends only over the northwestern portion of Oregon. Quarries massive tunnel systems, some up to 300' long and 6' deep; uses both its feet and its enlarged, outwardly angled incisors to dig. Conspicuous rounded mounds of excavated soil may be seen at or near the entrance; spends most of its life below ground, resting, sheltering, avoiding predators, burrowing, nesting, and eating. Does not hibernate; active all year. Establishes a home range of several thousand square feet, violently defending it against conspecifics.

A diurnal and nocturnal herbivore that leans heavily toward bulbivory, its diet is typical of the Rodentia order: seeds, fruit, grains, nuts, wood, flowers, grasses, bark, stalks, leaves, tubers, agricultural crops, and the item from which it derives its species name, bulbs. Its common name, camas, is a variant of the name of the camassia plant, or camas lily, an herb with edible bulbs upon which it feeds. Prefers vegetation that grows underground where foraging is safest. Caches surplus foods in specially made burrow larders, transporting it in its external, fur-lined cheek pouches or "pockets," from whence it derives its common name.

Its transient monogamous mating season takes place in spring. Breeding males and females form temporary pair-bonds that last for the length of one reproductive cycle. The following season they breed with new monogamous partners. The serially monogamous female gestates for nearly 3 wks., giving birth to an average of 3-4 altricial pups, which she weans within 6 wks. The male does not display paternal behaviors.

Communication and perception occur via normal Geomyidae channels: mainly touch, hearing, and smell, with little eyesight involved. Its natural enemies are reptiles, birds of prey, mustelids, wild felids, and wild canids. Beneficial as a soil aerator and recycler, as a host for parasites, as food for carnivores, and as a homebuilder for other animals. Considered an agricultural pest by many, both for the widespread damage its intense style of digging can inflict on cultivated and pastoral fields, and also for its destruction of croplands and orchards. The rarely seen camas pocket gopher probably lives an average of 1-2 yrs. in the wild; perhaps twice that long in captivity.

CANADA LYNX (*Lynx canadensis*): At around 2' in height, 3½' in total length, a 12" hind foot, and a weight of up to 40 lb., the Canada lynx is a medium-sized North American felid. It has large golden-green eyes and vertical black pupils. Gender dimorphism is present: the male is larger than the female.

Though often confused with its close relative the bobcat, the lynx has a number of special identifying characteristics, including a somewhat different diet. Its dorsal pelage is grizzled tawny-buff, intermixed with black and cinnamon-colored hairs (note that its summer coat is darker and shorter than its winter coat); the bobcat's pelage, however, is longer and tends more toward reddish-brown. The lynx's underparts are lighter than the bobcat's, and while some individual lynxes sport light or indistinct blackish or brownish spots, particularly on the legs, a bobcat's spots are usually darker and more conspicuous.

A lynx's facial or cheek ruffs are large, flared, and light in color, and flow downward, often forming two pointed beards at the throat; a bobcat's cheek ruffs are smaller and shorter. A lynx's tail is short, tannish, and tipped with black all the way around; a bobcat's tail is longer, banded across nearly its entire length, and has a bicolored tip: black above, white below. The lynx's ears, which have noticeable white spots on their backs (as do the bobcat's), are triangular, with long black tufts; the bobcat's ears are also triangular, but its ear tufts are shorter.

Canada lynx.

The lynx's spreadable toes and large furry paws act like snowshoes, while its long powerful legs aid in traversing deep snow and rough terrain; the bobcat's paws are smaller and its legs are shorter, an adaptation to life in more varied habitats. A lynx's hindquarters are higher than its shoulders (giving it a "raked" appearance), a result of the back legs being longer than the front legs; a bobcat's hips are more level with its shoulders. The Canada lynx, as its name denotes, generally lives in much colder climates (as far north as northern Alaska); the bobcat, however, is much more concentrated in the lower 48 U.S. states (their ranges overlap across southern Canada). Lastly, the 40 lb. Canada lynx is generally more mild-mannered, smaller in size, and more wiry than the pugnacious muscular bobcat, which can attain a weight of 70 lb., nearly twice that of its Canadian cousin.

Canada lynx.

A sedentary, cursorial nearctic native, the lynx's thick coat is well suited to its primary habitat: deep, mature, coniferous, boreal forests and windfalls with thick undergrowth, located in regions with long cold winters. It can also be found in tundra and rocky settings, and occasionally prairie with good hiding cover. Seeks out regions with a high density of hares, such as burnover and clearcut areas where regeneration is taking place after wildfires and logging. As mentioned, it ranges from Alaska and Canada (from British Columbia to Nova Scotia) south into the northern U.S., stretching down into the Rocky Mountain states and across to the upper Midwest and into the Northeast states. Dens are primitive and usually located in underbrush, snag, blowdown, hollow logs, stumps, deadfall, rock piles, or other similar microhabitats. Males establish home ranges that extend up to 100 sq. mi. or more, scent-marking trees and stumps along the boundaries with urine.

Nocturnal, secretive, and solitary (and thus rarely seen in the wild), it rests in a concealed location (under snag or rock ledges) during the day. Under cover of darkness it hunts silently using its powerful jaws, massive canines, and large retractable claws to grab and pin down prey. Though not built for long-distance running, it is an excellent climber, swimmer, short-distance sprinter, stalker, and ambush predator. A perch-hunter, sometimes it pounces down on its victims from a seat in a tree. It is easily able to traverse deep snow, which it uses to its advantage, catching fleeing animals that get bogged down. Its long ear tufts function like miniature aerial receivers, detecting the slightest sound over great distances. It possesses superior eyesight, and can see an object as small as a rodent from nearly the length of a football field away (300').

The lynx's favorite food is the snowshoe hare, which heavily populates its particular habitat, and with which its own population rises and falls—sometimes dramatically. An obligate meat-eater, this elusive carnivore also consumes fish, birds (grouse), mice, voles, squirrels, fox, beaver, sheep, and carrion; will go after larger game such as deer, caribou, and moose if it spots a young, sick, injured, or aged individual. Also feeds on gut piles left by hunters, and may take poultry and livestock on occasion. Hoards and caches surplus meat under forest litter or snow. Uses trees as scratching posts, which helps keep its claws sharp.

Canada lynx.

Its breeding season runs from February to March. The female gestates for about 2 mos., giving birth to 1-6 altricial kittens in late spring or early summer in a warm nesting den located in a dense thicket, stump, rock ledge, root tangle, shrub, or under logs and snag. The young are weaned at 5 mos. of age, but stay with their mother for another 5 mos., during which time they all travel together as she teaches them how to hunt. At 10 mos. they attain independence and disperse. The male does not aid in the rearing of his offspring.

Communication and perception channels are typical of the Felidae family: sight, touch, hearing, and smell are utilized during normal activities (hunting, scent-marking, mating, infant-raising). Vocalizations include hissing, growling, shrieking, yowling, mewing, wailing, purring, squeaking, chattering, and screaming (some of its sounds can be quite bizarre). A secondary consumer, this powerful wary mammal has few natural enemies, but it may occasionally be preyed upon by mountain lions, wolves, and bears. The Canada lynx can live 10-15 yrs. in the wild; 25 yrs. in captivity.

CANYON MOUSE (*Peromyscus crinitus*): Also known as the "canyon deer mouse," this small rodent grows to a total length of nearly 8" and a weight of around ½ oz. The long, soft pelage on the upper body is buffy-brown with a gray wash; the ventrum is white; the feet are light, usually whitish; the long 4½" tail is sparsely furred and has a tufted tip. The eyes are large; the ears are long; the muzzle is covered in tactile vibrissae.

This terricolous sedentary cricetid is a nearctic native who, contrary to that suggested by its common name, inhabits a variety of temperate biomes and ecosystems. Besides rocky canyons, this petraphile can also be found in and around deserts, scree, grassland, forests, shrubland, valleys, talus slopes, sand dunes, boulder fields, gravelly chaparral, rocky outcroppings, cliffs, and mountains. Has a preference for locations with bare, slick stone surfaces and rocky crevices.

A western species, its range covers portions of the states of Oregon, California, Nevada, Idaho, Arizona, New Mexico, Utah, Wyoming, and Colorado. Does not hibernate and is normally active all year; however, it will enter a state of light torpidity if conditions become harsh, helping conserve its energy. Establishes a home range of several thousand square feet, which it scent-marks and aggressively defends against conspecifics.

A shy, solitary, nocturnal omnivore, its diet differs little from most other rodents: seeds, leaves, grains, fruit, flowers, nuts, green foliage, and insects, is the common fare. Does not need to drink as its water requirements are met through its food; hoards and caches surplus edibles to carry it through the cold months. Its polygynous breeding season runs throughout the year. The highly fecund polyestrous female may produce 6 or more litters annually; she gestates for about 3-4 wks., gives birth to 2-3 altricial pups in the natal nest chamber, and weans them within 1 mo. or so.

Communication and perception occur via the normal Cricetidae channels: sight, smell, touch, and hearing. Engages in teeth-clicking, dirt-bathing, and foot-stomping. Its natural enemies are reptiles, birds of prey, mustelids, wild felids, and wild canids. An adept digger, sprinter, climber, and swimmer. Beneficial as a seed disperser, a host for parasites, food for carnivores, and as a subject of scientific research. The canyon mouse probably lives an average of 1-2 yrs. in the wild; perhaps twice that long in captivity.

Caribou.

CARIBOU (*Rangifer tarandus*): Familiar to Americans as Santa's reindeer ("reindeer" being the name by which it is known in Europe and Asia), this large common cervid stands up to 5' at the shoulder, is just over 7' in total length, and can weigh up to 700 lb. Its impressive, branched semipalmated antlers can reach spreads of over 60". Both genders grow antlers, a phenomenon unknown in any other deer species. Coat coloration, which depends on numerous factors (including geography and time of year), varies enormously, from brownish-black and bluish-gray to tawny-buff and white. Often, however, despite the color of the pelage, the neck and mane are whitish. Has a large muzzle, short rounded ears, and a short furry tail. It is gender dimorphic: caribou males are physically much larger than females.

It inhabits tundra, taiga, muskeg, and montane coniferous forest throughout its large range, which extends across the arctic, subarctic, and boreal regions of North America, from Alaska, the Yukon, British Columbia, and Washington in the west, to Quebec and Newfoundland in the east. Crepuscular, gregarious, and nomadic, it forms herds of up to 500,000 individuals that migrate 3,000 mi. each year (in the spring and fall). Traveling 15-30 mi. a day, they make the longest known treks of any terricolous animal on earth. Their snowshoe-like foot pads, which are large, flat, and rounded, become pliable in the summer, providing good grip on muddy ground, while in winter they become hard, the toughened hooves biting into the snow and ice, preventing slippage. Possessing a double-layered, shaggy coat with lightweight but stiff hollow hairs (that insulate the body and give buoyancy), caribou are aquatically adept and will swim across huge lakes and broad rivers, reaching speeds of up to 10 mph in the water.

This hardy herbivore feeds on lichens, grasses, roots, sedges, and fungi, as well as tree parts (leaves, wood, stems, twigs, bark). Also a ruminant, when necessary it will regurgitate its meal, re-chew it, and re-swallow it, a survival mechanism for procuring extra nutrients from its food.

The male's antlers can grow 52" long and weigh as much as 33 lb. Mature bulls begin to shed their antler velvet in late summer, just in time for the rut, which begins in the fall—when they begin fighting and competing over access to

Caribou.

estrus cows. These tournaments are quite violent and can leave a male battered and exhausted, and thus highly susceptible to predation. The female breeds once a year (usually in October), gestates for about 7½ mos., and gives birth to 1-2 precocial calves in late spring to early summer. A newborn reindeer can weigh up to 25 lb., and can suckle, walk, and run on its first day.

Communicates and perceives via standard Cervidae channels: sight, smell, hearing, and touch. Vocalizations include bellows, grunts, bawls, rattles, snorts, bleats, hoots, and barks; makes a clicking sound while walking (created by the snapping of its foot tendons). A primary consumer, its natural enemies are coyotes, wolves, and bears. Restless, wary, and fleet-footed, it can attain a speed of 50 mph. It is the only deer to have been domesticated, a process which began some 3,000 yrs. ago. The caribou lives an average of 5-10 yrs. in the wild; up to 20 yrs. in captivity.

CAVE MYOTIS (*Myotis velifer*): This large cave bat grows to 4½" in total length, its ears are about 1" long, and its wings can attain a span of up to 13". Its body is light brown to black, depending on the individual, as well as the geography of its colony. Its calcar lacks a keel.

Cave myotis at home.

A secondary consumer, it inhabits caves, bridge beams, rock crevices, mines, culverts, and the walls and ceilings of old buildings across the arid regions of the American Southwest, from Kansas and Texas west to southern California. Migrates between summer and winter roosts in order to hibernate, living off its stored body fat until spring.

This vespertilionid is a nocturnal insectivore that greatly benefits humanity. Flitting from its roost after dark, it feeds on a wide variety of insects, from weevils and beetles, to moths and butterflies. The female gestates for around 65 days, gives birth to one pup each spring, and raises it in a nursing colony of up to 20,000 individuals. The cave myotis lives for 10-12 yrs. in the wild.

CHIHUAHUAN GRASSHOPPER MOUSE (*Onychomys arenicola*): Also known as "Mearns' grasshopper mouse," it was once thought to be conspecific with (that is, identical to) the southern grasshopper mouse. Today, however, it is recognized as a separate species. Despite this, it continues to share nearly all of its ecology and life history with its close cousin, and thus, as their ranges overlap, the two can be difficult to distinguish in the field (absolute identification can probably only be achieved through genetic testing).

CHISEL-TOOTHED KANGAROO RAT (*Dipodomys microps*): This medium-sized rodent grows to a total length of about 12" and a weight of nearly 3 oz. The upper body is buff with gray-tipped hairs that give the dorsal pelage a gunmetal-like sheen; the underbody, from the lower jaw and throat to the belly and feet, is white. Some individuals have a thin white thigh stripe that curves over the upper leg. The body is small and ovoid; the head is large but narrow with a convex profile; the tapered face is lightly masked; the ½" ears are small, rounded, and erect; the eyes are small, beady, and dark; long tactile nasal vibrissae sprout from the muzzle; the front legs are short, the back legs are long and powerful; the feet have clawed toes; the long 7" tail is whip-like and bicolored, with a thin black stripe running along the dorsum and a thin white stripe running along the ventrum; there is a tuft of long wispy fur on the distal half of the tail. With the male being larger than the female, gender dimorphism is present.

Kangaroo rat.

This small heteromyid is a nearctic native that dwells in temperate desert environments (from low to high elevations) that may include chaparral, shrubland, piñon pine woods, brushland, rocky hills, sageland, juniper forests, sand dunes, and gravelly soil. Occupies the western U.S., with a range that covers parts of the states of Oregon, Idaho, California, Nevada, Utah, Arizona, and New Mexico. Highly fossorial, it spends much of its life underground in its well-constructed burrow system, with an entrance often located along a bank or beneath a bush. It scent-marks its burrow as well as its home range, agonistically defending them against intruders.

A solitary, sedentary, nocturnal omnivore, it is an ardent folivore whose feeding habits are concentrated on leaves, in particular the leaves of saltbush (*Atriplex*); it is, in fact, one of only a few rare mammals that can consume this highly saline plant. Its leafy diet is supplemented with seeds, nuts, fungi, and grains. Slightly carnivorous, it will prey on insects when necessary. It is also a coprophage, which allows it to live in regions where food supplies may be unstable during certain parts of the year. Does not need to drink water, as it derives all necessary liquid from its mostly folivorous fare. Like nearly all other congeneric species, it caches extra food supplies in its subsidiary burrows, transporting them in its massive cheek pouches.

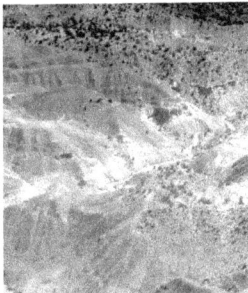

Hilly scrubland: chisel-toothed kangaroo rat habitat.

Depending on latitude, its breeding season generally runs from winter to summer. Bucks fight violently in the sand for access to estrus does, who gestate for about 1 mo., then give birth to 2-4 altricial pups. Weaning seems to take place at around 30 days or so. The male is non-paternal and plays no role in the parenting of his offspring.

A saltatorial mammal, as its common name implies, it is an excellent jumper; it is also a superb runner and swimmer. Engages in sand-bathing, which helps remove external parasites. Is active year-round; does not estivate or hibernate. Communicates and perceives via standard Heteromyidae channels: sight, touch, smell, and hearing. Its natural enemies are reptiles, birds of prey, wild felids, and wild canids, and, of course, other burrowing animals. The male engages in hind foot-drumming as a territorial signal, an alarm signal, and, during the breeding season, a courtship lure.

This species benefits the world ecosystem by spreading seeds, hosting parasites, and, as a primary consumer, providing a food base for carnivores; benefits humanity by serving as a subject of scientific research. The chisel-toothed kangaroo rat lives an average of around 1-2 yrs. in the wild; more than twice that long in captivity.

CINEREUS SHREW (*Sorex cinereus*): Also known as the "masked shrew," it grows to about 4" long, its 2" tail comprises half its body length, its hind foot measures ½", and it weighs up to ⅜ oz. This makes it our second smallest shrew species (after the pygmy shrew). Its coat on the dorsum is dull grizzled brown to gray; on the ventrum it is lighter in color, usually a light silvery-gray. Its head is cone-shaped; its eyes are tiny; its snout is long, slightly convex, and pointed; its teeth are sharp and brownish in color; its limbs and feet are small and delicate; its tail is bicolored: brown on the upper side, buff on the lower side.

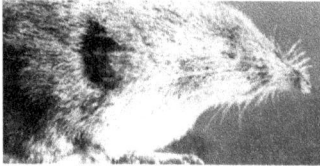
Cinereus shrew.

It lives in a wide variety of moist biomes and ecosystems, from marshes, woods, and lake shores, to meadows, bogs, and forests. Ranging across the entire northern tier of North America (from Alaska and Canada south into all of the northern states), it is the most widely distributed shrew in our region, hence its alternate name, the "common shrew." It shelters under logs and stumps, lining its nest with grass and leaves.

This active little soricid, like all of its kind, is a voracious omnivore that feeds primarily on invertebrates, including insect larvae, lepidopterans, coleopterans, arachnids, dipterans, chilopods, gastropods, and orthopterans. A granivore and a fungivore as well, it will also consume plant material, such as seeds and fungi. With its incredibly fast heart rate of around 1,000 beats a minute (common among most North American shrews), it must hunt almost constantly in order to meet its high energy demands. In pursuit of its prey it can run, climb, and dig, consuming more than its own body weight every 24 hrs.

The reproductive life of this solitary, secretive, nocturnal mammal is little known. A secondary consumer, its natural enemies are reptiles, birds of prey, mustelids, wild canids, and other shrews. The cinereus shrew lives an average of 1-2 yrs. in the wild.

CLIFF CHIPMUNK (*Neotamias dorsalis*): A member of the squirrel family, this medium-sized chipmunk grows to a total length of about 11", its hind foot measures nearly 2", and it can weigh up to 3 oz. Its pelage is grayish on the dorsum with indistinct alternating whitish, blackish, and brownish dorsal stripes running from shoulders to rump. Has a single dark medial stripe. The sides are grizzled-gray to rusty-orange; its brownish, blackish, and clear whitish face stripes echo the dorsal stripes on its back, running horizontally from nose tip to ears. The belly is whitish-tan; each ear has a white postauricular mark. The 5" tail is bushy and squirrel-like and runs from grizzled gray to longitudinally striped. Its coloration is both cryptic and protective, permitting it to blend in with its natural surroundings. Gender dimorphism is present: the female is larger than the male.

This energetic but timid sciurid is a nearctic native that inhabits temperate scrub forests made up of piñon, pygmy, juniper, Mexican white, and ponderosa pines; a palustrine species, it is also found in and around shrubland, sagebrush, rocky zones, bluffs, saltbrush, boulder fields, stump ranches, chaparral, talus slopes, lava fields, scree, desert, and, as its common name denotes, cliffs—preferably with snag, blowdown, deadfall, and logs. Its range is limited to the American West, occurring in varying density over the states of Idaho, Wyoming, Nevada, Utah, Colorado, Arizona, and New Mexico, as well as parts of northern Mexico. Terrestrial, sedentary, petraphilic, and fossorial, it constructs its den and nest in stone piles, rock crevices, bluffs, burrows, steep banks, and, being slightly arboreal, occasionally trees. Establishes a home territory of up to 5 acres.

A diurnal terricolous omnivore, its diet centers around the seeds, fruit, and nuts of the plants and trees in its environment. It is also fond of insects, amphibians, reptiles, birds' eggs, and birds. Hoards and caches surplus food in shallow holes, transporting it in its cheek pouches. May hibernate or employ torpor during extreme cold.

Its breeding season occurs throughout the spring and summer. The doe gestates for about 1 mo., giving birth to as many as a half dozen altricial pups in the nest chamber. Weaning and independence come at around 5-6 wks. of age. The buck does not contribute to the rearing of his offspring.

Communication and perception occur via typical Sciuridae channels: sight, smell, hearing, and touch. Vocalizations include chirps, chips, bird-like calls, chucks, chatter, and a fast repetitious bark accompanied by tail-twitching. Engages in dust-bathing. Its natural enemies are birds of prey, reptiles, mustelids, wild felids, and wild canids. Benefits its ecosystem by dispersing seeds, hosting parasites, and providing food for carnivores. The life span of the cliff chipmunk is unknown, but it probably lives no more than 3-4 yrs. in the wild.

COAST MOLE (*Scapanus orarius*): This eulipotyphlid grows to a total length of nearly 7", its tail is about 1½" long, and its hind foot measures ¾". Its pelage is dark gray to black in color (ideal camouflage for its subterranean environment), the fur is velvety (allowing it to slip more easily through its tunnels), and the nose, forelegs, and tail are hairless (increasing tactile sensitivity in its moist dark world). Its eyes are tiny (and thus it is almost functionally blind); its outer ears are inconspicuous; and the front feet are shovel-like and heavily clawed—all evolutionary designs that aid it in living underground. It is active both during the day and at night and has a home range about the size of a football field. It is gender dimorphic: the male is larger than the female.

This solitary talpid inhabits deciduous woodlands throughout its range: southwestern Canada, south (through Washington and Oregon) into northern California, and east to Idaho. A fossorial omnivore, vermivore, and insectivore, its diet is comprised primarily of worms, snails, and insects, though it also consumes fungus as well as vegetation, such as bulbs and tubers.

The breeding season of this secondary consumer takes place from midwinter to early spring. The female gives birth to as many as 4 altricial pups in a grass-lined nest chamber, weaning them in the summer. The male does not assist in the rearing of his offspring. Communication and perception run along standard Talpidae channels: touch and smell, and to a much lesser degree, sight and hearing. Its natural predators are reptiles (snakes) and birds of prey (mainly owls). Its massive tunnel systems and underground foraging helps aerate and recycle the soil, facilitates drainage, and provides homes for numerous other animals. It therefore plays an important role in North America's ecosystem. The coast mole probably lives from 3-4 yrs. in the wild.

COLLARED PECCARY (*Pecari tajacu*): Also known as the "javelina" (Spanish for "spear," a reference to its short, straight, pointed tusks), this medium-sized porcine reaches a shoulder height of 22", a total length of 40", and a weight of 65 lb. Its coarse fur is a grizzled brownish-black; a dark mane runs from the top of its head to its hindquarters, forming a blackish dorsal stripe. As its common name indicates, it has a tawny-whitish "collar" that runs around the neck, throat, and shoulder regions, one of its most distinguishing

Collared peccary.

characteristics. It has 4 toes on its front feet, 3 toes on its hind feet.

In our region it inhabits arid scrubland, cacti-rich desert, chaparral, canyons, wasteland, rangeland, and also urban areas, throughout the southwestern U.S.; mainly Texas, New Mexico, and Arizona. A social mammal, it forms herds of 10-30 individuals that are led by the dominant male. The herd defends its 5-square mile territory by scat-marking, as well as by scent-marking rocks and trees, using a pungent oil it emits from its rump gland.

Collared peccaries.

Like other members of the Tayassuidae family, this pig-like artiodactyl will feed at any time of day or night, and is therefore diurnal, nocturnal, and crespuscular. A terricolous omnivore, its diet consists of cactus (including the spines), grasses, roots, nuts, bulbs, agaves, fruit, eggs, carrion, invertebrates (insects), amphibians, reptiles, and birds. Though its eyesight is poor, this is offset by its excellent smell and hearing. Communication and perception are typical for a tayassuid: mainly touch and smell, and to a lesser degree sight and hearing. Vocalizations include grunts, squeals, cries, teeth-gnashing, and snorts.

Breeds year-round, the only wild hoofed mammal in our region to do so. The female gestates for about 5 mos., giving birth to 1-5 piglets at a nest site separate from the herd. Sometimes called the "stink pig" or "skunk pig" due to its odiferous scent glands, its natural enemies include coyotes and wild felids (mountain lions, bobcats, jaguars). This nomadic artiodactyl can swim, run 35 mph, and fight viciously when defending itself or its young. Considered by many to be harmless (even beneficial) to humans, this native species should not be confused with the wild boar, which is an invasive, highly destructive, and much more aggressive nonnative. The collared peccary lives about 8 yrs. in the wild; perhaps as long as 25 yrs. in captivity.

COLLARED PIKA (*Ochotona collaris*): Also known as a "cony," this solitary petraphilic mammal grows to a total length of nearly 8", its hind foot measures 1¼", and it weighs as much as 7 oz. Its dorsum is grizzled grayish-brown, the ventrum is white, ideal camouflage for the rocky environments it resides in. It has a light gray "collar" of fur on the neck and shoulders, the origin of its common name. The word pika (in North America pronounced pie-ka) comes from a Siberian species whose call sounds like "peeka."

Like most other members of the *Ochotona* genus, it has a small guinea pig-like shape, a mousy appearance, short semi-circular ears, long whiskers, short legs, furred foot soles, and a hidden tail. The collared pika's front feet have 5 toes, the hind feet have 4 toes. As with other pikas, it has an apocrine (scent) gland; in its case it is located on the cheek and is covered with a buff-colored fur patch (which helps distinguish it from other species). The males and females are similar in appearance and can only be differentiated by close physical examination.

It is found in mountainous rocky terrain and on talus slopes in alpine habitat, usually located near meadows. Its range covers southeastern Alaska south into coastal and northwestern British Columbia. A diurnal mammal, it does not hibernate or burrow.

This omnivorous ochotonid eats everything from lichens, leaves, and flowers, to animal matter such as birds. Also coprophagous, it consumes its own dung as well as that of other animals, one of Nature's many ways of optimizing

Collared pika.

nutrients—thereby enhancing an animal's chances of survival. An ecotone species, it collects copious amounts of plant cuttings (which it forms into haystacks throughout its 100 sq. ft. home range) to feed on during the long winter months.

Typically breeds once a year in late spring. The female gestates for around 1 mo., giving birth to 1-6 altricial (temporarily blind and hairless) young. When communicating this vocal lagomorph chatters, teeth-clicks, and squeaks; it is best known for its alarm call, a high piercing "meep" that can be heard over great distances.

A keystone species, this primary consumer is very sensitive to climate and food availability, and may be susceptible to human-induced changes in its environment. Its natural predators include birds of prey, mustelids, wild canids, wild felids, and ursids. The collared pika lives an average of 5-7 yrs. in the wild.

COLORADO CHIPMUNK (*Neotamias quadrivittatus*): Once believed by some to be conspecific with the Hopi chipmunk, today it is classed as a separate species. While it shares much of the same ecology and life history as its close cousin, there are 2 differences: its dorsal stripes are more vividly colored, and its geographic range covers the states of Colorado, Arizona, New Mexico, and Oklahoma (the Hopi's range extends only over Utah and Colorado).

Colorado chipmunk.

COLUMBIAN GROUND SQUIRREL (*Urocitellus columbianus*): One of the largest members of its genus, it grows to a total length of around 16" and it can attain a weight of almost 30 oz. Has distinctive coloration: its upper body is grizzled gray and black with white or buff flecks across the back, as well as a light wash of rust. The underfur is lighter, tawny or light gray. The upper cheeks are gray; the lower cheeks, chest, front legs, and the top of the muzzle are cinnamon-colored. The 4" tail is bushy, reddish, white, and black on top, dark-tipped, and edged in white. Has a typical ground squirrel morphology: husky body; top of head slightly convex; dark medium-sized eyes ringed in tan; small ears; shortish but strong limbs; and short dense fur. This species is gender dimorphic, with the male sometimes being slightly heavier than the female.

Columbian ground squirrel.

As with most of its kind, it inhabits a variety of landscapes and biomes, in its case from alpine grasslands and lowland meadows to arid subalpine brushland and savanna. A nearctic native, its range encompasses the Rocky Mountains, and more generally the northwestern corner of the U.S. (Washington, Oregon, Idaho, Montana), as well as portions of Alberta and British Columbia—the latter region giving it its common name.

An omnivorous sciurid, its diet includes seeds, grasses, stalks, bulbs, fruits, tubers, leaves, flowers, insects, birds, and small mammals. Diurnal and semi-social, it dwells in large colonies inside sprawling complex burrow systems that house nesting chambers, food storage compartments, and hibernacula. The latter is used for both hibernation during the cold months and estivation during the hot months. In total, it can spend up to 8 mos. (or nearly 67 percent) of the year sleeping, slowing down kidney function and lowering its body temperature to conserve energy.

Its breeding season takes place after hibernation, which ends in the spring. The female gives birth to an average of 4 altricial pups inside a grass nest, weaning them at around 1 mo. of age. Communication and perception channels are normal for a member of the Sciuridae family: sight, hearing, touch, and smell, the latter playing a particularly large role in a species that possesses multiple scent glands over its body. Its natural enemies are birds of prey and small to medium size carnivores. The Columbian ground squirrel probably lives 3-6 yrs. in the wild.

COLUMBIA PLATEAU POCKET MOUSE (*Perognathus parvus*): The largest member of its genus, it grows to a total length of about 8", its hind foot measures nearly 1", and it can weigh up to 1 oz. The dorsal pelage is pinkish-yellow with a wash of dark fur; its underparts are whitish or tawny; its 4" tail is long, furry, and bicolored; its hind legs are large, but smaller than those of its cousin the kangaroo rat. The body is ovoid; the eyes are large and dark; the ears are small and erect.

This crepuscular nocturnal heteromyid inhabits pine forests, dry short-grass steppe, chaparral, shrubland, grassland, moist meadows, desert, rangeland, hedgerow, montane sageland, and cropland, preferably containing loose sandy soil, rocks, and abundant hiding cover. Its range covers the western end of our region, extending over portions of British Columbia, Washington, Oregon, Idaho, California, Montana, Wyoming, Utah, and Arizona. May enter torpor during severely hot or cold weather. This fossorial rodent spends much of its time in its underground burrow and tunnel system. Seemingly solitary, it establishes a home range of several thousand square feet.

Omnivorous, it consumes mainly seeds, legumes, and grains, but also general plant material and insects. Hoards and caches surplus food, transporting it to its burrow larders in its external fur-lined cheek pouches. (It is from these "pockets" that all pocket mice derive both their common name and their genus name: *Perognathus* is from the Greek *pera*, meaning "pouch," and *gnathus*, meaning the "jaw.") Its breeding season peaks in spring and summer. The iteroparous polyestrous female may produce 2-3 litters annually. She gestates for nearly 1 mo., giving birth to an average of 4 pups in a subterranean nest chamber, weaning them some 3 wks. later.

Communicates and perceives via normal Heteromyidae channels: primarily touch and smell. Its natural enemies are reptiles, birds of prey, mustelids, wild canids, and other rodents. The precise life span of the Columbia Plateau pocket mouse in the wild is not known, but it is probably similar to that of other pocket mice: 1-2 yrs.

COMMON BOTTLENOSE DOLPHIN (*Tursiops truncatus*): The official state water mammal of Mississippi and South Carolina, it is also known as the "Atlantic bottlenose dolphin," the "bottlenosed dolphin" and the "bottlenose dolphin," and is the most familiar of North America's delphinids. It reaches a total length of 13' and an average weight of 700 lb., though much higher weights have been recorded—making it the largest of the beaked dolphins. Has a bluish-gray dorsum; a light bluish-gray ventrum; a blunt but prominent bottle-shaped beak (hence its common name); pointed flippers; and a tall hooked dorsal fin at the center of its back.

Common bottlenose dolphin.

Like its many dolphin relations, this apex predator uses echolocation to hunt, either alone or in cooperative teams. An opportunistic feeder, its preferred foods are aquatic crustaceans, cephalopods, and fish—which it grasps and tears with its many large teeth before swallowing them whole. Gregarious, it travels in pods of 10-1,000 individuals. Extremely communicative, it makes a host of sounds, from clicks and squeaks, to whistles and trills. One of the most fun-loving of all dolphin species, it readily rides bow and stern waves, lobtails (slaps the water with its tail), and leaps high into the air.

Both its habitat and range are broad. It is found in temperate to tropical waters, close to the coast (bays, rivers, lagoons, inlets, estuaries, reefs, harbors) and also far out in the deep ocean. Some populations migrate.

Females give birth to a single 3-4', 25 lb. calf every 2 yrs. or so. With its massive brain, it can learn, understand, and even mimic human language. (It is interesting to note that we do not yet fully understand its language.) A secondary consumer, its natural enemies are sharks and orcas. A regional heterotherm, it has the ability to alter the blood flow in specific areas of its anatomy in order to regulate its overall body temperature. The common bottlenose dolphin probably has an average life span of about 25 yrs. in the wild—though it may live as long as 60 yrs. As with nearly all wild marine mammals, it is impossible to know with certainty.

COMMON MINKE WHALE (*Balaenoptera acutorostrata*): Also known as the "minke whale," the "northern minke whale," the "little piked whale," and the "dwarf minke whale," this small slim balaenopterid grows to a total length of 33' and a weight of up to 10 tons. Its common name derives from the Norwegian word *minkevhal* (*minke* meaning "small" or "lesser," *hval* meaning "whale"—thus the "small whale"). It is dark gray-blue to blackish on the upper half, whitish on the lower half, with variable soft gray and white patterns flowing over its entire body.

It has a v-shaped upper jaw, a ridge that runs from the tip of the upper jaw back to its double blowholes, and at least one unique marking that makes it easily identifiable: a wide white band across the pectoral flippers. Its tiny curved dorsal fin is set far back on the dorsum. Its 300 whitish baleen plates, a biological feeding filter also called "whalebone," hang from the upper jaw, growing to 10" in length. Its 70 throat pleats expand and contract while feeding.

This streamlined, energetic planktivore can swim up to 25 mph. It feeds by straining small crustaceans and fish through its fibrous, comb-like baleen plates. It ranges throughout all of the world's oceans, from polar to tropical regions, and may be found in both pelagic waters and along shallow coastlines. In North America it spottily inhabits the entire Atlantic Ocean, from Nova Scotia south to Florida; it has also been sighted off the coasts of Texas, California, and Alaska.

Common minke whale.

Often solitary, it sometimes forms small pods of from 2 to several hundred individuals. It has been known to become trapped in ice fields. The female gestates for around 10 mos., giving birth to a 1,000 lb., 10' long calf. Quite inquisitive, it may seek out boats and follow them for several miles. A secondary consumer, its primary natural enemy is the orca. The common minke whale may live to around 50 yrs. in the wild.

COTTON MOUSE (*Peromyscus gossypinus*): Also known as the "cotton deer mouse," this large rodent grows to a total length of 8", its hind foot measures 1", and it can attain a weight of nearly 2 oz. Its upper body is grizzled buffy-tan with a gray wash; its lower body is white; its 3½" tail is dichromatic. This cryptic coloration provides excellent camouflage, helping it avoid predators. The body is stout; the profile of the head is convex; the eyes are large; the ears are tall and erect; the snout is pointed and covered in sensitive vibrissae; the limbs are short; the paws are delicate and pinkish.

This big riparian cricetid is a nearctic native that inhabits a wide variety of temperate biomes and ecosystems, including forests, deserts, beaches, brushland, rocky locations, hummocks, swamps, dunes, disturbed areas, bottomland, ridges, scrubland, sandy locations, and savanna, both in wilderness and urban settings. It is often found in and around moist habitats like marshes and bayous, and along waterways such as creeks and rivers. Its range covers much of the Southeastern and Midwestern U.S., extending over portions of Illinois, Oklahoma, Arkansas, Texas, Louisiana, Mississippi, Alabama, Georgia, Florida, Tennessee, Virginia, North Carolina, and South Carolina. Shelters and nests under logs, bushes, and scrub.

A shy, nocturnal, opportunistic omnivore, its diet consists of typical mouse fare: seeds, nuts, grains, flowers, fruit, fungi, mollusks, and insects. Its mating system is unknown, but like other similar species, it probably transits back and forth from polygamy to monogamy depending on such factors as population density, food abundance or scarcity, and climatic conditions. Its breeding season is hinged on geography, but generally it runs year-round, with the usual increase in activity during cold weather and a decrease during warm weather. The polyestrous female can bear 3-4 litters annually; she gestates for around 3 wks., giving birth to an average of 3-4 altricial pups, weaning them within 1 mo. If the male practices monogamy he will probably contribute to the raising of his offspring; if he engages in polygyny, this is unlikely.

Communication and perception occur via normal Cricetidae channels: sight, smell, touch, and hearing. Engages in foot-stomping; though terricolous it is also scansorial, and is an adept climber; has natatorial traits and is a good swimmer. Its natural enemies are reptiles, birds of prey, mustelids, wild felids, and wild canids. Beneficial as a seed disperser, a host for parasites, and food for carnivores; harmful as a carrier of deadly diseases and as a household pest. The cotton mouse rarely lives to 1 yr. of age in the wild, and its actual life span is probably closer to 6 mos. or less.

COUGAR (See mountain lion).

COYOTE (*Canis latrans*): The official state animal of South Dakota, it is also misleadingly called the "prairie wolf." One of the best known species in the wild dog family, it stands about 27" at the shoulder, grows to a total length of 53" long, has a 9" hind foot, and weighs up to 50 lb. The fur on its upper body ranges in coloration from white, tan, and tawny-gray, to reddish-brown, gray-brown, and grizzled. Its throat, abdomen, and legs are often lighter (buff), and its eyes have a yellow or golden iris.

Other distinguishing traits include a lean body, dense fur, narrow snout, a long 16" bushy tail (often with a black tip), black nose, and conspicuous pointed ears. May have a "collar" of whitish or darkish fur around the neck and/or shoulders. A digitigrade, it walks and runs on its toes (the heels do not touch the ground) rather than on the soles of its feet (as plantigrades do). Has powerful jaws with long canines (for grasping and tearing skin and muscle) and robust molars (for crushing and pulverizing ligaments and bone). A gender dimorphic species, the male is generally larger and heavier than the female.

Coyote.

Its common name comes from the Aztec word *coyotl*, the name bestowed upon it by the Nahuas, a Native American people of Mexico. It shares many of the characteristics of its canine cousins: wolves, foxes, jackals, and domestic dogs. Adaptive and versatile, and thus widespread, this crafty riparian inhabits almost every type of biome and ecosystem in North America, including prairies, plains, chaparral, sand dunes, rangeland, alkali sinks, savanna, scrubland, hedgerows, ranchland, farmland, steppes, arid deserts, tropical locations, pastures, coniferous and deciduous forests, railroad land, wetland, grassland, alpine meadows, cold tundra, sageland, woodlots, plantations, montane heath, taiga, riverbanks, brushland, disturbed areas, broken forest, orchards, and swamps; found in wilderness, suburban, urban, and industrial settings. It has been sighted both far from civilization and also in the middle of some of our largest metropolitan areas (for example, Central Park in New York City). Nearctic, its ever-expanding range covers nearly all of North America, from Alaska to Panama. Establishes a home range of as much as 35 sq. mi. Dens are up to 50' in length, have multiple entrances and tunnels, and are formed in hollow logs, rock crevices, sinkholes, thickets, steep banks, ledges, underbrush, dense vegetation patches, and hill slopes.

Though primarily nocturnal, and at times also crepuscular, the ever opportunistic, generalist coyote will also hunt during the day. It usually hunts alone, but sometimes also in small groups or large packs; occasionally hunts cooperatively with badgers, with whom it shares the same basic diet, making it both a mutualist and a symbiotic species. Extremely cunning and secretive, it seeks out high vantage points to scan for prey. An omnivorous carnivore (and also a part-time herbivore, folivore, invertivore, and vermivore), its diet consists of nearly everything edible: carrion, leaves, plants, fruit, vegetables, worms, insects, birds, amphibians, reptiles, rodents, and larger prey such as deer and pronghorn. If the opportunity arises, it will attack, kill, and eat both livestock (poultry, sheep, goats, cows) and domestic pets (such as cats and dogs). Has been known to snatch dogs as they are being walked by their unassuming human owners: quickly darting from nearby bushes, it viciously rips the pet from its leash and carries it off into the woods. As a synanthrope, it also enjoys human refuse and thus frequents garbage dumps. Will cache surplus food. Travels up to 12 mi. a day when foraging.

Its breeding season runs from midwinter to early spring. The bitch (female) gestates for about 2 mos., giving birth to 1-20 altricial pups in a carefully concealed and protected den. She nurses her babies with protein-rich milk until around 6 wks., when they are weaned. The dog (male) assists with the rearing of the young by defending the den and by, like the mother, regurgitating food for them to eat. While male pups leave the den site after about 6 mos. or so, female pups often remain with the main pack.

One of the most vocal of the canids (its Latin scientific name means "barking dog"), its familiar squeaks, yelps, barks, and howls help it communicate with, as well as locate, its companions. It also uses its sharp senses of smell (scent-marking its territory with urine), eyesight (it can see movement over long distances), and touch (play-fights with other individuals, as well as rubs and grooms). Runs with its tail down (wolves run with their tails straight out; dogs with their tails up). Though it is not a good climber, it is a superior swimmer and runner, and can leap 15' barriers, easily run 25 mph, and reach speeds of up to 40 mph over short distances—excellent defenses against this secondary consumer's few natural enemies: golden eagles, mountain lions, black bears, grizzly bears, and wolves (the latter which drive down the coyote population wherever their ranges overlap).

Despite being heavily hunted by humans, this truly "wily" canid is one of the most successful North American mammals and continues to extend its range across the U.S. and Canada. Occasionally breeds with the gray wolf (creating a hybrid known as the "coywolf") and the domestic dog (creating a hybrid known as a "coydog"). The coyote lives an average of 8-10 yrs. in the wild; up to 20 or more yrs. in captivity.

Coyote.

CREEPING VOLE (*Microtus oregoni*): Also known as the "Oregon vole," this short-haired rodent grows to a total length of about 6", its hind foot measures 2", and it can attain a weight of up to 1 oz. The pelage on the dorsum is grayish-brown to blackish; on the ventrum it is light gray; the 2" tail is bicolored. One of the smaller voles, its eyes are tiny, dark, and beady, the ears are large and rounded.

This nearctic native is a riparian cricetid that inhabits temperate grassland, brushland, chaparral, foothills, shrubland, herbaceous settings, disturbed habitat, and moist pine forests across the American Northwest, including portions of Washington, Oregon, California, and southwestern British Columbia. Prefers locations with good hiding cover and abundant forest debris. Found in both xeric and slightly hydric conditions. Excavates shallow mole-like burrows in friable soil; here it lives, sleeps, and seeks shelter from predators. Sets up grass-lined nests under stumps, fallen trees, or underground; will take over the abandoned tunnel systems of other fossorial creatures. Does not hibernate; active year-round; establishes a home range of several thousand square feet.

An endothermic herbivore, its diet consists primarily of plant matter (such as grass and forbs), along with fruit and fungi. Meets most of its water needs through the succulent foods it consumes. Its breeding season runs most of the year, with a peak in the spring. The iteroparous polyestrous female may become pregnant at just 3 wks. of age and is capable of producing up to 6 litters annually; she gestates for about 3-4 wks., giving birth to an average of 4 pups.

Communication and perception occur via normal Cricetidae channels: smell, touch, hearing, and to a lesser extent sight. Its natural enemies are reptiles (snakes), birds of prey (owls), mustelids (weasels), wild felids (bobcats), and wild canids (foxes, coyotes). The creeping vole probably lives no more than 6 mos. or so in the wild.

CUVIER'S BEAKED WHALE (*Ziphius cavirostris*): Also known as the "goose-beaked whale," it grows to a length of 28' and weighs up to 7,800 lb. (nearly 4 tons). It takes its common name from the French naturalist Georges Cuvier, who described the species in 1823. It has the typical robust cylindrical body of the beaked whale family, along with a prominent forehead, a short beak, small pectoral flippers, and two short conical teeth growing out from the end of the lower jaw (which juts out beyond the upper jaw or rostrum).

This ziphiid has a ridge running from its small sickle-shaped dorsal fin (which is set two-thirds of the way back on the dorsum) to the tail. There is great variation in coloration, with some individuals being gray or brown, others blueish or black. Its head, neck, and back get lighter (in some cases almost white) with age. Irregular spotting along the body can occur. Adults may have dark eye patches, possibly an evolutionary adaptation that helps reduce sun glare at the water surface.

Cuvier's beaked whale.

Travels alone, but also in groups or pods, which can contain a dozen or more individuals. A wide-ranging species, it is found in nearly every ocean and sea (except for those in the polar regions); it is common along the coast of the Atlantic Ocean, from New England to Florida.

Employing echolocation it feeds mainly on squid, but also on fish and aquatic crustaceans, using the suction method common to many odontocetes. Can dive to a depth of at least 6,000' (over 1 mile) and hold its breath for nearly 1½ hrs. Spends most of its life (over 75 percent) in the pitch black depths of the ocean; sometimes it strands on beaches. A secondary consumer, little else is known about this shy and elusive medium-sized cetacid. Cuvier's beaked whale may live to 25 yrs. in the wild.

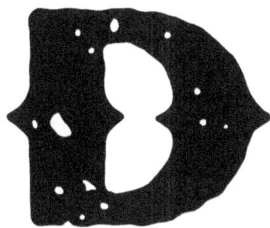

DALL'S SHEEP (*Ovis dalli*): The only species of a group called "thinhorn mountain sheep" (a common name by which it is also known), this medium-sized ovid stands up to 3½' at the shoulder, grows to 5' in total length, and weighs up to 225 lb. It is stocky, with thin legs and a short tail. Its coat runs from buff-white to pure white; may have gray markings or cinnamon highlights. The female's horns (which grow to about 8" in length) are small and pronged on the ends, while the male's (which can grow to over 3' in length) are massive, ringed, spiraled, and pointed, eventually curling into a large circle. (The rings on its horns, known as *annuli*, form once per year, and thus can be used to determine its age.) The ram's horns, which range in color from honey-beige to yellow-brown, and from grayish to transparent, grow only

Dall's sheep.

during the spring, summer, and fall. Gender dimorphism is present, with the males being much larger and more robust than the females.

Similar in many ways to its cousin the bighorn sheep, this petraphile inhabits steep cliffs, rugged rock shelves, and inhospitable mountainous areas in arctic and subarctic regions—a means of deterring predators. It has the distinction of being the world's most northern wild sheep, with a range extending from northern Alaska south into western and coastal Canada. In the warm months this herbivorous grazer searches out open meadows where it feeds on grasses, sedges, mosses, lichens, flowers, sage, herbs, and tree parts. Will migrate up and down mountainsides in search of food.

The ewe breeds in midwinter and gestates for around 5 mos., giving birth to 1-2 precocial lambs in the late spring or early summer in an isolated nursery set apart from the main flock. The young, which can nurse, stand, walk, and run within the first 24 hrs., rejoin the group after a few days, and are weaned at around 5 mos. of age. Rams engage in kicking, shoving and head-butting contests during the rut, but are protected from severe injury by their thick skulls, which sport a double layer of shock-absorbent bone. The crashing sound from these violent tournaments can be heard up to nearly a mile away.

Though highly social, like many other flocking mammals it tends to congregate in two separate main groups: a bachelorhood (in the Dall's case, made up of mature rams and juvenile males) and a matriarchy (made up of adult ewes, young adult females, and the yearlings of both genders). The playful exuberant young sometimes form their own yearling flocks within the matriarchal herd. A primary consumer, among its predators are bears, coyotes, lynxes, wolves, mountain lions, wolverines, and golden eagles (which feed on newborns). The Dall's sheep or "thinhorn sheep" lives 10-15 yrs. in the wild; as long as 20 yrs. in captivity.

DARK KANGAROO MOUSE (*Microdipodops megacephalus*): It grows to a total length of about 7", the hind foot measures nearly 1", and it can attain a weight of up to ⅝ oz. The tan pelage on its upper body has a brownish-gray to blackish wash, lending it its common name; its lower body is white; its 3½" tail is sparsely furred, has a black tip, and is thicker at the midpoint than at either the proximal or distal ends. Its countershaded cryptic coloring allows it to blend in with its shadowy surroundings, aiding its survival during nighttime foraging. The eyes and ears are large, as is its head, the latter which gives it its species name: *megacephalus* ("large head").

This nearctic native is a sedentary solitary heteromyid that inhabits temperate deserts, playa, sageland, scrubland, chaparral, salt flats, rangeland, alkali sinks, and arid sand dunes, where fine, friable, gravelly soil is abundant. Strictly a western rodent, its range extends over portions of Oregon, Idaho, California, Nevada, and Utah. A fossorial mouse, it excavates complex 4' long tunnel systems about 1' below ground, where it spends much of its time. Seals its entrances during the day; estivates during hot weather; may hibernate during cold weather. Establishes a home range of several thousand square feet, which it defends against conspecifics.

A saltatorial nocturnal omnivore, its diet consists of seeds, nuts, fruit, grains, and insects; enjoys the seeds of the desert star plant in particular. Does not need to drink as all of its water requirements are met through the solid food it eats. Probably caches surplus food in burrow larders, transporting it in its external cheek pouches. After waking from hibernation in the spring, its breeding season begins. The female gestates for 3-4 wks., giving birth to 4-6 pups, which she cares for, nurses, protects, and grooms until they are weaned soon after.

Communication and perception run along typical Heteromyidae channels: mainly smell and touch, and to a lesser extent, sight and hearing. Walks quadrupedally, hops bipedally (kangaroo-like, on its hind legs, from whence it derives its common name). Dislikes light, and even remains in its burrow during moonlit nights. Its natural enemies are snakes, owls, weasels, badgers, foxes, and other night creatures. Beneficial as a seed disperser, host to parasites, soil aerator and recycler, and as food for carnivores. The life span of the dark kangaroo rat probably averages 2-4 yrs. in the wild.

DEER MOUSE (*Peromyscus maniculatus*): Also known as the "North American deer mouse," this extremely variable species contains numerous nearly identical subspecies with only slight differences, the result of ecological and geographical influences. This has created a taxonomy that is complex, confusing, unsettled, and in need of further research. This entry then contains a general overview of the species as a whole, concentrating primarily on its two main forms: the woodland deer mouse (*Peromyscus maniculatus gracilis*), which dominates the West and the North, and the prairie deer mouse (*Peromyscus maniculatus bairdii*), which dominates the Midwest—the latter which has smaller ears and a shorter tail than its forest dwelling cousin.

Deer mouse.

It grows to a total length of nearly 9", its hind foot measures 1", and it can reach a weight of about 1 oz. Its pelage is soft, thick, and smooth; the coloration on the dorsum is buff brown with a gray to blackish wash; the ventrum is white; the feet are white; the 2½" tail is noticeably bicolored (brownish-gray on top, whitish on bottom), and is tipped with a sparsely haired, wispy tuft. Its eyes are large, black, and rounded, enhancing night vision and allowing for a wide field of vision; the ears are tall and erect; the muzzle tip is covered in sensitive vibrissae; the front limbs are shorter than the hind limbs.

The most widely distributed mouse in North America, this solitary, sedentary, riparian cricetid is a nearctic and neotropical native that inhabits nearly every temperate biome and ecosystem in our region, including savanna, desert, chaparral, brushland, prairie, forests, orchards, rangeland, scrubland, cropland, sand dunes, grassland, hedgerow, shrubland, mountainous regions, and mixed woodlands—in both wilderness

and urban settings. It thus has one of the largest ranges of any mammal in North America, one extending from Alaska and northern Canada south to central Mexico. It is found over most of the lower 48 U.S. states, but is less populous in the Southeast and may be completely absent from Florida.

Both the long-tailed forest dwellers and the short-tailed open-ground dwellers are terricolous and live on the ground. However, the former is semi-arboreal and generally nests above ground (in trees, rock crevices, buildings, hollow logs), while the latter nests in underground burrows. Highly territorial, it establishes a home range of up to ½ acre or so, which it readily defends against conspecifics.

A nocturnal and somewhat crepuscular omnivore, its diet entails standard *Peromyscus* fare: seeds, fruit, grain, nuts, flowers, fungi, cultivated crops (corn, wheat, etc.), annelids, mollusks, and insects. It is also a coprophage, enabling it to procure vital nutrients during times of food scarcity.

Its polygynandrous breeding season runs year-round, with a peak in the warmer months. The female gestates for about 1 mo., giving birth to 5-10 altricial pups, which she weans about 30 days later. Young females are able to begin reproducing within a few days. The male appears to play no role in the rearing of his offspring; however, he will kill young that he has not sired if he finds them unprotected.

Communication and perception occur via normal Cricetidae channels: sight, smell, touch, and hearing. May enter a state of daily torpidity in winter, cutting down on the need to eat in order to generate body heat. Its natural enemies are reptiles, birds of prey, mustelids, wild felids, and wild canids. Beneficial as a seed disperser, as a prey base for carnivores, and a subject of scientific research; harmful as a household and agricultural pest and as a carrier of deadly diseases, such as the hantavirus. The deer mouse lives an average of 6-10 mos. in the wild; at least twice that long in captivity.

DESERT COTTONTAIL (*Sylvilagus audubonii*): Also known as "Audubon's cottontail," this large rabbit grows to a total length of nearly 17", its hind foot measures almost 4" long, and it can attain a weight of up to 2½ lb. Shares many physical characteristics with other cottontails: has a grizzled brown, black, and white coat; large eyes; long ears; narrow feet; long back feet; a grayish or white "collar"; ringed eyes and ears; a "cotton" tail; and cinnamon fur patches at the nape of the neck and on the legs.

Though as its name suggests it usually inhabits arid regions (such as desert and chaparral), it is also fond of forests, prairies, rangeland, brushland, and scrubland, often taking shelter in thickets and brushpiles. It is also somewhat riparian and can be found near water corridors, where it feeds on marshy plants. Its range covers western North America, from Montana south to Mexico. Does not normally dig burrows, but will use the abandoned burrows of other animals on occasion. It is active all year long; does not hibernate.

This solitary, crepuscular herbivore feeds mainly on grass, but also on leaves, mesquite, fruit, cactus, bark, twigs, and vegetables. Will stand on its hind legs to reach food. One clear indication of its presence are twigs sheared off at a 45 degree angle, a typical sign of many leporids.

Desert cottontails.

Depending on the region, its breeding season generally runs from winter to summer. The polyestrous doe gestates for about 1 mo., giving birth to 1-5 altricial kittens up to 5 times a year in a grassy nest lined with her fur. Like some other cottontails, the mother does not live in the nest with her young, but instead visits them once a day for the purpose of short nursing sessions. The young are weaned and independent within a month or so, and are capable of starting their own families by the age of 3 mos. As with most lagomorphs, the buck is not involved in child-rearing.

Communicates and perceives using a variety of standard Leporidae channels and methods: scent, foot-stomping, squeals, and tail flashing (a sign of potential danger). The natural enemies of this primary consumer are reptiles, birds of prey, badgers, foxes, coyotes, and bobcats. It can run up to 15 mph and possesses a host of evasion tactics, from running (in a zigzag pattern) and freezing (for up to 15 mins.), to swimming and even climbing trees. The desert cottontail probably lives for about 1-2 yrs. in the wild; around 7 yrs. in captivity.

DESERT KANGAROO RAT (*Dipodomys deserti*): It grows to a total length of about 15" and a weight of up to 5 oz. It is nearly physically identical to many of its congenerics: the pelage on the upper body is light golden brown with a grayish metallic sheen; the lower body is white. The long whip-like 9" tail is dichromatic, with a dark horizontal stripe running along the top and a white horizontal stripe running along the bottom; the tail is tipped with a tufted crest of white fur and a black proximal stripe on the dorsal side.

Not a true rat, but a relative of the pocket gopher, its body is compact and ovoid; the head is large (in proportion to body size) with a convex profile; the eyes are large and dark, with a small white patch over the upper lids (some individuals have whitish eye rings); the nose is narrow, tapered, and pinched; long tactile whiskers grow from the end of the muzzle; a small dark patch of fur lies at the base of the vibrissae; the ears are small, rounded, and erect, with tiny white postauricular patches; there is a light, nearly indistinguishable face mask; a single white thigh stripe curves over and down the upper leg; the flank may have a large whitish spot; the front limbs are small and retrogressed; the back limbs are long, powerful, and 4-toed; the digits are clawed. Gender dimorphism is present: the male is larger than the female.

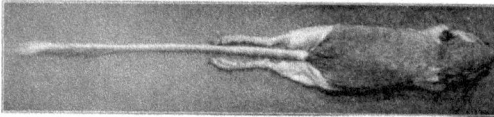

Desert kangaroo rat (museum specimen).

This desert-loving heteromyid is a nearctic native that inhabits arid, wind-blown sandy regions across some of the hottest and most inhospitable parts of southwest North America. Its terrain may include playa, shadscale scrubland, dunes, alkali sinks, creosote brushland, salt flats, gravelly desert pavement, and loose beach-like soil. Its range is limited, covering only small portions of southern California, Nevada, Utah, and Arizona, and perhaps a few areas of northern Mexico. Highly fossorial, it constructs elaborate underground burrow systems with multiple entrances and exits. Numerous pathways may radiate from these access holes, which are usually located in mounds of dirt situated beneath brush. Solitary and territorial, each individual violently defends it tunnels from conspecifics.

An herbivore, it specializes in a folivorous diet built around the consumption of dried leaves. Like others in its genus, it supplements this bland desiccated fare with granivorous, nucivorous, and legumivorous foods, such as grass seeds, pine nuts, and the beans of the mesquite plant. It is slightly carnivorous with an appetite for an occasional insect. Seldom drinks water since it extracts all the liquid it needs from its diet of plant matter. Hoards and caches surplus victuals in storehouse chambers underground, transporting the food in large cheek pouches.

Its polygynandrous breeding season runs from early winter to late summer, with bucks fighting over mating rights and estrus does. Strong dominant males have the advantage when it comes to reproductive success. The doe gestates for about 1 mo., gives birth to an average of 3 altricial pups in the natal nest, then weans them around 3 wks. later. The male is non-paternal and appears to play no role in the raising and care of his offspring.

Communicates and perceives in typical Heteromyidae fashion: sight, hearing, smell, and touch. Does not hibernate; is active year-round. Engages in dust-bathing by rolling in the sand, which aids it in removing unwanted dirt from its fur while protecting it from ectoparasites. Prefers hopping to walking; is an adept climber and swimmer; when alarmed it kicks sand at its target; though nocturnal it dislikes bright moonlight (when it is most likely to be seen by predators), and usually remains in its subterranean den during a full moon; practices hind foot-drumming; a saltatorial mammal (as its common name indicates), it can leap up to 9'.

Its natural enemies are reptiles, birds of prey, mustelids, wild felids, and wild canids—and potentially any burrowing carnivore. Benefits its ecosystem by aerating and recycling the soil, dispersing seeds, hosting parasites, and providing food for carnivores. The desert kangaroo rat lives an average of about 2-4 yrs. in the wild; probably twice that long in captivity.

DESERT POCKET GOPHER (*Geomys arenarius*): This muscular stocky rodent grows to a total length of almost 12", its hind foot measures 1½", and it can attain a weight of up to nearly 9 oz. Its pelage on the dorsum is dull brown with a light gray wash; the ventrum and feet are white; the 3½" tail is sparsely furred. Strongly fossorial, all 4 paws are big and heavily clawed, and the forelimbs are greatly enlarged compared to the hind limbs—adaptations to its life as an excavator. It has small eyes, while tiny ears allow for greater speed and dexterity underground. A diagnostic feature that distinguishes it from other pocket gophers in its range: bisulcate upper incisors. It is gender dimorphic: the male is significantly larger than the female.

Desert pocket gophers.

This sedentary solitary geomyid is a nearctic native that inhabits temperate river bottoms, deserts, grassland, farmland, sand dunes, hedgerows, scrubland, cropland, disturbed areas, and forest, with loose, well-drained, friable soil. Also a riparian species, it has a fondness for the sandy margins of irrigation ditches, streams, pools, and lakes. A western mammal, it has a limited range covering only small portions of Texas, New Mexico, and northern Mexico. Constructs elaborate tunnel systems up to 100' long, with multiple offshoots, side passages, and chambers; leaves large ejecta mounds outside its portals. Establishes a home range of perhaps 50-100 sq. ft., which it vigorously defends against conspecifics.

This herbivorous endotherm is an opportunistic feeder with a diet that focuses on subterranean fare: roots, tubers, and bulbs; it also consumes grasses, forbs, seeds, nuts, leaves, flowers, grains, pine needles, fruit, and woody material, such as bark, stalks, twigs, shrubs, and trees. As it meets its water needs from the solid food it eats, it rarely needs to drink. Hoards and caches surplus edibles in burrow larders, carrying them in its massive external fur-lined cheek pouches or "pockets."

Its 2-cycle, polygynous breeding season occurs in spring and summer. The female gestates for almost 3 wks., giving birth to an average of 3-4 altricial pups, which she weans around 1 mo. later; dispersal and independence take place at about 60 days. It is unlikely that the male contributes to the raising of his offspring.

Communication and perception run along standard Geomyidae channels: mainly touch and hearing, and to a lesser extent, sight and smell. Does not hibernate; active all year. Its natural enemies are reptiles, birds of prey, mustelids, wild canids, and various fossorial carnivores. Beneficial as a parasite host and a seed disperser; also, the biopedturbation resulting from its aggressive tunneling helps aerate and recycle the soil; it is harmful, however, as an agricultural and household pest; additionally, the microtopographical changes it leaves behind can lead to soil erosion. Its genus name, *Geomys*, comes from the Greek words *geo* ("earth") and *mys* ("mouse"). The desert pocket gopher lives an average of about 2-3 yrs. in the wild.

DESERT POCKET MOUSE (*Chaetodipus penicillatus*): Also known as the "sand pocket mouse" and the "tuft-tailed pocket mouse," it grows to a total length of about 8" and can attain a weight of up to ¾ oz. On the dorsum its coarse pelage is buff brown with a gray wash; on the ventrum it is white; its 5" tail is longer than its head and body, it is bicolored (brownish-gray on top, white on the bottom), and it has a crest and a tufted tip. The body lacks both a lateral line and the bristly posterior spines found in some other pocket mice species. Its genus name derives from the Greek words *chaeta* ("flowing hair"), *dis* ("two"), and *pous* ("feet").

This solitary riparian heteromyid is a nearctic native that inhabits dry deserts, chaparral, sloping hills, sand dunes, rangeland, streamsides, flat bottomland, and arid scrublands that include palo verde vegetation, loose sandy and gravelly soil, cacti, washes, creosote brush, and tough grasses; it avoids rocky settings. A southwestern rodent, it has a limited range that extends over portions of California, Nevada, Utah, Arizona, New Mexico, Texas, and northern Mexico.

Digs burrow systems under shrubbery using both its paws and teeth; here it nests, sleeps, hides from predators, and seeks shelter during inclement weather. Estivates during the summer months, closing its burrow entrance to block out the heat; may hibernate during frigid weather, but it is more likely that it enters a brief state of torpor. A territorial endotherm, it establishes a home range of several thousand square feet, which it aggressively defends against conspecifics.

A solitary nocturnal omnivore, its diet consists of seeds, nuts, forbs, leaves, grains, and insects. Hoards and caches surplus food in burrow larders, which it transports in its cheek pouches or "pockets." Needs little water, which it procures from its food. Its breeding season commences in spring and peaks in summer. The female of this prolific species is able to begin reproducing while still young. She gestates for around 3 wks., giving birth to 3-5 altricial pups in an underground nest chamber. She nurses, grooms, and protects her young without assistance from the male, who displays no paternal behaviors.

Arid desert: desert pocket mouse habitat.

Communication and perception follow typical Heteromyidae channels: mainly touch and smell, and to a lesser extent, sight and hearing. Its natural enemies are reptiles, birds of prey, mustelids, wild felids, wild canids, and various fossorial carnivores. Beneficial as a seed disperser, parasite host, and as food for carnivores. As nearly the entire population dies off each year, the desert pocket mouse probably only lives for an average of 6 mos. or so in the wild.

DESERT SHREW (*Notiosorex crawfordi*): Also known as "Crawford's gray shrew," it grows to a total length of about 3½", its hind foot measures ⅜", and it weighs about ⅛ oz. The short thick dorsal fur is silver-gray with brownish highlights. Countershaded, its ventrum is light gray. It has a cylindrical body; tapered head; small eyes; conspicuous rounded ears; 5-fingered toes; a 1" tail; and a long snout packed with tactile vibrissae.

It inhabits mainly arid regions, such as chaparral, sand dunes, and deserts, where sagebrush and cacti dominate; but, despite its common name, it is also occasionally found in swampy areas, grassland, and forests. Its range runs from southern California east through Arizona, New Mexico, Colorado, Texas, and Arkansas, and south into Central America. Marks off its small home range with a foul-smelling oil that it emits from scent glands located in its rump area.

Despite its minuscule size, like most shrews it is one of Nature's fiercest predators, hunting its prey, worms, insects, reptiles, birds, and other small mammals, using its

acute senses of hearing and smell. Cannibalistic, it will also consume carrion, pets, and other desert shrews if available. This highly active soricid can eat more than its own body weight each day, has a resting heart rate of 1,000 beats a min., and a resting respiratory rate of 800 breaths per min. With this type of extreme metabolic rate, it will die of starvation if it does not eat at least once every 2-3 hrs. It seldom needs to drink since it is capable of deriving most of its water from its diet; hoards and caches surplus food.

Desert shrew.

The female gives birth to around 4 altricial pups (sometimes twice a year) in a well concealed, fur-lined nest filled with shredded vegetation. The male does not participate in the raising of his offspring. Nocturnal and solitary, the primary enemy of this secondary consumer is the owl. The desert shrew probably lives no more than 1-2 yrs. in the wild.

DESERT WOODRAT (*Neotoma lepida*): It grows to a total length of nearly 16", its pes measures 1¼", and it can reach a weight of 12 oz. In coloration its upper body runs from golden yellow-gray to grizzled black; the underbody is typically light gray; the throat is dark; the 8" tail is bicolored (grayish on top, tan on the bottom); the hands and feet are pinkish-white. The body is compact and cylindrical; the head is small with a convex profile; the eyes are large and dark, and, as with many other prey animals, they protrude, permitting a wide arc of vision; the ears are large, broad, rounded, and erect; the nose is rather rounded, slightly pinched, and snubbed; the limbs are short; the front feet or hands have 4 clawed digits, the rear feet have 5 clawed digits; its long tail acts as a balancing weight, useful when climbing and leaping. Gender dimorphism is present: the male is larger than the female.

As its common name implies, this medium-sized cricetid is a nearctic native that is often found in temperate arid biomes, such as deserts, sand dunes, and coastal regions. However, it also inhabits piñon, Joshua tree, and juniper forests, chaparral, creosote scrubland, rocky locations, rangeland, sageland, and hillsides. It is strictly a western mammal, with a range that extends from Oregon and Idaho at its northern edge, Nevada and Utah at its eastern edge, and California and the Baja Peninsula at its southern edge. Often takes over the abandoned underground homes of burrowing animals; but will also construct its domicile in trees and boulder fields, building the haphazard-looking structure out of sticks, cactus spines, and stones.

Chaparral: desert woodrat habitat.

An omnivore with a strong granivorous, folivorous, nucivorous, frugivorous, florivorous, cactivorous, and lignivorous appetite, it prefers eating within the safety of its shelter (rather than out in the open), which helps protect it from predation. Has a cannibalistic streak, and will consume its own kind when the opportunity arises. Succulent plants, such as cacti, provide most of its water needs. Caches surplus food in dark subterranean storehouses to decelerate spoilage.

Nocturnal, but also occasionally crepuscular and diurnal, its breeding season occurs between midfall and midspring. The doe gestates for 1 mo. or so, giving birth to 2-3 altricial pups in an underground nest chamber. She weans them about 30 days later. As with most other members of the Rodentia order, the male is unlikely to provide any assistance in the rearing of his offspring.

This solitary, endothermic vertebrate communicates through ordinary Cricetidae channels: sight, smell, hearing, and touch. It scent-marks both during its mating season and as a territorial defense. Engages in foot-stomping and tail-rattling to communicate trouble and distress. An adept climber, it can expertly navigate through cactus spines without harm; it is non-hibernating and active year-round. Its natural enemies are reptiles, birds of prey, and wild canids. Benefits its ecosystems by dispersing seeds, aerating and recycling the soil, hosting parasites, and providing homes and food for other animals; detrimental to us as a potential transmitter of disease-causing viruses, which may be contracted through its urine, dung or aerosol (exhaled air). The desert woodrat lives about 1-2 yrs. in the wild.

DOUGLAS SQUIRREL (*Tamiasciurus douglasii*): Also known as the "chickaree," this tree squirrel grows to a total length of 14", the pes or hind foot measures around 2½", and it attains a weight of about 10 oz. The pelage coloration is variable, but is generally brownish with a reddish or grayish tinge on the dorsum, rusty to whitish on the ventrum. The throat, chest, and belly may be buff or cinnamon. A brownish band runs along the center of the back. Has a short small head; medium-sized ears (that become tufted during the cold months); a snubbed nose; dark eyes ringed in white or rust; orangish front teeth; shortish limbs; and strongly clawed digits.

The bushy 6" tail, which is grizzled and often tipped with a cinnamon or orange swash, serves a multitude of functions, from umbrella (during rain storms) and sun shade (during hot weather), to blanket (during cold weather), balancer (while jumping in trees), and warning flag (in times of danger). Some adults have a thin blackish flank stripe that

Douglas squirrel.

separates the upper half of the body from the lower half. During winter, the pelage becomes grayer and the flank stripe becomes more conspicuous. Sensitive vibrissae grow about the face, including around the eyes, nose, and chin, aiding it while foraging in dark woodlands.

This small sciurid inhabits temperate, old-growth coniferous and mixed forests on North America's Pacific coast, with a range that extends from southwestern British Columbia south through Washington and Oregon and into northern California. Establishes home territories up to 4 sq. mi. in size. Has no fear of people and will sound loud alarm calls to alert other forest dwellers of a human's presence.

An omnivore with a strong herbivorous and granivorous appetite, it consumes a wide variety of plant and animal matter, from sap, bark, leaves, cambium, wood, nuts, and fungi, to shoots, seeds, buds, stalks, grains, fruit, insects, birds' eggs, and birds. Hoards and caches surplus food. It is particularly fond of pine cone seeds, and thus throughout its range an overt indicator of its presence are large kitchen middens comprised of uneaten pine cones and discarded pine cone scales. Often used repeatedly by succeeding generations, a Douglas squirrel midden can reach a depth of 3' and a width of 6'. From its affinity for pine trees and pine cone seeds it derives one of its other common names: the "pine squirrel."

Its monogamous breeding system takes place from midwinter to midsummer. After a short courtship, in which the male pursues and isolates a single estrus female, she gestates for 5-6 wks., giving birth to as many as 8 altricial pups in a bolus nest or drey constructed out of shredded bark, twigs, lichens, and moss. This well-built maternal nest, set high up on tree limbs or in tree crevices, helps protect both

Douglas squirrel kitchen midden.

mothers and their newborns from predators and inclement weather. Sometimes the female will appropriate an abandoned bird nest. At 2 mos. the young are weaned, but they do not become fully independent until around 6 mos. of age. The father plays no role in caring for his offspring.

Also known as "Douglas's squirrel" or "Douglas' squirrel," this diurnal, lively, highly active, arboreal, solitary rodent does not migrate or hibernate; being scansorial, it is an adept tree climber, runner, and leaper. It communicates and perceives via normal Sciuridae channels: sight, touch, hearing, and smell. Its teeth never stop growing, so it must constantly chew on woody objects in order to keep them at the proper length. One of our noisiest small woodland mammals, its many loud vocalizations include various types of barks, trills, chicks, and chirps. Its natural enemies include birds of prey, mustelids, wild felids, and wild canids. It is beneficial due to its dispersal of pine cone seeds and fungi spore. The life span of the Douglas squirrel in the wild averages 5-6 yrs.

DULZURA KANGAROO RAT (*Dipodomys simulans*): This medium-sized heteromyid grows to a total length of about 13" and a weight of up to 2½ oz. Pertaining to its ecology and life history, it is almost identical to its larger cousin, the agile kangaroo rat. Outside of genetics (where there are a few differences), the main dissimilarity concerns their ranges. Though its range overlaps the agile's at its extreme northern edge, the Dulzura kangaroo rat's extends much further south, into Baja California.

Dusky-footed woodrat.

DUSKY-FOOTED WOODRAT (*Neotoma fuscipes*): This western packrat grows to a total length of about 19" and a weight of up to 10 oz. Its dorsal pelage is rusty-gray; the ventral pelage is gray-tannish-white; the long 12" tail is pinkish brown-gray. Its front and hind feet are a light smoky or sooty gray, from which it derives its common name. The body is long and cylindrical; the head is large with a convex profile; the eyes are smallish, dark, and protruding; the ears are large (almost bat-like), wide, rounded, and erect; the muzzle is somewhat tapered; the nose is pinched and snubbed; tactile nasal vibrissae emerge from the snout area; the limbs are shortish; the feet have clawed digits. Gender dimorphism is present, with the male being larger and heavier than the female.

This North American cricetid is a nearctic native that inhabits mainly dense forest communities (coniferous, deciduous, and mixed), though it is also fond of chaparral and shrubland. A riparian mammal, it requires a reliable water source and has a preference for standing snag, good dry hiding cover, forest floor detritus, hollow logs, and dead trees. Its range is narrow, extending from coastal Washington through Oregon and California, and south into northern Mexico.

Depending on the habitat it occupies, it builds its house on open ground or rocky slopes, in rock crevices, around tree bases, or high up in trees—normally in shady or cool locations with thick surrounding undergrowth. These massive elaborate structures (often used repeatedly by succeeding generations) can be 8' high and 8' wide, appearing as great mounds of sticks, bark, and debris that have been haphazardly thrown together. In fact, this woodrat's home is so large and contains so many "rooms" that commensals (from insects to small mammals) often take up residence within it.

A nocturnal herbivore, its diet focuses solely on various types of plant matter, from seeds, bark, and nuts, to fruit, fungi, and ferns. Derives its water requirements from the vegetation it eats. Hoards and caches surplus supplies in special storage compartments.

Its breeding season runs from fall to summer. In its semi-matriarchal society, the doe chooses which buck she will mate with. She gestates for about 35 days, gives birth to 2-4 altricial pups in a leaf-lined natal nest, then weans them at about 3 wks. of age. Though, unusually, this rodent seems to be briefly monogamous, the buck does not remain long with the female, for she drives him away after they have mated. Thus the male does not participate in the rearing of his offspring. Instead, like the males of most other mammalian species, he returns to a solitary way of life; in his case, taking shelter in a small stick tree nest nearby, where he awaits the arrival of the next breeding season.

Communication and perception occur via channels common to the Cricetidae family: sight, smell, touch, and hearing; it is known for clicking its teeth and beating the ground with its tail when alarmed. Engages in dust-bathing and self-grooming. Semicolonial and slightly social, it may live in close proximity to conspecifics—though each retains its own individual stick shelter. Its natural enemies are reptiles, birds of prey, mustelids, wild felids, and wild canids. Benefits its ecosystem by providing shelter (in its stick house) for other organisms. The dusky-footed woodrat probably lives 2-3 yrs. in the wild.

DWARF SHREW (*Sorex nanus*): One of the tiniest mammals in the world, it grows to a total length of 4", its tail is about 1½" long, its hind foot measures ⅜", and it weighs 0.1 oz. The pelage on its upper body is grayish-brown, its lower body is buff or gray; during the cold months its pelage becomes lighter.

Dwarf shrew.

It inhabits white fir, Douglas fir, and spruce forests, as well as prairie, alpine tundra, meadows, steppe, shrubland, montane grassland, scree, marshes, arid areas, talus, and rocky slopes, usually near a reliable fresh water source. Its range extends from Montana south through Idaho, Wyoming, South Dakota, Utah, Colorado, Arizona, and New Mexico.

As with many other aspects of this diminutive soricid, little is known about its diet, except that it is an insectivore, one that probably shares the same general eating habits as other members of the Soricidae family. Its breeding season may take place during early summer, with births occurring in late summer. A secondary consumer, its only known natural enemy is the owl. The dwarf shrew probably lives around 1 yr. or so in the wild.

DWARF SPERM WHALE (*Kogia sima*): Little is known about this little dolphin-sized physeterid. A secondary consumer, it is similar to its cousin the pygmy sperm whale (with which it is sometimes confused), but is slightly smaller, has less teeth in its lower jaw, grows to a total length of 9', and weighs about 600 lb. An inhabitant of the coastal waters of the southern U.S., it probably shares many of the pygmy's habits and traits.

A deceased dwarf sperm whale.

E

EASTERN CHIPMUNK (*Tamias striatus*): Our largest chipmunk, this familiar rodent grows to a total length of nearly 12", its hind foot measures about 1", and it can attain a weight of up to almost 5 oz. Its countershaded pelage coloration can change with geographic location; however, it is most commonly reddish-brown mixed with gray on the upper body, whitish on the lower body. Its most distinguishing features are its conspicuous ears and the 5 black dorsal and lateral stripes alternating with white and brownish-gray bands. The facial mask is light and somewhat indistinct; the 4" tail is dichromatic and bushy, grizzled brown-gray above, cinnamon-brown below.

This rust-colored sciurid is a nearctic native that prefers open deciduous forests and tree islands with good hiding cover, logs, snag, forest detritus, deadfall, stumps, blowdown, and rocks; it is also found in and around pine forests, bushland, chaparral, woodland edges, hedgerows, shrubland, cemeteries, fields, brushpiles, clearings, firewood stacks, houses, rock walls, and talus slopes, inhabiting nearly the entire environmental spectrum, from wilderness and farmland to suburbia and large cities. As its common name indicates, it is a citizen of the eastern half of our region, with a range that extends from southeastern Canada and New England in the north to Louisiana and Florida in the south, from North Dakota and Oklahoma in the west to Virginia and North Carolina in the east. It is the only chipmunk in most of its range.

Eastern chipmunk.

Fossorial, it digs vast underground burrow systems, some as long as 30'. Entrances lack ejecta mounds, for after excavation it carries away extra soil in its cheek pouches to prevent drawing attention to its home; entrances typically open up on hillsides or under objects such as stumps, wood heaps, roots, rocks, brushpiles, and tree bases. Estivates during hot weather; remains in its subterranean home during cold weather. Unable to store fat, it does not hibernate, but enters a state of torpor, waking up periodically throughout the winter months to forage on the surface. Though classified as a terrestrial animal, it has scansorial abilities and is quite capable of climbing trees. Establishes a home range of 100-200 sq. ft., which it defends aggressively against conspecific intruders.

A solitary diurnal omnivore, its diet revolves around seeds, nuts, grains, and fruit; it also feeds on fungi, slugs, worms, snails, insects, birds' eggs, and small mammals (such as juvenile mice). Both a larder hoarder and a scatter hoarder, it caches surplus foods in a central area, as well as in different locations throughout its home territory. Transports edibles in its large internal cheek pouches.

Has 2 polygynous breeding seasons: one in early spring, one in mid-summer. The polyestrous doe is thus capable of producing 2 litters a year. She gestates for around 4 wks., and gives birth to as many as 5 altricial pups in a leaf-lined underground nest chamber. Weaning occurs at about 5-6 wks., with dispersal following shortly thereafter. The buck does not exhibit paternal behaviors and does not participate in the rearing of his offspring.

Communication and perception take place via normal Sciuridae channels: sight, smell, hearing, and touch. A highly loquacious mammal, among the vocalizations it produces are whistle-like trills, clicks, chucks, bird-like chirps, chattering, clucks, barks, and chips—the latter the source of its common name. Its natural enemies are numerous and include birds of prey, reptiles, mustelids, procyonids, wild felids, and wild canids—and occasionally other larger sciurids (such as squirrels). Benefits the ecosystem by aerating and recycling the soil, supplying homes for other fossorial creatures, dispersing seeds, and serving as a prey base for carnivores. The eastern chipmunk lives an average of 1-2 yrs. in the wild; at least twice that long in captivity.

EASTERN COTTONTAIL (*Sylvilagus floridanus*): The most widely distributed member of the genus *Sylvilagus* and the most common rabbit in our region, this familiar lagomorph grows to a total length of nearly 19", its hind foot measures 4", and it weighs up to 4 lb. Its dense dorsal fur is grayish-brown with black tips; its ventral fur is lighter, tending to white. It has a conspicuous cinnamon-colored patch on the back of its neck. May have a whitish-gray "collar," as well as a white forehead patch. The underside of its tail is a soft cottony white, from which it derives its common name. Its tall 2½" ears, long body, short front legs, fur-covered feet, and large white-ringed eyes give it a distinctive appearance.

Eastern cottontails.

It is gender dimorphic, with the doe generally being larger than the buck. Despite this, the two genders may be difficult to distinguish in the field.

Highly adaptable, this leporid inhabits both rural and urban areas, seeking shelter in thickets, fields, pastures, orchards, prairies, deserts, and woods, and in particular shrubland and brushpiles. Its range is vast, covering southeastern Canada to South America. In the U.S. it extends over the entire eastern half of the country, from most of New England south to Florida, from Virginia west to Texas and beyond.

In summer this solitary crepuscular vegetarian feeds on clover, grasses, fruits, and vegetables. In winter it consumes tree bark, small branches, and buds. It produces 2 types of dung: one soft and edible, one hard and inedible. It eats the edible form in order to extract additional nutrients, a common survival practice among rabbits and hares (one known as coprophagy).

Mating occurs from February to September, at which time bucks drive off rival males and court females. The polyestrous doe may have from 1-8 litters a year. She gestates for roughly 27 days, giving birth to 1-10 altricial kits in a grass nest lined with fur she has plucked from her abdomen. She may mate again within hours after bearing young. One of the most prolific breeders in the animal kingdom, a single pair of eastern cottontails, as well as their offspring, can produce 350,000 descendants in just 5 yrs.

Communicates and perceives along standard Leporidae channels: sight, hearing, smell, and touch. Vocalizations include shrieks, wails, and snorts. Does not hibernate. It can hop, swim, stand on its hind legs, leap 15' into the air, and reach speeds of up to 20 mph, typically running in a zigzag pattern (to evade predators) while flashing its white tail as a danger signal to conspecifics. Other escape tactics include crawling close to the ground (with its ears laid back), jumping obliquely (to interrupt the scent of its trail), and freezing in place (to try and blend into its surroundings). Fun-loving, it is known to romp and play with conspecifics in snow. The natural enemies of this primary consumer are reptiles, birds of prey, weasels, bobcats, foxes, and coyotes. The eastern cottontail lives 2-5 yrs. in the wild; 10 yrs. in captivity.

EASTERN FOX SQUIRREL (*Sciurus niger*): The largest of our tree squirrels, its grows to a total length of 28", its pes measures 3", and it can attain a weight of 40 oz. Pelage coloration depends on geographical population, but generally the upper body is a grizzled gray-buff-rust, the lower body is buff, rust, or orangish. The body is stout; the head is medium-sized and boxy; the ears are erect and smallish; the eyes are medium-sized, dark, and ringed in buff; the forehead and muzzle are convex; the cheeks are buff colored; the limbs are short. The 13" tail is bushy with yellowish-tipped fur and orangish edging, otherwise it is highly variable in color, with the dorsum usually a grizzled buff-gray and the ventrum cinnamon. Both horizontal and vertical stripes are sometimes present along the upper part of the tail. Southern populations can be all black; other populations are sometimes rusty-brown with white ears, a white nose, and a white face blaze. Ears may grow tufts in winter.

A riparian, it prefers temperate, open, park-like, mixed forest environments with tree islands, thickets, and thin understory; thus it is found in piney woodlands, cypress wetlands, oak and hickory forests, longleaf pine groves, and mangrove swamps. It also enjoys savanna habitat and is at home in wild, rural, suburban, and urban settings, particularly along creeks and streams. A nearctic native, its range extends across the

Eastern fox squirrel.

eastern half of North America, from Canada in the north to Mexico in the south, covering most of the Southern and Midwestern states. It has been artificially introduced into some western states, such as Washington, Oregon, California, and Colorado.

This common colorful sciurid is an omnivore whose diet centers around acorns and hickory nuts; it also consumes seeds, fruit, buds, grains, fungi, insects, birds' eggs, birds, small mammals, carrion, and agricultural crops, especially corn. Sensitive vibrissae scattered over the face, muzzle, and upper forelegs aid in both foraging and navigation. Since it likes to eat on the same familiar feeding perches, one sign of its presence are large kitchen middens on the ground below. These are composed of food debris, such as the husks and shells of nuts and other discarded plant matter. Caches (buries) surplus food for the lean cold months.

Scansorial, it is well-designed for tree-climbing, with heavily clawed digits and strong leg, chest, and stomach muscles. In summer, when not foraging and eating, it relaxes on tree branches or rests in great leafy dreys (nests) that it constructs in tree craws; in winter it switches over to deep tree cavities for warmth.

Its polygynandrous breeding season runs year-round, though it crests around the summer and winter solstices. Males compete for estrus females and instigate mating chases; dominant males protect the females they breed with from other males. The female gestates for about 7 wks., giving birth to 2-5 altricial pups, often between late winter and late summer. The young are weaned at 2 mos. and become independent at around 3 mos. of age. The male does not assist in the rearing of his offspring.

This arboreal, diurnal, scent-marking rodent communicates and perceives mainly via sight, touch, and smell, and to a lesser degree hearing. Vocalizations include whines, clicks, screams, chits, barks, and chatter. By burying seeds and nuts it helps disperse plant life, a benefit to both us and the world ecosystem. Its primary natural enemies are reptiles, birds of prey, and wild canids. The eastern fox squirrel probably lives an average of 5-6 yrs. in the wild; twice that long in captivity.

EASTERN GRAY SQUIRREL (*Sciurus carolinensis*): The official state wild game animal species of Kentucky and the official state mammal of North Carolina, it grows to a total length of 20", its hind foot measures 2½", and it can attain a weight of up to 25 oz. The upper body of this common tree squirrel is, as its colloquial name indicates, mainly gray in coloration. The underbody is lighter, usually white or off-white, but may also run to grayish or fulvous. Despite its vernacular eponym, the dorsal fur can be sprinkled with reddish-brown, white, silvery, and black-tipped hairs, giving it a grizzled appearance; the face is sometimes washed in cinnamon; the 10" tail has a white edge and may be a mixture of black, brown, rust, and white hair.

Both melanistic (black) and albinistic (white) populations are known to occur, the former more common in northern regions (where its dark pelage absorbs heat), the latter more common in southern regions (where its light pelage deflects heat). The body is stout; the head is large and convex from forehead to nose tip; the eyes are dark and ringed with buff fur; the ears are large and erect; the limbs are short and strong; the feet are well-developed with clawed digits. The male and female are generally indistinguishable in size and color.

As with other arboreal squirrel species, its long, bushy, flattish tail serves a myriad of utilitarian functions: a stabilizer (while leaping); steering gear (while swimming); an alarm signal (when excited); a shade (when sunny); an umbrella (when raining); and a wing, aileron, spoiler, or flap (when ascending and descending in midair).

It does not hibernate and is active all year long. It is crepuscular during the hottest months, but it is diurnal on days with clement weather. Constructs massive, permanent, leafy, bolus nests for denning in winter (protection against the cold), but more loosely constructed nests in summer (protection against the heat). A conspicuous sign of its presence are the scattered remnants of discarded nut shells and husks covering the ground in its home range.

One of North America's most popular game animals, the favored habitat of this familiar rodent is extensive, uninterrupted, mature, mixed forest (preferably composed of oak or hickory) with plenty of dense ground vegetation. A highly adaptable riparian that is closely associated with bottomland and heavily forested biomes, nonetheless it can be found in wild, rural, suburban, and urban environments (especially enjoys city parks), preferably with a permanent fresh water source, standing snag, stumps, and logs. Its range, where it is our mostly commonly seen mammal, covers the entire eastern half of the northern tier of North America. This includes parts of Canada as well as all of the

Eastern gray squirrel.

U.S., from just west of the Mississippi River east to the Atlantic Ocean, from North Dakota to Texas, from Maine to Florida. It has been introduced into several areas, such as British Columbia, Alberta, Washington, Oregon, Montana, and Wyoming.

This opportunistic omnivore is primarily a nucivore, but it is also a fungivore, granivore, insectivore, invertivore, frugivore, avivore, ovivore, florivore, folivore, amphibivore, mammalivore, carcassivore, and an osteophage. Also a cannibal that is known to kill and eat its own kind, it is fond of agricultural plants, such as wheat and corn, and will often construct its dreys near farms to take advantage of the readily available food supply. A well-known and sometimes comical acrobat, it can easily climb or leap onto bird feeders, and is the bane of many a birdwatcher. A scatter hoarder, it digs through snow using its keen sense of smell to find foods that it cached underground the previous year.

The polygynandrous breeding season of this archetypal sciurid occurs twice a year: the first in early winter, the second in late spring. Bucks fight for mating rights to estrus does, with the most dominant males usually winning access. The female gestates for approximately 45 days, giving birth to as many as 4 altricial pups in a leafy drey located in a tree cavity or an abandoned woodpecker hole. The mother may occasionally transport her babies from one nest site to another in order to prevent predation, weaning them at around 8 wks. of age. The father does not contribute to the rearing of his offspring.

Communication and perception occur via normal Sciuridae channels: sight, smell, touch, and hearing. Tail-flicking accompanies alarm-barking. Its natural enemies are birds of prey, mustelids, wild felids, and wild canids. Beneficial due to its practice of storing seeds underground, which helps disperse vegetation. It also consumes creatures we dislike, and provides food for carnivores as well. It is considered a pest by many farmers and homeowners, however. The eastern gray squirrel lives to an average age of 8-10 yrs. in the wild; twice that long in captivity.

EASTERN HARVEST MOUSE (*Reithrodontomys humulis*): It grows to a total length of nearly 6", its hind foot measures about ⅝", and it can reach a weight of up to ½ oz. Its dorsal pelage is buffy-brown with a gray wash; its ventral pelage is white to tan; its 2½" sparsely-furred tail is indistinctly bicolored; has pale flanks and a distinct lateral stripe; the dorsal stripe is dark brown. Has large, dark protruding eyes; the ears are large and erect; the snout is covered in sensitive vibrissae; the feet are pinkish-white and delicate.

This small nocturnal cricetid is a nearctic native that inhabits brushland, brackish ditches, sedgeland, briar fields, roadside drainage trenches, weedy bottomlands, and wet grassy meadows in low-lying areas across the southeastern corner of the U.S. Its temperate and subtropical range covers portions of, or all of, the following states: Texas, Louisiana, Mississippi, Alabama, Tennessee, Kentucky, Ohio, West Virginia, Pennsylvania, Virginia, Delaware, North Carolina, South Carolina, Georgia, and Florida.

An endothermic omnivore, its diet consists of seeds, nuts, fruit, sprouts, green vegetation, and insects. Caches surplus food in larders located within its home range. It breeds in the spring and autumn. The female gestates for about 3 wks., giving birth to 2-6 altricial pups, which

Snake hunting eastern harvest mice.

disperse and become independent 1 mo. later. Its natural enemies are reptiles, birds of prey, mustelids, wild felids, and wild canids. On average the eastern harvest mouse probably lives no more than 6-12 mos. in the wild; perhaps twice that long in captivity.

EASTERN MOLE (*Scalopus aquaticus*): Also known as the "common mole" it grows to a total length of nearly 9", its hind foot measures about 1", its short tail is 1½" long, and it weighs up to 5 oz. Its stout body is covered with a dense, short, soft coat that is generally grayish, cinnamon, or almost charcoal black in color. The individual hairs are sometimes glossed with silvery tips. Though it has no external ears and no visible eyes, it does indeed possess the latter: its tiny eyes are each covered with a thin membrane of skin, which is itself obscured by fur, leading some to mistakenly believe that it is eyeless and blind. In reality, these sensitive, visual organs are able to ascertain subtle changes in the intensity of light, an asset in its dark subterranean world. This solitary species is gender dimorphic: the male is larger than the female.

Eastern mole.

A fossorial eulipotyphlid, its massive front paws are hairless and spade-shaped with webbed toes, giving it its Latin species name, *aquaticus* ("water dweller"). This webbing, along with its large front feet and claws, make it both an excellent excavator and an adept swimmer. The nostrils at the end of its slender muzzle are located on top of its nose, functioning a bit like a "snorkel" for traveling underground. Its plush velvety fur allows it to easily change direction while moving through its burrows. Though it does not leave tracks or scat behind on the earth's floor, its presence can be detected by its tunnels, which often break the surface, pushing up the soil beneath into "molehills" (the nemesis of many a homeowner). Establishes a home range of about 300 sq. ft.

It inhabits moist open meadows, fields, pastures, gardens, lawns, cropland, waste areas, and forests, preferring soft loose earth, such as well-drained sand or loam. Avoids clay and wet conditions, as well as gravelly and rocky soil. It ranges over the Central states, Midwest, and the South, from Wyoming to New England, and from Canada south to Florida, Texas, and Mexico.

Active from dusk to dawn, as well as all year-round, this omnivorous talpid has a high metabolic rate. An avid insectivore and vermivore, its diet includes vegetation, crop foods, grubs, worms, slugs, snails, and various insects, as well as spiders, centipedes, and millipedes. Can consume its own body weight every 24 hrs.

Its mating season seems to occur sometime between late winter and early summer. The sow may gestate for about 5-6 wks. She then gives birth to 2-4 altricial pups in a well concealed grass-lined nest deep underground—typically situated beneath a log or large rock. The boar displays no paternal instincts and does not participate in the rearing of his offspring.

It contributes to the world ecosystem by aerating and recycling the soil and by providing underground homes for other wildlife. It aids us by consuming large quantities of creatures we

Eastern mole.

consider harmful. As it spends nearly its entire life below the earth's surface, this secondary consumer is usually protected from predation—though snakes, birds of prey, procyonids, wild felids, and wild canids will take it if given the opportunity. The eastern mole may live from 2-5 yrs. in the wild.

EASTERN PIPISTRELLE (*Pipistrellus subflavus*): The smallest bat in the eastern U.S., this little vespertilionid grows to a total length of 3½", its forearm is around 1½", its wingspan measures about 9", and it weighs as little as ⅛ oz. It tragus is short; its calcar is not keeled. Employing countershading, its upper body is reddish- or rust-brown, the underbody slightly lighter. It has tricolored hairs, with blackish tips, light mid-sections, and a blackish base. The ears are narrow and rounded; the wings are usually darker than the body.

Somewhat solitary (known to hang in its roost alone, or at least apart from conspecifics), this chiropterid shelters in caves, mines, quarries, hollow trees, buildings, crevices, and various forms of vegetation. It ranges over the entire eastern U.S., from Minnesota and New England, to Texas and Florida, south to Mexico.

A secondary consumer, it migrates infrequently; hibernates during the winter months. An obligate carnivore, it feeds exclusively on flying insects. Using echolocation, it snatches its prey out

Head and skull of a pipistrelle bat.

of the air as it flits along the edges of forests and waterways. The female delays fertilization, gestates for around 55 days or so, and gives birth to twins in June at her maternity colony. The eastern pipistrelle may live up to 10 yrs. or longer in the wild.

EASTERN SMALL-FOOTED MYOTIS (*Myotis leibii*): Named for its hind feet, which tend to be smaller than other members of the genus *Myotis*, this tiny flying chiropterid grows to a little over 3" in total length, its forearm is not quite 1½", and it weighs a mere ¼ oz.—about the same as 2 pennies. This makes it the smallest of the mouse-eared bats, with a wingspan of only 8" or so. Its dorsal pelage is beige to cinnamon brown; the ventral pelage is grizzled tan to whitish. Its ears, wings, and interfemoral membrane are dark brown to black. Its most distinguishing traits are its black face mask and a tail that extends past the end of the uropatagium. It has a slender tragus; a shortened muzzle; a strongly keeled calcar; and small wide wings that help it maneuver through thick vegetation and tree branches while hunting.

It shelters in typical myotis habitats: caves, mines, buildings, tunnels, rocky cliffs, and under tree bark, stone slabs, and bridges. Prefers various types of forest in its range: Ontario, Canada, through New England to the southern U.S. states, west to Missouri and Oklahoma.

Eastern small-footed myotis.

This elusive slow-flying vespertilionid hibernates in winter (often alone), sometimes hiding in the crevices of cave floors and walls. Nocturnal, it feeds after dark, fluttering erratically through woodlands—typically just a few feet above the ground—using echolocation to track down its insect prey, mainly moths, beetles, and mosquitos.

The female practices delayed fertilization, bearing one pup in early summer at her maternity roost, a colonial nursery comprised of several dozen adult females and juveniles of both genders. A secondary consumer, its natural enemies are birds of prey (such as owls), reptiles (such as snakes), and small mammals (such as rodents). The rare eastern small-footed bat lives 10 yrs. or more in the wild.

EASTERN SPOTTED SKUNK (*Spilogale putorius*): This small mephitid grows to 23" in total length, its hind foot measures 2", and it weighs nearly 2½ lb. It is jet black with white spots and irregular dashed lines and stripes over its body, and has a brushy (not bushy) black and white tail, often white-tipped. It has a white forehead spot and 2 white preauricular spots. The fur is silky; the body is stout and compact; the head is small; the nose is rounded; the legs are short; the paws have 5 claws each. Those on the forefeet are adapted for digging and defense and are therefore longer, sharper, and more curved than those on the hind feet. With the male being larger than the female, gender dimorphism is present.

This endothermic mammal is a nearctic native that inhabits a number of different biomes, ecosystems, structures, and terrains, including open fields, woodpiles, scrubland, fencerows, farmland, savanna, gullies, prairie, thickets, outbuildings, rocky areas, and woodlands, typically near a fresh water source. Its range extends from Wyoming, Colorado, and the northern Midwest, east to Pennsylvania and south to Texas, Florida, and Central America. It dens in hollow logs, brushpiles, and the dens of other mammals, sometimes sharing its home with conspecifics.

Eastern spotted skunk.

This agile carnivore is nocturnal and omnivorous, and, depending on the season, feeds on vegetables, fruit, insects, grubs, amphibians, birds (including poultry), eggs, reptiles, and small mammals (mice, rabbits); its diet also includes carrion and, as a synanthrope, human refuse.

Its breeding season starts in early to midspring and may last as late as August. The female does not employ delayed implantation like her cousin the western spotted skunk (proving that they are separate species). Polyestrous, she has 1-2 litters a year, gestates for about 5-6 wks., and gives birth to 4-9 altricial kits in a concealed den. The young become independent at around 4 mos. of age.

Communication and perception occur via standard Mephitidae channels: smell and touch, and to a lesser degree sight and hearing. The natural enemies of this secondary consumer are many and include birds of prey, wild felids, and wild canids, as well as nearly every other medium- to large-sized carnivore. If a predator disregards its aposematic "warning sign" (that is, its conspicuous black and white coat), this highly skilled climber will run up a tree for safety. If cornered, however, it stands on its front feet, lifts its tail, and emits a foul-smelling sulfuric musk from its rump glands, shooting it accurately up to 12'. The eastern spotted skunk lives an average of 2 yrs. in the wild; as long as 10 yrs. in captivity.

EASTERN WOODRAT (*Neotoma floridana*): Also known as the "Florida woodrat," this large rodent grows to a total length of about 18", its pes measures nearly 2", and it can attain a weight of up to 1 lb. Its dorsum is a gunmetal gray mixed with rusty brown fur; the ventrum is light gray to white; the 8" tail is dichromatic (grayish on top, whitish below); the feet are pinkish. As with most other woodrats, adults molt annually, while its countershaded pelage is long, soft, and flexible, allowing it to easily reverse directions in tight spaces. Though its cryptic coloration is quite uniform and lacks overt markings, this helps it blend in with its mostly gray-brown environment.

The body is compact and cylindrical; the head is small, narrow at the crest, broad at the lower jaw, and has a convex profile; the eyes are large, dark, and protruding, supplying a wide angle of night vision; the ears are large, ribbed, erect, wide, rounded, and bat-like, allowing for excellent sound detection; the muzzle is robust; the snout is covered in long, curving, dark, sensitive vibrissae; the nose is snubbed; the limbs are short; the digits are clawed; the long tail provides balance while running and leaping.

This stout cricetid is a nearctic native that inhabits many types of biomes and ecosystems, from wilderness to urban areas. These include forests, hedges, grassland, cliffs, houses, coastal marshes, farmland, mountains, bogs, bluffs, lowlands, intermontane regions, woodlands, bottomland, caves, savanna, barns, and boulder fields. As its common name denotes, its range covers much of the eastern half of the U.S.; but it also includes parts of the Midwest and most of the South, from South Dakota, Colorado, and Texas, east to Pennsylvania and New York, and south to Georgia and Florida. Outside its mating season this solitary species is quite antisocial and highly territorial, and will defend its turf aggressively; a violent agonist, it is known to kill conspecifics on sight.

Eastern woodrat.

As with many of its cousins, it is known for constructing large dome-like middens from various natural and human-made materials, including sticks, branches, stones, feathers, leaves, grass, bark, fecal pellets, bones, and pollen, as well as human refuse. Its midden—which may appear to the casual onlooker as little more than a disorderly pile of randomly assembled debris—is typically located in a concealed area, such as dense vegetation, abandoned buildings, logs, stone ledges, caves, tree craws, branches, ground holes, or rock crevices. It is used for multiple purposes, from denning, nesting, foraging, sleeping, and protection from predators, to resting, storage, hunting, eating, and shelter from the weather. This enormous complex structure is often so rich in resources that it is like a miniature, self-contained world. As a result, some individuals never venture outside of it, and instead remain within its confines for their entire lives.

Despite the midden's massive size, only one woodrat occupies it at a time (though it does provide shelter for numerous other smaller animals, such as invertebrates and reptiles). Since a woodrat's midden is used by succeeding generations, it can grow to gargantuan proportions over the years, with some reaching a width of 10' and a height of 6'. Under the proper conditions a woodrat midden can survive for thousands of years, preserving many layers of rat deposits that provide scientists with a unique view into the past.

A generalistic feeder, this omnivorous rodent eats primarily seeds, ferns, tubers, leaves, nuts, wood, fungi, bark, fruit, and grains; when available it will supplement its diet with cultivated crops (such as wheat and corn) and insects. Hoards and caches surplus mast in storage areas within its midden. Does not need to drink water if it has enough plant material at hand. Despite being a terricolous mammal, it is a good climber and will forage both on the ground and in trees. Rather sedentary, it does not usually wander far from its midden, maintaining a home range of about 5,000 sq. ft.

Depending on geography, its polygynous breeding season runs all year long, with peaks between winter and summer. The polyestrous doe gestates for about 35 days, gives birth to an average of 2-4 smoky-gray, altricial pups in a feather- or grass-lined nest. Weaning comes about 1 mo. later. The mother nurses, grooms, cares for, and protects her neonates until they are ready for independence at about 3 mos. of age. The non-paternal buck plays no role in the care of his offspring.

This nocturnal packrat is an ardent collector of the unusual, and will stash away whatever items catch its fancy: sticks, cans, newspaper, stones, string, and golf balls, for example, and more particularly shiny human-made objects, like keys, rings, bullets, silverware, eyeglasses, earrings, broken glass, belt buckles, screws, and coins. Will sometimes leave behind an object to replace the one it has taken, a trading behavior for which it—as well as all other members of the genus *Neotoma*—has been nicknamed "trade rat."

Communication and perception take place via standard Cricetidae channels: sight, smell, hearing, and touch. Like many other rat species, it makes a variety of sounds, from squeaks and teeth-chattering, to squeals and back feet-drumming. Engages heavily in grooming activities and will take dust-baths when necessary. Its natural enemies are reptiles, birds of prey, mustelids, procyonids, wild felids, and wild canids. It benefits its ecosystem by supplying other animals with shelter (in its midden), providing food for carnivores, and acting as a host to numerous parasites. Though common in parts of its range, in others it seems to be on the decline. The eastern woodrat lives to an average age of 1-2 yrs. in the wild; twice that long in captivity.

ELK (*Cervus elaphus*): The official state animal of Utah, it is also known as the "North American elk," "red deer," and "wapiti" (a Native American word meaning "white rump"). It is the second largest member of the deer family (after the moose), attaining a shoulder height of 5', a total length of nearly 9', and a weight of around 1,000 lb. (a half ton). Its summer pelage is light brown, its winter pelage is grayish-brown to dark brown. This is contrasted by blackish-brown fur that covers the head, neck, abdomen, and legs.

Elk (male).

Its body is robust with thin legs. It has a thick neck, a short tail, and a white or cream-colored rump patch. Its ears are large and slightly pointed, and a scraggly mane runs from the top of the head around the neck and down onto the chest. The male sports a set of massive multi-tined antlers that can spread to 5' in width, weigh up to 40 lb., and grow as much as 1" a day or more in the hot summer months. Gender dimorphism is present: females do not grow antlers and are smaller and weigh far less than males.

This nearctic and palearctic native inhabits open wooded slopes, swampy areas, coniferous forests, and elevated mountain savanna, taiga, and grassland. Its once immense historic range (which covered most of the northern part of our region) has been drastically reduced by habitat loss and hunting. Today it is found primarily in parts of southern Canada south into the Rocky Mountain states, extending down to New Mexico (elk were recently reestablished in Michigan as well).

This imposing artiodactyl migrates into higher areas during summer, lower elevations during winter. It is both nocturnal and crepuscular, and can move stealthily

Elk (male).

through brush and trees in near total silence. An herbivorous grazer, it is also a lignivore and a folivore, with a diet consisting of grasses, leaves, lichen, plant stems, forbs, sedges, roots, and woody vegetation (wood, bark, twigs). Also a ruminant, it regurgitates its food, re-chews it, then re-swallows it—Nature's way of helping maximize an animal's nutrient intake.

Gregarious, it lives in a variety of herd types, depending on the season. The male loses his antlers in March. In summer wapiti congregate in large herds of up to 500 members, led by an alpha female. By the time the breeding season begins in autumn, bulls have regrown new antlers. They then lose their antler velvet and begin to compete for females. They "trumpet" loudly, snort, challenge, paw the ground, and joust with one another, digging wallows into which they urinate, then roll their bodies. The resulting pungent "cologne" attracts cows, which the bulls gather into small all-female herds known as "harems." The injuries a bull sustains while defending his harem can be quite serious, even resulting in death.

Cows gestate for nearly 9 mos., giving birth to 1-2 precocial calves in the spring. The mother and her young live apart from the main herd for about 2 wks., rejoining it when her baby (or more rarely, babies) are strong enough to function safely in a large group. During calving, bulls separate from females, forming bachelor herds comprised of mature males and young adult males. Around the same time, females form maternal herds made up of new mothers and their young. Stags (males) do not participate in the rearing of calves. In summer the herding cycle begins again.

Elk, known as the most vociferous of the world's cervids, have a wide range of vocalizations, from the familiar "bugle" to all manner of squeals, grunts, roars, bellows, barks, and bleats. A primary consumer, the natural enemies of this powerful deer include bears, wolves, coyotes, bobcats, and mountain lions. A healthy adult male may avoid predation

Elk stag and herd.

in a number of ways: it can swim rivers and lakes, jump up to 8' high, trot at 25 mph over long distances, and run up to 45 mph in short bursts. If these fail, as a last resort it can stab and toss with its long pointed horns and kick and slice with its sharp-edged front hooves, making it a formidable opponent. Elk may live up to 20 yrs. in the wild; as long as 25 yrs. or more in captivity.

Ermine (summer coat).

ERMINE (*Mustela erminea*): Also known as a "stoat," it grows to a total length of 1', its hind foot is nearly 2" long, and it weighs up to 7 oz. It has a long slim body; short legs; a pointed face; alert dark eyes (that protrude for broader vision); rounded ears; long whiskers; sharp non-retractable claws; and a black-tipped nose and tail. In summer the dorsal fur is cinnamon to dark brown while its ventral fur is creamy to white from the throat to the rump; in winter its entire coat turns snowy white, giving it its common name, ermine, a Middle English word that is related to the Old High German word *harmo*, meaning "weasel." Simply put, the *ermine* is a weasel in its *winter* pelage; a *stoat* is a weasel in its *summer* pelage. Thus this particular weasel species has been given the name of its winter coat. A gender dimorphic species, males are larger than females.

This small, sedentary, petraphilic mustelid is a nearctic and palearctic native that inhabits many types of biomes, ecosystems, and terrain, from taiga, mixed forest, shrubland, tundra, and grassland, to beaches, open fields, farmland, rock fields, ice fields, and wetlands. A northern riparian mammal, it is found across Alaska, Canada, and most of the northern states into California, the Rockies, and as far east and south as Pennsylvania and Maryland. Dens have multiple burrows and are constructed under stumps, logs, brushpiles, rock piles, and roots. Establishes a large home range, which it marks with its scent glands, aggressively patrolling its boundaries for intruders.

A scrappy nocturnal carnivore, it feeds readily on insects, crayfish, fish, amphibians, birds (including domestic fowl), and small mammals (shrews, voles, mice, hares, lemmings, rats, opossums, rabbits), which it kills with several swift savage bites to the throat or neck. It will also consume carrion and, as a synanthrope, human garbage. Can travel over 10 mi. a night, moving rapidly over and under snow, in search of prey. This highly active mammal requires large amounts of nutrients to meet its energy demands, which it helps fulfil by storing surplus food for emergencies. Though also partly crepuscular, the solitary territorial ermine is sometimes seen during the

Ermine (winter coat).

day. A competent swimmer and climber (that can run down a tree trunk headfirst), it also possesses excellent vision (which it uses to hunt aquatic prey), smell (which it uses to hunt other mammals), and hearing (which it uses to hunt arthropods).

Its breeding season, the only time males and females interact socially, starts in early summer. The bitch (female) employs delayed implantation, gestating for about 9 mos., then giving birth to as many as 18 altricial kits the following spring. The dog (male) does not contribute to the rearing of his offspring.

Like other members of the weasel family the ermine is highly vocal, and uses hisses, chatter, and grunts to communicate. It also relays information through scent, body language, and touch. A secondary consumer, its natural enemies include birds of prey as well as the fisher, marten, badger, and fox. Its luxurious coat has long been highly prized, and is the traditional fur used to embellish the robes of European royalty. The ermine lives an average of 1-3 yrs. in the wild, but may live as long as 8 yrs. under favorable circumstances.

EUROPEAN HARE (*Lepus europaeus*): Also known as the "brown hare," this large nonnative leporid grows to a total length of 28", its hind foot is nearly 6" long, and it weighs up to 10 lb. In the warm months its upper body is a grizzled cinnamon-brown; in the cold months its upper body is a grizzled gray; its lower body is whitish all year-round; the tail is black dorsally, white ventrally. Its nearly 4" ears are black-tipped and its eyes are ringed; its pupil is black; its iris is gold.

This endothermic lagomorph shelters in forms made of grass or bushes, inhabiting open fields, cropland, pastureland, savanna, and woodland. Though as its name denotes it is a native of Europe, it is now a North American species as well, having been introduced to the northeastern U.S. in the late 1800s. Since then it has spread across New England, north into southern Ontario and the Great Lakes region, and south along the coast of the mid-Atlantic states. It does not hibernate and is active all year. Solitary, crepuscular, and

European hare.

nocturnal, by day this herbivorous coprophage rests in a form, coming out on summer nights to feed on grasses, herbs, agricultural crops, and other green vegetation. During the cold winter months it grazes on tree parts: twigs, stalks, buds, and bark.

Its mating season runs from winter to summer. The polyestrous doe may bear up to 4 or more litters a year. She gestates for about 37 days, giving birth to as many as 6-8 precocial leverets, which become independent at around 1 mo. of age. She scatters her young into different nests (enhancing their survival), returning quietly each night to nurse them. The buck, who boxes (often violently) with other males during breeding season, does not contribute to infant care.

The natural enemies of this primary consumer are birds of prey (owls, hawks), wild felids (bobcats), and wild canids (coyotes, foxes, wolves). Its evasion tactics include hiding in brush, freezing in place, swimming, bounding (up to 12'), leaping (up to 5' in the air), and running 30 mph (up to 45 mph in short bursts). The European hare lives from 6-12 yrs. in the wild.

EUROPEAN RABBIT (*Oryctolagus cuniculus*): Also known as the "San Juan rabbit," it is the ancestor of every variety of domestic rabbit and the sole member of its genus. It grows to a total length of 24", its hind foot is nearly 5" long, and it weighs up to 5 lb. Though its pelage is typically grizzled gray or brown, it can be found in almost any color, from white to black. Indeed, as one would expect from a species that has undergone domestication in numerous countries for over 1,000 years, there is an almost infinite variability in coloration, markings, and size. The underside of its 3" tail is white. At nearly 4", its ears are somewhat small for its body size.

European rabbit.

Introduced to North America from Europe as a game animal (and thus also sometimes called the "Old World rabbit"), this invasive leporid can now be found in almost every part of our region, from temperate to tropical climates. Its habitat ranges from brushland, savanna, fields, and dry forest, to sand dunes, farmland, and suburban and urban areas. A highly social, extremely territorial species, it forms large hierarchical groups (of up to 30 rabbits) in massive underground warrens (intricate tunnel and burrow systems). Like others of its kind it communicates with conspecifics through the use of scents, foot-stomping, and various vocalizations.

A nocturnal herbivore (plant-eater) and coprophage (dung-eater), its diet also classes it as a folivore (leaf-eater), a lignivore (wood-eater), and granivore (seed-eater). Such a diet includes roots, grains, fruit, herbs, garden vegetables, nuts, tree bark, flowers, and agricultural crops.

Essentially its breeding season is year-round, with the polyestrous doe bearing up to 6 litters annually. She gestates for around 35 days, giving birth to as many as a dozen altricial kittens. The mother does not live with her offspring; she attends to them in their own nest (lined with her fur) once every 24 hrs., nursing them with nutrient-rich milk. These once-a-day visits help avoid attracting the attention of predators. At around 4 wks. they are weaned and become independent. The buck is non-paternal and is not involved in the care of his offspring.

The natural enemies of this primary consumer include birds of prey, mustelids, wild felids, and wild canids. It protects itself and evades predators by standing on its hand legs (to observe its surroundings), tail-flashing, and altruistic foot-thumping (which alert other members of the warren of danger).

Considered an alien (nonnative) pest by many, this lagomorph remains an economic boon to North Americans: it is hunted extensively for its meat and skin, it is widely bought and sold as a pet, and it is used by researchers as an animal model in laboratories across our region. Though the European rabbit's life span in the wild may be as long as 10 yrs., in most cases it probably lives for 1 yr. or less.

EVENING BAT (*Nycticeius humeralis*): It grows to a total length of nearly 4", its forearm is 1½" long, its wingspan is around 11", and it weighs ¼ oz. Its fur is dark brown above, grizzled or buff below. Its head is wide; its snout is broad; its ears are black; the tragus is curved; the calcar is not keeled. A vespertilionid (as its common name indicates), it shelters in trees (under loose bark or in holes) and buildings. Avoids caves (but may occasionally swarm near the entrance). Prefers open areas near fresh water sources. Its range extends from Michigan south to Florida, covering most of the Midwest and Southern states west to Texas.

Evening bat.

Social, migratory, and polygynous, this chiropterid moves south in the fall, at which time it also mates. The female, however, postpones fertilization until spring when ovulation takes place. She gives birth to as many as 4 altricial, temporarily blind, hairless pups in her maternity roost—comprised of as many as 1,000 female colony members (and their young). If a baby falls from its roost, the mother is able to locate it by its particular squeak and return it to the nest site. Lactating females will sometimes act as "wet nurses," and provide milk for each others' pups.

This slow-flying nocturnal insectivore leaves its roost after sundown, tracking its quarry (moths, flies, ants, beetles) using echolocation. A secondary consumer, its natural enemies include reptiles, birds of prey, and small mammals. The evening bat lives from 2-5 yrs. in the wild.

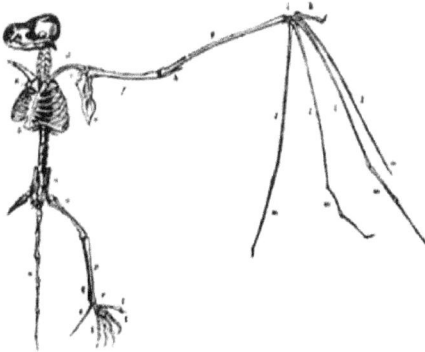

Bat skeleton showing its elongated left arm, hand, and finger bones, over which its wing membrane stretches.

Muskox.

FALLOW DEER (*Dama dama*): Introduced to North America from Europe, its common name derives from the Old English word *fealu*, which means "pale," a reference to its most common fur color: yellowish-brown. This dappled cervid stands about 40" at the shoulder, grows up to 70" in total length, and can weigh as much as 200 lb. The color of its pelage runs from white to black, and from yellow to gray, making it the most chromatically-varied of any deer species. However, generally in summer its coat is fulvous with white spots; in winter its coat is brownish-gray without spots. Its abdomen is nearly always white to tan. The tail has a single black stripe down the mid-section, which sometimes extends up the back and also around the whitish rump patch. Gender dimorphism is present: the male has semipalmated antlers, a conspicuous "Adam's apple," and is larger than the female—who does not usually have antlers.

Fallow deer.

This terricolous artiodactyl inhabits many types of biomes and ecosystems, including: grassland, meadow, brushland, savanna, fields, forest (deciduous and coniferous), scrubland, and hilly terrain. In our region its range is spotty, with populations in parts of British Columbia, California, Texas, Oklahoma, Alabama, Kentucky, Georgia, and Maryland.

A small crepuscular herbivore, it forms herds of up to 200 individuals comprised of young, old, and both genders (though all-male bacherlorhoods and all-female matriarchal groups are common around the rut). Graminivorous, granivorous, and frugivorous, in the warm months it feeds mainly on green grasses, fruits, seeds, nuts, and herbs; also a lignivore and a folivore, during the cold months it consumes tree parts, such as leaves, bark, wood, and twigs.

At the beginning of the breeding season, which runs from early fall to midwinter, bucks compete with one another over territory and females. Violent shoving, bush-thrashing, ground-pawing, and horn-clashing ensues, sometimes resulting in serious injury. The doe gestates for 6-7 mos., searches out a secluded spot away from the main herd, and gives birth to 1 precocial fawn in late spring.

Communicates and perceives in typical Cervidae fashion, using scent, body language, and a wide variety of vocalizations, such as bleating, barking, and grunting. A primary consumer, its natural enemies are numerous: wild felids, wild canids, and ursids.

Fallow deer.

Though able to run at tremendous speed, this shy, short-legged, Old World deer will sometimes employ stotting, a form of bounding in which all four limbs are kept stiff and straight and the tail held high—probably a way of bewildering predators.

Well-known as the traditional "park deer," due to its docile nature it is often kept in parks and zoos and on estates and farms as a semidomesticated pet. Its famous Megacerine cousin, the gigantic Irish elk, which had antlers that spread up to 13' from tip to tip, went extinct some 8,000 yrs. ago. The life span of the fallow deer in the wild is unknown, but it may live up to 20 yrs. in captivity.

A pod of false killer whales.

FALSE KILLER WHALE (*Pseudorca crassidens*): This large ocean-going dolphin grows to a total length of 20' and a weight of up to 3,000 lb. It has a rounded head; a small dorsal fin; small elbow-shaped pectoral flippers; and a groove running above the upper jaw. Lacks the bulbous forehead found in many cetaceans, and instead has a gracefully rounded snout that extends slightly out over the lower jaw.

It is not closely related to the true killer whale (the orca). Where then did its common name come from? Its binomial name answers the question: *Pseudorca crassidens* means "imitation orca, heavy-toothed"; in other words, with their similarly shaped conical teeth, the jaws of the two species resemble one another.

In our region this pelagic delphinid prefers the warm waters of subtropical and tropical oceans; may be found in open water and along the coast; it is attracted to oceanic islands. A formidable carnivore, it uses echolocation while searching for its prey: from crustaceans and cephalopods, to fish, sea lions, and other cetaceans (mainly the young of whales and other dolphins). Pods cooperate in hunting expeditions and also share captured prey. It communicates with conspecifics using clicks, grunts, squeals, and body language. Highly social and acrobatic, like many other marine mammals it enjoys breaching, bow-riding, and playing with objects.

False killer whale.

This polygynandrous cetacid has no set breeding season. The female only mates about once every 7 yrs. She gestates for around 13 mos., giving birth to 1 precocial calf, which is weaned by 2 yrs. of age. Females pass through menopause with age. In size one of the largest of the world's dolphins, it is known to engage in mass strandings. A secondary consumer, its only known natural enemy is, ironically, the orca—after which it is named. The false killer whale may live up to 65 yrs. in the wild.

FIN WHALE (*Balaenoptera physalus*): The second largest cetacean, as well as the second largest animal, in the world (after the blue whale), it reaches a total length of nearly 90', weighs up to 140,000 lb. (70 tons), and can eat 2 tons of food a day. Its dorsum is gray to black; its ventrum is white. Its body is long, slender, and torpedo-like; the head is v-shaped and flat; and its small 2' long dorsal fin is set far back on the spine. A striking trait is the coloration of its lower jaw: the left side is dark, the right side is light. Like other baleen whales it possesses a double row of blowholes, as well as some 100 throat grooves: long accordion-like pleats in the skin that allow its throat to expand and contract

Fin whale.

while feeding. A grooved ridge extends from the tip of the upper jaw back to the blowholes, from which it can spout up to 20' into the air. Its large tail fluke is centrally notched with pointed tips. As the male and the female grow to be the same size and weight, this species is not gender dimorphic.

Fin whale.

In contrast to the toothed whales (suborder *Odontoceti*), the fin whale (also known as the "finback whale" or "common rorqual") is a member of the baleen whales (suborder *Mysticeti*), a cetacean family named for the grayish-purple-white, comb-like plates (also known as "whalebone"), whose fibers hang in drapery-like rows from the two sides of its upper jaw. Composed of a protein called keratin (the same substance out of which our fingernails and hair are made), the roughly 400, 3' long "combs" act like a giant food strainer while feeding: as the whale opens its mouth and seawater pours in, these strong but flexile bristles filter out its prey—zooplankton (such as krill and copepods), as well as various other creatures (including squid and small schooling fish)—while allowing the seawater (which it cannot consume) to spill back into the sea.

The fastest swimmer of all the whales, it can reach speeds of 25 mph, earning it the nickname "greyhound of the sea." This planktivorous balaenopterid pursues its prey (as noted, mainly planktonic crustaceans) down to a depth of 1,000' or more while holding its breath for 15 min. The female gestates for about a year and gives birth to a 22' long, 4,000 lb. calf. Though often solitary, this secondary consumer sometimes forms pods of up to several hundred individuals. It inhabits all oceans and is found both in open deep waters and along shallow coastlines. The fin whale may live as long as 140 yrs. in the wild.

FISHER (*Pekania pennanti*): Also known as a "fisher cat," this large lithe member of the weasel, badger, and otter family grows to 40" in total length, its hind foot measures about 6" long, and it weighs up to 18 lb. Long and slender, its pelage is highly variable from individual to individual, but it is generally chocolate brown, often with grizzled cinnamon tinting around the head, face, and back. The head is wide; the face is tapered; the ears are small, rounded, and ringed; the eyes are tiny and dark; the muzzle is compact and dog-like; the feet have 5 toes with retractable claws; and its short legs and long 17" tail are black. Often there are several light stripes or rings running around the neck. This species is gender dimorphic: males can grow almost twice the size of females.

Fisher.

A riparian mammal, this terricolous, sedentary nearctic native prefers closed canopy areas under large trees, readily inhabiting taiga, mountainous regions, wetlands, and most especially dense, contiguous, old-growth coniferous, deciduous, and mixed forests—usually near water and steep sloping terrain. Found only in North America, its range covers southern Canada and the northern tier of the U.S., from California and the Rocky Mountains to Appalachia and New England. Its home range can extend to over 100 sq. mi. Rests, shelters, and nests in hollow stumps, snag, slash piles, crevices, holes, burrows, deadfall, wood debris, and logs, and often in living trees. In winter it digs subnivean dens with multiple tunnels under the snow. A secondary cavity user, it will sometimes take over bird and squirrel nests.

Solitary, omnivorous, and both diurnal and nocturnal, its diet is comprised of fruit, fungi, nuts, seeds, insects, reptiles, birds, and other mammals, such as shrews, voles, mice, chipmunks, squirrels, muskrats, porcupines, racoons, rabbits, and hares. When available it will readily eat carrion, and, as a synanthrope, it also consumes farm animals, pet food, and human refuse. Cannibalistic, it is known to occasionally prey on members of its own kind. An excellent swimmer and powerful climber that is fast on its feet, this scansorial mustelid is quite capable of capturing prey both underground and high up in trees. Despite its common name it does not hunt or eat fish. In fact, the origins of its common name are not known (though theories abound).

Its breeding season takes place in early spring. The female mates once a year and employs delayed implantation. Gestating for up to 12 mos., she gives birth to 1-5 altricial kits the following spring in a maternal den, often located in a treetop cavity. They are weaned at 2-3 mos. and become independent at 4-5 mos. of age. The male does not contribute to the rearing of the young.

Fisher.

This secretive mammal, the only known living member of its genus, communicates and perceives using traditional Mustelidae channels: smell (via scent-marking), sight (body language), sound (hissing, chattering, screeching, growling, snarling), and touch. A secondary consumer, its natural enemies are birds of prey (mainly hawks, but also eagles), wolverines, wild felids (lynx, mountain lion, bobcat), and wild canids (fox, coyote). Despite being heavily impacted by habitat loss (logging), habitat fragmentation (roads), and trapping (its luxurious pelt is highly prized), it is making a comeback in some regions. The fisher lives 5-7 yrs. in the wild; a few years longer under optimum conditions.

FLORIDA MOUSE (*Podomys floridanus*): Also known as the "Florida deer mouse," this large rodent grows to a total length of nearly 9", its hind foot measures about 1", and it can weigh up to 1½ oz. On the dorsum its long, soft, countershaded pelage runs from buff to brown in color; the ventral pelage is white; the 3¾" tail is sparsely furred and pinkish-brown. Has orangish cheeks and shoulders; a distinct lateral line; large bulbous eyes; large thin erect ears; large feet; and a tapered, vibrissae-covered muzzle. Diagnostic trait: most deer mice have 6 or 7 pads on each hind foot; the Florida mouse has 5.

This terrestrial fossorial cricetid is a nearctic native that inhabits open grassland, shrubland, chaparral, xeric hummocks, mosaic locations, arid scrubland, coastal settings, flatwoods, burned out woodlands, uplands, and mixed forests, with a preference for high sandy ridges and loamy hills, along with abundant oak, blackjack, pine scrub, and palmetto. As its common name indicates, its range is limited to Florida, and more specifically to the central and southern regions of the Sunshine State (it is sometimes found as far south as the Gold Coast and Miami). Excavates large subterranean burrows, or appropriates the abandoned burrows of other creatures, such as the gopher tortoise (a propensity that may have contributed to another one of its nicknames, the "gopher mouse"). Spends most of the day in its burrow, only coming out at night to forage. Establishes a sizeable home range of several thousand square feet.

A nocturnal omnivore, its diet includes seeds, fruit, fungi, nuts, and insects. Its breeding season runs year-round, with a peak in the cooler months. The female gestates for about 4 wks., giving birth to 1-4 pups in a moss-lined underground nest. Weaning comes at about 1 mo. of age. Communication and perception take place via the usual Cricetidae channels: touch and smell. Engages in foot-drumming. Its natural enemies are reptiles, birds of prey, wild felids, and wild canids. Beneficial as a host for parasites, a prey base for carnivores, and a subject of scientific research. As most of the population dies each year, on average the Florida mouse probably only lives 6-12 mos. in the wild; somewhat longer in captivity.

FRANKLIN'S GROUND SQUIRREL (*Poliocitellus franklinii*): This ground-loving rodent grows to a total length of around 16", its pes measures about 2", and it weighs up to 25 oz. The pelage is short and stiff. The body coloration is a washed mixture of brown, black, gray, cinnamon, and white fur; around the head and tail the coat becomes a grizzled gray; the underfur is usually lighter, often tan or rust colored; the feet are grayish. The body is long and supple; the head is large with a convex profile; the eyes are dark and surrounded by a buff ring; the ears are small, erect, and trimmed in buff; the pedes are powerful; all 4 feet are clawed; the 6" gray tail is somewhat bushy, has a white border, and is occasionally striped. Though it is not strictly gray and it is not a gopher, its somewhat gopher-like appearance has lent it the nickname the "gray gopher." Gender dimorphism is present: the male is slightly larger and more robust than the female.

A nearctic native, it inhabits dense, tall and short grassland, shrubland, hedgerows, meadows, woodland, fields, and brushy areas; thus, it is fond of savanna, parks, edgeland, tall grass prairies, wetland, and old pastures, often making its home along ditches, railroad beds and grades, rock inclines, and riverbanks. May be found in almost any temperate ecotone with thick vegetation on one side and woody plants on the other, from central Canada south through the Midwestern states of the U.S. (North Dakota, Indiana, Iowa, Illinois, Kansas, Wisconsin, Minnesota, and Michigan). Digs extensive tunnel systems with numerous entrances and exits, all well hidden amidst thick vegetation or under logs and rocks. Sets up a home range of around 300 sq. ft.

Franklin's ground squirrel.

Though smaller than many other squirrels, it is the biggest ground squirrel in its range in North America. It is an expert tree climber and a skilled swimmer, and yet it lives nearly its entire life underground: 7 mos. are spent hibernating, 5 mos. are spent in its burrows. It is most likely to emerge from its subterranean home on sunny days when there is abundant light. Though primarily solitary, it is also semicolonial and will sometimes congregate in a small group of conspecifics, which, as with other squirrels, is called a "scurry."

This omnivorous sciurid has the typical highly varied diet of its kind, feeding on seeds, fruit, clover, thistle, and numerous agricultural products (such as oats, wheat, and corn). Nearly half of its food, however, is comprised of animal matter, such as insects (beetles), amphibians (frogs), birds' eggs, birds (ducks), small mammals (mice), and carrion. Hoards and caches surplus food in its tunnels.

Depending on its geographical location, this facultative hibernator enters winter torpidity from midfall to early spring, when the breeding season begins. At this time bucks begin establishing territories, fighting one another, and chasing estrus does. The latter gestate for about 1 mo., and give birth to 1 litter of 4-10 altricial pups once a year in late spring or early summer. The young are able to venture above ground by 1 mo. of age, and become independent at 6 wks., dispersing to begin life on their own. Males do not seem to participate in the rearing of their offspring.

Communication and perception channels are standard for members of the Sciuridae family: sight, touch, smell, and to a lesser degree, hearing. Vocalizations include musical chirps, bird-like trills, metallic-like pinging sounds, and whistle alarms, the latter for which it has been given the alternate common name the "whistling ground squirrel." Its main natural enemies are birds of prey, reptiles, mustelids, and wild canids. Shy and wary, this once common species is on the decline. Male Franklin ground squirrels live 2-3 yrs.; females live 3-4 yrs.; captive individuals live 6-7 yrs.

FRINGED MYOTIS (*Myotis thysanodes*): This large chiropterid grows to nearly 4" in total length, its forearm measures about 2", and it can attain a weight of up to ⅓ oz. Employing countershading, its dorsum is cinnamon brown to dark brown, its ventrum is buff, whitish, or grizzled gray. It is the only member of the myotis family with a fringe of hairs lining its uropatagium or interfemoral membrane, hence its common name.

A social petraphilic species, this nocturnal bat forms colonies of several hundred individuals, taking shelter in caves, rocky outcroppings and fissures, canyons, mines, tunnels, and buildings located in scrubland, sageland, and oak forests. Its range covers British Columbia south to Mexico, with some density in the southwestern U.S. An agile nocturnal flyer, this vespertilionid leaves its roost after sunset, and, often foraging over water, uses echolocation to hunt down its prey: beetles, moths, and spiders—which it trawls, "nets," or scoops up using its fringed membrane.

Like most members of the Vespertilionidae family, it consumes hundreds of pounds of insects during its life, making this secondary consumer of great benefit to humanity as a natural pest controller. The female gestates for about 55 days, giving birth to a single pup in early summer in a female-only nursery. It shares its roosts with other bat species. The fringed myotis lives for 10-15 yrs. or more in the wild.

Myotis perched on a post.

FULVOUS HARVEST MOUSE (*Reithrodontomys fulvescens*): This small rodent grows to a total length of nearly 8", its hind foot measures almost 1", and it can attain a weight of up to 1 oz. Its common name comes from its dorsal coloration, which is reddish-yellow with a gray or blackish wash; the underparts are white; the pinkish-gray 4½" tail is furry, slightly bicolored, and longer than the body and head combined; its dark eyes protrude; its ears are large and erect; the feet are small, pinkish, and delicate.

This nocturnal cricetid is a nearctic native that inhabits sandy savanna, arid grassland, rocky outcroppings, cactusland, mixed woodland, hedgerows, weedy bushland, cropland, shrubland, conifer-sedge ecotones, and rocky valleys with loose soil and hiding cover. Its range covers parts of the southern and southwestern regions of the U.S., including the states of Arizona, New Mexico, Texas, Oklahoma, Kansas, Missouri, Arkansas, Louisiana, and Mississippi, and extends as far south as Central Mexico.

Constructs small bolus-like nest-shelters—sometimes in shrubs and trees several feet off the ground—out of sticks, grass, sedges, leaves, and other natural debris. There

is only one portal (for entering and exiting), located near the bottom of the nest. Occasionally it will appropriate the underground tunnel systems of fossorial animals; establishes a home range of several thousand square feet.

This motile cricetid is a nocturnal omnivore with a diet consisting of seeds, grains, fruit, insects, and green vegetation in general. Its breeding season runs all year long, but crests in or around early and late summer. The female gestates for about 3 weeks, giving birth to 2-4 altricial pups. Weaning comes within 2 wks., with dispersal and independence arriving a few weeks later.

Communication and perception occur via standard Cricetidae channels: touch and smell, and to a lesser extent sight and hearing. Its natural enemies are reptiles, birds of prey, mustelids, wild felids, and wild canids. The

Nest of a fulvous harvest mouse.

fulvous harvest mouse lives an average of 6-8 mos. in the wild; considerably less under harsh conditions, slightly longer under advantageous conditions.

G

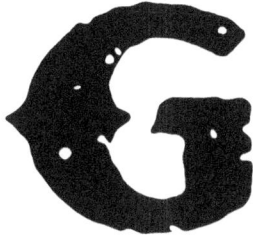

GERVAIS' BEAKED WHALE (*Mesoplodon europaeus*): Also known as the "European beaked whale" and the "Gulf Stream beaked whale," its common name derives from its discover, Victorian French zoologist Paul Gervais. It grows to a total length of 17' and attains a weight of 2,600 lb. Has delicately curved jaws; a narrow beak; a gently rounded forehead; and a triangular, recurved dorsal fin set far back on the body. Coloring is dark blue to grayish-black on the upperside, with the underside being lighter and blotched. Males and females have a dark circle around the eyes. Adult males produce two visible teeth at the front of the lower jaw.

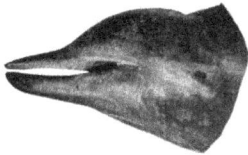

Head of a beaked whale.

This secondary consumer is a carnivorous ziphiid that inhabits warm waters in subtropical and tropical regions; ranges over the Atlantic Ocean, Caribbean Sea, and Gulf of Mexico. A pelagic species, it is a deep diver that relies heavily on sound for communication, hunting, and traveling. A suction-feeder, it eats by sucking in its prey, which includes shrimp, squid, and fish. Stranding is common. Estimated life span is about 30 yrs., though it may live for as long as 50 yrs. As with most other fully marine mammals, sightings of Gervais' beaked whales are rare, making it difficult to study. Thus little else is known.

GHOST-FACED BAT (*Mormoops megalophylla*): The only North American bat belonging to the Mormoopidae family (ghost-faced bats), it is about 3½" in total length, its forearm measures 2½", it weighs around ⅔ oz., and it has a wingspan of 15". Its body is sandy-red to grizzled brown; its short skull is highly arched on top; it has an upturned nose; it has large rounded ears with ridges that connect above the face; the chin sports leafy facial ornamentation; and its short tail is located on the dorsal side of the uropatagium or interfemoral membrane (the thin skin that stretches between a bat's hind legs). Unique among bats are its folds of skin that run from one ear across the chin to the other ear, a cup-like design for enhancing the ability to capture returning sonar echoes.

In our region this mormoopid inhabits mines, tunnels, and caves in the hot desert and humid scrublands of southern Texas and southern Arizona, south into Mexico. Shares caves with other colonial bat species. A powerful and fast flyer, this nocturnal hunter uses echolocation to find its food, primarily moths, as it swoops just above the ground or water in complete darkness.

Bat in flight.

Roosts during the day in colonies of up to 500,000 individuals; does not cluster tightly together (like many other bat species), but keeps a distance of about 6" from conspecifics while roosting. The female has one pup a year in the late spring. A secondary consumer, its main natural enemies are birds of prey and snakes. The ghost-faced bat may live to 15-20 yrs. in the wild.

GIANT KANGAROO RAT (*Dipodomys ingens*): As its common name suggests, it is the largest of all the kangaroo rats, growing to a total length of about 14" and a weight of 6-7 oz. Not a true rat, its closest relative is the pocket gopher. Like most other kangaroo "rats" in our region, its back is buff and gunmetal gray in color, its belly is white, and its tail is bicolored, with a dark stripe on the dorsum, a white stripe on the ventrum. A thin white thigh stripe crosses up and over the upper back leg, and there is a white supraocular spot over each eye and a subauricular one below each ear—which sometimes connect. The body is compact and ovoid; the head is large with a convex profile; the eyes are large, dark, and

Giant kangaroo rat.

protruding, permitting a wide arc of vision; the ears are small, rounded, and erect; the muzzle is short and robust; the nose is pinched, snubbed, and covered in long tactile nasal vibrissae; the front limbs are small and retrogressed; the hind legs are large and kangaroo-like; the hind feet are 5-toed; the long 8" tail (which is used for balance and stability) is whip-like and ends in a bicolored tuft of fur. Gender dimorphism is present, with the male being larger than the female.

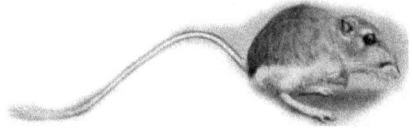

This large heteromyid is a nearctic native that inhabits open desert and sand dune terrain. It is also found, however, in dry grassland, sloping foothills (known as "piedmont"), valleys, ranchland, and chaparral, and also along roadsides and around grazing livestock. Prefers well-drained barren landscapes and sandy, loamy, or gravelly soil with little or no brush. It has one of the smallest ranges of any North American mammal, extending only over parts of several counties in southwestern California. Solitary and territorial, it excavates its own individual burrow system with numerous entrances and also multiple separate chambers: one for eating, one for sleeping, one for waste, and one for food storage. Establishes a home range of about 3,000 sq. ft.

A nocturnal omnivore with a granivorous appetite, its diet centers mainly around grains and seeds, both wild and cultivated. Dries cached food in little holes in the ground or in small piles above ground known as "haystacks." Supplements its vegetable diet with insects when necessary; transports edibles in its large cheek pouches. Derives its water requirements from the food it eats. It has highly efficient, specialized kidneys that can flush waste from the body with little loss of water. Instead of sweating, this desert dweller conserves bodily fluids by lowering its metabolic rate, and by remaining in its subterranean home during the heat of the day.

Its breeding season occurs between winter and summer. The doe gestates for approximately 1 mo., giving birth to an average of 3-4 altricial pups in the natal nest chamber. Weaning takes place at 1 mo. of age; dispersal comes at about 3 mos.

Communicates and perceives along regular Heteromyidae channels: sight, hearing, touch, and smell. Engages in foot-tapping and dust-bathing. Although it can scurry along on 4 legs, more often it hops on its hind legs (like a kangaroo), particularly when it is in a hurry. Can reach a top speed of nearly 7 mph, leap up to 9', and change direction in midair (known as "ricochetal locomotion"). Its natural enemies are reptiles, birds of prey, wild felids, and wild canids, as well as any burrowing carnivore. Benefits its ecosystem through seed dispersal. Once quite common, this elusive rodent is now endangered and on the decline due to habitat loss from agricultural spread, mining development, and urban expansion. The giant kangaroo rat lives an average of 3-4 yrs. in the wild; perhaps twice that long in captivity.

GOLDEN-MANTLED GROUND SQUIRREL (*Callospermophilus lateralis*): This chipmunk-like rodent grows to a total length of around 12", its pes or hind foot measures about 1½", and it can weigh up to 10 oz. Its pelage is brownish-gray with grizzled black-tipped hairs; the underside and feet are buff. Its most conspicuous features are a single whitish lateral stripe on each side (both bordered by 2 black stripes), and its cinnamon-colored head and shoulders, a coppery-red or golden "hood" from which it derives its common name.

The body is small and supple; the eyes are dark with buff eye rings; the head is small and slightly concave at the nose bridge; the ears are medium-sized, erect, and fringed in black; the nose is snubbed; the limbs are short; the feet are clawed; the 4½" tail is bicolored (dark on the dorsum, light on the ventrum). Though it is often mistaken for a chipmunk, it is actually a squirrel with a chipmunk-like appearance, and can be differentiated by a few simple identifiers: it does not have facial stripes (chipmunks do) and it is larger than a chipmunk. Gender dimorphism is present: the male has a bigger head and brighter coloring than the female.

A petraphilic nearctic native, its favored habitat is moist pine and mixed forest up to and beyond the timberline. It also haunts scrubland, sageland, scree, meadows, chaparral, talus, tundra, boulder-strewn areas, rock inclines, rock outcroppings, mountain slopes, campgrounds, and especially ecotone landscapes. Its range covers the montane regions of west and northwestern North America, from British Columbia and Alberta south through Washington, Oregon, California, and Arizona, and east through parts of Idaho, Montana, Wyoming, Nevada, Utah, New Mexico, and Colorado. It constructs elaborate burrow systems that serve as rest areas, protective shelters, hibernacula, and natal nest sites. Though shallow, its tunnels can be up to 100' long and contain multiple entrances and exits, openings that usually emerge from beneath roots, thickets, rock crevices, stumps, cabins, roots, streambanks, or fallen timber.

Golden-mantled ground squirrel.

This diurnal omnivorous sciurid is a primary consumer with a highly variable diet typical of its kind, one that includes shoots, tubers, bulbs, leaves, forbs, fungi, seeds, flowers, nuts, grains, fruit, insects, reptiles, birds' eggs, birds, small mammals, and carrion. Large cheek pouches allow it to carry food to its burrow, where it is eaten or cached.

It hibernates and estivates from midfall to midspring, after which its polygynous breeding season begins. Bucks clash with other males, set up territories, and mate with the later-waking does. The female gestates for approximately 1 mo., giving birth to 2-8 altricial pups by early summer (or later, depending on geographic location) in a grass-lined nest chamber. Weaning takes place after 1-2 mos. The father displays no paternal inclinations, and thus does not provide for his offspring.

Also known as the "copperhead squirrel," it communicates via standard Sciuridae channels: sight, touch, and smell, and to a lesser degree, hearing. Vocalizations include squeals, growls, chicks, and chirps. Though largely solitary and asocial (it is usually quite hostile toward conspecifics), when danger is imminent it gives out alarm calls as a warning to those nearby, a sign of altruistic behavior. Cleans itself by taking dust-baths. Of benefit to both humans and forest ecology by aerating and recycling the soil and spreading conifer seeds. Its primary natural enemies are reptiles, birds of prey, mustelids, wild felids, wild canids, and ursids. The precise life span of the golden-mantled ground squirrel is unknown, but it probably lives from 3-5 yrs. in the wild.

GOLDEN MOUSE (*Ochrotomys nuttalli*): This medium-sized mouse grows to a total length of nearly 8", its hind foot measures around ¾", and it can attain a weight of up to 2 oz. It receives its common name from the color of its dorsal pelage: bright cinnamon-yellowish-brown; the ventral area is white to creamy-white; the 3½" tail is sparsely furred and pinkish-gray. Its fur is soft, dense, and fine; its eyes are large, dark and protruding; the ears are thin, large, and erect; the tapered muzzle is covered in long tactile vibrissae; the limbs are short; the feet are small, pinkish, and delicate.

This terricolous cricetid is a nearctic native that inhabits temperate, mixed woodland, swamps, moist lowland forests, hedgerows, floodplains, chaparral, and boulder-laden hillsides, preferably containing one or more of the following: shrubs, brier patches, leaf litter, vines, and brushy thickets. In particular it enjoys hot humid wetland. Strictly a rodent of the southeastern U.S., its range encompasses the states of Texas, Oklahoma, Arkansas, Missouri, Louisiana, Mississippi, Alabama, Tennessee, Kentucky, Illinois, West Virginia, Virginia, North Carolina, South Carolina, Georgia, and Florida. Builds large 4"-12" globular nests in trees or concealed on the ground; materials used in house construction include grass, bark, pine needles, cotton,

Golden mouse.

leaves, fur, and Spanish moss. Highly social, it is semi-communal and will share its feeding platform, nest, and turf with conspecifics. Establishes a home range of several thousand square feet.

This crepuscular nocturnal omnivore feeds on seeds, nuts, grains, fruit, sumac, poison ivy, and insects. Depending on locality, its breeding season runs all year long, with crests in spring and fall. The prolific polyestrous female is capable of mating once a month, and thus can produce up to a dozen litters annually. She gestates for about 1 mo., giving birth to an average of 2 altricial pups in a nest hidden in shrubbery.

Communication and perception take place via standard Cricetidae channels: primarily touch and smell, and to a lesser degree, sight and hearing. Though terrestrial, this mammal is quite arboreal, with excellent tree-climbing abilities; will ascend to a height of 30' or more; uses it prehensile tail for support and balance, and to suspend itself from branches. Its natural enemies probably include reptiles, mustelids, birds of prey, and wild felids. Gentle, shy, and rarely seen, the golden mouse can live 1-2 yrs. in the wild; perhaps twice that long in captivity.

GRAY-COLLARED CHIPMUNK (*Neotamias cinereicollis*): This smallish chipmunk grows to a total length of about 10", its hind foot measures around 1½", and it can attain a weight of up to 3 oz. The dorsum is light to dark gray, the ventrum is whitish. Distinguishing markings are its 5 dark dorsal and lateral stripes (the median stripe is black), interspersed with white and fulvous stripes. As both its common and scientific names denote, it has a gray "collar," one that also flows down over the nape of the neck; the hind feet are yellow-brown. Gender dimorphism is present: the female is larger than the male.

This smoky gray sciurid is a nearactic native that inhabits temperate, mature, coniferous montane forests with plenty of ground cover, fallen logs, soft soil, rocks, forest debris, and clearings. Like most of its kind, however, it is quite adaptable, and thus can also be found in agricultural, suburban, and urban areas. Its range is extremely narrow, covering only small swaths of the states of Arizona and New Mexico.

Gray-collared chipmunks.

A petraphilic fossorial mammal, it excavates elaborate burrow systems that can reach 30' in length. Here it rests, nests, sleeps, stores food, and seeks protection from predators. Entrances are carefully concealed by removing excess dirt and locating them under the bases of trees, rock walls, and boulders. It is often seen on logs, rocks, and stumps chipping and tail-twitching. Establishes a home territory of about 2,000 sq. ft., which it defends against intrusive conspecifics.

A diurnal, solitary, terricolous omnivore, it feeds primarily on seeds (Douglas fir), nuts (acorns), bark, wood, fruit (berries, currants, cherries), stalks, tubers, leaves, and fungi. These foods are supplemented with carnivorous items, such as worms, insects, birds' eggs, birds, small mammals (mice), and carrion. It is both a larder hoarder and a scatter hoarder. Estivates during extremely hot weather; enters den hibernation during the coldest months, though it will experience periods of arousal, for example, on mild winter days, at which times it briefly ventures to the surface to forage.

Its breeding season occurs twice annually, once in spring and once in fall; the iteroparous polyestrous female may thus produce 2 litters a year. Males compete for rights to estrus females. The doe gestates for about 1 mo., giving birth to as many as 5 altricial pups in her underground nest chamber. They are weaned at around 5-6 wks., with dispersal and independence taking place shortly thereafter. The buck does not display paternal behaviors and thus does not participate in the rearing of his offspring.

Communication and perception run along normal Sciuridae channels: sight, smell, hearing, and touch. Its natural enemies are reptiles, birds of prey, mustelids, wild felids, and wild canids. Though terrestrial, it is adept at tree-climbing; its tail stretches out horizontally while running. Benefits the world ecosystem by dispersing seeds, aerating and recycling the soil, supplying homes for other fossorial creatures, and providing food for carnivores. Harmful to humans via garden and crop destruction, the spread of diseases (such as the hantavirus), and the weakening of structural foundations. The gray-collared chipmunk lives an average of 2-3 yrs. in the wild.

GRAY-FOOTED CHIPMUNK (*Neotamias canipes*): Once thought to be a subspecies of the gray-collared chipmunk, based on recent genetic studies it is now recognized as a separate species. Morphologically, however, it remains nearly identical to the gray-collared, with roughly the same ecology and life history (including coloration, size, weight, habits, habitat, reproductive life, diet, and longevity). They also share overlapping ranges, though the gray-footed is also found in Texas, whereas the gray-collared is not. One other difference is that while the gray-collared has dark dorsal stripes, the dorsal stripes of the gray-footed tend to be a lighter grizzled cinnamon-gray color.

GRAY FOX (*Urocyon cinereoargenteus*): The official state wildlife animal of Delaware, this slender, medium-sized canid stands 15" at the shoulder, it grows to about 3½' in total length, its hind foot measures nearly 6", and it can attain a weight of up to 14 lb. It gets its common name from its grizzled, almost silvery gray coat, which is mixed with coarse white, red, and black fur. The top of the head, the outer ears, the back of the neck, the chest, the abdomen, and the shortish legs are often cinnamon-red. Its throat is usually white or cream. Its long brushlike tail is gray laterally and reddish on the ventrum, with a black stripe running along the dorsum, ending in a black tip. Its eyes are amber-

Gray fox.

brown; its ears are large, sharply tapered, and edged in buff fur; its muzzle is long and pointed; it has prominent vibrissae, small cheek ruffs, and a black nose. A narrow black line runs from the side of the face to the outer side of the eyes, and from the inside of the eyes down onto the snout, giving it a slight "bandit" look. Gender dimorphism is present: males are larger and more robust than females.

This terricolous nomadic troglophile is a nearctic native that inhabits forests, woodland, brushland, and farmland, usually near reliable fresh water sources. Also a neotropical species, its range extends from southern Canada to South America—though

Gray fox.

it is not generally found in parts of the northern Rocky Mountain states or the Great Plains states. It dens in rocky crevices, tree cavities, wood piles, caves, slash piles, hollow logs, dense brushy thickets, and on boulder-strewn slopes. Will sometimes take over the abandoned burrows of woodchucks. Establishes a home range of 1-2 sq. mi.; scent-marks its territory using numerous glands located over the body.

Like most dogs, this homeotherm is an endothermic carnivore, an insectivore, an herbivore, a frugivore, and a granivore; in other words it is an omnivorous hunter and scavenger, one whose diet includes everything from vegetables, nuts, eggs, grains, roots, and fruit, to insects, birds, rodents, small mammals, and carrion; will take poultry when available. Hoards and caches surplus food in loose piles of dirt or moss. Scansorial and possessed of retractable claws, it is the only North American canid that climbs trees, which it does while searching for resting places, protection, and food. Built for speed, it can run 20 mph and chase its prey up and down trees. It is also an excellent swimmer and will drive its quarry into rivers and lakes.

Its breeding season runs roughly from December through March, depending on geography. A temporary monogamous bond is formed between the mating pair at this time. The vixen gestates for about 6 wks., giving birth to 1-10 altricial pups in the spring. Weaning takes place in 4-6 wks.; the young begin hunting for themselves by 4 mos. of age; independence occurs at around 1 yr. or so. Though he does not den with her, the dog (male) assists the female with infant-rearing, bringing food to the burrow, defending the nest site, and teaching the pups how to hunt. After dispersal of the offspring the male and female return to living solitary lives.

Communication and perception take place via typical Canidae channels: body language, vision, hearing, touch, and smell. Vocalizations include yelping, barking, whining, howling, chattering, yapping, growling, and "coughing." With its catlike claws and compact body structure, this fox may be more primitive than other fox species (its ancestry dates back at least 10 million yrs.). A secondary consumer, its natural enemies include coyotes, wolves, mountain lions, bobcats, and birds of prey, such as owls and eagles. Quite capable of defending itself, it escapes most predators by hiding, running, or climbing trees. The gray fox lives an average of 6-8 yrs. in the wild; up to 12 yrs. in captivity.

GRAY MYOTIS (*Myotis grisescens*): Though medium-sized, it is the largest species of the *Myotis* genus, growing to nearly 4" in total length, with a forearm that is nearly 2" long, and a weight of ¼ oz. Unusual in chiropterids, its entire upper body is unicolored, in its case—as its species name indicates—gray (though during molting it is sometimes cinnamon-brown). The lower body is lighter in color. It has well-developed somewhat pointed ears; a short rounded tragus; a keel-less calcar; and a wing membrane that connects to its ankle (rather than its toe, as in some other bats).

Its primary distinguishing characteristics are its preference for caves and water. Thus, as a true speleophile and aquaphile it is most often found inhabiting caves with abundant water; or caves near water, such as rivers and lakes—particularly in karst areas. Avoids human structures. Lives year-round in caves, hibernating in them in winter, roosting in them in summer. This means that its habitat

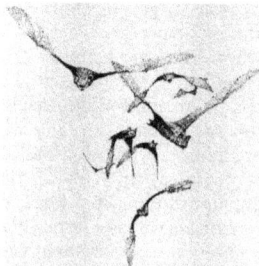

A *Myotis* species.

is extremely limited—and on the decline, as more and more caves are opened up to recreation, exploration, and development (pesticides, water pollution, and human-made flooding also take their toll). As a result, nearly the entire population is now restricted to several caves in only a few Southern and Midwestern states, such as Tennessee, Arkansas, Kentucky, and Indiana.

Nocturnal and carnivorous, this riparian vespertilionid leaves its roost at nightfall, seeking out its prey (nocturnal aquatic insects) over water using ultrasonic sounds (echolocation). Eating thousands of insects a night, it is one of nature's great pest controllers. As a member of the Chiroptera order, it communicates with other colony members vocally, and also through touch and smell. Highly social, in order to hibernate it migrates annually between summer and winter caves, in the latter forming massive colonies with over 150 bats packed into 1 sq. ft.

The female gives birth to a single altricial pup in May-June, raising it to maturity in a female-only maternity cave (the adult males and male and female juveniles roost in a separate cave). A secondary consumer, its main natural predators are reptiles (such as snakes), birds of prey (such as owls), rodents (such as mice), and small mammals (such as foxes). The gray myotis lives to about 15 yrs. in the wild.

GRAY SEAL (*Halichoerus grypus*): This large, rotund North American mammal can reach a total length of 12' and a weight of up to 800 lb. Depending on the population, its upper body can be brown, grayish-green, silvery, or black, while the lower body is lighter and usually mottled or spotted. Occasionally the entire body, including the squarish head, can be either solid or splotched. It has a fusiform morphology, typical of members of the Phocidae family. Its eyes are large, dark, and watery; its neck skin is wrinkled (on the male the thick neck is often scarred from fighting); the snout is long and rather bulbous (particularly in the male) with large flaring nostrils. The latter trait gives it its scientific name: *Halichoerus grypus* is Latin for "hooked-nose sea pig." Gender dimorphism is present, with the male being larger and differently shaped and marked than the female (the former is often dark with light patterns, the latter is often lighter with dark patterns).

This portly petraphilic phocid is an inhabitant of coastal marine environments, ranging from temperate to polar conditions. Its favored habitat is a rocky beach, though it also frequents icebergs and islands; it is often found in strong currents amidst cliffs,

Gray seals.

reefs, and rocks. The North American range of this Atlantic Ocean native extends from Labrador south to New England. (Outside our region it can be found from Europe to Russia.)

A natatorial mammal, it is an opportunistic piscivore whose diet centers around pelagic, midwater, and benthic fish species. Also a molluscivore with a penchant for marine crustaceans, it can consume up to 5 percent of its body weight each day. It has been known to hunt and kill larger prey, such as harbor seals and porpoises, holding them underwater to drown them; it may also practice cannibalism on occasion. Gregarious, it often hunts in groups, diving to a depth of nearly 500' and remaining submerged for as long as 20 mins. while foraging. A gray seal colony can have as many as 1,000 members.

Because it is connected to a population's geographical location, the gray seal's breeding season varies, occurring sometime between winter and fall. The polygynous bulls establish territories and fight over mating rights, forming harems of up to 10 cows. Both the adult males and females abstain from food during this 4-6 wk. period. Cows gestate for around 11 mos., giving birth to a single 30 lb. pup on ice or onshore, sometimes far inland. For the next 2 wks.

the mother nurses and aggressively guards her youngster, after which it is weaned and she abandons it in the rookery. Within a month or so the pup sheds its white lanugo and takes to the water to forage and hunt.

Communication and perception are typical for a member of the Phocidae family: sight, smell, touch, and hearing. Vocalizations include barking, hooting, mooing, yapping, howling, and hissing. Its primary natural enemies are sharks and orcas. Slaps the water when distressed. This massive, agile, carnivorous predator can be quite dangerous if cornered or threatened, and despite its bulk and cumbersome appearance, it can move swiftly on land. The male gray seal lives an average life span of about 20 yrs. in the wild; the female up to 35 yrs. Captive individuals have survived into their 40s.

GRAY WHALE (*Eschrichtius robustus*): The official state marine mammal of California, the female grows to a total length of about 50' and can weigh up to 80,000 lb. (about 40 tons). Its slender, dark slate gray body is tapered at both ends, and is scattered with white, silver, or light gray patches, scars, and mottling—the result of hundreds of pounds of whale lice and barnacles which attach themselves to its skin. It has 2 blowholes and lacks a dorsal fin. In its place there is a dorsal hump set toward the back half of the body. Behind the hump is a line of some 10 dorsal "knobs"

Gray whale.

that grow smaller toward the end of the tail. Its flippers are broad and stout; the upper jaw overshoots the lower jaw slightly; the mouth curves downward and under the eye then back upward; the horizontally flat tail is notched. Gender dimorphism is present: the female is larger than the male.

This nearctic and palearctic native inhabits temperate, tropical, and polar waters along North America's Pacific coast. A benthic eschrichtiid, it haunts shallow waters with soft sandy and muddy bottoms, often where seaweed and eelgrass grow. Migrates to warm waters off the coast of Mexico in the fall, and to cool waters in the Arctic in the spring, traveling up to 14,000 mi. a year—among the longest migrations of any mammal on earth. Not particularly social, it may travel in pods of 2-5 adults and their young, always staying within 5 mi. or so of shore.

This carnivorous mysticete is a planktivore, a piscivore, and a marine invertivore, vermivore, and molluscivore. An opportunistic feeder, its diet ranges from zooplankton, worms, and shrimp, to eggs, mollusks, and fish. It feeds in several ways. It may group-hunt fish, driving schools into large silvery balls, then smashing up through them with its mouth agape. Another method, probably unique among whales, is bottom feeding. This it does by swimming along the seabed on its side, sucking in large gulps of sediment (mud, sand, and water), in which its food (crustaceans) lies buried. This slurry is filtered through some 300 short, fringed baleen plates that hang from the roof of its mouth. The water is then forced back out. The trapped food is rubbed off the plates with its tongue and swallowed. There are 4-5 throat pleats located under its lower jaw that expand and contract, facilitating the feeding process. It can consume over 1 ton of food a day, and can often be located by the muddy water it stirs up while foraging, the origin of another one of its nicknames: "mud whale."

This shallow water, continental shelf-hugging, natatorial mammal is an agile swimmer that can hold its breath for up to 30 mins. and dive to a depth of 500'. Like others of its kind it enjoys spyhopping and breaching, and, being curious and generally friendly, will often approach boats and even interact with people. Its heart-shaped spout can reach a height of 20'. Though a medium-sized whale, it is massive and powerful by human standards, giving it its species name: *robustus* ("robust," "strong," "vigorous"). Small hairs on its snout betray its prehistoric descent from a distant land mammal. Communicates and perceives via typical Eschrichtiidae channels: body language, touch, and especially hearing. Vocalizations include low frequency grunts, moans, bangs, pulses, croaks, and knocks.

The polygynandrous gray has no set breeding season. The cow gestates for around 13 mos., giving birth to a single 16', 1,500 lb. precocial calf underwater. Birthing takes place on the calving ground: a shallow, highly saline coastal area (such as an estuary or lagoon) that has an average depth of about 12'. She nurses and aggressively protects her youngster until it is weaned at around 7 mos. of age. Due to the mother gray's violent ferocity in defending her baby, early whalers gave her the nickname "devilfish" (though she is neither devilish or a fish). The non-paternal bull does not assist in the rearing of its offspring.

A secondary consumer, the primary enemy of this cetacid is its cousin the orca, which hunts both adults and the young. Once thought to be extinct, it is popular among whale-watchers, providing benefits to us by way of the ecotourism industry. The average life span of the gray whale in the wild is probably 50-70 yrs.

GRAY WOLF (*Canis lupus*): Also known as the "timber wolf," it is the largest member of the over 30 living wild canid species, with a shoulder height of 3', a total length of around 6½', and a weight of up to 175 lb. Though it is named for its dense, grizzled gray coat, depending on the individual, its geographic location, and its color phase (it has 3),

Gray wolf.

the fur color on the upper body can range from white to black, with many variations of gray and brown in between. The underbody is lighter, usually white, buff, cream, or light silvery gray, and there is often cinnamon coloration on various parts of the body, particularly on the snout, neck, sides, and legs. The tail is medium length, bushy, and typically black-tipped. While pups are born with blue eyes, the adult's eyes are usually amber or brown, but they may also be green or yellow. Its ears are triangular and pointed; the jaws are powerful; the canines are large; and its muzzle is robust, long, and tapered. Unlike coyotes and domestic dogs, while running it carries its tail horizontally. A gender dimorphic species, the male is larger than the female.

Highly adaptable, and both nearctic and palearctic, it inhabits a myriad of biomes and ecosystems, from tundra, taiga, broken terrain, rangeland, and chaparral, to prairie, scrubland, ranchland, forest, mountains, and intermontane regions. This North American wild dog once lived over most of our region, but due to habitat loss it is now found mainly in the northern Rockies, Great Lakes region, Pacific Northwest, the Southwestern U.S., Canada, and Mexico. It dens in an underground burrow up to 30' long, often leaving a fan-shaped ejecta mound at the entrance, along with bones, dung, and paw prints. Outside the breeding season, timber wolves usually sleep out in the open—even in the middle of winter.

A well designed carnivorous hunter and scavenger that can run up to 40 mph, this mainly nocturnal canid feeds on fruit, insects, fish, birds, rabbits, hares, beavers, bighorn sheep, deer, elk, bison, moose, caribou, muskox, livestock, and carrion. A synanthrope, it will also consume our garbage and our pets. Culls the weak, sick, and aged from healthy animal groups, which serves to strengthen the prey's gene pool. Pack hunting, which usually takes place at night, involves complex tactics, problem-solving, and strategical thinking, which displays immense cognitive skills. After ambushing

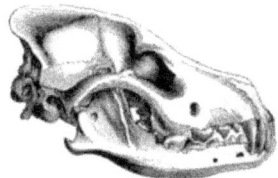

Skull of a gray wolf.

their prey, they test it for vulnerabilities. If a successful kill seems probable, the pack gives chase, working together in an attempt to cut the tendons of its back legs. In summer individuals often hunt by themselves or in pairs. Each wolf can consume 20 lb. of meat at one sitting, after which little is left of the carcass but hair and scattered bones.

The gray wolf is highly intelligent, highly social, and highly territorial. It lives in dominance hierarchy-governed packs of up to 15 related individuals, led by an alpha male and an alpha female (usually these are the two physically strongest, or the two most psychologically dominant, of the group). Pack ranking is based on complex, ever-shifting dynamics, with betas (secondary leaders), subordinates (neutrals), and omegas (lowest position) rounding out the group's social structure. It is the alpha pair, however, the pack leaders, who direct

Gray wolf.

hunts, select rest areas and home sites, and fix the group's territory—an area that can extend up to 1,200 sq. mi. or more, and which the resident pack vigorously defends against other wolf packs. Trespassers may be killed or accepted, depending on a host of variables. Nomadic and motile, this long distance traveler can cover up to 125 mi. a night.

Adult males and females form lifelong monogamous pair-bonds with one another. The dominant alpha couple are the only ones allowed to mate, however. Depending on a pack's geographic location, the breeding season normally runs from January to April. The alpha female gestates for about 2 mos., giving birth to 4-14 altricial pups in late spring or early summer in a carefully concealed den—usually located in a hole or cave. The entire pack assists in raising the young. For example, up until about 6 wks. of age, pups consume food regurgitated by pack members returning from the hunt. After this, they are weaned, becoming fellow pack members and hunters at 9-10 mos. When the maternal den site is no longer needed, the pack may decide to begin laying up in a "rest area," typically an expansive field where members can meet, socialize, recuperate, and cache food after the hunt. By around 2 yrs. of age or so, some of the youngsters will desert the pack to find a mate, or start their own pack and set up their own territory.

Communication and perception among timber wolves is standard for most members of the Canidae family: vocalizations (yelping, barking, growling, whining, whimpering, and communal howling—the latter which solidifies the pack and serves as a location beacon); body language (dominance and submission postures); touch (rubbing, licking, rolling); sight (facial expressions, tail positions); smell (scent-marking with urine and dung). The natural enemies of this daring secondary consumer are few, mainly coyotes and others of its own kind. Since it has been domesticated several times over the last few thousand years, many consider it the original ancestor of the domestic dog (a topic still debated). Beneficial as a population controller of prey animals and as a subject of ecotourism; detrimental as a farm

Wolf skeleton.

and ranch pest. The gray wolf lives an average of 5 yrs. in the wild; more than twice that long in captivity.

GREAT BASIN POCKET MOUSE (*Perognathus mollipilosus*): The largest member of its genus, it grows to a total length of almost 8" and can weigh up to 1 oz. or so. The soft pelage on the dorsum is pale buff with a drab gray or blackish wash; the ventrum pelage is white to light tan; the lateral line is an indistinct olive-green; the 4" tail is long and dichromatic (dark on top, lighter below), with a barely distinguishable crest near the tip. The body is ovoid; the eyes are dark and protruding; the ears are small, erect, and have a lobed antitragus; the snout is covered with sensitive vibrissae; the hind legs are long for a pocket mouse; it lacks the bristle-hairs found in some other species.

This endothermic heteromyid is a nearctic native that inhabits arid and semiarid deserts, montane grassland, open savanna, steppe, sand dunes, sageland, shrubland, pine forests, brushland, and sandy woodlands; prefers adequate hiding cover and loose soil. It is often found with the following plants: creosote, bitterbrush, saltbush, pigweed, greasewood, wild mustard, shadscale, and fescue-wheatgrass. A western rodent, it derives its common name from part of its range, which includes much of the Great Basin, a large geographical area that extends over portions of Oregon, California, Idaho, Nevada, and Utah. This mammal also occurs in British Columbia, Washington, Wyoming, Colorado, and Arizona.

Pocket mouse.

Excavates extensive burrow systems with multiple portals located under shrubbery; ejecta mounds are usually present near the entrance; uses its underground home to nest, sleep, feed, and avoid predators; to conserve energy it estivates (during extreme heat) and hibernates (during extreme cold). Establishes a large home range averaging some 20,000 sq. ft.

A nocturnal omnivore, its eats seeds, forbs, grasses, and various other types of plant material; feeds on insects when available. Does not drink water, which it derives from its diet. A scatter hoarder, it caches surplus food in small holes around its home range and also in burrow larders, transporting it in its external, fur-lined cheek "pockets" or pouches. After emerging from hibernation in spring the breeding season begins, cresting in late summer. The iteroparous polyestrous female may produce 2 litters annually. She gestates for about 3 wks. or so, giving birth to an average of 5 pups.

Communication and perception channels are typical of the Heteromyidae family: primarily touch and smell, and to a lesser extent, sight and hearing. Its natural enemies are reptiles, birds of prey, mustelids, wild felids, and wild canids. Beneficial as a seed disperser and as a prey base for carnivores. The Great Basin pocket mouse probably lives less than a year in the wild.

GRIZZLY BEAR: See Brown bear.

GUADALUPE FUR SEAL (*Arctocephalus townsendi*): This rare otariid grows to a total length of a little over 6' and can weigh up to 350 lb. It is grizzled brownish-gray on the dorsum, brownish-black on the ventrum. The head area is a silvery-brown; a yellowish "mane" runs down the back of the neck; the front and rear flippers are long and narrow. Gender dimorphism is present: the male is much larger, heavier, and more robustly built than the female.

This social but territorial petraphile is found along rocky shorelines, particularly those on coastal islands. Its range is narrow, only extending from California's Channel Islands in the north to the Baja Peninsula, Mexico, in the south.

Fur seal (male).

Like others in the Otariidae family it is a carnivore with a piscivorous- and molluscivorous-centered diet. Using its abilities to dive to a depth of 55' and stay submerged for nearly 3 mins., it hunts in open water for fish and squid. It has a sleek fusiform body. Nimble underwater, it can remain at sea for nearly a week at a time.

Its breeding season begins and ends in summer. The polygynous bulls, who establish temporary territories and fight other males for access to females, form harems of up to 10 cows. Almost unique among eared seals, the males guard their harems from both onshore and in the water. The female gestates for about 12 mos., giving birth to a single pup at her favored breeding site, a rocky cave. After 11 mos. of care, nursing, and protection, the pup is weaned.

Communicates and perceives via standard otariid channels: touch and smell, and to a lesser degree sight and hearing. This marine mammal came close to extinction in the 1800s. Its natural enemies are sharks and orcas. The male Guadalupe fur seal lives about 12 yrs. in the wild; the female may live up to 22 yrs. in the wild.

GUNNISON'S PRAIRIE DOG (*Cynomys gunnisoni*): Closely related to North America's other 3 prairie dog species, this large ground squirrel grows to a total length of nearly 16"and can attain a weight of up to 2 lb. On the dorsum its cryptically-colored coat is tannish-cinnamon with black-tipped hairs; lateral coloration is sandy; the ventrum is white; its brushy 2½" tail is gray-tipped and edged in white. As the male is larger than the female, gender dimorphism is evident.

Gunnison's prairie dog.

This motile, endothermic sciurid is a nearctic native that inhabits elevated chaparral, grassland, plateaus, brushland, savanna, sparsely-treed woodland, shrubland, valleys, and forest. Its range is restricted to portions of 4 Rocky Mountain states: Utah, Arizona, Colorado, and New Mexico. This robust rodent forms small, dense, informal colonies that are based around family clans and close relations. Burrow systems are excavated in small hills or on hillsides. In the northernmost parts of its range it may hibernate in winter; probably estivates during extremely hot weather.

A diurnal omnivore, its diet consists of forbs, grains, sedges, weeds, flowers, grasses, leaves, and insects. Its breeding season occurs in the spring. The female gestates for about 1 mo., giving birth to as many as 8 altricial pups in an underground nest chamber.

Communication and perception run along typical *Cynomys* channels: touch and smell, and to a lesser degree sight and hearing. Vocalizations include barking, growling, and chattering. Its natural enemies are reptiles, birds of prey, procyonids, mustelids, wild felids, and wild canids. Its eye placement, at the sides and top of the head, gives it a wide field of vision, permitting it to see long distances in nearly all directions—an added defense against predators. "Sentinels" sitting atop ejecta mounds constantly survey their surrounding home range for signs of danger, altruistically alerting the entire colony with sharp alarm calls.

A prairie dog mound.

Beneficial as an insect pest controller and as food for carnivores; detrimental as a farm and ranch pest, and as a carrier of deadly diseases (such as sylvatic plague, which is transmitted through flea bites). The life span of Gunnison's prairie dog in the wild is probably around 4 yrs. or so.

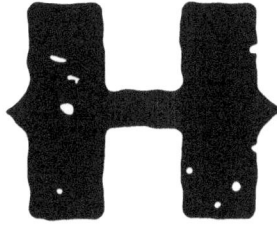

HAIRY-LEGGED VAMPIRE (*Diphylla ecaudata*): One of only 3 species of blood-eating bats, this medium-sized phyllostomid grows to a total length of nearly 3½", its forearm measures about 2", and it weighs in at around 1½ oz. Its dorsum is reddish- or grizzled-brown, its ventrum is light brown. It has large eyes; small rounded ears; a blunt snout; and more teeth than the other 2 vampire bat species. Other distinguishing traits: it has short fur; long pointed wings; it lacks a tail; and it has hair on its interfemoral membrane, contributing to its common name. It has several unique abilities: it can creep along as well as hop like a frog, talents that allow it to take flight from the ground (this it does by jumping several feet into the air before winging away).

Vampire bat.

The habitat of this secondary consumer is varied, ranging from subtropical and tropical forests to arid regions, where it roosts in caves, tunnels, old buildings, hollow trees, and mines in small groups of 2-10 individuals. Native to the New World, it is found as far north as southern Texas and as far south as South America.

A nocturnal obligate hematophage (an animal that feeds solely on blood—in the case of this species, warm-blooded vertebrates), it sleeps all day and awakens at dusk, flying from its roost in search of birds (chickens, geese, ducks) and mammals (pigs, cattle, horses) to prey upon. Using sonar to navigate, it gently crawls onto (or lands lightly on) its usually sleeping victim. Heat sensors located in its face help it find a blood-rich area beneath fur and feathers. After licking the skin to cleanse and loosen it, the bat uses its razor-sharp upper incisors to cut a tiny, shallow, and thus painless, v-shaped slit. A strong anticoagulant in its saliva creates a continuous flow of blood, which it laps up with its tongue through a small notch in its lower lip. The loss of blood is modest (only a few teaspoons), and the next day its prey is usually unaware of the bat's stealthy night attack.

Reclusive, highly intelligent, and intensely social, this shy chiropterid is known to engage in allogrooming, even caring for fellow colony members who may be in need (this includes food sharing). Will share its roost with other bat species. The female gestates for about 7 mos., typically giving birth to 1 altricial pup a year. Though to some it is frightening in appearance (and it will occasionally feed on sleeping people), it is not aggressive and actually flees from the presence of humans. It is thought that the rare sanguivorous hairy-legged vampire bat lives to around 10 yrs. in the wild.

Vampire bat.

HAIRY-TAILED MOLE (*Parascalops breweri*): Also known as "Brewer's mole," it grows to a total length of nearly 7", its tail is 1½" long, its hind foot measures about 1", and it weighs close to 2 oz. The soft silky pelage on its dorsum is gray to black; its ventrum is light gray. It lacks external ears and has a short muzzle; its tiny eyes are concealed under fur; it has upward-facing nostrils on its red-tinted nose. Typical of its family (Talpidae), it has large shovel-like front feet with turned-out palms and toes with long claws—adaptations to life as a fossorial mammal. Its hind feet are mouse-like. It is the only mole in the eastern part of our region with a short, scaly, hairy tail, hence its common name. Gender dimorphism is present: the male is larger than the female.

This excavating eulipotyphlid inhabits well-drained meadows, cultivated fields, woodland edges, forests, mountainous areas, open pastures, lawns, and golf courses with light dry soil. A North American native, it ranges from Ontario, Canada, down through much of New England, west into parts of the Midwest, and south into Tennessee and North Carolina.

Hairy-tailed mole.

This omnivorous talpid is an insectivore and vermivore that preys primarily upon insects and worms; but it will also consume vegetation (such as roots) if meat is scarce. Can eat up to 3 times its body weight per day. Solitary and chiefly diurnal, it can also be active at night, at which time it leaves its complex subterranean tunnels and forages above ground.

Its breeding season takes place in early spring. The sow gestates for approximately 38 days, gives birth to 4-5 altricial pups, and weans them at 4 wks. of age. The boar does not appear to assist in the rearing of his young. It destroys thousands of insect pests every year making it of great benefit to us. A secondary consumer, a host of natural enemies prey on this nearctic mammal, including amphibians, reptiles, birds, wild felids, and wild canids. The hairy-tailed mole lives 2-3 yrs. in the wild; 4 yrs. or longer in captivity.

HARBOR PORPOISE (*Phocoena phocoena*): Also known as the "common porpoise," at from 4'-6' in total length and a weight of roughly 135 lb., it is one of the smallest of all cetaceans. A member of the Phocoenidae family, its dorsum is gray, the ventrum is light gray to whitish. There is often some lateral mottling. Has a gently sloping forehead, small flippers and tail (for its size), and a wide, slightly curved dorsal fin located near the center of the back.

As its common name indicates, this cetacid is a frequent inhabitant of inland waters, and is a familiar sight in bays, estuaries, and fjords, along the coasts, and in rivers many miles from the ocean. It prefers cool water, and in our region ranges throughout the North Atlantic Ocean down to New England.

Somewhat solitary, this natatorial phocoenid forms small pods of a half dozen members. Will hunt alone or in groups, diving to a depth of 700' for up to 5 min., in search of its favorite foods: cephalopods, fish, and benthic crustaceans. A secondary consumer, its natural enemies are sharks and orcas—though bottlenose dolphins and gray seals are also known to attack them. The harbor porpoise's average life span is around 12 yrs. in the wild, relatively short compared to other marine mammals.

Harbor porpoise.

HARBOR SEAL (*Phoca vitulina*): The official state marine mammal of Rhode Island, its grows to a total length of 6', its tail measures about 4", and it can attain a weight of up to 300 lb. It pelage coloration is quite varied, ranging from yellowish-cream, gray, silvery, or dark brown on the upper body, to off-white, buff, or silver-gray on the lower

Harbor seals.

body. Fur markings and color patterns are also highly variable depending on the individual, population, and coat phase. The body can thus be plain, spotted, mottled, dappled with rings, or two-toned (dark dorsum, light ventrum). With multiple layers of blubber, or subcutaneous fat, the body is portly, though cylindrical and aquadynamic. The head is dog-like; the eyes are large, dark, and watery; and all 4 flippers have 5 webbed "fingers." Each side of the snout is covered with sensitive whiskers that are able to pick up the sound vibrations of prey items underwater. Gender dimorphic, the male is larger than the female.

Also known as the "hair seal," this familiar nearctic and palearctic phocid inhabits the shallow littoral waters of North America's Pacific and Atlantic Coasts. Also found in freshwater, including rivers and inland lakes, it is most commonly seen hauled out on rocks, ledges, and beaches; frequents saltwater piers, as well as quiet islets, rocky shores, bays, and estuaries. More widely distributed than any other pinniped, in the West its range extends from the subarctic to Southern California; in the East its range runs from Greenland to New England, and south to the Carolinas.

Like others of its kind, this shy carnivore is also a piscivore and molluscivore, one that feeds on, among many other aquatic creatures, flounder, mackerel, salmon, herring, rockfish, octopus, crayfish, and shrimp. Hunts along the bottom of shallow waters during high tide; basks and rests during low tide. While searching for its food it can dive to a depth of nearly 1,500' and stay submerged for 30 mins. Will swim hundreds of miles up rivers pursuing fish runs.

Diurnal and nonmigratory, its breeding season occurs between spring and summer, depending on geographic location. The polygynous (sometimes monogamous) males establish territories and fight violently with other males for mating rights. Females employ delayed implantation, gestating for a total of around 10½ mos. The cow gives birth to a single 25 lb. pup which sheds its lanugo before birth. Born with an adult-like coat, it is able to take to the water and swim within a few days. The mother nurses her youngster with nutrient-rich milk, 50 percent of which is fat. At about 6 wks. the pup is weaned and becomes independent.

Though largely solitary, it will join groups that sometimes casually form onshore. The playful but ever attentive harbor seal communicates and perceives in traditional Phocidae fashion, using sight, hearing, touch, and smell. Common vocalizations include barks, grunts, and yelps. Its natural enemies are polar bears, orcas, and sharks, which feed on adults, and coyotes and eagles, which feed on the young. The life span of the wild male harbor seal is probably about 25 yrs.; females perhaps live to 30 yrs. in the wild; in captivity this common marine mammal may attain an age of 45 yrs.

Harbor seal (juvenile).

HARP SEAL (*Pagophilus groenlandicus*): It can grow to nearly 7' in total length and weigh up to 400 lb. Adult males are generally yellowish to gray on the dorsum; light grayish and spotted on the ventrum. Most distinguishing is a large, dark, lyre- or harp-shaped, dorsolateral marking over its back and sides, the origin of its common name. It is the only light-colored seal with such a saddle marking. Females are lighter in color and have more spots. Adults have dark brown or black facial fur. Newborns enter the world with yellowish or white lanugo, which they molt within 3-4 wks. as they begin to take on the spotted markings and thicker silvery pelage of juveniles. Gender dimorphic, the male is larger, more generously colored, and more robustly constructed than the female.

This holarctic phocid is a cold water species that is found in polar marine habitats, chiefly around and on drifting pack ice. It is also known to enter streams. Its range covers the Arctic Ocean and the northern section of the Atlantic Ocean, extending from northern Canada east through Scandinavia, Europe, and Russia. Highly migratory, it travels an average of 1,500 mi. north in summer, and 1,500 mi. back south in winter. The total distance of its seagoing migrations may reach as much as 6,000 mi. a year.

Harp seal.

A carnivorous marine mammal, it is chiefly a piscivore and an invertivore, with a diet centered on fish (cod, sculpin, herring, redfish, capelin, halibut) and invertebrates (shrimp, prawns, krill). It can dive up to 900' deep and stay underwater for 15 mins.

The gregarious harp seal swims in huge herds, congregating in densely packed colonies in winter in preparation for the breeding season, which runs from midwinter to early spring. Bulls battle for access to estrus cows, who gestate for 11-12 mos., employing delayed implantation so that parturition takes place at the optimum time of year, namely spring. Near the edge of the ice pack the female gives birth to a single, precocial, 12 lb., lanugo-covered pup, which she nurses with high-fat milk, weaning it within 2 wks. Having attained its independence, the mother leaves the now 100 lb. pup on the ice to fend for itself. Over the following few weeks it is unable to eat and loses up to 50 percent of its body weight. By about 4 wks., however, its juvenile pelage grows in and it is able to swim and hunt on its own. Up to 30 percent of pups die of natural causes within their first year. The non-paternal father does not assist in the raising of his young.

Harp seal.

(Note: It is during the 3 mos. following weaning that Canada's massive, government-sponsored commercial seal hunt takes place on ice floes off the country's east coast. Off-season fishermen, known as "sealers," wielding guns, picks, and clubs, kill 60,000-80,000 harp seal pups for their pelts and oil each year. Intense controversy continues to surround the hunt. Proponents argue that culling the seal population is needed to protect fish stocks; opponents maintain that the hunt is nothing but an inhumane and unnecessary slaughter.)

Communication and perception take place via normal Phocidae channels: sight, touch, sound, and smell. Vocalization occurs mainly underwater, where individuals trill, roar, chirp, "hoh," and click to one another. Its natural enemies are orcas, walruses, polar bears, and sharks. The harp seal lives an average of 30 yrs. in the wild.

HARRIS' ANTELOPE SQUIRREL (*Ammospermophilus harrisii*): Also known as the "Yuma antelope squirrel," this small chipmunk-like ground squirrel grows to a total length of 10", its pes measures about 1½", and it can weigh up to 4 oz. The upper body is generally grayish or light brown, the underbody is whitish, tan, or light gray. The top of the head and the outer legs are often washed with cinnamon-colored hair. Its predominant marking is a thin white lateral stripe on each side of the torso that runs from shoulder to rump. Its thinnish 3" tail—which is grizzled on top and gray below—is brushy rather than bushy and is carried arched over the back to shade it from the sun. Its coloration provides ideal camouflage in its preferred habitats: dry, sandy, rocky areas.

Often misidentified as a chipmunk, the pelage is coarse; the head is small and rounded; the nose is snubbed; the humanlike ears are small and erect; the eyes are dark and ringed in white; the neck is long; the body is cylindrical and supple; the limbs are short; the feet are clawed; sensitive vibrissae cover the muzzle. Like others of its kind it likes to sit upright on its back feet. This species is not gender dimorphic.

A Southwestern petraphilic native, it inhabits arid shrubland, cactusland, playa, dunes, valleys, sandy plains, rocky slopes, foothills, salt flats, and canyons, usually in or near sparsely vegetated, gravelly, desert environments and xeric zones generally. Its range is limited to the dry parts of southern California, southern Nevada, southern Arizona, and southern New Mexico, south to northern Mexico. Creates burrow systems for protection, shelter, and nesting; these often open and exit below thickets and shrubs.

Antelope squirrel (juvenile).

This omnivorous sciurid has a varied diet that includes seeds (particularly from cacti), roots, stalks, grains, nuts, fruit, insects, small mammals (such as mice), and carrion. Transports food to its burrow in large cheek pouches; hoards and caches surplus food. It is diurnal, does not hibernate, and is active year-round, even on the hottest and coldest days.

Normally solitary, it is social only during its polygynandrous breeding season, which runs for a few weeks between midfall and midsummer—depending on latitude. The doe gestates for about 1 mo., giving birth to as many as a dozen or more altricial pups in a grass-lined, underground nest chamber. By 2 mos. of age the young are weaned, dispersing as independent individuals for the first time. The father plays no role in rearing the offspring he sires.

This hot weather rodent communicates and perceives primarily via sight, touch, and smell, and to a lesser extent, hearing. When disturbed or distressed it stomps its front feet and utters a loud trill; a petraphile, it often lives near rock or boulder fields, which it uses to avoid predation. Runs with its tail held straight up; dashes up and down cactus plants, often perching at the top to check for predators; it heat dumps by laying belly-down in the shade. Though it is beneficial to us as a seed disperser, it can be detrimental due to its habit of crop raiding. Its main natural enemies are birds of prey, reptiles, wild felids, and wild canids. Harris' antelope squirrel lives from 3-5 yrs. in the wild; up to 10 yrs. in captivity.

HEATHER VOLE (*Phenacomys intermedius*): Also known as the "western heather vole," this small, short-tailed arvicolinid grows to a total length of about 6", its hind foot measures ¾", and it can attain a weight of up to 1½ oz. Though its coloration is geographically-dependent, the dorsal pelage is generally grayish-brown with a grizzled or silver-tipped wash over the fur; the ventral pelage is grayish-silver; the feet are white; the 1½" tail is bicolored. The body is cylindrical and stout; the eyes are small; the ears are furred and erect; the limbs are short; the feet are small, pinkish, and delicate. It is often mistaken for the meadow vole.

A fossorial cricetid, it is a nearctic native that inhabits open, grassy, ericaceous (heather) fields, hence its common name. It is also found in aspen and conifer forests, spruce stands, chaparral, taiga, mossy fields, blueberry patches, clearings, shrubland, montane meadows, and tundra, at elevations from subalpine to alpine. Prefers settings with rotting logs, blowdown, and abundant duff. Its range extends over much of central and southern Canada, south into Washington, Oregon, California, Utah, Idaho, Wyoming, Colorado, Montana, and New Mexico. Constructs its house out of grass, moss, lichen,

Heather vole.

and other debris, locating it under shrubbery, stumps, rocks, and logs. In winter it burrows beneath the snow, forming a round subnivean stick and grass shelter. Though not particularly social, during cold weather it may cluster into transitory groups for warmth; does not hibernate.

A solitary, nocturnal (and sometimes diurnal) herbivore, it is a primary consumer of plant matter, in particular, seeds, fruit, leaves, and fungi, as well as woody material, such as twigs, buds, and bark. Hoards and caches surplus victuals around its burrow.

Its breeding season takes place from midspring to late summer, at which time males compete over polyestrous females, the latter who may produce up to 3 litters per year. The female gestates for 3 wks. or so, giving birth to an average of 5 altricial pups in the natal nest; they are weaned 3 wks. later. The male does not participate in the rearing of his young.

Communicates and perceives via standard Cricetidae channels: primarily smell and touch, and to a lesser degree, sight and hearing. Its natural enemies are reptiles, birds of prey, mustelids, wild felids, and wild canids. Beneficial as food for carnivores; detrimental as a carrier of deadly diseases. At most, the heather vole probably lives an average 6-12 mos. in the wild; perhaps longer under optimum conditions.

HEERMANN'S KANGAROO RAT (*Dipodomys heermanni*): This medium-sized rodent grows to a total length of about 14" and a weight of up to nearly 4 oz. In common with most other kangaroo rats in our region, it has long smooth fur that is dusky gray-olive and orangish-brown on the dorsum and white on the ventrum, while the flanks are buff colored and the feet are white. Used for balance while running and hopping, its long 9" tail is whip-like and bicolored (dark on top, light on the bottom), and ends in a bicolored tuft.

Not a true rat, but a relative of the pocket gopher, its body is compact and ovoid; its face is wide and its head is large with a convex profile; the eyes are large, dark, and protruding, each with a white supraocular spot; the face is lightly masked; the ears are small, rounded, and erect, each often possessing a white subauricular mark; the muzzle is stout; the nose has a black edge-band on top and is pinched, snubbed, and covered in sensitive nasal vibrissae; the front limbs are retrogressive; the hind limbs are large, kangaroo-like, and well designed for hopping—the trait from which it derives its common name. The back feet are 5-toed; a narrow white thigh stripe runs laterally over the upper leg. Gender dimorphism is present: the male is larger than the female.

This fossorial heteromyid is a nearctic native that inhabits a number of different biomes and ecosytems, including well-drained piedmont, shrubland, mixed forest, knolls, chaparral, hills, grassland, ridges, valleys, and plains. Prefers dry, friable, loose, fine, sandy, or gravelly soil in areas with sparse vegetation. Its range is very narrow, extending only through parts of southern, central, and coastal California. It is active all year and does not hibernate; but it spends up to 95 percent of its life underground in its large burrow system. When excavated in soft soil, its tunnels may attain a length of 30'-40'. Each burrow contains multiple tiers,

Kangaroo rat mound, showing entrances, runway, and subsidiary portals.

tunnels, entrances, exits, and compartments—the latter which are used for a variety of purposes, such as eating, sleeping, nesting, protection, waste, and storage. Will appropriate the subterranean homes of other animals when necessary (such as when the soil is too hard to dig through). A secretive and reclusive rodent, it dislikes inclement weather, sunlight, and even moonlight. Its home range is small, probably only 100'-200' in diameter.

An endothermic omnivore, essentially it has a granivorous and graminivorous diet surrounding seeds, forbs, and grasses; it supplements this fare, however, with animal matter (such as insects). Stores surplus victuals; transports its food in large cheek pouches. With highly effective kidneys that require little fluid to metabolize nutrients, it is able to satisfy most of its water needs from the food it eats; it also reduces moisture loss and maintains its body temperature by remaining in its den during the day.

Its breeding season runs from spring to fall. The doe gestates for about 1 mo. and gives birth to 2-4 altricial pups in a grass-lined natal nest chamber. She weans them at around 1 mo. of age, at which time they become independent and disperse. When food is abundant the polyestrous females may bear 2-3 litters a year.

This nocturnal, aggressive, solitary "rat" communicates and perceives via established Heteromyidae channels: sight, hearing, smell, and touch. Engages in dust-bathing and hind foot-thumping. It is plantigrade and quadrupedal when walking; it is saltatorial and bipedal when hopping. When pursued it engages in ricochetal locomotion: short fast hops that allow it to change direction quickly, even when in midair. Benefits its ecosystem by hosting parasites and by providing food for carnivores; aids humans by serving as both a pet and as a subject of scientific research. Its natural enemies are reptiles, birds of prey, mustelids, wild felids, wild canids, and other burrowing animals. An agile and speedy sprinter, it can run (hop) 10-12 mph. Heermann's kangaroo rat may live 4-5 yrs. in the wild.

HISPID COTTON RAT (*Sigmodon hispidus*): This medium-sized vole-like rodent grows to a total length of about 15" and a weight of up to 7 oz. Cryptically colored, its upper pelage is grizzled brown and black, its lower pelage is tan to light gray. Derives both its common name and its species name from the stiff, coarse (that is, "hispid") appearance of its long fur. Like other sigmodonts, there is an s-shaped pattern on the tops of its molars, the origin of its genus name. The body is small, compact, and cylindrical; the head is small and narrow with a convex profile; the eyes are small, dark, and protruding; the ears are small, rounded, and erect; the muzzle is short and tapered; the nose is snubbed with radiating vibrissae; the limbs are short; the feet are small and delicate; the 6" tail is shortish, scaly, sparsely furred, and lighter on the ventral side. Gender dimorphism is present: the male is larger than the female.

This terricolous cricetid is a nearctic native that inhabits a variety of biomes and ecosystems, from temperate grassland, seeps, moist meadows, prairies, salt marshes, weedy farmland, palm savanna, canal banks, and pastures, to scrubland, field edges, brackish ponds, roadsides, levees, oak hummocks, ditches, mesquite bosque, and dry cacti-rich areas.

Hispid cotton rat.

Prefers warm temperatures, along with good hiding cover and shrub overstory. Its range is quite large, extending from Nebraska in the north, New Mexico in the west, Virginia in the east, and South America in the south. Constructs its house under objects such as rocks and logs using grass and natural litter; as it clears and maintains its runways, it leaves behind small irregular piles of cut grass. Excavates an attached subsidiary burrow system which it uses for warmth and protection; will take over the abandoned homes of other small and medium-sized mammals (such as skunks). Does not hibernate; it is active all year, and at all hours of the day and night. Establishes an average home range of 100-200 sq. ft.

A solitary diurnal and nocturnal omnivore, its diet is composed of seeds, roots, leaves, nuts, grains, woody material, stems, tubers, foliage, and fruit. It supplements these foods with insects, aquatic crustaceans, birds' eggs, and small birds and mammals, as well as carrion. It does not hoard or store surplus food; acquires most of its water needs from the foods it consumes.

Hisbid cotton rat.

Depending on geography, the breeding season of this fecund reproducer runs year-round. The polyestrous doe may produce 5-6 litters a year. She gestates for about 1 mo., gives birth to as many as a dozen or more altricial pups in a round grass-lined nest, and weans them within a few weeks. The young are capable of mating and reproducing at a mere 6 wks. of age—though they do not become adults until they are 5-6 mos. old. As in most polygynous species, the male is not paternal and does not contribute to the rearing of his offspring.

This endothermic eutherian communicates and perceives via standard Cricetidae channels: sight, hearing, touch, and smell. Its natural enemies include reptiles, birds of prey, mustelids, procyonids, wild felids, and wild canids. Benefits the world ecosystem by hosting parasites and by providing a prey base for carnivores. Harms humans by carrying disease (such as rabies) and by destroying crops (such a sugarcane, corn, sweet potato, carrot, peanut, and cotton). The hispid cotton rat has a short life span, probably not living more than an average of 4-8 mos. in the wild.

HISPID POCKET MOUSE (*Chaetodipus hispidus*): This large chaetodipid grows to a total length of nearly 9", its hind foot measures 1", and it can attain a weight of 2 oz. Its fur is coarse, hence its common name and its species name: *hispidus* is Latin for "rough," "bristly," "stiff," "shaggy," or more generally, "spine-like hair." Its dorsal pelage is buffy with an olivaceous-gray to blackish wash; the ventral pelage is white to tan; the clear-cut lateral line is yellowish; the 4" tail is bicolored (brown above, off-white below) and sparsely furred.

This solitary, petraphilic, abundant heteromyid is a nearctic native that inhabits prairies, grassland, rocky settings, chaparral, rangeland, slopes, shrubland, desert, pastures, cultivated fields, mixed forests, piñon-juniper woodland, and savanna, usually with light vegetation and gravelly soil. A Western and Southern rodent, its range extends from Montana and North Dakota in the north to Texas and northern Mexico in the south, from Arizona and Wyoming in the west to Kansas and Louisiana in the east.

Pocket mouse.

A burrow maker, it spends much of its time underground, where it nests, feeds, and sleeps; its tiny burrows—which are less than 1" in diameter—reach a depth of about 1' and are located under shrubbery or rocks; burrow systems have multiple side branches and portals, the latter which occur with small ejecta mounds around them; entrances are stopped up during the day (to keep humidity in and predators out). This species estivates during hot weather; enters torpor during the coldest months; may hibernate during winter in its northernmost range. The diet of this nocturnal omnivore consists of seeds, leaves, forbs, cacti, and insects; caches food in subterranean larders, transporting it in its fur-lined cheek "pockets" or pouches.

Its breeding season runs year-round, peaking at the start of warm weather. The polyestrous female can produce up to 2 litters annually. She may gestate for 3-4 wks., giving birth to as many as 8 pups in a grass-lined natal nest. Communicates and perceives along standard Heteromyidae channels: mainly touch and smell, and to a lesser degree, sight and hearing. Its natural enemies are reptiles, birds of prey, mustelids, wild felids, and wild canids. Beneficial as a consumer of the seeds of pestilent plants; detrimental as a consumer of the seeds of agricultural plants. The hispid pocket mouse probably lives less than a year; perhaps as long as 1-2 yrs. under optimum conditions.

HOARY BAT (*Lasiurus cinereus*): This mouse-like chiropterid has the distinction of 1) being the largest bat in the eastern half of North America, 2) the most widespread of all American bats, and 3) the only bat known to inhabit the Hawaiian Islands. It grows to a total length of 6", its forearm is just over 2" long, its wings span up to 17", and it weighs a little over 1 oz. It has a blunt nose; rounded black-rimmed ears; small eyes; a short blunt tragus; and a keeled calcar. Its large skull is filled with the robust crushing teeth of an insectivore. Its soft silky pelage is grizzled gray-brown, while each individual hair is heavily frosted with a white tip, giving it its common name: hoary means "old" or "gray-haired." Uniquely, its furry tail membrane is also grizzled gray, giving it its scientific name: *Lasiurus* means "hairy tail," *cinereus* means "ash-colored."

It shelters in both deciduous ad coniferous trees, from dense forests to cleared land, from rural to urban settings. Has also been found in caves, buildings, and squirrel nests. Wrapped snugly in its furred tail membrane and clinging to a twig by a single tiny foot, it cleverly camouflages itself as a pine cone or dry dead leaf as it hangs in tree foliage. In our region this broadly distributed, intercontinental vespertilionid ranges from Canada all the way to Central America, and is found is all 50 U.S. states. Despite being the most widespread bat in the Americas, it is rarely seen, for it is a solitary creature that does not usually form large colonies—generally occurring in groups only when it is migrating and hunting for food.

Hoary bat.

Some populations seem to hibernate. Depending on the latitude, it may simply enter torpor during cooler periods. A strong, fast nocturnal flyer, it echolocates to find its main prey, moths, but it also consumes flies, wasps, and beetles. While in the air it can fly 15 mph and soar to heights of nearly 2 miles (making it one of our highest flying bats). Though this species mates in the fall or early winter, the female delays fertilization until spring when conditions are more advantageous to birthing and raising young. She then gestates for about 55 days, giving birth to as many as 4 altricial pups in the spring or early summer, which she nurses at one of her 4 mammary glands. A secondary consumer, this flying mammal shares the primary predators of many other North American members of the Chiroptera order: birds of prey and reptiles. The hoary bat lives to at least 2 yrs. or more in the wild.

HOARY MARMOT (*Marmota caligata*): Also known as the "mountain marmot" and the "rockchuck," this is our largest ground squirrel. It grows to a total length of up to 33" and can attain a weight of 20 lb. Its base pelage on the dorsum is buffy-cinnamon brown and black with grizzled silver hairs washing over the back, shoulders, and sides; this unusual fur color gives it its common name: hoary, meaning "grayish-white" from "old age." The ventrum runs from white and gray to tan and rusty; the 10" tail is bushy, cinnamon colored, and quite long for a marmot; the posterior is tannish-brown; the head, nose, and shoulders often have white, black, or white and black markings and patches; the feet are black, a trait that gives it its species name: *caligatus* is Latin for "booted" (and was thus an ancient term for the "common soldier").

Its body is stout and bulky; its head is robust and broad; the eyes are dark and small; the ears are small, rounded, and erect; the muzzle is banded (in black) and covered in sensitive vibrissae; the limbs are short and muscular. Its dense silvery-gray fur not only acts as an insulating layer against cold, it also supplies good camouflage for the rocky locations it haunts. It molts once a year. Gender dimorphism is present: the male is larger than the female.

This terricolous petraphilic sciurid is a nearctic native that inhabits temperate talus slopes, treeless alpine meadows, scree, bare rocky outcroppings, and open grassy tundra located in mountainous regions—both subalpine and alpine. One of North America's most widely distributed mammals, its range extends from Alaska south through British Columbia and into Washington, Idaho, and Montana. Fossorial, it spends most of its life underground, where its extensive burrow system protects it from harsh weather and predators. Its powerful front limbs and paws are well-clawed and specially adapted to digging; tunnels are over 1' in diameter; ejecta mounds are present at burrow entrances. Estivates during hot weather; enters hibernation with the onset of cold weather, living off its fat reserves until the arrival of spring. Establishes a home range of over 30 acres.

A diurnal herbivore, its diet is composed primarily of grasses, but it also consumes mosses, flowers, herbs, lichens, leaves, vetches, and sedges. Its reproductive season begins shortly after breeding bucks and does emerge from their hibernacula in spring. The female produces 1 litter every other year. After a gestation period of about 1 mo., she gives birth to 3-5 altricial pups in a subterranean maternity chamber, weaning them within a few weeks. The young do not disperse for another 1-2 yrs., by which time they are reproductively mature and ready to be independent. The male provides marginal assistance to the female in the rearing of their offspring.

Communicates and perceives along standard Sciuridae channels: sight, touch, hearing, and smell. Vocalizations include chirps, trills, growls, chucks, scream-like calls, and whistles, the latter a very humanlike sound from which it derives one of its many nicknames: "whistler." Highly social, it lives and eats in large communal groups led by an alpha male and several alpha females; each colony possesses as many as 40 individuals and includes neonates, yearlings, and juveniles up to 2 yrs. old or so. Energetic and playful, it enthusiastically engages in shoving, mock-fighting, and wrestling matches with conspecifics; spreads out on rocks to sunbathe; though quite wary, it is not particularly fearful of humans. Among its natural enemies are birds of prey, mustelids, wild canids, and ursids. Beneficial by providing food for carnivores and meat and pelts for humans. One of the longest-lived of our rodents, the hoary marmot has an average life span of 6 yrs. in the wild.

Hoary marmot.

HOODED SEAL (*Cystophora cristata*): The most common seal in the North Atlantic Ocean, it attains a total length of 10' and a weight of nearly 900 lb. The male's pelage is countershaded, with a whitish-blue or silvery-gray to blackish upper body and a lighter lower body. His entire torso is covered with irregular dark spots and splotches. The female is more lightly colored, and her markings are not as distinct. The head is short; the flippers are smallish with tough claws; the nostrils are extremely large; the mouth contains 30 teeth. Both the male and female sport dark gray faces, and often darkish throats and necks as well. Gender dimorphism is present, with the male being larger, much heavier, and more robustly built than the female.

The most identifying characteristic in both genders provides its common name: a rubbery, hood-like, nasal air-bladder located on top of the head, unique among all seal species. (It is much smaller and less developed, and hence not as noticeable, in the female.) For clarification's sake, I have given this appendage the name the "primary air bladder." In the male, when inflated, this 2-lobed enlargement of the nasal cavity—situated on the forehead and extending down to the nose tip—grows quite bulbous, making the head look much larger than it actually is. When deflated the "hood" becomes trunk-like and crinkly, hanging down over the side of the muzzle. The male possesses a second odd piece of body ornamentation as well, a specialized septum that I have named the

Hooded seal.

"secondary air bladder": a pinkish-red mucus membrane that is located inside the nose. When inflated, this bright crimson, balloon-like nasal sac protrudes from the left nostril, growing larger than a football.

A holarctic native of the northern Atlantic Ocean, it inhabits deep water on the edge of drifting pack ice, with a range extending from eastern Canada to Scandinavia. A highly migratory animal, it travels great distances in pursuit of moving pack ice as it retreats and advances each year. Occasionally it is spotted as far south as Florida (and, outside our region, Portugal).

In keeping with its kind, this carnivorous phocid is both a piscivore and a molluscivore, with a diet centered around fish (mackerel, cod, halibut, flounder, herring, redfish) and marine invertebrates (mussels, shrimp, starfish, squid, octopus). When hunting, this powerful swimmer and deep-diving mammal can go down to a depth of 3,300' and remain submerged for 30 mins. or more.

Its mating season runs from midspring to early summer. Though normally solitary, at this time both bulls and cows become gregarious, congregating at their breeding grounds where the polygynous males joust and fight over receptive females. The males' primary and secondary air bladders now come into play: the former is inflated to scare off would-be male rivals, while the latter is inflated to draw the attention of breeding females—who are attracted by their pink "balloons" or nasal sacs.

Cows gestate for about 8 mos., giving birth to a single, precocial, slate-blue, 40 lb. pup the following spring on the ice. The pup sheds its lunago *in utero* and is able to crawl and swim shortly after birth. The mother nurses it with milk that has a fat content of nearly 70 percent. This enables the pup to gain weight quickly, helping ensure its survival. Female hooded seals have the shortest lactation period of any mammal on earth: in as little as 3 days the pup is weaned, becomes independent, and is abandoned on the ice by its mother. The male plays no role in parenting his young. As with most mammals, the nuclear family of this species is comprised only of the mother and her offspring—brief as it is.

Hooded seal.

Communication and perception take place via normal Phocidae channels: sight, touch, hearing, and smell. Vocalizations include bellows, growls, and roars. Males can also communicate through their 2 air bladders, emitting various sounds, such as pings and pulses, that carry long distances. Its primary natural predators are orcas, sharks, and polar bears. The average life span of the hooded seal in the wild is probably 25-30 yrs., with females generally living longer than males.

HOODED SKUNK (*Mephitis macroura*): Also variously known as the "southern skunk," the "long-tailed Mexican skunk," and the "white-sided skunk," it grows to 32" in total length, stands about 8" at the shoulder, its hind foot measures nearly 3" long, and it weighs up to 2 lb. It may appear in any one of a number of color phases, including a white upper body (with a black lower body), a black upper body (with white stripes on its sides and a black lower body), or an all black body. It can sport a myriad of other black and white patterns as well. It has a slender body; long soft fur; a broad head; a tapered face; small eyes; a slightly upturned nose; rounded medium-sized ears; shortish limbs; and a long whisk-like tail usually comprised of mixed white and black hair (its species name, *macroura*, is Greek for "long tail"). Sometimes there is a single vertical white stripe down the middle of the forehead running onto the nose.

It gets its common name from the aforementioned first phase, in which a patch of

Hooded skunks.

longish white fur covers the top of its head and neck, giving it a hood-like appearance. Though it is often confused with its close cousin the striped skunk, it is this head or neck ruff that distinguishes the hooded from the striped skunk—the latter which has coarser fur, a less bushy tail, and is generally larger in size. The hooded skunk is gender dimorphic, with the males being larger than the females.

This secretive riparian mephitid inhabits numerous biomes and ecosystems, from scrubland, arid deserts, talus, swamps, canyons, grassland, chaparral, and deciduous forests, to marshes, pastures, coastal plains, mountains, rainforests, scree, and rocky outcroppings, always near a fresh reliable water source. Its range covers Mexico and parts of the extreme U.S. southwest, primarily New Mexico, Arizona, and Texas. Paludal, nocturnal, and solitary, its favored den site is in a rocky ledge along a stream bank or wash with plenty of hiding cover. Its home range may extend up to 2 sq. mi.

A generalist omnivore, it is primarily an insectivore whose diet also includes vegetation (mainly prickly pear cactus), seeds, nuts, grains, fruit, worms, wild bird eggs, fish, amphibians, small mammals, carrion, and, as a synanthrope, pet food and human refuse. It is usually found foraging along streams with dense tangle and vegetation, but will visit farms and take chicken eggs if the opportunity arises. The hooded skunk population consumes thousands of pounds of pestilent insects each year, a great benefit to us.

Its breeding season takes place in late winter to early spring. The doe gestates for about 2 mos.; parturition takes place between late spring and early summer with the birth of 2-8 kits. The buck does not assist in rearing or protection of the young. Though little is known about this skunk, it is assumed that it communicates and perceives in a similar fashion to other members of the Mephitidae family, using primarily vocalizations, smells, and body language.

A secondary consumer, it has few natural enemies, except birds of prey and wild canids. Its defenses include its aposematic black and white markings (a warning signal to predators), running, hiding, and, as a last resort, its chemical spray: when threatened it turns its posterior toward its attacker and emits a noxious oil from its rump glands. These organs contain controllable tips or nipples, allowing it to aim the smelly fluid accurately 10'-15'. The life span of the hooded skunk remains a mystery, but it is probably similar to that of other mephitids: 2-3 yrs. in the wild; perhaps twice that long (or more) in captivity.

HOPI CHIPMUNK (*Neotamias rufus*): Once thought to be a subspecies of, or even conspecific with, the Colorado chipmunk, due to recent genetic studies it is now recognized as a separate species. While it shares much of the same ecology and life history as its close cousin, there are 2 differences: the Hopi's dorsal stripes are less vividly colored, and its geographic range is restricted to the states of Colorado and Utah (the Colorado's range extends over Colorado, Arizona, New Mexico, and Oklahoma).

HOUSE MOUSE (*Mus musculus*): Also known as the "common mouse," this invasive pervasive rodent grows to a total length of nearly 8" and can attain a weight of up to ¾ oz. The coloration of the dorsal pelage runs from tan to blackish; the ventral pelage is lighter, from white to grayish-tan; the long 4" tail is sparsely furred, scaly, and dichromatic (dark above, light below). The body is cylindrical; the head profile is convex; the eyes are dark, small, and beady; the ears are large, broad, and erect; the muzzle is tapered and covered in tactile vibrissae; the limbs are short; the feet are small, delicate, and pink.

This terricolous murid is a palearctic native, making it, for us, an alien species. It originated in Asia, spread to Europe, then, as a stowaway on ships, was introduced to North America by early Spanish, French, and English explorers in the 16th and 17th Centuries. It is now one of the most widely distributed mammals in our region, occurring in all of the lower 48 U.S. states and most of Canada.

House mouse.

Largely a commensal mammal, it is usually never far from humans, inhabiting primarily anthropogenic structures and locations, such as houses, buildings, barns, orchards, and agricultural fields. Though it is capable of living in almost any type of temperate biome or ecosystem (including savannas, deserts, beaches, and forests), it is seldom found in wild areas—unless they are near human society. Quite social, it lives in hierarchical colonies of up to 500 individuals or more; the colony in turn is comprised of separate family groups, each one led by a territorial alpha male, who dominates his small harem and their offspring. This species shares burrows, tunnels, and houses with conspecifics, though it nests singly; will migrate en masse to new territory if local conditions deteriorate. Its home range is probably less than 500 sq. ft.

A cursorial nocturnal omnivore, it is first and foremost a granivore, with a strong affinity for seeds and grains. It also consumes leaves, roots, fruit, tubers, and nuts, along with woody material, such as stalks, bark, and wood. When available it will readily feed on carrion as well as insects (like caterpillars) and arachnids (such as spiders). Its ungrooved incisors are designed for quick, neat cutting—dental tools it skillfully uses while foraging and feeding; it also uses them, however, for more destructive habits, such as chewing up furniture, walls, and flooring. Hoards and caches surplus edibles, often in the same places humans store their food.

When indoors its polygynous breeding system takes place year-round. The prolific polyestrous female may produce over a dozen litters annually. She gestates for about 3 wks., gives birth to an average of 8 altricial pups in a hidden natal nest chamber, then weans them about 3 wks. later. The young become reproductively mature at just 5 wks. of age, helping to insure an abundant population. The father does not contribute to the rearing of his offspring.

Communication and perception occur via typical Muridae channels: sight, smell, hearing, and touch. Engages in gregarious allogrooming. Besides having extremely keen physical senses, it is an adept climber, swimmer, jumper, hider, and runner, characteristics that have greatly contributed to its success as a species. Its natural enemies are reptiles, birds of prey, mustelids, wild felids, and wild canids.

Beneficial as a pet, parasite host, consumer of pestilent plants, and a subject of scientific research; detrimental as a destroyer of cultivated crops, contaminator of granaries, ruiner of clothing, buildings, appliances, wiring, etc. (it is capable of initiating house fires), and a carrier of deadly diseases (such as bubonic plague, hantavirus, typhus, and salmonella). Even breathing its aerosol (exhaled air) can be dangerous to us. It has well-earned its species name: *musculus* is from the Sanskrit word *musha*, "the thief" (from *mush*, "to steal"). The house mouse probably does not live more than 1 yr. in the wild; perhaps 2 or 3 times that long in captivity.

House mouse.

HUMPBACK WHALE (*Megaptera novaeangliae*): The official state marine mammal of Hawaii, it grows to a total length of 55' and a weight of 45 tons. It has blackish baleen plates (with dark bristles) that grow to 3'; its small triangular dorsal fin lies back of center; its massive knobbed flippers stretch out to ⅓ the length of its body; it has double blowholes from which it can spout some 20' into the air.

Named for both the hump it makes before diving and the actual hump on its back, it is the only whale in this group with such a protuberance. It can also be easily identified by its scalloped pectoral flippers and tail flukes (often used to identify individual whales), which are also unique traits to this family. Gender dimorphism is present: the female is larger than the male.

Humpback whale.

An opportunistic carnivore, this cetacid feeds mainly on small schooling fish, but also on zooplankton, sieving the microscopic animals from the water through its baleen bristles. It possesses nearly 50 throat pleats, accordion-like furrows in the skin which expand and contract, aiding in the feeding process. The humpback may use various complex techniques during the hunt, such as entrapping its prey inside large foam circles (made by churning the water with its tail), or corralling its food inside massive bubble "nets" (made by releasing a cloud of bubbles from its mouth while swimming in a spiral up toward the surface).

For a rorqual, it has extremely thick blubber relative to its size (only the blue whale's blubber is thicker). Sometimes solitary, it normally forms pods and migrates back and forth to its feeding and breeding grounds with the seasons. One of the most familiar of all whales, the humpback is best known for its haunting "songs," complex vocalizations of repetitive pulsed and tonal sounds that it rearranges each year. The precise meaning and exact purposes of these songs are not yet known (though theories abound).

This natatorial migratory balaenopterid enjoys breaching, often leaping completely clear of the water, then twisting around in midair before crashing back into the sea. It inhabits all oceans, from the poles to the equator, from the coast to deep water. In our region it ranges the Atlantic Ocean from Canada southward. A shallow diver, it can swim 18 mph at top speed and hold its breath for up to 15 min.

Humpback whale.

Known as one of the more aggressive whales, if alarmed or threatened this gigantic megapterid may lash the water with its huge tail, a signal to keep away. It sometimes forms friendly associations with other cetaceans, such as minke whales. Females gestate for around 11 mos., giving birth to (normally) a single 15', 2-ton calf. The male does not assist in the rearing of its young. A secondary consumer, its primary natural predator is the orca. The humpback whale is believed to live around 50 yrs. in the wild.

IDAHO POCKET GOPHER (*Thomomys idahoensis*): Once believed to be conspecific with the northern pocket gopher, it is now classed as an individual species. Nonetheless, it possesses much of the same ecology and life history as its close cousin. The main difference is their ranges: the northern's extends over much of the American Northwest, while the Idaho's range is limited to the states of Idaho, Montana, Wyoming, and Utah.

Indiana myotis.

INDIANA MYOTIS (*Myotis sodalis*): Its scientific name means "mouse-eared comrade," and indeed it forms close bonds with other members of its colony, giving rise to its alternate common name: the "social myotis." This small chiropterid is nearly 4" in total length, its forearm is almost 2" long, its wingspan is nearly 11", and it weighs up to ¼ oz. Its soft fine fur is generally grizzled gray to brown in color; its underside is lighter than the upper body—excellent camouflage that helps protect it, both during its daytime rest periods and its nighttime hunting forays. Its tragus is short; its lips and nose may be reddish-pink; its calcar is keeled. This species is gender dimorphic, with the female being slightly larger than the male.

It mainly inhabits cool but humid limestone caves, but will also house under peeling tree bark and bridge beams, as well as in abandoned human-made structures (such as mines), particularly those situated near lakes, rivers, and streams. It lives only in the U.S., with a range covering New England, parts of the Midwest, and the upper South. Also known as the "Indiana bat," this highly gregarious vespertilionid migrates between summer and winter roosts, hibernating with thousands of other individuals in precise, tightly packed rows. It may travel as many as 400-500 mi. during the coldest months for this purpose, taking shelter in carefully selected hibernacula.

Feeding between sunset and sunrise, this crepuscular hunter searches in the dark for its prey, beetles, wasps, flies, and moths, using echolocation. Consuming up to half its body weight in insect pests each night, it greatly benefits humanity. It mates in the fall, but the female delays fertilization, gestating for around 65 days, giving birth to a single altricial pup in early summer. The polygynous male does not contribute to the rearing of his offspring. A secondary consumer, it has numerous natural enemies, including birds of prey, reptiles, and mammals. The Indiana myotis lives for 15-20 yrs. in the wild.

INSULAR VOLE (*Microtus abbreviatus*): This rodent is probably an individual species, one closely related to the singing vole. This taxonomic status has not been accepted by all, however. Also known as the "Hall Island insular vole," many systematists now consider it to be a subspecies (*Microtus abbreviatus abbreviatus*) of the singing vole.

Polar bears.

JAGUAR (*Panthera onca*): The biggest cat in North America and the third largest in the world (only the tiger and lion are bigger), this muscular mammal stands 30" at the shoulder, grows to a total length of nearly 8', and weighs up to 350 lb. Though its coat color can range from white and tan to brown and black (the latter is a melanistic form), it is typically a tawny yellow with black rosettes over the body that turn to spots on the head, legs, and tail. The base coat under its speckled chest, belly, and inside legs is usually cream-colored. These colors and markings serve as wonderful camouflage for life in its usual habitat: sun-dappled forest environments.

Jaguar.

It sports a lithe but compact body; a large head; massive eyes (in relation to body size, the largest of any carnivore) with round pupils and amber-yellow irises; small rounded ears; broad cheeks; robust square jaws; massive canines; a thick chest; large paws; and a medium-length spotted tail. Its cryptic pelage, short limbs, and powerful build reveal it to be designed as an ambush predator rather than a pursuit predator (like the African cheetah). Though when viewed in photos and illustrations it is sometimes confused with the African leopard (*Panthera pardus pardus*), this would not be possible in the wild since the two cats live on different continents. The jaguar is gender dimorphic: the male is larger than the female.

Its highly varied nearctic and neotropical habitats include jungle, rainforest, floodplains, warm deciduous forest, brushland, pampas grassland, swamp, mountains, lagoons, woodlands, rocky areas, mesquite scrubland, and sometimes arid regions. A riparian, it prefers biomes with abundant food, thick hiding cover, and at least one water source (slow-moving rivers are a favorite) in which it plays, bathes, and cools off, using the area as a hunting ground as well. Though during the Pleistocene it flourished north into what is now Washington and California and east to Tennessee, Maryland, and Florida, it is now rare in the U.S. and is seen only in southern Texas, New Mexico, and Arizona, and from there south into Mexico and Central and South America. Its home range or territory is usually about 15 sq. mi., but may extend up to 50 mi. A territorial beast, it marks off its turf with urine, dung, and tree-clawing.

This crepuscular stocky felid is a strict carnivore that feeds on eggs, fish, snakes, frogs, turtles, caiman, birds, rabbits, tapirs, peccaries, capybaras, monkeys, bighorn sheep, deer, and livestock (sheep, cows, horses). A solitary hunter, it will wade in water or swim to catch aquatic prey; it will sometimes fish from shore using its tail as a fishing lure. Its sharp teeth deliver a single crushing blow to the skull of its prey (piercing the brain); or it may deliver a suffocating bite to the neck. Its jaws have the most powerful bite force of any cat in the world, and it is easily able to crack open bones and cut through tough caiman hide and hard turtle shells with its conical saber-like teeth. It can drag, lift, and carry the carcasses of animals much bigger than itself. Caches surplus food.

Jaguar on the hunt.

A New World pantherine with Old World roots, it is believed to have crossed over from Eurasia into North America via the Bering land bridge during the Pleistocene. Its common name derives from the Tupi word: *jaguara*. It is the only feline in our region that roars, and is thus sometimes called the "roaring cat." Natatorial and cursorial, it is an excellent swimmer and runner (it can reach a speed of 50 mph for short distances). It generally rests in trees during the heat of the day, coming out in low light to hunt; it is, however, sometimes active during the day as well.

The polygynandrous jaguar has no set breeding season; that is, mating can occur anytime throughout the year. The female only mates once every 2 yrs. Males compete for mating rights during the females' short estrus cycle (1-2 wks. long), forming a temporary pair-bond with one another that ends shortly after mating. The female

Jaguar.

gestates for about 100 days, giving birth to 1-4 blue-eyed, altricial cubs in a secluded nesting den, usually located under tree roots or in caves, thickets, or rocky outcroppings. Weaning takes place at around 6 mos. of age, but the young remain with their mother for up to 2 yrs. or more (at which time she teaches them to hunt) before becoming independent and dispersing. The non-paternal male does not assist in the rearing of his offspring.

It communicates and perceives via standard Felidae channels: vocalizations (snarling, grunting, roaring, mewing, growling), body language, and scent. Both an umbrella species and a keystone species, this secondary consumer is at the top of the food chain and therefore has no natural predators. The elusive stocky jaguar lives 12-15 yrs. in the wild; up to 23 yrs. in captivity. Thus it has one of the longest life spans of any cat species.

JAGUARUNDI (*Puma yagouaroundi*): It stands about 14" at the shoulder and weighs nearly 20 lb. It grows to 55" in total length, its long thick panther-like tail (which helps it balance) making up nearly half its body length. While it is spotless and uniform in color, it has three distinct color phases: all black (to brownish), all gray (to grizzled), and all red (to tawny). Each of these three morphs can occur in a single litter, and in adulthood may be found in the same or different regions. With its elongated, slender, muscular body, small flattish head, long neck, short legs, and rounded ears, this unusual feline species has a sleek otter-like or weasel-like appearance,

Jaguarundi.

lending it the nicknames "otter cat" and "weasel cat." Males are larger and heavier than females, making it gender dimorphic.

This solitary terricolous feline is a riparian nearctic and neotropical native that inhabits rainforest, arid lands, thorny thickets, chaparral, deciduous woodland, swamp, scrubland, mesquite, coniferous woodland, grassland, and savanna, often near running water. Its natural range covers the southern U.S. (mainly southern Arizona and southern Texas), where it is rare, south to the tip of South America. Its home range can extend to 65 sq. mi., which it controls through scent-marking, ground-scratching, tree-scratching, and urination.

While this shy carnivore is actually an omnivore that includes leaves and fruit in its diet, its main source of nutrition comes from animals: arthropods (insects and crustaceans), fish, amphibians, birds, reptiles (lizards, whiptails, iguanas), and small and medium-sized mammals (mice, rats, guinea pigs, rabbits, opossums, armadillos, marmosets). A skilled tree climber and an expert swimmer, it will dive underwater for its food; will sometimes take poultry, but this problem is offset by its consumption of many animals that we consider agricultural pests. Athletic and built for stealth and speed, it can leap over 6' into the air and easily chase down prey whatever the habitat.

Closely related to the mountain lion and the African cheetah, the secretive jaguarundi is atypical in that, unlike many other cats, it spends most of its time on the ground and, being both diurnal and nocturnal, it is afoot at night and during the day (the latter period when it is most active). Though primarily solitary, males and females form short pair-bonds during the mating season.

Jaguarundi.

Almost nothing is known about its reproductive habits, and it seems to have no set breeding season. The iteroparous polyestrous female gestates for about 2 mos., giving birth to as many as 2 litters of 1-4 spotted, altricial kittens a year in a concealed nest, typically located in underbrush, a hollow log, or a dense thicket. The young are weaned by 3-4 wks. and probably become independent within a year or so. During this time the mother nurses, feeds (solid food), protects, grooms, and teaches her kittens hunting techniques and possibly survival skills. It is doubtful that the father plays any role in the rearing of his offspring.

A secondary consumer, its senses of sight, hearing, and smell are acute. Extremely communicative, its vocalizations include over a dozen sounds, from chirps, yaps, peeps, spits, and whistles, to cries, chatters, purrs, hisses, and screams. It also communicates via scent (urination, flehmen), body language, and touch (head- and neck-rubbing). The life span of the jaguarundi in the wild is unknown, but it may average 5-7 yrs.; perhaps twice that long (or more) in captivity.

Sea lions.

KEEN'S MYOTIS (*Myotis keenii*): Named after Anglo-Canadian missionary and natural historian John Henry Keen (who discovered it in the late 1890s), this medium-sized chiropterid is a member of the night bat group known as the Vespertilionidae family. It is 3½" in total length; its forearm measures 1½"; it has a 10" wingspan; and it weighs ⅜ oz. It has long ears; a long narrow tragus (pointing up); and its calcar is not keeled. Countershaded, it has silky dark brown fur on the dorsum, light brown fur on the ventrum. It so closely resembles other bats in its class that in many cases only a detailed scientific examination (of skull, dentition, DNA, etc.) could definitively determine this species.

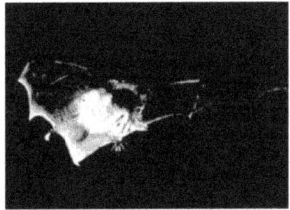

A member of the *Myotis* genus.

A secondary consumer, it inhabits the dense coastal forests of the Pacific Northwest (from Alaska south through British Columbia into Washington), roosting in caves, hollow trees, mines, attics, and snag, as well as under bridges, tree bark, and roofs. Does not migrate, but hibernates in special caves (known as hibernacula) in the fall. Both crepuscular and nocturnal, this insectivore actively hunts its prey (spiders, midges, flies, moths) between sunrise and sunset.

The female gestates for about 50 days, giving birth to one altricial pup a year in the summer. In the maternity roost she nurses (with protein-rich milk), grooms, cares for, and protects her offspring, teaching it to fly and echolocate until it is ready for independence about 5 wks. later. This social vespertilionid shares its roosts with other bat species. A beneficial pest controller, its natural enemies are reptiles (snakes), birds of prey (hawks), rodents (mice), and small mammals (cats and racoons). Keen's myotis may live to about 10 yrs. in the wild; 20 yrs. in captivity.

Kit fox.

KIT FOX (*Vulpes macrotis*): Looking somewhat similar to a gray fox in miniature, it is 1' tall, 20" in total length, and can attain a weight of up to 5 lb. Its coat is reddish-yellow-gray, and often grizzled. The brushy tail is black-tipped and a small black facial swash often runs from the lower inside eye forward along the snout. The ventrum (from throat to rump) is white to cream-colored; the eyes are dark amber; the nose is black; the pelt is coarse and dense. Its robust head, large eyes, massive bat-like ears, pointed concave muzzle, small lean body, long legs, and dainty paws give it a distinctive appearance.

Though it is the smallest fox species (about the size of a domestic house cat), its ears are the largest of all the foxes. Lined with numerous blood vessels that lie near the skin's surface, their size not only increases its hearing capacity, but they also act as heat-radiators, helping to keep it cool in the hot regions it occupies. The kit fox should not be confused with its close cousin the swift fox, which is a larger, separate, distinct species that lives further east (the two sometimes interbreed). Gender dimorphic, the male is larger than the female.

Active year-round, this terricolous sedentary canid is a nearctic native whose habitat affinities include warm open desert, prairie, playa, alkali sink, dune, mixed-grass shrubland, grassland, sagebrush, savanna, foothills, salt flat, rangeland, piñon-juniper woodland, chaparral, both rural and urban areas, and plant communities characterized by shadscale, saltbrush, greasewood, sagebrush, and creosote. Its range extends across the western U.S.—from the Pacific Coast (Oregon, California) east to Idaho, Nevada, Utah, and Colorado, and south to Arizona, New Mexico, Texas, Baja California, and Mexico. Establishes a home range of 1-4 sq. mi.

An opportunistic carnivore with omnivorous inclinations, its diet consists of seeds, fruit, cacti, insects, fish, reptiles (snakes, lizards), birds, shrews, rodents (mice, rats, voles, squirrels, gophers), pikas, rabbits, and carrion; its preferred prey is the prairie dog; hoards and caches surplus food. Though nocturnal and crepuscular, it is commonly seen on cool days. When the temperature rises, however, it dens up and waits for dusk, when it leaves its burrow to forage, scavenge, and hunt. It may be solitary or social, depending on the region, climate, and available food.

Semifossorial, it usually constructs its own den sites, but will also use human-made structures, or take over the burrows of other animals, often enlarging and customizing them. When these are abandoned they provide homes for medium-sized animals, making the kit fox a mutualist species. Dens are dug in soft soil and have multiple portals and chambers.

Kit fox.

Its breeding season runs from early to midwinter. The dog (male) and vixen (female) typically form a lifelong monogamous bond. After mating, the monestrous female gestates for roughly 2 mos., giving birth to 1-6 altricial pups in an underground whelping den in late winter to early spring. Weaned at 2 mos., the pups attain independence and disperse at around 6 mos. of age. Unlike most other mammalian fathers, the male kit fox participates in the feeding and protection of his offspring.

Uses typical communication and perception channels endemic to the Canidae family: sight, body language, vocalizations, scent, and touch. The smallest wild dog in our region, the natural enemies of this secondary consumer include birds of prey, badgers, bobcats, coyotes, and other fox species. The kit fox probably lives 3-5 yrs. in the wild; it may live 10-15 yrs. in captivity.

LABRADOR COLLARED LEMMING (*Dicrostonyx hudsonius*): Also known as the "Ungava collared lemming," it shares much of the ecology and life history of its close cousin the arctic lemming. One difference is that the Labrador has a much more limited range, occurring in only 2 northeastern Canadian provinces: Quebec and, as its common name indicates, Labrador.

LEAST CHIPMUNK (*Neotamias minimus*): The smallest of our chipmunks (hence both its common and scientific names), it grows to a total length of nearly 9", its hind foot measures about 1½", and it weighs around 2 oz—the weight of about 8 pieces of typing paper or 24 U.S. dimes. Pelage coloration may change with geographic location, but generally it is dullish cinnamon-brown on the upper body, whitish-gray on the lower body. Lighter in color than other western chipmunks, it has 5 distinct dark dorsal and lateral stripes, alternating with 4 whitish stripes. The median stripe, on the spine, is the darkest and longest. Sharply defined facial stripes mimic its dorsal stripes. Outer legs may be grayish. The 4" tail is bushy and bicolored (grizzled tan on top, lighter on the bottom). By breaking up the outline of its body, this cryptic protective coloring provides superior camouflage in its habitat, where sticks and branches cast line-like shadows. Gender dimorphism is present, with the female being larger than the male.

This tiny petraphilic sciurid is a nearctic native that favors subalpine taiga; that is, open, boreal coniferous forests. It is particularly fond of tree islands and woodland edges. It is also found in and around rocky locations, tundra, river banks, scrubland, moist pinelands, aspen woods, cliffs, meadows, and sage deserts. Its range is extensive, covering much of the northern and western portions of North America, from Canada in the north to New Mexico in the south, from Washington and California in the west to Wisconsin and Michigan in the east.

Least chipmunk.

Builds several different types of houses, depending on the season: summer homes are constructed of light woody materials in trees; winter homes are more substantial and are manufactured underground. With the start of cold temperatures (October) it enters hibernation in its subterranean nest; actually a state of torpor, it experiences frequent arousals, at which times it rises to forage outside when weather permits. Territorial and sedentary, it establishes a home range of up to 10,000 sq. ft., which it defends aggressively against conspecifics.

A diurnal solitary omnivore, its diet centers around seeds, grasses, tubers, flowers, nuts, fruit, roots, leaves, and fungi; these are supplemented with the occasional worm, insect, bird, and small mammal. Also known to consume carrion, it hoards and caches surplus food, transporting it to its den in its expandable cheek pouches.

Its polygamous breeding season begins in the spring with the awakening of hibernating males, who are quickly followed by the females. Bucks compete over estrus does, who gestate for about 1 mo., then give birth to as many as 7 altricial pups in a well concealed, grass-lined, maternal nest chamber. Weaning and dispersal occur at about 2 mos.

Communicates and perceives via normal Sciuridae channels: sight, smell, hearing, and touch; it is known for its bright, fast, repetitious "chip," "chip," "chip" call. Though a terricolous mammal, it is also somewhat arboreal and is a proficient tree climber; its tail is held vertically while running. Its natural enemies are birds of prey, reptiles, mustelids, wild felids, and wild canids. Benefits its ecosystem by aerating and recycling the soil, hosting parasites, dispersing seeds, and providing food for carnivores. The least chipmunk probably lives an average of to 2-3 yrs. in the wild; twice that long in captivity.

LEAST SHREW (*Cryptotis parva*): The body of this nearctic soricid is around 3" long. Its tail adds another ⅞", making it nearly 4" in total length. Its hind foot measures ½" and it weighs up to ¼ oz. This makes it one of our region's most diminutive mammals. Its coat is countershaded, with the upper body a grizzled gray-brown, the lower body light gray to tan. Smaller than a man's thumb, its head is conoidal, its nose is pointed, its eyes are small, and its short tail is dark on the dorsum,

Least shrews.

light on the ventrum. Its ear holes are concealed beneath its short, thick fur, thus it does not have visible external ears. Its tiny nose and feet are delicate and pinkish.

It inhabits grassy areas, scrubland, fields with tall brush, marshes, cropland, savanna, moist woodlands,, and forests with rotting logs. Aboveground it uses the pathways of mice, rats, and other rodents; underground it digs its own shallow runways and burrows, often tunneling beneath logs and boulders. Its complex burrow systems are excavated a few inches under the surface and may extend up to 5' in length. Its geographical range covers much of the midwestern and eastern U.S., from North Dakota and the Great Plains east to New York, and south to Texas and Florida, extending further south to Mexico and Central America.

An aggressive carnivore, its diet is comprised mainly of beetle and moth larvae, snails, earthworms, centipedes, sow bugs, spiders, grasshoppers, and crickets. Also a folivore and a fungivore, it sometimes devours leaves, fungi, and other vegetable matter. Will attack and even appropriate beehives in order to feed on bee larvae—lending it the misleading nickname the "bee mole." Possessed of an extremely high metabolism, it must eat almost constantly, which is why it is diurnal, crepuscular, and nocturnal: using its stiff haptic whiskers to help detect its prey, it spends most of its waking hours hunting various microhabitats, such as beneath litter layers and soft soil, consuming its own body weight every 24 hrs. Probably caches surplus food; known to cannibalize its fellows if necessary.

Far more social than many other shrews, it will sometimes establish colonies of several dozen individuals who live, hunt, work, and sleep together. Its breeding season runs throughout much of the year, from winter to fall. The female gestates for around 3 wks., giving birth to as many as 6 altricial young in a carefully hidden nest chamber lined with shredded vegetation. She weans them at 3 wks. of age, at which time they become independent. The male assists in the rearing of his offspring by guarding the nest chamber.

Also known as the "North American least shrew," this tiny secondary consumer communicates with conspecifics using typical Soricidae channels: sight, touch, smell,

and sound. One of the more vociferous of its kind, like some other shrew species its many vocalizations include high-pitched squeaks (ultrasonic sounds), a bat-like form of echolocation that it uses to help navigate in the dark, and also possibly to hunt prey. Its natural enemies are reptiles (snakes), birds of prey (owls, hawks), and small to medium-sized mammals (skunks, foxes). Like other fossorial mammals, its burrows help aerate and recycle the soil while providing homes for other creatures. It benefits us more directly by consuming a host of harmful insects. The least shrew probably lives about 1 yr. or less in the wild.

LEAST WEASEL (*Mustela nivalis*): With a total length of only about 7" (including its tail), a hind foot only ¾" long, and a weight of only 1½ oz., this tiny, slender mustelid is the smallest carnivore in our region (the origin of its common name). It has a long tube-shaped body; a small head; long neck; tapered snout; large dark eyes; short tail; long whiskers; small rounded ears; and short limbs. Protruding eyes give it a wide arc of vision. In the summer the dorsal pelage of northern residents is a rusty-cinnamon brown while the ventral pelage (from throat to back legs) is white or cream. During the winter its coat turns completely white. Southern residents, however, tend to remain brown above and white below all year. Gender dimorphism is present, with the males being larger than the females.

Terricolous, sedentary, and motile, the habitat of this nearctic native is quite varied and includes a number of widely differing biomes and ecosystems: grassland, brushland, scrubland, woodland, marshland, desert, sand dunes, rocky areas, prairie, taiga, tundra, rainforest, and mountainous regions. Its circumboreal range covers Alaska, most of Canada, and the American Midwest into Appalachia. This petraphilic cousin of the mink is comfortable both below ground and under snow. It uses the dens of other animals, preferring to hole up in microhabitats like tree bases, in grass clumps, in rock piles, and under logs, tree roots, stumps, and the foundations of barns, carefully lining each burrow with feathers and fur.

Least weasels.

Has a home range of about 2 acres, which it scent-marks (with a noxious smelling secretion emitted from its rump glands) to ward off intruders. Males agonistically defend their turf against other males.

Though minuscule, it is a fearless and vicious predator that can take down prey nearly twice its size. An ambush hunter, it relies on various microhabitats, such as leaf litter and dense vegetation, as a cover from which to surprise its favorite quarry, mice. After springing upon its victim, it despatches it with a quick bite to the neck. Its tiny jaws hold 34 razor sharp teeth, as well as powerful muscles for clamping down on its prey. Few animals escape: it can run 15 mph, is able to leap several feet into the air, and its lithe, highly flexible body allows it to quickly turn around in even the smallest spaces.

Both diurnal and nocturnal, this voracious carnivore and opportunistic scavenger also feeds on invertebrates, fish, frogs, birds, eggs, shrews, voles, rabbits, moles, carrion, and particularly lemmings (with whose density its own population rises and falls). Stockpiles and caches surplus food.

It has no set breeding season, thus polygynandrous mating can take place at any time throughout the year. Depending on the availability of prey, the polyestrous doe may bear up to 3 litters a year by different bucks. She gestates for about 35 days, giving birth to an average litter of 6 altricial kits. Weaning starts at about 1 mo.; independence and dispersal begin at around 2 mos. of age. The male is non-paternal and therefore does not contribute to the rearing of his offspring.

Its means of communication and perception are standard for most members of the

Mustelidae family: sight, touch, hearing, and smell. Vocalizations include hissing, screeching, shrieking, chirping, twittering, whistling, barking, purring, trilling, squeaking, squealing, and snarling. Benefits humans (in particular farmers) by acting as an efficient rodent killer and overall pest controller. Solitary, secretive, excitable, and highly energetic (it is always on the move), the natural enemies of this secondary consumer are birds of prey, reptiles, minks, wild felids, wild canids, and fellow weasel species. The least weasel lives an average of 18 mos. in the wild; 6-8 yrs. in captivity.

LESSER LONG-NOSED BAT (*Leptonycteris yerbabuenae*): Also known as "Sanborn's long-nosed bat," it grows to a little over 3" in total length, its forearm is just over 2" long, its wingspan measures around 10", and it weighs about 1 oz. Its upper body is reddish-brown in color, its underbody is light brown. Has large eyes (for increased night vision); a minuscule tail (that is often difficult to observe); and an arrowhead-shaped nose-leaf (which helps amplify and detect returning sonar echoes).

This small phyllostomid inhabits mines, caves, rock shelters, abandoned buildings, trees, and shrubs in the arid desert, scrubland, woodland, and oak-forested regions of southern Arizona and southwest New Mexico, south to Mexico and beyond. Prefers locations with agave plants and cacti; is known to haunt streams and lakes.

A nocturnal omnivore, insectivore, frugivore, and nectarivore, it drops from its perch after sunset, flying up to 20 mi. a night to feed on insects (making it a significant pest controller), fruit (making it a significant seed disperser), and the nectar and pollen of night-blooming flowers (making it a significant plant pollinator). As it hovers over a flower bloom, its lengthy 3" tongue (tipped with hairlike papillae) extends from its long muzzle (the source of its common name), lapping up its food. It does not need to drink (or be near) water since it is able to procure all of the fluid it needs from its fruit and nectar diet.

Its daytime roost may consist of tens of thousands of individuals. Females gestate for roughly 6 mos., give birth to a single altricial pup each year, and care for it in large maternity colonies. The male does not assist in the rearing of his offspring. Restricted to the warm South, it does not hibernate or enter torpor. Shares its daytime roost with other bat species. A long-distance flyer, it

Southwestern cave: lesser long-nosed bat habitat.

migrates south to Mexico in the fall, back north in the winter, reaching speeds of up to 15 mph during flight. A secondary consumer, its natural predators are snakes, owls, and bobcats. The lesser long-nosed bat may live 10-20 yrs. in the wild.

LITTLE BROWN BAT (*Myotis lucifugus*): Also known as the "little brown myotis," it is nearly 4" in total length, its forearm measures about 2", it weighs ½ oz., and it can attain a wingspan of nearly 11". Employs countershading: the coloring on the upper body is golden-brown to reddish-brown, on the underbody it is lighter, from buff to white. Its fur has a somewhat iridescent sheen; the snout is bluntish; its ears are small and pointed; its tragus is short and rounded; the calcar lacks a keel.

This extremely common North American vespertilionid is cosmopolitan in its habits, and is found from desert to grassland, from forests to mountains; from rural to urban, and from arid to riparian environments. It inhabits caves, mines, wood piles, trees, rock ledges, and the roofs and attics of buildings, ranging over most of our region, from Alaska and Canada south through the American West, upper Midwest, and Northeast, south into Mexico. (Its current range does not include much of the American Southeast, the plains, and southern California.)

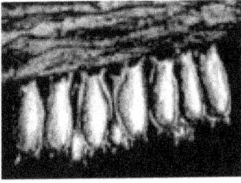

Little brown bats.

It shelters in different types of roosts depending on the time of day or night, the weather, and the season. Nonterritorial, little brown bat colonies may contain several hundred thousand individuals. A crepuscular and nocturnal hunter, this energetic little troglophile leaves the roost at sunset to forage over water, using echolocation to search out its main prey, aquatic insects, which it trawls by using its wings and tail membrane as a "net." It also feeds on non-aquatic insects such as mosquitos, gnats, moths, wasps, midges, and beetles, eating up to 1,200 bugs an hour. Consuming half or more of its body weight each night in arthropods makes it highly beneficial to humans as a pest controller.

Communicates with conspecifics via sight, hearing, touch, and smell. It hibernates in warm humid roosts from fall to spring, sometimes waking up during brief thaws to fly about outside. Swarms cave entrances. The female gestates for around 55 days. Then, while hanging right-side up, she gives birth to a single altricial pup which clings to her as she flies. A secondary consumer, it has numerous carnivorous enemies, including birds of prey, reptiles, wild felids, and other mammals, such as rodents, weasels, racoons, martens, and fishers. The little brown bat may live as long as 10 yrs. in the wild; up to 30 yrs. in captivity.

LITTLE POCKET MOUSE (*Perognathus longimembris*): In its range it is the smallest of our pocket mice, growing to a total length of slightly less than 6", with a hind foot that measures about ¾", and a weight of only ¼ oz. or less. The soft dorsal pelage is buff with a gray or blackish wash; the ventral fur varies from white to tan; the long 3⅜" tail is monocolored: buffy-brown. The body is ovoid; the head profile is convex; the eyes are dark and beady; the ears are small and erect; the muzzle is tapered and covered in vibrissae; the forelimbs are short and small; the hind limbs are long; the paws are small and delicate. At the base of each ear there is a small white subauricular patch; it lacks the bristly guard hairs found in some other pocket mouse species.

Pocket mouse.

This small heteromyid is a nearctic native that inhabits temperate open chaparral, shrubland, grassland, coastal sageland, dry deserts, rangeland, and valleys with sandy gravelly soil, often dotted with rough brush, cactus, and creosote bush. Its narrow range is restricted to portions of the western states of Oregon, California, Idaho, Nevada, Utah, and Arizona. Constructs underground burrow systems where it rests, shelters, sleeps, feeds, nests, estivates, and hibernates.

A solitary nocturnal herbivore, its diet is comprised mainly of seeds; also consumes general green plant matter when available; caches surplus food in its burrow larder, transporting it in its fur-lined cheek "pockets" or pouches (from whence it derives its common name). Its breeding season runs year-round, cresting in spring. The female gestates for 3-4 wks., giving birth to an average of 5-6 altricial pups in an underground natal nest; she weans them 30 days later.

Communicates and perceives via normal Heteromyidae channels: sight, smell, touch, and hearing. Dislikes extremely hot or cold weather; in winter it hibernates for over 6 mos. to conserve energy. Its natural enemies are reptiles, birds of prey, mustelids, wild felids, and wild canids. The little pocket mouse may live for an average of 2-4 yrs. in the wild; longer in captivity.

LODGEPOLE CHIPMUNK (*Neotamias speciosus*): This retiring rodent grows to a total length of nearly 10", its hind foot measures almost 2", and it attains a weight of up to 2 oz. It is vivid cinnamon-brown on the dorsum and sides, grayish-brown on the outer legs, top of head, and hindquarters (and sometimes the shoulders), and tan, gray, or white on the ventrum. The number of dark and light alternating dorsal and lateral stripes may vary from individual to individual; dark dorsal stripes are not as distinct as in other species; however, its outermost lateral stripes are bright white. The 4" tail is bushy and grizzled rusty-gray, with a blackish tip on the underside. The facial mask is comprised of prominent alternating black, brown, and white stripes that mimic those on the back and sides; each eye and each ear has a black mark in front of it and a white mark behind it. Gender dimorphism is evident: the female is larger than the male.

This medium-sized sciurid is a nearctic native that, as its common name denotes, inhabits montane stands of lodgepole pine. It also thrives in chaparral, meadows, shrubland, and mixed woodlands of fir and oak, particularly areas rich in manzanita (an evergreen shrub found along the Pacific Coast). Prefers areas with plenty of tree cavities, snag, logs, rocks, deadfall, forest detritus, blowdown, and pliable soil. It has an extremely limited range, extending only over small portions of eastern central California and western Nevada.

Estivates during extremely hot weather; enters a state of torpor (or what some consider a form of hibernation) during the winter months, from which it frequently awakens to forage; digs burrows for safety, resting, and maternity nesting. Establishes a home range of about 1,000 sq. ft., which it readily defends against conspecifics and other intruders.

A western chipmunk subspecies.

A sedentary diurnal omnivore, its diet is comprised of the usual chipmunk fare: seeds, nuts, fruit, forbs, grains, leaves, pollen, flowers, fungi, woody material, insects, non-insect arthropods, birds' eggs, birds, small mammals, and human refuse. A hoarder, it caches surplus food, carrying it to its larders (usually in ground holes or under rocks and logs) in its large cheek pouches.

Its polygamous breeding season takes place in late spring shortly after emergence from its winter sleep. The doe gestates for around 1 mo., giving birth to as many as 6 altricial pups, which she weans within 20-30 days. The buck does not seem to provide any care in the raising of his offspring.

Communication and perception occur via standard Sciuridae channels: sight, hearing, smell, and touch. Vocalizations include chipping, chattering, whistling, and chucking. Like others of its kind, it engages in tail-twitching as a visual display when alarmed or during courtship. Both terricolous and scansorial, it is a proficient runner and tree climber with terrestrial and arboreal abilities. Benefits its ecosystem by acting as a parasite host, a seed, spore, and pollen disperser, and as a prey base for carnivores. As a predator itself, it helps keep the populations of smaller creatures in check. May be harmful to humans as a carrier of diseases like rabies, the plague, and possibly the hantavirus and salmonella (even its aerosol may be hazardous). Its natural enemies are reptiles, birds of prey, mustelids, wild felids, and wild canids. Rarely seen due to its timid nature, the lodgepole chipmunk lives an average of 1-2 yrs. in the wild; perhaps twice that long under optimal conditions.

LONG-BEAKED COMMON DOLPHIN (*Delphinus capensis*): One of two species of common dolphin (the other is the short-beaked common dolphin), it grows to a total length of around 8' and can weigh up to 300 lb. It shares much the same ecology and life history as its close cousin. This delphinid has a thinner more gracile body contour, however, with markings and colorations that are not as prominent. Additionally, the long-beaked has a more limited range than the short-beaked, extending only along the coasts of California and Mexico.

LONG-EARED CHIPMUNK (*Neotamias quadrimaculatus*): This medium-sized reddish chipmunk grows to a total length of about 10" and can attain a weight of up to nearly 4 oz. As with most other members of this genus, it is rusty-brown on the upper body, whitish-gray on the lower body. Summer coloration is brighter than winter coloration. The body striping (5 dark stripes, 4 light ones), however, can be somewhat indistinct. Its premier distinguishing features are its long, pointy, erect, slightly tufted 1" ears, from which it derives its common name. Additionally, each ear possesses a dark subauricular stripe and a white postauricular patch. Sharply defined facial stripes mimic the dorsal and lateral stripes; the 5" tail is brushy and washed with grizzled gray-brown fur on the dorsum, buff on the ventrum. Gender dimorphism is present: the female is larger than the male.

This pert sciurid is a nearctic native that inhabits temperate pine forests. It is also found in mixed woodlands, chaparral, edgeland, scrubland, and brushland, preferably with rocks, snag, deadfall, decaying logs, forest floor detritus, blowdown, and healthy thicket growth. Its range is very limited and is restricted to small portions of eastern central California and western Nevada. May nest above or below ground. Enters hibernation with the onset of cold temperatures (October); experiences frequent arousals, at which time it may feed on stored food supplies or journey above ground to forage—particularly on mild winter days.

A sedentary diurnal omnivore, its diet consists of seeds, fruit, grains, flowers, nuts, pine cones, fungi, and insects. Forages both on the ground and also in shrubbery and trees. Leaves behind kitchen middens of empty pine cones and seed husks on the ground below its feeding trees. Caches surplus edibles in preparation for periods of cold and food shortages.

Its polygamous breeding season takes place in early spring. Males emerge from hibernation before females and compete with one another for mating rights, often engaging in violent encounters. The doe gestates for about 1 mo. and gives birth to 3-4 altricial pups in a leaf-lined maternity chamber set up in a subterranean burrow; she weans them about 30 days later. The buck seems to play no role in the raising of his offspring and, as is true of nearly every North American mammal, the nuclear family of this rodent species is made up solely of the mother and her young.

Communication and perception fall along standard Sciuridae channels: sight, hearing, smell, and touch. Engages in tail-twitching and various body postures when alarmed or excited, and during courtship. Its natural enemies include reptiles, birds of prey, mustelids, wild felids, and wild canids. Benefits its ecosystem by aerating and recycling the soil, hosting parasites, dispersing seeds, providing food for carnivores, and checking the populations of the flora and fauna it feeds upon. Commonly seen perched on a stump or darting along a log, the long-eared chipmunk lives an average of about 1-2 yrs. in the wild; 3-4 yrs. under optimum conditions.

LONG-EARED MYOTIS (*Myotis evotis*): It grows to nearly 4" in total length, its forearm measures about 2", its hind foot is about ½" long, it has a wingspan of 10", and it weighs around ¼ oz. Its tragus is long and narrow; its calcar often has no keel; and at 1" in length it has the longest ears of any myotis species in our region. Its dorsum is grizzled cinnamon to yellowish brown; its ventrum is lighter in color; the ears are dark brown to black. Externally it so closely resembles other members of the *Myotis* genus (such as *Myotis keenii*) that only a detailed scientific examination (of, for example, its teeth) can positively identify it.

This petraphilic vespertilionid roosts in rocky outcroppings, cliffs, caves, snag, downed trees, deadfall, tree stumps, mines, blowdown, ground crevices, abandoned buildings, bridge beams, and standing trees, preferring riparian zones in coniferous forests (whether coastal or mountainous). It ranges from the Dakotas and Wyoming to

Long-eared myotis.

British Columbia, and south to Arizona and New Mexico, hence one of its other common names: the "western long-eared myotis."

A beneficial, crepuscular, nocturnal, slow-flying but highly maneuverable insectivore, it uses both its sensitive hearing and echolocation to track down its prey, moths, beetles, termites, flies, and spiders, hunting mainly at sunrise and sunset when insects are most active and abundant. Besides being able to perform the usual aerial acrobatics of its kind, it can also hover and is able to snatch up both airborne and stationary creatures—feeding techniques known as "hawking" and "gleaning" respectively.

As with a number of other vesper bats, though it mates in the fall, the female delays fertilization until spring when conditions are more favorable for parturition and raising young. Altricial pups are born in early summer and cared for by their mothers in large maternity roosts. Animals that prey on this secondary consumer include reptiles and small and large mammals. The life span of the long-eared myotis is not known; it probably lives for 10-20 yrs. in the wild.

LONG-FINNED PILOT WHALE (*Globicephala melas*): Once known as the "common pilot whale," both this name and its current common name are misleading, for it is not a whale, it is a dolphin. It grows to a total length of 28' and weighs up to 6,000 lb. (3 tons). It is unique among dolphins in that it has both a bulbous globe-like forehead and is nearly all black. Another of its outstanding characteristics are its pectoral flippers, which give it its common name: long, narrow, and pointed, they can reach one-fifth of the animal's body length.

As noted, the head bulges; it is also rounded on top and flat at the front, presenting a rather squarish appearance. Has a short blunt beak and spouts some 5' into the air from its single blowhole. Concerning coloration, the dark gray to black body has several lighter patches, one behind the eyes, one on the throat, and one on the underside. Its body is

Long-finned pilot whale.

cylindrical and sinuous, with a curved, triangular dorsal fin set slightly forward of the center of the back. Its genus name refers to one of its most distinctive and already mentioned features: *Globicephala* means "sphere head." Gender dimorphism is present, with the male being larger than the female.

Like most other delphinids, the long-finned is both social and energetic. It spends most of its life traveling in large pods, or even superpods (of 1,000 or more), in the deep cold waters of the world (though some populations prefer temperate coastal waters). In our region it inhabits the coast of the Atlantic Ocean.

Mainly a nocturnal hunter, this natatorial nomadic carnivore uses echolocation to find its prey, cephalopods and fish, eating up to nearly 100 lb. of food a day. Can dive to nearly a ½ mile and hold its breath for about 15 min. Communicates using pops, whistles, and clicks. This secondary consumer is often confused with its close cousin, the short-finned pilot whale. Strandings are not uncommon, as this closely bonded species will unhesitatingly follow its pod members into dangerously shallow waters. The male long-finned pilot whale may live up to 50 yrs. in the wild; the female's life span can reach 60 yrs. or more.

LONG-LEGGED MYOTIS (*Myotis volans*): This large bat—whose scientific means "mouse-eared" and "flying"—grows to over 4" in total length, its forearm is nearly 2" long, its hind foot measures about ½" long, and its wings extend to around 11". It has a short snout; the tragus is long and slender; the ears are shortish and rounded; its upper body is brown, reddish, or grizzled gray; and its underbody is buff to whitish, sometimes with dark sections. Unusual for a member of the genus *Myotis*, its interfemoral membrane is furred from knee to elbow, while its calcar has a conspicuous keel. Its extra long shinbone (tibia) gives it its common name: long-legged myotis.

A secondary consumer, in the warmer months it shelters in trees, rock crevices, buildings, and along waterways in forests throughout its range: western North America, from Alaska through British Columbia, south to Mexico. It migrates and hibernates.

A nocturnal hunter, this widespread vespertilionid comes out at night, and, using echolocation, tracks down airborne prey (mainly moths) as it flits through the forest. In a female-only nursery made up of hundreds of other pregnant mothers, the female bears one pup in early summer. By eating creatures we consider pests, this insectivorous slow-flying bat is one of our great mammalian benefactors. The long-legged myotis has been little studied, and thus not much else is known.

LONG-TAILED POCKET MOUSE (*Chaetodipus formosus*): This medium-sized pocket mouse grows to a total length of 8½", its hind foot measures ¾", and it can attain a weight of just under 1 oz. The upper body coloration is buffy with a gray to blackish wash; the lower body is white with golden-tipped hairs; the long 4½" tail is bicolored (dark above, light below) and has both a distal crest and a terminal tuft; lacks the bristly rump hairs found in many other similar species.

This terricolous heteromyid is a nearctic native that inhabits arid desert scrubland with abundant sagebrush and hardpacked rocky soil. Also occurs in and around valleys, bare rocky outcroppings, lava beds, boulder fields, canyons, sand dunes, scree, mesquite zones, talus slopes, and open sandy wasteland, as well as along dry river beds and pebbly stream beds. Its range is restricted to the western part of the U.S., extending over portions of the states of California, Nevada, Utah, and Arizona, and south into northern Mexico.

Arid Arizona hills: long-tailed pocket mouse habitat.

Fossorial, it excavates a network of tunnel systems under shrubbery, with special sections for nesting, sleeping, storage, and waste. May enter a state of torpidity during extremely cold weather; probably estivates during very hot weather. Solitary, it establishes a small home range which it vigorously guards against conspecifics.

The diet of this sedentary nocturnal omnivore consists of seeds, grains, fruit, nuts, leaves, stems, general plant matter, and insects. Does not require water as this is metabolized from the succulent solid food it consumes. Caches surplus edibles, transporting it in its external cheek pouches or "pockets." Its breeding season runs from spring to early fall. The female gestates for roughly 1 mo., giving birth to an average of 3-4 altricial pups in the natal nest chamber.

Communication and perception take place via normal Heteromyidae channels: sight, smell, touch, and hearing. Its natural enemies are reptiles, birds of prey, and wild canids. Beneficial as a seed disperser, parasite host, soil aerator and recycler, and as a prey base for carnivores; detrimental as a crop destroyer and carrier of deadly diseases. The long-tailed pocket mouse probably lives an average of 1-2 yrs. in the wild; twice that long under optimum conditions.

LONG-TAILED SHREW (*Sorex dispar*): It grows to a total length of 5", its thick 2½" tail makes up half its body length, its hind foot measures ⅝", and it weighs as much as ¼ oz. Generally longer and more slender than other shrews, its coat is dark slate gray year-round; the venter is sometimes lighter in color. The head is long and cone-shaped; the eyes are tiny; the ears are hidden under its fur; the elongated snout is pointed and whiskered; the teeth are narrow, sharp, and rust-brown in color; the limbs are small; the feet are delicate. It displays gender dimorphism, with the male being slightly heavier than the female.

A rare solitary creature that is both diurnal and nocturnal, this riparian mammal inhabits mainly cool moist forests (deciduous and coniferous) in rugged mountainous regions, usually near streams and bogs with a predominance of roots, humus, moss, leaf litter, and leaf mold. Extremely stenotopic and petraphilic, it is found almost solely near or in deep dark crevices on rocky slopes and sheltered canyons, in loose scree and talus fields, and around boulders and rockslide debris. This preference has earned it the alternate common name, the "rock shrew." Its range extends from northeastern Canada down through the Northeastern states, south to Tennessee, North Carolina, and Georgia, with some density in the Appalachian Mountains. It establishes a home range of about 1-3 acres, violently defending it against conspecifics.

Shrew belonging to the *Sorex* genus.

In addition to being extremely active, the minuscule body of this asocial soricid has a small surface to volume ratio, both traits requiring it to feed round the clock in order to maintain body heat and meet its high daily energy requirements. A carnivorous invertivore that is primarily an insectivore and a vermifore, it can eat twice its body weight each day, consuming a myriad of prey, from spiders, crickets, and beetles, to earthworms, centipedes, and flies; will eat vegetable matter when necessary; it risks dying if it goes without food for more than a few hours.

Difficult to study in its natural habitat, we know only a few details about its mating system. Its breeding season runs from spring through summer. Like many other types of shrews, the polyestrous female of this short-lived species bears several litters a year. The gestation period is unknown. She gives birth to an average of 2-5 altricial pups in a nest chamber fitted out with dried cut grass. The young are weaned within a month or so, becoming independent by 8 wks. of age. The male does not appear to help in the rearing of his offspring.

A secondary consumer, its natural enemies include reptiles, birds of prey, and small mammals, such as mustelids. It benefits humans by consuming large quantities of pestilent insects. The life of a long-tailed shrew is fast and furious, with an average duration in the wild of 1-1½ yrs.

Long-tailed vole.

LONG-TAILED VOLE (*Microtus longicaudus*): It grows to a total length of nearly 11", its hind foot measures about 1", and it can reach a weight of up to 3 oz. Its pelage is countershaded: buffy-tan with a grayish-blackish wash on the dorsum, whitish-gray on the ventrum; the body is stout and ovoid; the eyes are smallish, dark, and beady; the ears are large and furry; receives both its common name and its species name from its long 5" tail (*longicaudus* is Latin for "long tail"), which is bicolored (dark on top, light on bottom) and sparsely furred.

This reticent, riparian, terricolous cricetid is a nearctic native that inhabits temperate savanna-woodland ecotones, sedgeland, dry grassy slopes, chaparral, alpine meadows, shrubland, willow and alder zones, mountain marshes, brushland, dense mixed forests, disturbed land, wetlands, and stream beds surrounded by thickets. Its range extends over most of the western side of Canada and the U.S., from Alaska south to Arizona and New Mexico, and east into Montana, Wyoming, Colorado, and South Dakota. Does not hibernate; active all year; establishes a home range of 5,000-10,000 sq. ft.

A diurnal, crepuscular, and nocturnal omnivore, its diet consists of standard Cricetidae fare: seeds, roots, grains, nuts, stems, fruit, tubers, leaves, fungi, wood, and insects. Its breeding season runs from spring to fall. The female gestates for about 3 wks., giving birth to an average of 4 altricial pups in a grass-lined maternity nest chamber.

Communicates and perceives via standard Cricetidae channels: sight, hearing, smell, and touch. Its natural enemies are reptiles, birds of prey, mustelids, wild felids, and wild canids. Beneficial as food for carnivores; detrimental as a crop destroyer and as a carrier of dangerous diseases. The long-tailed vole lives for an average of 8 mos. in the wild.

LONG-TAILED WEASEL (*Mustela frenata*): Also known as the "American weasel," it grows to 22" in total length, its hind foot measures 2", and it weighs up to 12 oz. Like many other members of its family, the pelage of southern populations is normally cinnamon-brown on the dorsum and cream on the ventrum all year-round. In northern populations, however, the coat molts to white in winter, a product of photoperiodicity. It has a long, slender, supple body; a small head; long whiskers; small round ears; short legs; short soft fur; and a long, bushy, black-tipped tail that makes up half its total body length (and from which it derives its common name). A gender dimorphic species, the male is larger than the female.

A riparian mustelid, the habitat of this nearctic and neotropical native includes brushland, wetland, farmland, swamp, scrubland, chaparral, hedgerows, field edges, rangeland, marsh, wooded areas near water, and suburban environments. Prefers friable soil for digging, and forest debris and litter layers for hiding and hunting. The most wide-ranging weasel in our region, it flourishes throughout southern Canada and most of the lower 48 U.S. states, south to South America. Dens in a myriad of microhabitats, such as rock piles, inside decomposing logs, hollow stumps, and tree roots, and under human structures; a secondary cavity user, it also appropriates the abandoned burrows of other animals, often lining it with the hair of its prey (mainly from mice). Relies on standing snag, blowdown, and deadfall for food and protection. Establishes a home territory of up to ½ sq. mi.

Ferocious and aggressive, it will attack animals many times its own size, and, when aroused by the smell of blood or a tiny movement, it is known to go on killing

Long-tailed weasel.

sprees (a behavior known as "surplus killing")—even on a full belly. Extremely nimble and alert, when running it leaves a 20" stride; will dig underground, dive into water, or climb into trees in pursuit of its food: fruit, crayfish, snakes, birds, shrews, rats, chipmunks, rabbits, and, as noted, mice.

When hunting, it corners its prey, then performs a "war dance" comprised of wild hopping, twisting, and leaping. (Since it also executes this frenzied romp when it is not hunting, its purpose is not yet fully understood.) It then snatches its victim up into its powerful jaws, wraps its body around it (to prevent it from escaping), and delivers a crushing bite to the back of its head or upper neck—puncturing its brain or severing the spinal cord. With its very high metabolic rate, the insatiable long-tailed weasel must consume at least half its own body weight each day. Caches extra food in cool underground larders outside its main den site.

A polygynous species, its breeding season takes place during the summer (July-August). The female mates only once a year, then employs delayed implantation (postponing attachment of the egg on the wall of the uterus) in order to synchronize the birth of her young with the most favorable conditions (that is, during warm months). Her total gestation period is around 10 mos., after which she gives birth to an average of 6 altricial kits in the spring in a concealed natal nest. The young are weaned at about

Long-tailed weasel.

5 wks. The mother then teaches them how to hunt, and by 2-3 mos.—by which time gender dimorphism is already becoming apparent—they attain independence and disperse. The solitary male does not seem to aid in the rearing of his offspring.

Though primarily nocturnal, it is also diurnal and crepuscular, and thus may be active anytime day or night. Possessed of excellent senses, it communicates and perceives like other members of the Mustelidae family, using sight, hearing, smell, and touch. Its vocalizations include hissing, chattering, squeaking, twittering, squealing, whistling, barking, and purring. A secondary consumer, its natural enemies are snakes, birds of prey (owls, hawks, eagles, buzzards), fishers, martens, badgers, wild felids (bobcats), and wild canids (foxes, coyotes). Among its defense mechanisms is, as its scientific name denotes, the ability to emit an offensive musky oil from its rump glands when threatened or excited. Beneficial to humans as a natural pest controller; detrimental as a farm and ranch pest. The long-tailed weasel lives 1-3 yrs. in the wild; over twice that long in captivity.

MARGAY (*Leopardus wiedii*): Standing 12" at the shoulder, its body can reach a total length of 70" (though 50" is more typical), its tail grows to 20", its hind foot measures about 4", and it can attain a weight of nearly 11 lb. This makes it the smallest spotted wild cat in North America. With its tawny-gray body, white or cream-colored belly, chest, chin, and throat, and dark mottled rosettes and irregular striping, this tiny graceful felid could easily be mistaken for a small version of its larger cousin the ocelot—which is why it is sometimes called the "little ocelot."

Margay.

The margay has soft thick fur; the back of each ear is black with a white spot on it; and its base pelage can range in color from light tan to pure black (a rare melanistic form). Its golden brown eyes are rimmed in black, trailing off to form a vertical stripe down each cheek from the inner eye, and a horizontal stripe from the outer eye along each side of the face. Two more facial stripes run upward from the inner eye to the top of the head, all which lend it a "bandit-like" appearance.

Such cryptic markings, along with a ringed, black-tipped tail, provide superb camouflage in its main habitat: sun-speckled, often humid forests, from subtropical, tropical, and coniferous, to deciduous, gallery, and montane. This neotropical native is also sometimes found in temperate savannas, marshy areas, and disturbed forests, and on agricultural land and deserted farms. Its range once extended into the southern U.S. (it was last seen in Texas in 1852), but today it is found only from northern Mexico south to Argentina. Its home range is about 10 sq. mi., with a core area about half that size.

A scansorial nocturnal hunter, this sedentary solitary carnivore feeds on fruit, insects, amphibians, reptiles, birds (including poultry) and their eggs, and small terrestrial and arboreal mammals, including sloths, porcupines, and monkeys. Largely arboreal itself, the gymnastic margay is specialized for life in the forest canopy, where it rests, hunts, sleeps, and bears its young: its large dark eyes possess excellent night vision; its broad soft paws and flexile toes provide good traction and grip; its long heavy tail aids in stability and balance while running and jumping from tree limb to tree limb; flexible digits and extra long hind claws give it the ability to hang from branches with one foot; highly limber ankles allow the feet to rotate outward 180 degrees (a trait known as "hind foot reversal"), permitting it to run under branches or walk slowly down the sides of trees headfirst (the only known North American cat able to do so). Nimble and light, it can jump 20' into the air and leap a distance of 30'.

The reproductive system of the margay has yet to be studied in depth in the wild, but it is probably similar to the ocelot. There seems to be no set breeding season, thus mating can take place at any time of the year. The female gestates for about 3 mos., giving birth to 1-2 altricial kittens that are weaned at about 2 mos. or so. She has only one pair of mammary glands with which to nurse her young (relatively rare among terrestrial mammals), an indication of small litter size. An asocial feline, the male probably does not assist in the rearing of his offspring.

Uses typical communication and perception channels endemic to its genus *Leopardus* (small American felines): touch, hearing, smell, and sight. The natural enemies of this secondary consumer are not known but, again, they are probably the same or similar to those of the ocelot: snakes, birds of prey, and other felids, such as jaguars and mountain lions. Secretive, rare, and elusive, this cryptic creature of the night is rarely seen by humans. The margay, also known as the "tiger cat," lives 10-15 yrs. in the wild; nearly twice that long in captivity.

MARSH RABBIT (*Sylvilagus palustris*): This medium-sized solitary mammal grows to a total length of 18", its hind foot measures nearly 4", and it weighs up to 3½ lb. It has a small head, small feet, small broad ears, and a small tail. The sparsely furred coat on the dorsum is cinnamon to dark brown with buff, gray, or black mottling; the ventrum is lighter, tending to white. The underside of its gray-brown tail, however, is dark, a trait unique among cottontail rabbits.

As its common name indicates, this is a riparian species whose lifestyle requires water. Indeed, it is a semiaquatic leporid with excellent swimming and diving abilities. Thus, when fleeing a predator it will jump into a pond or swamp before running over ground. Safe from its tormentor, it bobs motionless at the surface with only its nose and eyes showing, until it is safe to return to shore. When forced to run on land, it will make quick zigzags in order to elude its pursuer.

Marsh rabbit.

It shelters in dense thickets or logs, making a large nest of grass, rushes, and leaves that it lines with its own fur (which it plucks from its belly). Its usual habitat is temperate, low-lying bottomland with plenty of brackish and freshwater sources, such as lakes, marshes, estuaries, canals, flooded fields, intra-coastal waterways, and wetlands generally. Its range covers the American Southeast, from eastern Virginia south to the Florida Keys.

This water-loving, nocturnal herbivore feeds mainly on marshy plants, such as wet grasses, cattails, and hyacinth. It will also eat fruit, herbs, leaves, bulbs, twigs, and tree bark. Its breeding season lasts for about 8 mos., from February to September (longer further south). The polyestrous female may have up to 5 litters (or more) during that time. She gestates for about 1 mo., giving birth to 2-5 altricial kits that are independent by 3-4 wks. of age.

Like all members of the Lagomorpha order, this primary consumer has a host of natural predators: reptiles, birds of prey, wild canids, and wild felids. Due to its preferred habitats, it is particularly susceptible to alligators, marsh hawks, and water snakes. Will sometimes walk on its hind legs, leaving a strange track that can baffle even some of the most experienced mammalogists. The marsh rabbit lives 3-4 yrs. in the wild; 8 yrs. in captivity.

MARSH RICE RAT (*Oryzomys palustris*): This smallish mouse-like rodent grows to a total length of 12", its hind foot measures a little over 1", and it can attain a weight of nearly 3 oz. Countershaded, its coloration is grizzled brownish-gray on the dorsum, whitish or tan on the ventrum; the long, annulated 6" tail is dichromatic, with sparse dark fur on top, sparse light fur on the bottom. The body is compact and ovoid; the head is small with a convex profile; the eyes are small, dark, beady, and protruding; the ears are medium-sized, rounded, and erect; the muzzle is tapered; the nose is slightly snubbed and covered in sensitive nasal vibrissae; the limbs are short; the whitish feet are small and delicate.

As both its common name and its species name imply, this cricetid is a semiaquatic nearctic native that haunts swamps and marshy habitats, both freshwater and saltwater. However, it can also be found in and around grassland, dikes, hedgerows, forests, cropland, sedgeland, levees, and wetland borders. Its range covers much of the southeastern U.S. and parts of the Midwest, from Illinois, Pennsylvania, and New Jersey on the northern edge, to Missouri, Oklahoma, and Texas on the western edge, to Georgia down to the Florida Keys on the southern edge. It often constructs its round

Marsh rice rat.

15" nest near water, weaving it out of shredded terrestrial grass and aquatic material; elaborate runways radiate out through its home territory; feeding stations overhang at the water's edge. Other signs of its presence are the vestiges of arthropod meals it leaves along the shoreline. Does not hibernate; active all year.

A nocturnal omnivore that is strongly granivorous, it gets its genus name from its predilection for rice (*oryza* is Greek for "rice," *mys* is Greek for "mouse"). But it also consumes grass, nuts, sedges, aquatic plants and terrestrial fruit and fungi. On the carnivorous side it enjoys dining on insects, snails, crab, fish, birds' eggs, chicks, and carrion.

Its breeding season runs year-round. The prolific polyestrous doe can produce up to 6 litters a year, nursing recently born offspring while pregnant with her next litter. She gestates for nearly 1 mo., giving birth to about a half dozen altricial pups that she weans over the following few weeks.

Communication and perception take place via normal Cricetidae channels: sight, smell, hearing, and touch. Adeptly natatorial, it can swim underwater and also at the surface, easily crossing an area of open water longer than a football field; will dive underwater to escape its natural enemies: reptiles, birds of prey, mustelids, procyonids, wild felids, and wild canids. Debate continues over the taxonomic position of several closely related members of *Oryzomys*, including Coues' rice rat and the silver rice rat. The marsh rice rat probably lives an average of 6-8 mos. in the wild.

MARSH SHREW (*Sorex bendirii*): Also known as the "Pacific water shrew," its grows to a total length of nearly 7", its tail is 3⅛" long, its hind foot measures nearly 1", and it weighs up to ½ oz. These statistics make it the largest shrew in North America. Its velvety pelage is dark brown on both the dorsal surface and on the ventral surface. Its head is typically cone-shaped; the muzzle is long and pointed; the snout is convex with a flexible nose tip; the chocolate brown tail makes up nearly half the length of the body. The eyes are tiny; the feet are delicate; the tips of the teeth are rust-colored.

Marsh shrew.

This massive soricid is a riparian that inhabits moist mixed forests, stream banks, marshes, grassland, muddy woodlands, skunk-cabbage wetland, and scrubland dominated by slow-moving creeks, forest litter, dense canopy, and mossy logs. These habitat preferences give it its common name, "marsh shrew." Its range covers the coast along the Pacific Northwest, extending from southwestern British Columbia south into Washington, Oregon, and northwestern California.

A rapacious invertivore that is well adapted to a semiaquatic life, this secondary consumer is an expert swimmer and diver, feeding primarily on aquatic insects, gastropods, fish, and amphibians, but also on annelids and various terrestrial invertebrates. Air bubbles trapped in its fur help insulate it from the cold water in which it spends much of its time. Detects prey using its sensitive hearing as well as its tactile vibrissae, which act as sensory antennae. Kills its victims with a series of quick violent bites; caches and hoards surplus food; does not hibernate; it is active all year and, due to its high metabolism, must feed almost constantly.

The breeding season of this rare mammal runs from midwinter to midsummer. Males emit a strong musk from their flank glands, which may act as an attractant to females. As is typical of shrews in our region, the female gestates for about 21 days, producing an average of 2 litters a year, each containing about 5 altricial young. The marsh shrew lives about 1½ yrs. in the wild.

MEADOW JUMPING MOUSE (*Zapus hudsonius*): It grows to a total length of about 10", its hind foot measures 1½", and it can attain a weight of 1 oz. Cryptically colored, a band of dull, coarse, grizzled, dark brownish fur washes over the dorsum; the sides are olive-yellow; the ventrum is tannish-white; the very long 6" tail is whip-like, nearly hairless, bicolored (dark on top, lighter on the bottom), and, unlike the tails of many other mice, lacks a white tip. The body is small and compact; the head is small and narrow with a convex profile; the eyes are small and dark; the ears are large and erect; the muzzle is short, pointy, and covered in haptic vibrissae; the limbs (especially the hind legs) are long for its body size; the feet are tiny, delicate, and pinkish. Diagnostic identifier: this rodent is the only known mammal possessing 18 teeth. Gender dimorphic, the female is slightly larger than the male.

This terricolous solitary dipodid is a riparian nearctic native that inhabits temperate moist meadows (from whence it derives the first half of its common name), montane savanna, brushland, marshy fields, scrubby wetland, and the weedy banks of creeks, rivers, and swamps edged with thick vegetation. Its massive range stretches from the Pacific Coast to the Atlantic Coast, covering mainly the upper tier of North America, from Alaska east to New England, and from Labrador south to Arizona and Georgia. Occurs across the Midwest, the Great Plains, and much of the Rocky Mountains, but is absent from Washington, Oregon, California, Idaho, Nevada, Utah, Texas, Louisiana, Arkansas, and Florida. Fossorial, it builds its burrow system under shrubbery and stumps or in the sides of hills and riverbanks. Estivates during extremely hot weather; enters hibernation in the fall, living off its built up body fat until the arrival of warm weather.

Meadow jumping mouse.

The diet of this crepuscular, nocturnal, nomadic omnivore consists of seeds, nuts, leaves, buds, grains, fruit, stems, fungi, and insects. To reach grass seeds, it cuts down the stems at their base, tossing the remnants into little piles; cuts leaf buds off of their stalks, consuming them on the spot; does not cache food. Its mating season begins when breeding adults emerge from their grass-lined hibernacula in spring. The iteroparous polyestrous female may produce 2 or more litters annually. She gestates for about 3 wks. or so, gives birth to an average of 4 altricial pups in an underground nest chamber, then weans them approximately 1 mo. later. The male does not appear to participate in the rearing of his offspring.

Communicates and perceives via normal Dipodidae channels: sight, touch, hearing, and smell. Saltatorial, with its elongated hind limbs it is an excellent jumper, leaper, and hopper, the source of the second half of its common name; can jump up to 4'. Its natural enemies are birds of prey, mustelids, and wild canids. Beneficial as a seed disperser and as food for carnivores. The meadow jumping mouse lives an average of 6-12 mos. in the wild; twice that long in captivity.

MEADOW VOLE (*Microtus pennsylvanicus*): Also known erroneously as the "field mouse," this stout rodent grows to a total length of nearly 8", its hind foot measures almost 1", and it can attain a weight of up to 2 oz. Its pelage coloration, as well as its coat's color placement, cover a wide spectrum, and depend on population, location, season, and the individual—making it difficult to positively identify. The dorsum can range from palomino, yellowish-gold, or rusty, to silvery, reddish-brown, or blackish; the ventrum is tan, whitish, or grayish; the long 2" tail is bicolored (dark on top, lighter below); unlike many other rodents it has dark brown or blackish feet. It is easily confused with other voles in its range, and often can only be distinguished by a thorough tooth examination. There is some gender dimorphism, with the male being a little larger than the female; but the difference is so slight that it would not be noticeable in the field.

This abundant, terricolous, sedentary cricetid is a riparian nearctic native that inhabits temperate savanna, grassland, moist fields, orchards, chaparral, sedgeland, hedgerows, marshy settings, shrubland, cultivated farmland, grassy bogs, overgrown pastures, thick forests, and wet woodlands, from valley to mountaintop. Prefers locations with one or more of the following: friable soil, forest duff, blowdown, rocky outcroppings, logs, deadfall, and snag. Its enormous range

Meadow vole.

covers the northern tier of North America, extending from portions of Alaska and nearly all of Canada, south to the Rocky Mountains and east to New England, with occurrences in parts of the Great Plains and the Midwest, running south to Kansas, New Mexico, South Carolina, and Georgia.

A fossorial mammal, it constructs large subterranean burrow systems that are connected to paths, grassy nests, and runways at the earth's surface. Solitary and territorial, after establishing a home range of several thousand square feet, it agonistically guards it against conspecifics, only sharing its turf in extreme cold weather conditions.

A diurnal and nocturnal omnivore, its diet mainly surrounds grass seed; but it also feeds on other customary items on the *Microtus* menu: grains, leaves, fruit, roots, fungi, nuts, flowers, tubers, vegetables, stems, bark, wood, insects, and carrion; will sometimes resort to cannibalism. A voracious eater, each day it can consume its own weight; leaves piles of grass clippings around its domicile.

Its polygynandrous breeding season takes place all year, but crests between spring and fall. The female gestates for approximately 3 wks., giving birth to an average of 7 altricial pups in a globular leaf-lined nest. She nurses, grooms, and protects them for about 2 wks., after which weaning and independence occur. The male displays no paternal behaviors and thus plays no part in the rearing of his offspring.

Communicates and perceives via standard Cricetidae channels: sight, touch, smell, and hearing. Engages in scent-marking (using pheromones), foot-stomping, teeth-clicking, and ultrasonic sounds. Its natural enemies include: reptiles (snakes), birds of prey (owls, hawks), mustelids (fishers), mephitids (skunks), procyonids (ringtails), wild felids (bobcats, lynxes), wild canids (foxes, coyotes, wolves), and ursids (black bears, brown bears). Beneficial as a seed disperser, parasite host, and food for carnivores; detrimental as a pestilent crop destroyer. The meadow vole lives for an average of 4-8 mos. in the wild; twice that long in captivity.

MERRIAM'S CHIPMUNK (*Neotamias merriami*): Named after American naturalist Clinton Hart Merriam, this large ground-dwelling rodent grows to a total length of about 11", the hind foot measures nearly 2", and it can weigh up to 4 oz. Coloration on the upper body is grayish-brown; on the lower body it is whitish; palish sometimes indistinct brownish-gray and brownish-white stripes run along the back and sides; facial stripes echo the dorsal and lateral stripes; the sides run from bright cinnamon to dull buff; the outer legs are grayish; the long nearly 6" tail is bushy, grizzled gray-brown, and edged with whitish-sienna fur. Each ear has a brownish preauricular patch and each eye has both brownish-black preocular and postocular patches. Sometimes confused with the similar California chipmunk, the female is larger than the male, thus Merriam's chipmunk is gender dimorphic.

A solitary, petraphilic, western sciurid, this shy nearctic native inhabits temperate, subalpine coniferous forests, along with chaparral, scrubland, rocky locations, woody foothills, particularly in mountainous regions. It prefers areas with thickets, fallen and rotting logs, deadfall, snag, creek banks, and plenty of forest duff. Its range is narrow, extending only over parts of southern and central California. Does not normally hibernate, unless the temperature drops precipitously; estivates during hot weather. May construct its shelter either above ground (under a log or stump) or below ground (in a rocky hole or burrow). Will appropriate the abandoned dens of fossorial mammals, such as pocket gophers. Establishes a home range of about 1,000 sq. ft., which it scent-marks and defends against conspecifics.

Merriam's chipmunk.

A terricolous diurnal omnivore, its diet encompasses the typical fare of its kind: seeds, nuts, woody vegetation, grains, fruit, forbs, insects, reptiles, birds' eggs, and small birds and mammals. A scatter hoarder, it caches surplus food throughout its home range, generally burying it or hiding it in trees and logs for later use.

The breeding season takes place between midwinter and late spring. The doe gestates for roughly 1 mo., giving birth to an average of 6 altricial pups in the maternal nest chamber. To prevent detection by predators, like many other chipmunk species, the mother may periodically move her young to a new nest site. At between 2 and 4 wks. of age the neonates are weaned, with independence and dispersal occurring shortly thereafter. The buck does not appear to assist in the rearing of his offspring.

Communication and perception are standard for this energetic member of the Sciuridae family: sight, touch, smell, and hearing. Vocalizations include chips, chucks, and trills. Its natural enemies are reptiles, birds of prey, mustelids, wild felids, and wild canids. Benefits its ecosystem by hosting parasites, dispersing seeds, and providing food for carnivores. Merriam's chipmunk probably lives an average of 2-4 yrs. in the wild; longer under optimum conditions.

MERRIAM'S GROUND SQUIRREL (*Urocitellus canus*): Once thought to be a subspecies of Townsend's ground squirrel, based on recent genetic studies it is now recognized as a separate species (for one thing, it has more chromosomes than Townsend's). Morphologically, however, it is identical to Townsend's ground squirrel, with the same ecology and life history (including coloration, size, weight, habits, habitat, reproductive life, diet, and longevity). One dissimilarity: Merriam's ground squirrel has a different geographical range, one covering portions of Oregon, California, Idaho, and Nevada.

MERRIAM'S KANGAROO RAT (*Dipodomys merriami*): The smallest of our kangaroo rats, it is neither a true rat or a relative of the kangaroo. It is a cousin of the pocket gopher, one that grows to a total length of 10" and a weight of nearly 2 oz. Its dorsal pelage is buff with a gunmetal grayish wash; its soft ventral pelage is white or tannish; the long 6" tail is whip-like and has dark and light lateral stripes and a dark tufted tip. It has a dark nose line on each side of its muzzle. The body is ovoid and compact; the head is large; the eyes are large, dark, and protruding, allowing for excellent panoramic night vision; the ears are medium-sized, hairless, and erect; the front legs are small; the hind legs are long and kangaroo-like, the origins of its common name. The white lateral stripe is distinct; the sensitive nasal vibrissae are prominent; the feet are 4-toed; walks quadrupedally, jumps bipedally.

Named after the American naturalist Clinton Hart Merriam, this solitary heteromyid is a nearctic native that inhabits temperate, dry, shrubby, low-lying deserts, salt flats, sageland, savanna, grassland, rangeland, sand dunes, playa, chaparral, and alkali sinks; can tolerate a wide variety of soil types, from soft and sandy to hard and rocky. Its range is limited to the American Southwest, covering portions of the states of California, Nevada, Utah, Arizona, New Mexico, and Texas, extending south into northern Mexico. Fossorial, it spends the day inside its burrow system, which opens under bushes or tufts of scrub grass; comes out to forage at night; territorial, it establishes a home range of at least 5,000-10,000 sq. ft., vigorously defending it against intruders.

Merriam's kangaroo rat.

A nocturnal omnivore, it feeds mainly on seeds from plants such as purslane, mesquite, creosote bush, ocotillo and grama grass; consumes insects when necessary; transports its food in its external fur-lined cheek pouches. Its highly efficient kidneys need little fluid to flush bodily wastes, permitting this species to live in areas where there is little or no water. Its breeding season runs from late summer to early spring. The female gestates for approximately 1 mo., gives birth to an average of 3-4 altricial pups in an underground nest chamber, then weans them some 3 wks. later. This prolific rodent can begin reproducing at 2 mos. of age.

Communication and perception take place via traditional Heteromyidae channels: touch and smell, and to a lesser extent, sight and hearing. Its natural enemies are reptiles, birds of prey, mustelids, wild felids, and wild canids; survival skills include the ability to kick sand in the eyes of predators and leap great distances. Beneficial as a seed disperser and a soil aerator and recycler. Merriam's kangaroo rat probably lives 1-2 yrs. in the wild; longer under optimum conditions.

MERRIAM'S MOUSE (*Peromyscus merriami*): Also known as "Merriam's deer mouse" and the "mesquite mouse," this pale gray cricetid was once thought to be conspecific with the cactus mouse. Though now regarded as a separate species, the 2 share much of the same ecology and life history, making them nearly impossible to distinguish in the wild. The primary noticeable difference is their overlapping ranges: Merriam's is much smaller and is restricted to Arizona and northern coastal Mexico, while the cactus' range extends over 6 southwestern states and both northern and central Mexico.

MERRIAM'S POCKET MOUSE (*Perognathus merriami*): Systematists disagree on whether this rodent is a distinct species, a subspecies of the silky pocket mouse, or is conspecific with the silky pocket mouse; some consider the 2 to be sister species. The taxonomic status of this mammal thus remains unsettled. Whatever the outcome of this debate, it appears to be most closely related to the silky, with the 2 sharing much of the same ecology and life history.

MERRIAM'S SHREW (*Sorex merriami*): Its grows to a total length of 4" or so, its tail is 1½" long, its hind foot measures ½", and it weighs up to ¼ oz. It is countershaded, with a pale brown-gray upper body, buff flanks, and a white to tan belly. The tail is dichromatic: grayish on the dorsum, lighter on the ventrum. It has a short, broad conoidal head; tiny eyes; a slightly convex muzzle; sharp rust-colored teeth; a pointed whiskered nose; and small delicate limbs and feet.

This secondary consumer inhabits somewhat arid biomes and ecosystems, including shrubland, chaparral, savanna, grassland, woodland (mixed), and prairie dominated by sagebrush and generally dry vegetation. It ranges across the western part of our region, covering British Columbia south through Washington, Oregon, and California, to New Mexico and Arizona, east to Idaho, Montana, Wyoming, Nebraska and North Dakota. Does not hibernate; is active all year. Digs its own burrows only rarely; prefers taking over the abandoned runways and tunnels of other animals, such as voles.

This carnivorous soricid is an avid insectivore, arachnivore, and vermivore, consuming large quantities of insects (wasps, caterpillars, crickets, beetles), spiders, and earthworms. Due to its small size (and consequent small surface to volume ratio), it must eat nearly nonstop in order to maintain its energy level and body heat. As a natural pest controller it serves humanity well.

Nocturnal and fossorial, it is difficult to study. Thus little is known about many aspects of the ecology and life history of Merriam's shrew, including its mating system. The female gestates for about 1 mo., giving birth to 4-6 altricial young, who are weaned and independent by 1 mo. of age.

Communication with conspecifics is through typical Soricidae channels: sight, hearing, touch, and smell. The latter is particularly important to males, who posses flank glands that emit a powerful scent that seems to be connected both to attracting mates and to repelling its primary enemy: the owl. Merriam's shrew probably has an average life span of around 1½ yrs. in the wild.

MEXICAN GROUND SQUIRREL (*Ictidomys mexicanus*): It grows to a total length of about 15", its hind foot measures 2", and it can weigh up to 12 oz. Its pelage is grayish-brown on top, light brown or pinkish on the underside from chin to rump. The outer legs are sometimes rust-colored. Most conspicuous are 9 dorsal stripes over the back (from neck to posterior) made up of off-white squarish spots. The body is slender and cylindrical; the head is large; the eyes are dark and ringed in tan; the ears are rounded and small; the nose is somewhat tapered and tinged with reddish-yellow; the limbs are short; the 6" tail is rather flat and brushy. It is often mistaken for a chipmunk, though it lacks that mammal's familiar malar stripes. Gender dimorphism is present, with the male being larger than the female.

Examples of ground squirrel feet.

It inhabits well-drained brushland, savanna, creosote scrubland, cactus flats, and arid grassland, as well as open terrace and desert habitats with sandy or gravelly ground; found in both wild and suburban areas, including pastures, yards, and roadsides. Its range is limited to southern Texas, southern New Mexico, and northern Mexico. A fossorial ground squirrel, it constructs multiple subterranean burrow systems in which it eats, sleeps, rests, nests, hoards food, escapes predators, and possibly hibernates (in cold regions). Often appropriates the abandoned tunnels of pocket gophers. Individuals may share tunnels, making this species semicolonial.

An omnivorous sciurid, it feeds primarily on plant matter (seeds, bulbs, roots, stalks, leaves, nuts, grains, fruit, and agricultural products); also a rabid meat-eater, it readily supplements its herbivorous diet with animal matter (insects, birds' eggs, small mammals, and carrion, such as roadkill). Cannibalistic, it occasionally kills and eats its own kind. Transports food to its den in large cheek pouches.

The breeding season of this solitary, chipmunk-like rodent runs for only a few weeks in early spring. The female forms a grass-lined nest underground, gestates for about 3-4 wks., and gives birth to 4-5 altricial pups, who attain independence within 2-3 mos.

Shy, diurnal, and quiet, it communicates and perceives via channels typical of the Sciuridae family: sight, touch, smell, and hearing. Its vocalizations include hissing and a trill-like alarm call. Its natural enemies are birds of prey, and most likely reptiles, wild felids, and wild canids. It benefits us by consuming pestilent insects; it harms us by digging up golf courses, cemeteries, and cultivated crops. The life span of the Mexican ground squirrel is not known, but it is most likely brief, possibly 2-4 yrs.

MEXICAN LONG-NOSED BAT (*Leptonycteris nivalis*): It is nearly 4" in total length, its forearm measures roughly 2½", and it weighs about ¾ oz. Its dorsum is grayish to brownish in color, the ventrum is light tan to whitish. Its most outstanding characteristics (of many) are its nose-leaf and a minuscule tail that is barely detectable.

It inhabits desert scrubland and montane woodland, preferring arid areas where century plants, cacti, and creosote bushes flourish. It is found primarily in the more temperate regions of the southwestern U.S., including Texas, Arizona, and New Mexico, ranging south into Central America. A colonial bat, it roosts with thousands of conspecifics in caves, tunnels, mines, deep caverns, abandoned buildings, and rocky shelters, usually not far from the entrance. It migrates north in spring, south in fall.

Desert agave plant.

Rare among chiropterids, this primary consumer, like its cousin the Mexican long-tongued bat, is a nocturnal nectarivore, using its 3" tongue—with tiny nipple-shaped organs on the tip—to lap up protein-rich nectar from night-blooming flowers, such as agave, as well as the blooms of various species of cacti and trees. Its long muzzle (from whence it gets its common name) acts as an extension when nectar-feeding.

This noncarnivorous phyllostomid is a powerful and agile flyer which, like hummingbirds, can hover while feeding. Also a pollenivore and a pollinator, it both consumes vitamin-rich pollen and is instrumental in pollinating several plants, including the tequila plant—a symbiotic relationship between bat and plant known as chiropterphily. Females give birth to one or two altricial pups in late spring. The Mexican long-nosed bat probably lives from 10-20 yrs. in the wild.

MEXICAN LONG-TONGUED BAT (*Choeronycteris mexicana*): It grows to a little over 3" in total length, its forearm measures nearly 2", and it weighs less than 1 oz. This gray to brownish medium-sized chiropterid has pale fur on its undersides. It is distinguished by its large eyes, short ears, long slender snout, erect leaf-like projection on the tip of its nose (used to enhance reception of sonar echoes), and a tiny tail that grows only halfway across its interfemoral ("inner thigh") membrane.

This secondary consumer inhabits caves, rock shelters, old buildings, and mines in the deserts, grassland, scrubland, montane woodland, and canyons of the southwestern U.S. It shelters by day in small colonies of 10-50 individuals, hanging by one foot (allowing it to rotate 360 degrees for optimum viewing) near the roosting entrance.

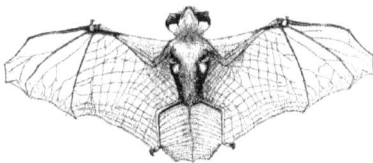

Body of a typical bat species.

Primarily an herbivore and nectarivore, its common name derives from another special characteristic: a long tubular tongue (which extends up to one-third of its body length) with small brushy projections on the tip, used to lick up the protein-rich nectar of night-blooming flowers, as well as the juices of various fruits. Also a frugivore (that consumes catus fruit), a pollenivore, and an insectivore, it feeds on pollen (making it an important pollinator) and insects (making it a useful pest controller), and can hover in midair momentarily. Will visit and take food from hummingbird feeders.

The female gives birth to a single, fully furred, altricial pup in early summer, which she carries as she flies. Does not hibernate. Built for long-distance flying, this phyllostomid migrates annually, following nectar corridors south in the fall, north in the spring. The life span of the gentle, nectar-feeding, Mexican long-tongued bat in the wild is unknown.

MEXICAN SPINY POCKET MOUSE (*Liomys irroratus*): This medium-sized member of the *Liomys* genus grows to a total length of nearly 12" and can attain a weight of 1¾ oz. Its upper body coloration is a grizzled grayish-black with rusty orange highlights; the lower body is cream white; the 6" tail is sparsely furred and dichromatic (brownish above, whitish below). The fur is shiny and coarse while the individual hairs are bristly, flat, grooved, and pointed—the source of the second word in its common name. Its upper incisors lack grooves. Diagnostic identifier: has an unusual shovel- or scoop-shaped rear claw. Gender dimorphism is present: the male is slightly larger than the female.

Pocket mouse.

This petraphilic heteromyid is a neotropical native that inhabits rough brushland, montane scrubland, terraces, rocky foothills, rangeland, ridges, river banks, chaparral, humid forests, and dense thickets of cactus. Its range extends over portions of southern Texas south into Mexico, from which it derives the first word of its common name. It excavates underground burrow systems, which it uses for sleeping, nesting, shelter, and protection; leaves small ejecta mounds near the entrance.

A nocturnal herbivore, its diet consists almost solely of plant seeds, which it gathers and caches in underground larders; it transports its food in its external fur-lined "pockets" or pouches, the origin of the third word in its common name. Little studied, almost nothing is known about the reproductive life of this mammalian member of the Rodentia order.

Communicates and perceives via standard Heteromyidae channels: primarily touch and smell, and to a lesser extent, sight and hearing. Its natural enemies are reptiles, birds of prey, mustelids, wild felids, and wild canids. The life span of the Mexican spiny pocket mouse is unknown, but it is likely that it lives an average of 1-2 yrs. in the wild.

MEXICAN VOLE (*Microtus mexicanus*): It grows to a total length of 6", its hind foot measures ¾", and it can weigh up to 2 oz. The upper pelage is gray-brown with a rusty tint; the lower pelage is whitish to silvery gray; its 1½" tail is bicolored (dark on the dorsum, light on the ventrum). The body is stout; the head is small; the eyes are small and dark; the ears are large, broad, and erect; the muzzle is small; the feet are tiny and delicate.

Mexican voles.

This diurnal cricetid inhabits dry savanna, forest clearings, and wetlands. Its range extends over portions of Arizona, New Mexico, Utah, Colorado, and Texas. Its grassy nests may be located on the surface or underground. An herbivore, it seems to feed primarily on green vegetation. Its breeding habits are little known. Communication and perception channels are probably typical of Cricetidae. Its natural enemies are reptiles, birds of prey, mustelids, procyonids, mephitids, wild felids, and wild canids. The life span of the Mexican vole has not been studied. Note: the classification of this mammal (and its subspecies) is unsettled and contradictory, with ongoing disagreement over taxonomic issues.

MEXICAN WOODRAT (*Neotoma mexicana*): This medium-sized member of the *Neotoma* genus grows to a total length of nearly 17", its hind foot measures 1½", and it can reach a weight of almost 7 oz. The dorsal pelage is a grayish-brown with a buff wash; the ventral pelage is white; the long 8" tail is sparsely furred and bicolored (brownish on top, whitish below). The body is robust and muscular; the head is large with a convex profile; the eyes are large, dark, and protruding; the ears are large, broad, and erect; the muzzle is thick; long sensitive nasal vibrissae are present; the limbs are strong; the feet are pink and delicate.

This petraphilic terricolous cricetid is a nearctic and neotropical native that inhabits scrubby mountainous rocky regions, as well as coniferous and deciduous forests. Habitations also include cliffs, chaparral, talus slopes, stone outcroppings, boulder fields, shrubland, brushland, discarded human structures, rock piles, and scree, preferably with logs, snag, and forest duff. A southwestern rodent, its range covers portions of the states of Utah, Colorado, Arizona, New Mexico, Oklahoma, and Texas, extending south into Central America. Builds stick and grass nests in tree cavities, debris mounds, or rock crevices; establishes a home range of about 200 sq. ft.; does not hibernate, active year-round.

A solitary nocturnal herbivore with generalized feeding habits, its diet consists of seeds, grains, leaves, cacti, nuts, fruits, and fungi. Hoards and caches surplus food; like other woodrats it is considered a "packrat," so named due to its predilection for accumulating odd items such as stones, animal remains, wood, and

Woodrat.

human refuse. Its breeding season runs from late winter to late spring. The iteroparous polyestrous female may produce up to 2 litters annually. She gestates for about 1 mo., gives birth to an average of 3-4 altricial pups, and weans them some 40 days later.

Communication and perception run via normal Cricetidae channels: sight, smell, touch, and hearing; engages in foot-stomping, scent-marking, and tail-vibrating. Its natural enemies are reptiles, birds of prey, mustelids, wild felids, and wild canids. Beneficial as a seed disperser and as food for carnivores; detrimental as a carrier of deadly diseases. The life span of the Mexican woodrat is unknown.

MOJAVE GROUND SQUIRREL (*Xerospermophilus mohavensis*): This monochromatic rodent grows to a total length of around 9", its pes measures 1½", and it weighs up to 5 oz. In summer the upper body is grayish-brown, the abdomen is white or tan. The cheeks are cinnamon-brown. Unlike many others in its family it lacks spots, streaks, flecking, mottling, blotches, and stripes. The winter pelage is slightly redder, with a washed rusty cast over the fur. The body is small and cylindrical; the head is small and somewhat tapered; the eyes are dark with a white ring; the nose is snubbed; the ears are small, rounded, and erect; the limbs are short; the paws are clawed; the short 2½" tail is furred, thinnish, and dichromatic (dark on top, light underneath).

The plain but cryptic coloration of this solitary, diurnal, nearctic native serves as suitable camouflage for its arid habitat: level temperate desert, creosote scrubland, sandy flatland, sand dunes, sparse sageland, gravelly terrain, alkali sink, and friable sun-soaked woodland. It is especially fond of biomes containing Joshua trees, whose seeds it relishes. As its common name indicates, it has a limited distribution. In fact, it is endemic solely to California, with a tiny range covering only the Mojave Desert in the southern part of the state.

This omnivorous sciurid feeds on seeds, fungi, nuts, leaves, flowers, grains, and insects; also a scavenger that consumes carrion, it hoards and caches food in its burrow system. Estivates and hibernates from midsummer to early spring. The tail is usually held aloft over the back, particularly when running.

After hibernation its polygynous breeding season begins. Males awaken first, enabling them to establish mating territories before the females emerge from their winter naps. The doe gestates for about 1 mo.,

Mojave ground squirrel.

giving birth to an average of 6 altricial pups in her natal chamber (usually constructed beneath thickets), who are weaned about 5 wks. later. The male does not assist in the raising of his offspring.

Communicates and perceives in typical Sciuridae fashion using sight, touch, smell, and hearing. Vocalizations include whistling, shrieking, squeaking, and various bird-like sounds. Its main natural enemies are reptiles, birds of prey, mustelids, wild felids, and wild canids. Benefits the Southwest ecosystem by dispersing seeds and aerating and recycling the soil. This species is on the decline. Because it spends most of its time underground, the life span of the Mojave ground squirrel has not been researched. However, it probably lives from 3-4 yrs. in the wild; twice that long in captivity.

MONTANE VOLE (*Microtus montanus*): Also known as the "Montana vole" and "mountain vole," this small rodent grows to a total length of around 7", its hind foot measures 1", and it can attain a weight of up to 3 oz. The dorsal pelage is muddy grayish-brown with a rusty undertone; the ventral pelage is silvery gray; the 2½" tail is long and bicolored; the feet are light to dark gray.

This riparian arvicolinid is a nearctic native that inhabits wet temperate meadows, alpine valleys, stream sides, thick forests, mesic savanna, and herbaceous grassland in mountainous regions, hence its common name. Also found in or around cropland, hedgerows, lakes, fence lines, and pastures. Prefers settings with friable soil, forest duff, stumps, deadfall, snag, blowdown, and logs. Its range covers portions of the western U.S. as well as southwest Canada, and includes the states of Washington, Oregon, California, Idaho, Montana, Wyoming, Nevada, Utah, Colorado, Arizona, and New Mexico. Excavates underground burrow systems with grassy pathways leading to and from the surface entrance. Does not hibernate; active year-round.

Montane vole.

An herbivore, its diet consists of seeds, sedges, tubers, roots, stems, fungi, rushes, leaves, forbs, and grasses. Its breeding season runs from spring through fall. The polyestrous female may produce 2 or more litters annually, each with an average of 4-5 pups. Communicates and perceives via standard Cricetidae channels: mainly touch and smell, and to a lesser degree, sight and hearing. Its natural enemies are reptiles, birds of prey, mustelids, wild felids, and wild canids. Beneficial as prey for carnivores; detrimental as an agricultural pest. The life span of the montane vole has not been verified.

MONTANE SHREW (*Sorex monticolus*): This North American native grows to nearly 5" long, its hind foot measures ⅝", and its 2¼" tail makes up half the length of its body. Its summer pelage is silvery rust-brown above, light rust below; its winter pelage is dark reddish-brown above and below. This coloration gives it its alternate common name, the "dusky shrew." Has a cylindrical body; conoidal skull; pointed snout; tiny eyes; conspicuous rounded ears; and tiny limbs.

Hard to identify in the wild (as it has few if any obvious features that sharply distinguish it from other shrews), it flourishes in the surface litter and debris of meadows, scree, tundra, bogs, prairies, talus, fen, and alpine forests, usually in, near, or around riparian zones (creeks, rivers, wetlands) and corridors; it is also found in mountainous regions, hence its common name, the "montane shrew." This secondary consumer frequents moist grassy areas and mossy stream banks. Will burrow in soft soil and under fallen logs. One of our most common shrews, its range runs from western Alaska and Canada south to New Mexico, with some density in the Rocky Mountain states.

This tiny solitary soricid is a round-the-clock, voracious insectivore with a high metabolic rate (burns calories quickly) and a small surface to volume ratio (loses heat quickly), making it necessary for it to hunt and feed almost constantly. Its diet consists primarily of insects, though it will also consume seeds, fungi, lichens, worms, and amphibians. Easily eats more than its own body weight in a 24-hr. period.

Mountains: montane shrew habitat.

The reproductive biology of this mammal has not yet been studied, thus its breeding season is unknown; however, it probably takes place in mid to late winter. The female may gestate for about 3 wks., giving birth to a litter of altricial young up to 4 times a year. Birthing takes place in early spring in a well hidden vegetation-lined nest. The male does not assist in caring for his offspring. The montane shrew probably lives for about 1½ yrs. in the wild.

MOOSE (*Alces americanus*): This massive cervid, also known as the "American moose," is the official state land mammal of Alaska and the official state animal of Maine. It grows to nearly 8' in height (at the shoulder), 9' in total length, and a weight of 1,800 lb. (almost a ton). The male's imposing palmated antlers reach an average spread of 5' (60") from tip to tip—though they have been known to grow to 6¾' (81"). A calf can weigh as much as 35 lb. at birth. These statistics make the moose the world's largest and heaviest deer species. Its thick insulating coat, made up of hollow hairs, varies in color from tawny to grizzled beige, from reddish brown to blackish.

Along with the male's antlers (the largest of any mammal in the world), distinguishing features of this species include a long head, large ears, humped shoulders, long muzzle, flexile nose and upper lip, long lanky legs, and a furry dewlap or pendulous "bell" located beneath the throat (this appendage, whose purpose is still not known, is bigger on males than females). The bull sheds its antlers in winter and regrows them throughout the summer. Gender dimorphic, males are larger than females and the females do not grow antlers.

A water-loving artiodactyl with wide hooves designed for walking in mud and snow, it prefers mosaic habitat that includes boreal wetlands, tundra, marsh, bogland, muskeg, elevated taiga, mountains, and shady coniferous forests in northern regions with plenty of water sources (streams, rivers, ponds, lakes). This last requirement is important, in great part, because a moose has no sweat glands: unable to perspire, it needs water to help keep its gigantic body cool. Thus it is one of the most lacustrine of all North American mammals. In our region its range covers all of Alaska, most of Canada, the Rocky Mountain states, and the northern tier across the American Midwest and New England.

Mainly a folivore and a lignivore, its winter diet consists mainly of tree parts (wood, bark, leaves, twigs, roots, shoots, buds, pine needles, pine cones), while in the summer it prefers shrubs, herbs, and aquatic plants (such as horsetails, pondweed, water shield, water lilies). While feeding on water vegetation this riparian mammal may wade in over its head and completely disappear under the surface.

Moose (male).

Natatorial, it is a powerful swimmer that is well adapted to its watery environment, and can swim up to 6 mph for several hours. When running, its stride (the distance between footsteps) can reach 8'. Though enormous in size, the moose can move noiselessly through a forest at 35 mph. Both crepuscular and diurnal, this solitary deer spends the majority of the day feeding in the same areas (though it will occasionally migrate latitudinally when looking for food), as well as searching out mineral licks (from which it obtains much-needed sodium).

Its breeding season runs from early fall to midfall. Cows seek a mate using calls and scents. Like elk, during the rut bulls create ground wallows: after urinating in them, they roll around in the muck in order to coat their fur with the resultant female-attracting aroma. To mark their territory and challenge rivals, males tear up vegetation with their antlers—masculine displays of strength and aggressiveness. Male-on-male battles can become quite violent, with injuries sometimes leading to death. The cow gestates for about 8 mos., giving birth to 1-2 light-colored, precocial calves in the summer. At around 6 mos. they are weaned, but remain with their mother until the birth of the next young.

Moose (males).

A primary consumer, the natural enemies of this gargantuan vertebrate are wolves and bears, which take mainly calves, the injured, the ill, and the aged. A healthy adult, who can run, swim, and kick with its sharp front hooves, is usually safe from predation. Unpredictable and peevish, it will aggressively charge even inanimate objects without provocation, making it highly dangerous. Females with calves are particularly agonistic. The male moose lives an average of 5-10 yrs. in the wild; the female, whose life is far less hazardous than the male's, may live as long as 25 yrs.

MOUNTAIN BEAVER (*Aplodontia rufa*): Despite its common name, it is not specifically a mountain dweller nor is it a beaver (though it shares several activities with its distant cousin, such as blocking streams and gnawing trees). The only member of the family Aplodontiidae, it is a primitive muskrat-like rodent also known by the vernacular names "aplodontia" (from the Greek words *aplos* and *donti*, "simple tooth") and "sewellel" (from a Chinook word for "mountain beaver pelt").

Medium-sized, it grows to a total length of 19" and attains a weight of up to 3 lb. Coloration on the dorsum is a dark muddy reddish-brown (hence its Latin species name, *rufa*, "red"); the ventrum is whitish-tan; and, unlike the beaver and the muskrat, its 2" tail is both small and furry. Its coat is thick and coarse; the body is stout; the head is robust; the eyes are small; the ears are small, rounded, and erect; the muzzle is blunt; the end of the snout is covered with long sensitive nasal vibrissae; the limbs are short; the 5-toed feet are strongly clawed; each ear bears a small whitish subauricular patch.

Mountain beaver.

This solitary aplodontid is a riparian nearctic native that inhabits damp deciduous and coniferous forests, preferably with moist friable soil, ground duff, shrubbery, seepage slopes, and creeks; it is found at nearly all elevations below the timberline.

Its range is limited to the western part of our region, covering portions of the states of Washington, Oregon, California, and Nevada, as well as southwest British Columbia. Fossorial, it excavates shallow but complex underground burrow systems nearly 20" in diameter, with separate chambers for sleeping, nesting, and waste; entrances are camouflaged with a teepee-like structure using leaning sticks and draped plant material; ejecta mounds, bits of cored earth, and grassy runways are usually present near portals; leaves haystacks of cut plants on nearby logs and rocks. Establishes a home range of several hundred sq. ft.; will create a casual colony of conspecifics under certain circumstances.

A crepuscular herbivore, its diet surrounds green plant matter, particularly ferns, but also includes grass, herbs, tree bark, roots, seeds, and forbs; will consume vegetation that other animals consider inedible, such as stinging nettle. It is also a coprophage, which allows it to procure additional nutrients from its food. Though it does not hibernate, it caches surplus edibles for use during cold weather. Its breeding season takes place in spring. The female gestates for about 1 mo. or so, gives birth to an average of 4 altricial pups in a grass-lined nest chamber, and weans them at around 7 wks. of age.

Mountain beaver.

Communication and perception occur via normal Aplodontiidae channels: primarily smell and touch, and to a lesser extent, sight and hearing. Vocalizations include chirps, whistles, squeals, and tooth-grating; it also makes a "booming" noise, lending it another one of its nicknames, the "boomer." Spends most of its life underground, seldom roaming far from its subterranean den. The primary natural enemies of this unusual rodent are birds of prey, mustelids, wild felids, and wild felids. Can swim, run, and climb quite adroitly. Beneficial as a prey base for carnivores; detrimental as an agricultural pest and tree destroyer. The misnamed mountain beaver lives an average of 4-5 yrs. in the wild; longer under optimum conditions.

MOUNTAIN COTTONTAIL (*Sylvilagus nuttallii*): Also known as "Nuttall's cottontail," this medium-sized leporid grows to a total length of nearly 16", its hind foot measures almost 4", and it can attain a weight of up to 2½ lb. Its dorsum is brownish-gray, with black, white, and cinnamon mottling; its ventrum is tan to white; its whiskers are white. Like other cottontails it has large eyes, rounded ears, long hind legs, and a tail with a white cottony underside.

It inhabits forest, mountain slopes, brushland, grassland, riverbanks, and sageland throughout its range: the western U.S., from Montana west to Washington, and south to New Mexico and Arizona. Solitary and secretive, this crepuscular coprophage is a folivore, lignivore, and herbivore that feeds on grasses, sagebrush, juniper berries, and parts of trees (bark, stems, leaves, twigs). Is active year-round; does not hibernate.

Depending on the region, its breeding season lasts from March to July. The polyestrous doe gestates for about 1 mo., giving birth to 1-8 altricial kits in a grass nest chamber lined with fur she has plucked from her body. She may have up to 5 litters a year. As with leporids in general, the mother's milk is rich in fat and protein.

Like other lagomorphs, it communicates primarily by scent, secreting oils, in its case, from glands near its throat and rump area. The natural enemies of this primary consumer are reptiles, birds of prey, mustelids, wild felids, and wild canids. To evade predators it will dart away, jump into water and swim, or freeze in place. The mountain cottontail probably lives for about 1-3 yrs. in the wild, but it may live as long as 7 yrs. in captivity.

Mountain cottontail.

MOUNTAIN GOAT (*Oreamnos americanus*): Also known as the "Rocky Mountain goat," it stands 3½' high at the shoulder, grows to 5½' in total length, and weighs up to 300 lb. Distinguishing traits are its stocky compact body; black eyes, nose, and hooves; cream to pure white thick shaggy coat; 5" "beard" or throat-mane; and thin dagger-like horns (12" long on the male, 9" long on the female), which range in color from tan to black. Gender dimorphic, the males are slightly larger and more muscular than the females, with horns that curve backward more sharply than the female's (whose horns are almost straight).

Mountain goat.

Despite its common name, it is not a true goat, but instead belongs to the goat-antelope group. True goats, for example, have transversely ridged or spiraled horns, while the mountain goat has relatively smooth horns, like its close European cousin, the goat-antelope known as the chamois (*Rupicapra rupicapra*).

With its muscular forequarters and specialized hooves (which, with their sharp outer rims and spongy soles, provide excellent grip and traction), it is an expert climber, a petraphile well-suited to its alpine habitat: high, steep rocky cliffs, bluffs, ledges, and mountains. A nearctic native, its range extends from southern Alaska through western Canada, south into Washington, Idaho, Montana, South Dakota, and Colorado.

Like many other bovids, it is an herbivore, a lignivore, and a folivore, whose diet centers around green and woody plants, grasses, herbs, sedges, mosses, lichens, and tree parts (bark, wood, stems, leaves). It migrates elevationally (higher in summer, lower in winter) in search of food and mineral-rich salt licks. A social mammal, it forms large herds during the cold months, smaller herds during the warm months, seldom straying outside its 10-15 sq. mi. home range. It is both diurnal and fond of dust-bathing. It spends its mornings and early evenings feeding; in the afternoon it rests in 2"-deep forms that it scrapes into the ground.

Its breeding season runs from midfall to early winter. During the courtship phase the ram rubs the ewe with musk oil emitted from a gland located between its horns, then fights other males in competition over mating rights. They do not engage in head-

Mountain goat.

butting, however. Instead they stab at one another's sides with their horns, contests that can lead to serious injury and death. The female gestates for nearly 5 mos., giving birth to 1-3 altricial kids in late spring at a safe and secluded rocky shelter. The young are weaned at about 3½ mos., but do not become completely independent until around 9 mos. old. As protectors, rams sometimes contribute to infant-rearing.

This rugged artiodactyl communicates and perceives using body language, scent, and an assortment of vocalizations along the lines of sheep and true goats. A primary consumer, its natural enemies are mountain lions, bobcats, bears, wolves, coyotes, and birds of prey (the latter which feed on the young by driving them off cliffs). Besides its ability to scale high sheer rock faces, it can also defend itself by kicking out its sharp front hooves. The mountain goat lives an average of 10-15 yrs. in the wild; around 15-20 yrs. in captivity.

MOUNTAIN HARE (*Lepus timidus*): Also known as the "Northern hare," this large lagomorph grows to a total length of 27", its powerful snowshoe-like hind foot is about 7½" long, and it weighs up to 10 lb. In response to photoperiodicity, it passes through 2 main molts: in the summer its dorsal coat turns grizzled grayish-brown to cinnamon-brown (the ventral coat is whitish); as the days grow shorter, its upper coat lightens, turning white. Such sharp alteration in the color of its pelage, which matches the changing seasons, serves as good camouflage for this prey animal. Has white eye-rings and black ear-tips year-round.

In our region this leporid inhabits mainly tundra, clearings, moorland, and boreal forest, with a range extending across northern and western Alaska and northern Canada. A nocturnal herbivore, it spends the day alone napping and grooming in a form. At night it feeds on grasses and lichen in the warmer months; heather, leaves, stems, bark, twigs, and shoots in colder months. It is thought that it eats snow instead of drinking water.

Its breeding season runs roughly from January to September. The polyestrous doe, who may bear up to 3 litters a year, gestates for about 50 days, giving birth to 1-4 precocial leverets who are weaned about 1 mo. later. The young grow up in their nest by themselves, the mother briefly visiting the burrow once a day to nurse them. The buck plays no role in the raising, protection, or feeding of his offspring.

Mountain hare.

Though solitary, it may temporarily congregate in groups of up to 70 individuals under various conditions (such as an overabundance of food or extreme weather). This arctic hare with fur-covered paws communicates through scent, as well as a series of vocalizations, such as crying and hissing. A primary consumer, its natural enemies are birds of prey, wild felids, and wild canids. Saltatorial, it sometimes uses hooking to avoid detection by predators. Can run nearly 40 mph in short bursts. The mountain hare typically lives around 5 yrs. in the wild, though some reports range up to 9 yrs.

MOUNTAIN LION (*Puma concolor*): The second largest cat in the Americas (after the jaguar), it stands up to 3' at the shoulder, grows to a total of 8' in length, and weighs up to 275 lb. Widely distributed, and thus known by some 100 different monikers across our region, this furtive carnivore has rightly been called "the cat of many names." Variously referred to as a "cougar," "el león," "puma," "panther," "painter," and "catamount," the most common name in North America is mountain lion—all representing the same slender but muscular animal.

Though its binomial or scientific name means "lion of one color," its coat is actually multicolored, and even varies from gray to reddish depending on the season and the individual's location. (All-black, or melanistic, forms have also been reported.) Generally speaking, however, its short coarse fur is fulvous, with lighter buff-coloring below and on the chest, sometimes tapering into white. The back of its short rounded ears, as well as its whisker patches, are blackish; the tail is black-tipped. This cryptic coloring allows the mountain lion to stealthily blend into shadowy landscapes where various shades of brown, black, and gray predominate.

The head is small and wide; the slightly convex forehead slopes sharply back and upward; the snout is short and broad; the striped muzzle is robust; the upper lip and chin are large and cream-colored. Its powerful jaws carry massive molar and premolar teeth (for shearing flesh), long canines (for tearing flesh), and strong incisors (for cutting flesh). Its long thick 36" tail, which makes up one-third of its total body length, helps with balance; large amber eyes with round pupils, and ringed in black, provide enhanced vision, both at night and during the day. Its well padded paws are large with curved, retractable claws, permitting quiet locomotion. This species is gender dimorphic: the male is larger than the female.

Mountain lion.

Mountain lion (cub).

Highly adaptable, it inhabits a myriad of biomes and ecosystems. Though it prefers mountainous regions, this nearctic and neotropical native is also found in semiarid areas, dense scrubland, woodland, grassland, hilly terrain, savanna, desert, swampland, chaparral, sand dunes, rangeland, and boreal, subtropical, and tropical forests—at all altitudes. With its natural habitat shrinking daily, it is seen more frequently in suburban areas, with interactions between cat and human increasing. This solitary, elusive felid once flourished across the entire lower 48 U.S. states, and at one time had the largest range of any mammal in the Western Hemisphere (which, as noted, is why it has been given so many different names over the centuries). Today, however, it is limited to the Western states and Florida (sightings are occasionally reported elsewhere). Yet, despite the reduction in its range it remains the most widely distributed cat in North, Central, and South America.

Seeks shelter under cliffs and ledges, and in rocks, caves, tunnels, mines, and thick hiding cover. Some individuals may establish both summer and winter home ranges, migrating back and forth with the changing seasons. Requires large corridors of undisturbed land with dense vegetation and abundant game. Extremely territorial, it marks off its 100-500 sq. mi. home range with urine, dung, scrapes, and scratching posts.

Largely a crepuscular and nocturnal hunter, this obligate carnivore can also be active during the day, and will traverse over 2 dozen miles (at a constant pace of 10 mph) in a 12-hr. period in search of prey. Though as an opportunistic hunter it will eat insects, snails, fish, and birds, its diet is mainly comprised of fellow mammals large and small, from mice, marmots, raccoons, rabbits, squirrels, skunks, and hogs, to hares, porcupines, beavers, bobcats, coyotes, deer, and elk. Will resort to cannibalism when necessary. Averages one deer kill a week; known to prey on livestock.

Few alpha species are the equal of this agile ambush predator: it is not only an expert climber and swimmer, it can sprint 50 mph (easily outpacing deer for short distances), leap a 40' gap in a single bound, and spring 15' into the air from a standing position. A part-time ridge-walker, it generally stalks from behind, and also sometimes from above (may jump from rocky overhangs). Nearing its victim, it pounces quickly, delivering a traumatic 200-300 lb. body-slam, while simultaneously digging into the flesh with its sharp claws and crunching down on the neck with its saber-like canines. This bite either severs the trachea (windpipe) or breaks the cervical spine (neck) of its prey. A bite to the head punctures the brain, also leading to a rapid demise. Caches surplus food under brushpiles made of forest litter (sticks, leaves, branches, grass, pine needles), usually sleeping within 150 feet of its stash. It will return to feed on the carcass over several succeeding nights.

Has no specific mating season (though northern populations seem to breed during the winter). The queen and the tom form a temporary pair-bond during her brief estrus cycle, hunting and sleeping together. This relationship dissolves after about 2 wks., the two returning to their preferred life of solitude. The pregnant female gestates for around 3 mos., giving birth to 1-6 altricial kittens in a moss-lined den in a concealed thicket, cave, or rock crevice. The young are covered in camouflage markings (black spots and stripes) that fade with time. They are weaned at

Mountain lion.

around 5 or 6 wks., becoming independent by 2 yrs. of age. The polygynous male does not help with the rearing of his offspring, and in fact, like African male lions, it will kill kittens that are not its own (which stimulates the female to enter estrus again).

Communication and perception channels are typical for a member of the Felidae family: sight, hearing, touch, and smell. Vocalizations include growls, hisses, whistles, purrs, yowls, chirps, squeaks, shrieks, meows, and a humanlike scream called a "caterwaul." When necessary, it uses piloerection to make itself look larger and more fearsome. A secondary consumer and an apex predator, its natural enemies are few. Only wolves, bears, and those of its own kind are equipped to hunt and kill it.

Secretive, usually silent, and living at low density, this feline night-stalker is seldom seen—though it may only be a few feet away. By helping keep the population of ungulates in check, this keystone species is beneficial to us; it is detrimental, however, as a large deadly cat that occasionally attacks and kills both livestock and humans. The mountain lion lives up to 10 yrs. in the wild, though 5 is more typical (failing eyesight and worn teeth make it harder to catch and kill prey; weight loss leads to malnutrition and an ultimately death). In captivity it may live up to 20 yrs.

MOUNTAIN POCKET GOPHER (*Thomomys monticola*): This stocky rodent grows to a total length of 11", its hind foot measures about 1", and it can reach a weight of up to 3 oz. Coloration on the upper body is generally rusty brown with a dull gunmetal gray wash; the lower body is whitish, gray, silvery, or tawny; the 3½" tail is thick, grayish, and sparsely furred. The body is stout; the head is robust; the eyes are small and dark; the ½" ears are medium-sized, erect, and somewhat pointed; each ear has a dark postauricular mark; the muzzle is short, blunt, and darker than the face; long tactile nasal vibrissae are present; the limbs are short; the muscular front legs and strongly clawed front paws are designed for digging.

Pocket gopher.

This solitary geomyid is a nearctic native that inhabits temperate open meadows in and around montane coniferous forests. Also an edge species, it often frequents the borders between fields and woodlands, as well as between rocky hillsides, intermontane regions, and mountains; prefers damp, loose, pebbly soil. Its range is limited to small portions of California and Nevada. Fossorial, it excavates extensive tunnel systems where it spends much of its life; in winter it lives in subnivean burrows; plugs up the entrances as protection against predators and to control the climate underground. Does not hibernate; active year-round. Establishes a home range of up to several thousand square feet.

A strict herbivore, its diet consists of various types of vegetation, from roots, weeds, and grasses, to forbs, shrubbery, and trees. Caches surplus food, transporting it in its external cheek "pockets" or pouches. Its breeding season takes place in summer. The female gestates for approximately 3 wks. and gives birth to about 4 pups.

Both diurnal and nocturnal, it communicates and perceives via standard Geomyidae channels: primarily touch and smell, and to a lesser degree, sight and hearing. Its natural enemies are reptiles, birds of prey, mustelids, wild felids, and wild canids. Beneficial as a soil aerator and recycler, as a driver of plant diversity, as food for carnivores, and as a provider of subterranean homes for other animals. The mountain pocket gopher probably lives about 2-3 yrs. in the wild; perhaps twice that long in captivity.

MULE DEER (*Odocoileus hemionus*): (Note: This entry includes the black-tailed deer, *Odocoileus hemionus columbianus*, a smaller subspecies of the mule deer that shares the same general traits, habitat, and range.) The "mulie," as it is commonly known, stands 3½' at the shoulder, grows to 6½' in total length, and weighs up to 450 lb. The male's antlers (which females lack) begin to grow in the spring, are shed in early winter, and can reach a spread of 4'. Its dorsal coat varies widely in coloration, from reddish-brown in summer to grayish-brown or blue-gray in winter. In all seasons its ventral coat, including the inside leg, throat, and rump patch, is creamy-tan to white. A grayish dorsal stripe may run down the back to the rump. This species is easily distinguished by its large 6" mule-like ears (from which it derives its common name), its light-colored face, its stocky body, and its short, all-white, black-tipped tail. Gender dimorphism is evident: adult bucks are heavier and larger than adult females.

Mule deer (male).

This highly adaptable cervid inhabits nearly every biome in the western part of our region, from arid desert, dune, brushland, rangeland, foothills, and chaparral, to savanna, riparian areas, grassland, plains, rainforest, and boreal mountain forest. Water sources, southern exposure, mineral licks, adequate fawning cover, and thermal cover are important. It seeks out both dense hiding cover and obstacle-rich environments (gorges, windfalls, broken terrain, sandy areas, steeply sloped mountains) that are apt to discourage predators. Its range is extensive, covering much of western Canada, almost all of the western half of the U.S., and much of northern Mexico.

A crepuscular browser, grazer, and ruminant, this large familiar artiodactyl rests during the day in a depression bed, feeding and moving about at dawn and dusk. An herbivore that is more specifically a graminivore, lignivore, granivore, frugivore, and folivore, it forages mainly on grasses, herbaceous plants, shrubs, forbs, fruit, seeds, nuts, and cultivated crops in the summer, and tree parts in the winter (wood, twigs, needles, leaves). Migrates as much as 100 mi. seasonally: to higher elevations in summer, lower ones in winter.

Males and females are social, but, depending on the season, tend to congregate in their own small gender-specific groups (some individual mature bucks tend to be solitary). Due to survival instinct they are more likely to "yard up," or herd, during the harsh winter months than at any other time of year. Home ranges run from 1-25 sq. mi., and are bigger among bucks than does.

Its breeding season depends on the region and ecosystem it inhabits. However, generally it runs from late fall to early winter. Bucks engage in jousting contests with their great racks, the most dominant individuals procuring the most mating privileges. Occasionally, however, antlers become locked and both bucks die of starvation. The doe gestates for about 6½ mos., then seeks out a secluded thicket where she gives birth to 1-2 white-spotted, odorless fawns in the summer. Communication and perception occur via standard Cervidae channels: sight, body language, touch, and a variety of specific scents that are produced by glands located over the body.

Mule deer (females).

A primary consumer, its natural enemies are wild canids, wild felids, and ursids. It employs a number of evasion tricks, relying on visual cues (using its excellent binocular eyesight), olfactory cues (using its keen sense of smell), and aural cues (using its big, independently moving, radar-like ears). Defensive tactics include its cryptic coloration, freezing in place, concealment, swimming, leaping, and running (up to 45 mph) in a zigzag pattern. It also employs stotting (springing up into the air and landing on all four feet), a bounding maneuver that confuses and deters predators—particularly when it stotts uphill. The mule deer lives 5-10 yrs. in the wild, and to at least 20 yrs. in captivity.

MUSKOX (*Ovibos moschatus*): Though it is a member of the ox or cow family, as its genus name indicates, this hoofed arctic artiodactyl has traits of both sheep (*ovis*) and ox (*bos*), and is actually more closely related to the former than the latter. It is, in fact, both a caprid (a member of the goat-antelope family) and a bovid (a member of the cattle family).

Muskox.

It grows 6'-8' in total length, stands 5½' tall at the shoulder, and can weigh up to 900 lb. It has a stocky, bison-like body; a large broad head; thin lips; humped shoulders; short neck; small ears; heavily built legs; and a short tail. Its coat is made up of 2 types of fur: an inner layer of soft thick hair known as qiviut, and a protective outer layer comprised of rough guard hairs, providing insulation against the cold in winter and protection against insects in summer. Its shaggy coat—which at 23" long, reaches to its knees—covers nearly the entire body and is dark brown to black, with a light "saddle" on the back and pale or white lower legs.

The most distinguishing trait among both males and females is a set of broad massive horns, which curve down alongside the head, then flare out and up, ending in sharp hooked tips. Light brown to dark gray in color, the horns merge on the forehead, creating a hefty 8" thick "boss" at the center. Large wide hooves aid in navigating through snow. Gender dimorphism is present, with the males (and their horns) being larger than females.

Primarily a circumpolar inhabitant of alpine tundra and arctic coastlines, in summer this nearctic native migrates into meadows, river valleys, and lakeshores where green food is plentiful; in winter it moves onto exposed hills, plateaus, and slopes, where wind clears the snow off vegetation. One of the few large animals on earth capable of living in arctic conditions all year long, its range in our region is today quite narrow, spanning only from northern Alaska into northern Canada. Depending on the season and terrain, its home range extends from 25-150 sq. mi.

In summer this large herbivorous grazer feeds on grasses, forbs, sedges, lichens, fruits, grains, mosses, seeds, nuts, flowers, and herbs; also a lignivore and a folivore, in winter its diet consists mainly of lichens and woody vegetation (wood, leaves, stems, bark). A social species, it may be found in mixed herds of 10-100 individuals. Some males, however, are solitary or choose to live in bachelor herds. During the summer rut, dominant bulls form small harems, which they defend in terrific head-butting battles. Speeding headfirst at one another at 25 mph, the crash of skulls and horns (cushioned by their heavy forehead bosses) can be heard over a mile away.

The cow, who has a relatively low reproductive rate for a mammal (often becoming pregnant only once every other year), gestates for about 8½ mos., giving birth to a single precocial calf in the spring. It is weaned and becomes independent in 1-2 yrs. It is not known if the male contributes directly to the raising of the young. When endangered, the main herd forms a living circular barrier around the calves, at which time several bulls may launch an attack on their pursuer (a type of indirect care-taking of newborns and yearlings).

Muskox use acoustic forms of communication that are typical of bovids: snorting, grunting, roaring, and bleating. They also use body language (such as head-swaying), and tactile cues (such as aggressive shoving) and chemical cues. The latter form of communication refers to its "musk," from whence it derives its common name. This eponym, however, is misleading: while the male muskox indeed exudes a musky odor during breeding season, it is not musk, for it has no

Muskox (juvenile).

musk glands. The powerful smell derives from its urine, which it dribbles on itself as a dominance display to other bulls.

A primary consumer, its natural enemies are wolves, grizzly bears, and polar bears. Its thick body, heavy fur, and short legs usually make outrunning predators impractical. Thus its main methods of defense are its aggressive nature, hooked horns, and nearly impregnable circle formations. Due to its harsh and violent lifestyle the male muskox probably lives only 10 yrs. or less in the wild; the female may live 20-25 yrs. in the wild.

MUSKRAT (*Ondatra zibethicus*): Also known as the "common muskrat," this medium-sized, beaver-like, semiaquatic rodent grows to a total length of 25", its hind foot measures nearly 4", and it can attain a weight of up to 4 lb. Coloration on the upper pelage is dark brown and is occasionally mixed with a blonde to rusty tint; the lower pelage is light brown; the long 1' tail is blackish, vertically flat, scaly, sparsely furred, and pointed. The outer guard hairs are long, oily, and rough, while the undercoat is short, thick, and soft. The body is large and stout; the head is robust; the eyes and ears are small; the muzzle is massive; long gray vibrissae cover the snout; the nose and nostrils are large; the limbs are short; the large, heavily clawed front feet are designed for digging; the large, slightly webbed rear feet are designed for swimming and diving. Takes its common name from the strong musky odor it emits from its scent glands and its rat-like appearance.

Muskrat.

Smaller than the beaver and the nutria (both with which it is often confused), this territorial, sedentary, riparian cricetid is a nearctic native that inhabits temperate watery biomes and herbaceous ecosystems, from forested marshes, swamps, muskeg, bogs, and ponds, to streams, rivers, canals, and lakes; it can be found in freshwater, brackish water, or saltwater wetlands; prefers shallow water with a depth of not more than 5'-6'. Its range is extensive, covering nearly all of North America, from Alaska to Labrador, from California to New England, from North Dakota to Texas, from the Midwest to the Midsouth and on into northern Mexico. It is rare in the extreme Southeast states, however, particularly Florida. Establishes a home range of about 300 sq. ft.; it is usually solitary, though individual home ranges may overlap and den-sharing may occur during the cold months. Does not hibernate; active year-round and at all times of day and night.

Builds massive, complex, round, beaver lodge-like houses out of mud, sticks, logs, and cut plant material in and around watery areas; its house is positioned on a strong support, such as a mat of aquatic weeds, the pond bottom, or a stump; some muskrat homes may grow to nearly 10' in diameter and reach a height of 6'; each dome-shaped lodge contains multiple chambers, subsidiary tunnels, and portals that are located both above and below water; this species digs dens into stream banks and shorelines; feeding platforms (elevated piles of mud and weeds), scent mounds, and abandoned reed cuttings will be found nearby. Spends most of its life in water.

Muskrat.

A crepuscular and nocturnal omnivore, its diet centers mainly on the roots, stalks, and leaves of aquatic vegetation, such as rushes, sedges, cordgrass, cattails, and water lilies; it also consumes land plants (such as ferns), as well as insects, snails, fish, frogs, crayfish, and clams; it is known to engage in cannibalism, with a preference for neonates. Its breeding season generally runs from winter to fall. The iteroparous

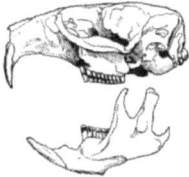

Muskrat skull.

polyestrous female can produce up to 5 litters annually; she gestates for approximately 1 mo., giving birth to as many as 15 altricial pups in the nest chamber; within a few weeks they can swim; by 4 wks. they are weaned; by 6 mos. of age they are reproductively mature; and within a year they are fully independent.

Communication and perception take place via standard Cricetidae channels: mainly touch and smell, and to a lesser degree, sight and hearing. One of our most skilled mammalian swimmers, it paddles with its hind feet, steers with its tail, can remain submerged for nearly 20 mins, and can swim backward as effortlessly as it can swim forward; bristly, fringed toe-hairs aid in propelling it through the water. Its lush coat captures air bubbles, serving as an additional layer of protection against cold while simultaneously operating like a scuba diver's buoyancy vest. As with its cousin the beaver, its mouth closes behind its protruding front teeth, permitting it to both eat and transport building material underwater. A regional heterotherm, it can change the blood flow in specific parts of its anatomy, aiding in regulating its overall body temperature.

Its natural enemies are reptiles, birds of prey, mustelids, procyonids, and wild canids. Beneficial as food for carnivores and as a provider of meat and valuable pelts for humans; detrimental as an agricultural pest, as a disease carrier, and as a destroyer of human-made structures. The muskrat seldom lives more than 4 yrs. in the wild; it can live twice that long in captivity, however.

Muskrat house.

Sperm whales.

NARROW-FACED KANGAROO RAT (*Dipodomys venustus*): Shares much of the same ecology and life history as its close cousin, Heermann's kangaroo rat. The primary differences are that it is darker and has larger ears than Heermann's, while its range is much smaller, extending only along the montane coastal regions of central California.

NARROW-HEADED VOLE (*Microtus gregalis*): Little is known about this rodent. The only member of its subgenus, *Stenocranius*, it grows to a total length of nearly 5" and a probable weight of 1 oz. to 1½ oz.; its bicolored tail measures approximately 1" or so. Depending on the season, its pelage coloration on the dorsum runs from buff to grayish to reddish; on the ventrum it is light buff.

This cricetid inhabits open elevated grassland, meadows, and alpine steppe, with a range spreading across Alaska and, outside our region, throughout northern Europe and Asia. Fossorial, it excavates intricate burrow systems with multiple portals, chambers, and subsidiary tunnels. Its narrow skull, from which it derives both its common name and its subgenus name (*steno* means "narrow," *cranius* means "skull"), is an adaptation to its subterranean lifestyle. Diurnal, crepuscular, and nocturnal, it is an herbivore that feeds on seeds, grass, grains, green vegetation generally, and probably fruit.

Its breeding season occurs during summer, with the iteroparous polyestrous female producing as many as 4 litters annually, each with 10 or more pups. Its natural enemies are likely those typical of the smaller members of the Rodentia order: reptiles, birds of prey, mustelids, wild felids, and wild canids. The life span of the narrow-headed vole has not been researched.

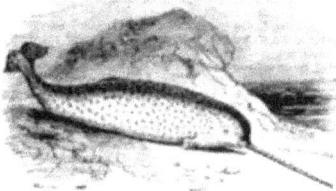

Narwhal.

NARWHAL (*Monodon monoceros*): One of only two cetacean species belonging to the family Monodontidae (the other is the beluga, with which it sometimes interbreeds), the narwhal grows to 18' in total length and weighs up to 3,500 lb. The dorsum is a mottled blue, gray, or black, while the ventrum is white. Males become more white with age.

It has only two fully formed teeth, both canines in the upper jaw. In the male, normally one of these grows into a long spiraled tusk that projects forward from the left side of the mouth (only rarely does it grow from the right). This enlarged hollow tooth can reach a length of 10' and weigh up to 25 lb.; the second canine remains rudimentary. (On rare occasions a male will grow 2 tusks. Even more uncommonly, a female will grow 1 or 2 tusks, though these are usually smaller than the males.) This odontological phenomenon gives it its scientific name, which is Greek for "one-tooth one-horn." The tusk's exact function is not known, but it would seem to serve several purposes: a weapon (males use them to "fence" and possibly "joust"); a sensory organ (containing millions of nerve endings) for collecting data from the narwhal's environment; and a hunting tool (used to strike and stun fish).

Preferring the deep cold waters of the Arctic Ocean, its habitat and range are quite restricted, mainly to Canada, Iceland, Greenland, Norway, and Russia. Closely related to dolphins and porpoises, this medium-sized monodontid migrates inland toward the coast in the spring; in winter it migrates back out to the deeper waters beneath the pack ice. It will gather in mixed- and sometimes gender-specific pods of 500 individuals or more.

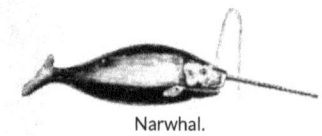

Narwhal.

Like its cousin the beluga (as well as land mammals), this cetacid has jointed rather than fused neck vertebrae, giving the head great flexibility of movement, aiding in hunting, navigation, and defense. It also shares with the beluga a large melon (a rounded sensory organ set high in the forehead), a short snout, small flippers, and a missing dorsal fin, which Nature has replaced with a dorsal ridge (in both species) that runs along the back.

A suction-feeding carnivore, this pelagic polar homeotherm swims slowly, using echolocation to find its favored foods: cod, halibut, squid, redfish, shrimp, flatfish, and wolffish. Will spyhop (lift its head out of the water to look around) and slap the surface with its flippers. Can dive down to a depth of 5,000', and hold its breath for nearly 30 min. A secondary consumer, its main natural predators are sharks, polar bears, orcas, and walruses. Vocalizations include clicks, knocks, squeaks, and whistles. The narwhal, known as the "unicorn of the sea," can live up to 50 yrs. or more in the wild.

NELSON'S ANTELOPE SQUIRREL (*Ammospermophilus nelsoni*): This chipmunk-like ground squirrel grows to a total length of about 10" and a weight of around 6 oz. Its ecology and life history are nearly identical to its close relation the white-tailed ground squirrel, except that it is slightly larger and occupies a smaller range: central California.

NELSON'S POCKET MOUSE (*Chaetodipus nelsoni*): Named after American naturalist Edward William Nelson, this medium-sized pocket mouse grows to a total length of almost 8" and reaches a weight of up to nearly 1 oz. The base fur coloration on the upper body is tan to rusty and is overtopped by a grizzled gray-brown-blackish wash; the lower body is white to tan; the long 5" tail is dichromatic (dark on top, lighter below) and has a crest near its tufted tip. Like many other pocket mice, this rodent has bristly dark-tipped hairs or spines on its posterior. The fur is coarse; the body is thinnish and cylindrical; the head is large (for its body size); the eyes are dark; the ears are small, rounded, and erect, and each one has a subauricular mark; the front feet are small; the hind feet are large; all 4 paws have black soles. Gender dimorphism is present, with the male being slightly larger than the female.

Cactus scrubland: Nelson's pocket mouse habitat.

This terricolous heteromyid is a riparian nearctic native that inhabits temperate rocky desert, shrubland, hillsides, grassland, sand dunes, sand flats, stone mounds, scrubland, chaparral, and piney woodland; has a preference for living in and around pebbly soil, chino grass, cactus, and creosote bushes, the latter which it uses as hiding cover while foraging at night. Its range is limited to small portions of New Mexico, Texas, and northern Mexico. Fossorial, it excavates burrows among rock-strewn settings with entrances concealed under shrubbery. Does not hibernate; active year-round. Establishes a small home range, seldom wandering far from its subterranean shelter.

A nocturnal omnivore, its diet consists of standard pocket mouse fare: seeds, grains, roots, wood, nuts, tubers, bark, stalks, leaves, and insects; caches surplus food in its burrow larder, transporting it in its external, fur-lined cheek "pockets" or pouches—the source of its common name. Its breeding season begins in late winter and ends in midsummer. The female gestates for about 1 mo., giving birth to 2-4 altricial pups in a grass-lined underground nest chamber; weaning occurs at around 1 mo., with reproductive maturity and independence occurring shortly thereafter. The father does not appear to contribute to the rearing of his offspring.

This chaetodipid communicates and perceives via normal Heteromyidae channels: sight, touch, smell, and hearing. Its natural enemies are reptiles, birds of prey, wild felids, and wild canids. Beneficial as a seed disperser, host for parasites, soil aerator and recycler, and as prey for carnivores; detrimental as a house and agricultural pest. Nelson's pocket mouse seems to live no more than 1-2 yrs. in the wild; slightly longer under optimum conditions.

Edward William Nelson.

NEW ENGLAND COTTONTAIL (*Sylvilagus transitionalis*): This medium-sized lagomorph shares many physical traits with its cousin the eastern cottontail, and from a distance the two may be difficult to distinguish. It grows to a total length of 19", its hind foot measures 4", and it weighs up to 3 lb. Its pelage is tawny to cinnamon brown with black mottling. The eyes are large; the ears are short and fringed in black; the underside of the tail is white (the "cottontail"); and there is often a black mark on top of the head.

It inhabits forests and brushy areas throughout, as its common name indicates, New England. Secretive and wary, it is rarely seen in the open, preferring the protection of the underbrush in its ½ acre home range. In the warmer months this herbivorous leporid feeds mainly on grasses and forbs; during cold months it consumes bark and twigs. Like many other rabbits and hares, it engages in coprophagy, a means of procuring additional nutrients.

Its breeding season ranges from January to September. The doe gestates for about 1 mo., giving birth to 3-8 altricial kits. Polyestrous, she may have as many as 3 litters in 1 year. The young are weaned at around 2 wks. of age.

New England cottontail.

Communicates using a wide variety of sounds, from foot-thumping, screaming, and purring, to huffing, crying, and wailing. The enemies of this primary consumer are identical to the eastern cottontail: reptiles, birds of prey, weasels, bobcats, foxes, and coyotes. The New England cottontail is a short-lived rabbit, with a life span of only 3 yrs. (or less) in the wild. However, it can live up to 10 yrs. in captivity.

NINE-BANDED ARMADILLO (*Dasypus novemcinctus*): The official state mammal of North Carolina and the official small state mammal of Texas, this solitary North American cingulatid grows to 2½' in total length and can weigh up to 17 lb. Its body is generally tan in color, with many variations, including brownish to yellowish skin. It is sparsely furred (with bristly hair); the nose is narrow and pig-like; the ears are soft and erect; the long 16" tail makes up half the total body length. Its legs are short; the feet are heavily clawed, with 4 toes on the front feet and 5 toes on the hind feet. Its long

Nine-banded armadillo.

sticky tongue is well adapted to foraging for insects in small spaces. Its teeth are small, rudimentary, and peg-like.

It is easily identifiable from its suit of hard bony plates or scutes, which cover the head, body, and tail—the only mammal in North America with such living armor. Despite their rigid appearance, the plates are quite elastic, allowing the animal a wide range of movement. Their flexibility derives from the fact that they are made of a keratinous tissue similar to our fingernails. Contrary to popular belief, however, the nine-banded armadillo does not, and cannot, roll itself up into a ball. Only 1 species, the three-banded armadillo (*Tolypeutes tricinctus*) is capable of this. And despite its common name, the nine-banded armadillo does not always have 9 bands of plates; it may have as few as 8 or as many as 12. It is gender dimorphic: the male is larger than the female.

This secondary consumer shelters in 6' deep burrows, which it digs in clay, or preferably in soft sandy soil along waterways, throughout its habitat: temperate to tropical forest, bottomland, scrubland, hardwood forest, savanna, brushland, and grassland. Primarily an inhabitant of the American Southeast, it is a riparian that can be found near water holes and streams. Nevertheless, its range is expanding further northward and westward each year (though its dislike of extreme cold weather has thus far prevented it from moving into the most northern parts of our region).

A fast runner, an adept hopper, and a powerful digger (when trying to escape danger it can create a small shallow burrow in seconds), it can also swim, as well as walk underwater along the bottoms of rivers and ponds, holding its breath for up to 6 mins. Using its keen sense of smell to track down its prey, this crepuscular, diurnal, nocturnal omnivore consumes a wide variety of foods, from fungus, tree bark, seeds, fruit, and farm crops, to earthworms, insects, birds' eggs, amphibians, and reptiles. Also an opportunistic scavenger, it will eat carrion and even small mammals, many of them considered pests and vermin—making it of benefit to us. A geophage, it sometimes consumes dirt while eating its prey (it is not known if this is accidental or intentional). Does not hibernate.

The male and the female form a temporary pair-bond during the breeding season in early summer. The female delays implantation (of her fertilized egg in the uterine wall) for nearly 4 mos., then gestates for another 4 mos., bearing up to 4 identical precocial pups in the spring. This delay means that the weather will be milder and food more plentiful for both mother and neonates. She nurses her young for as long as 3 mos., after which they are weaned and ready for independence.

Nine-banded armadillo.

This dasypodid can carry the bacteria associated with leprosy, as well as a host of other diseases that affect humans; even its aerosol is potentially lethal. Its natural enemies are birds of prey, alligators, wild felids, and wild canids. The nine-banded armadillo lives an average of 15 yrs. in the wild.

NORTH AMERICAN PORCUPINE (*Erethizon dorsatum*): Also known as the "Canada porcupine," this big stocky mammal is the second largest rodent in our region (after the American beaver). Its name derives from the Latin words *porcus* ("pig") and *spina* ("spine"), loosely meaning "the pig-like animal with bristles." Its grows to a total length of 3' and it can attain a weight of up to 40 lbs. The overall pelage is dark brown to black. On the back and sides this coloration is mixed with long, shaggy, pale yellow- to rusty-tinged guard hairs.

Most notably, some 30,000 sharply pointed modified hairs known as "quills" run along the back to the tail, making it the only mammal in North America with this type of spiny defensive armor. Semi-hollow with hard, microscopically barb-covered tips, these are lightly affixed (by a ligament) to a follicle embedded in a layer of voluntary muscles beneath the skin. There are often several white stripes crossing over the back.

North American porcupine.

Comprised of white-tipped hairs, these somewhat indistinct markings function as aposematic coloration, a warning signal to would-be predators.

The limbs are muscular, the paws are heavily clawed, and the sole pads are covered in coarse bumpy skin, arboreal and scansorial adaptations for life in the trees. The forelegs are much shorter than the hind legs, giving the back an arched appearance. The 11" tail serves as a balance while climbing and also as a defensive weapon. Gender dimorphism is present: the male is larger than the female.

This terricolous sedentary erethizontid is a riparian nearctic native that occupies a wide variety of biomes, habitats, and ecosystems. Although primarily a mixed forest dweller, it is also found in grassland, desert, shrubland, rangeland, tundra, sand dunes, taiga, scrubland, chaparral, and savanna, from valleys to mountains, from warm climates to cold ones. Prefers locations with snag, deadfall, logs, stumps, forest duff, blowdown, and hiding cover. As its common name suggests, it is the only porcupine in North America, and as such its range reaches further north than any of the world's other 25 or so porcupine species. Its distribution is extensive, covering much of Alaska and Canada, as well as most of the 48 contiguous U.S. states, south into northern Mexico. It is rare or nonexistent, however, in the American Southeast and parts of the Midwest. Does not hibernate; active year-round.

Primarily nocturnal, it does not build nests; instead it dens inside hollow trees, rock crevices, subterranean burrows, or beneath logs and tree roots. Mostly solitary and territorial, this expert climber establishes a home range of at least 5,000 sq. ft., which it defends by scent-marking and aggressive behavior. Like some other mammals, it may occasionally form loose colonies with conspecifics under duress, such as extreme weather, food shortages, or predator danger.

Unlike specialized herbivores, the North American porcupine is a generalist herbivore with a wide-ranging but rigid vegetarian appetite. Despite this, it is primarily a lignivore, one with a specific and keen taste for tree parts, from buds, twigs, roots, fruit, stems, to phloem, leaves, seeds, nuts, and pine needles. One sign of its presence are piles of dung, which it leaves at the bases of trees as it forages. It can inflict tremendous damage on both wild and cultivated trees: it not only clips off small branches, but it gnaws through bark, then strips it from the tree until it reaches the nutritious cambium (inner bark) underneath. Debarking exposes trees to disease and insect infestations, posing a threat to both delicate forest ecosystems and the lumber industry. This waddling docile mammal also feeds on grains, flowers, grasses, herbs, and tubers.

North American porcupine.

Its polygynous breeding season takes place between fall and early winter, with vocal vigorous rivalry between males for dominance and mating rights. The female gestates for almost 7 mos. (long for a rodent), giving birth to a single precocial pup in the spring. She nurses, grooms, and protects her youngster (known as a "porcupette") for the next 4-5 mos., after which it becomes independent. The father does not display paternal behaviors, and thus does not contribute to the rearing of his offspring.

Nonaggressive and poor-sighted, it communicates and perceives along standard Erethizontidae channels: hearing, touch, and smell, with vision being its weakest sense. One of our most loquacious mammals, its vast array of vocalizations include squeaking, moaning, wailing, grunting, whining, groaning, shrieking, howling, and teeth-clacking. When disturbed it may rattle its tail quills, a form of acoustic aposematism that can be quite intimidating to its natural enemies: birds of prey, mustelids, wild felids, and wild canids.

North American porcupine.

Its primary defense against predators is that from which its common name derives: quilling, which utilizes the thousands of long light-weight spines that normally lay flat against its body. It does not "throw" its quills, as is commonly believed; rather, when alarmed, it raises them by retracting the muscle tissue they are attached to. It then lowers its head, emits a foul odor, and backs up toward its opponent with its tail swinging. If the predator brushes up against any of the erect spines, or if the erected tail spines make contact, they easily detach from the porcupine and puncture the foe's skin. After entering the epidermis, the victim's body heat causes the barbed bacteria-covered quills to expand, while at the same time its muscle movements pull them ever deeper into its flesh. Due to their tiny, backward-facing, fishhook-like barbs, the spines are difficult if not impossible to remove. Infection soon sets in, triggering a multitude of serious physical effects, from pain, irritation, trauma, and abscesses, to blindness, joint pain, puncturing of vital organs, and swelling of the throat (usually accompanied by the inability to eat). Starvation and death are often inevitable.

This familiar endotherm is beneficial as a parasite host and food for carnivores, but it is detrimental as an agricultural pest (it readily eats farm and orchard products such as corn, grains, vegetables, and fruits), a tree destroyer (trees that have been attacked and girdled by porcupines often have a gnarled "spooky" appearance), and a household pest. Besides eating ornamental plants, to satiate its craving for sodium chloride it will gnaw on anything containing salt, including furniture, wooden house walls and siding, and the handles of wooden tools; it will also chew on vehicles that have accumulated road salt. One of our longest-lived rodents, the gentle slow-moving North American porcupine lives for an average of 8 yrs. in the wild; at least twice that long in captivity.

NORTH AMERICAN RIVER OTTER (*Lontra canadensis*): Also known as the "Canadian otter" or "northern river otter," it grows to total length of 5', its hind foot measures nearly 6", and it can attain a weight of up to 30 lb. Its dorsal coat is dark chocolate brown, its ventral coat is light brown, and its throat and chest may be sienna, grizzled gray, white, cream, or cinnamon (color variations are common).

This highly energetic semiaquatic mammal is North America's only river otter. As a riparian it is specially adapted to life in water: it has a sleek torpedo-shaped body, a small broad head, short rounded ears, and a long neck (all enhancing its aquadynamics). It eyes and nose are high-set (for swimming at the surface); it has long stiff vibrissae (that act as sensory antennae in dark forests and in the murky depths of rivers and ponds—the latter environment which restricts its hearing); short muscular legs (well adapted for paddling); water-repellant fur (shuts in heat and shuts out cold); ear- and nostril-valves that seal tight when underwater;

North American river otter.

webbed feet (that increase its speed and agility when submerged); non-retractable claws (for digging and running); and a muscular, tapered, flattened 20" tail that makes up nearly half its body length (used as a rudder and for propulsion).

Possesses specialized teeth designed for gripping, puncturing, and crushing. Both eyes possess a nictitating membrane, or third eyelid, which, like swimmers' goggles, allow it to see clearly when beneath the surface. (Its excellent underwater vision has resulted in it being somewhat nearsighted on land.) The male is larger than the female, thus gender dimorphism is present.

This natatorial nearctive native inhabits unpolluted, undisturbed riparian zones made up of freshwater, saltwater, or brackish water systems. These include: streams, rivers, marshes, ponds, tidal flats, swamps, muskeg, wetlands, bogs, reservoirs, estuaries, lakes, watersheds, canals, isthmuses, and coastal and littoral zones (though it will sometimes wander about in waterless areas). Its common name, "river otter," is thus not entirely accurate. Motile and highly adaptable, it can live in low or high altitudes and in cold or hot regions. Its range covers most of Alaska and Canada, through the American Northwest, upper Midwest, New England, and down along the eastern coast to Florida. It occurs as far south as the Rio Grande and Mexico.

North American river otters at play.

Its home range can extend up to 40 sq. mi. Sedentary, it permanently dens in riverbanks with entrances on both land and underwater. Prefers holt locations with rocks and dense vegetation. A secondary cavity user, it will take over the burrows of other mammals (such as muskrats and beavers, with which it often shares territory), customizing them with leaves, moss, sticks, reeds, grass, bark, fur, and other debris. Uses stump cavities, bank undercuts, hollow logs, and root overhangs as resting spots.

A diurnal, crepuscular, and nocturnal piscivore, its diet is comprised primarily of fish; but it also feeds on aquatic plants (tubers, algae), insects, invertebrates (crustaceans, mollusks, clams, snails), amphibians (salamanders, frogs), reptiles (turtles), birds (and their eggs), and small mammals (mainly rodents, such as mice and squirrels). This fearless ambush predator is known to attack pets. Will travel up to 20 mi. a day foraging and hunting. Due to its active lifestyle and high metabolism, it must spend much of its time locating food and eating.

This species does not hibernate, and is active all year-round. Normally nocturnal during warm months, it becomes increasingly diurnal as the cold months approach. Both social and solitary, depending on a number of factors (population density, food availability, time of year), it may live alone or with its family, or sometimes form large temporary groups called "romps," made up of unrelated individuals.

Dexterous and agile on both land and in the water, with its heavily built limbs and thick powerful tail, this sleek mustelid can swim forward or backward, swim under ice, propel itself up to 8 mph through the water, dive to a depth of 60', swim over 1,000' underwater (without coming up for air), and stay submerged for up to 8 min. While all mammals possess a wide range of emotions, the river otter is one who displays an overt sense of playfulness and fun. It thoroughly enjoys, for example, wrestling, teasing, chasing, and rolling about in water, mud, and snow with family or fellow romp members.

Favorite recreational activities include tobogganing. This it does by creating 1' wide slides (in winter on snowbanks; in summer on muddy riverbanks), some as long as 25', on which it races up and down with obvious delight. It will also roll on the ground, dive into the water from shore, and body surf (atop fast flowing currents) for sheer pleasure (sometimes river otters slide simply as a convenient and efficient mode of travel, such as when descending hills and mountains). These social activities (which can include traveling and hunting together, as well as allogrooming) are not just agreeable to the otter, they also serve to solidify family or group bonds while simultaneously marking off an individual's home range.

Its breeding season takes place in late spring to early summer. The sow is a delay-implanter. Thus, even though she gestates for only 2 mos., her babies can be born as much as a year after mating. Around this time, the following spring, she gives birth to 1-5 altricial pups in a concealed stick nest located near water. When they are 2 mos. of age the mother begins teaching her young how to swim and hunt; they are weaned by 4 mos., and begin dispersing after 6 mos. Normally sociable, during the mating season boars fight over estrus females and territory, the latter which they scent-mark with scratches, urine, and dung, as well as with a malodorous musk they emit from two specialized glands under their tails. The polygynous male seems to play no role in the rearing of his offspring.

Communication and perception take place through the usual Mustelidae channels: primarily sight (body language), but smell, hearing, and touch as well. Vocalizations include growling, snorting, chuckling, chirping, purring, buzzing, chattering, whistling, twittering, grunting, and screaming (the latter a piercing alarm call that can be heard nearly 2 mi. away). A

North American river otter.

secondary consumer, its natural enemies are birds of prey, reptiles, wild felids, and wild canids. Because it consumes undesirable fish and pestilent rodents, it is beneficial to humans. The North American river otter lives to an average age of 12 yrs. in the wild; up to 27 yrs. of age in captivity.

NORTH ATLANTIC RIGHT WHALE (*Eubalaena glacialis*): The official state marine mammal of Georgia and Massachusetts, and the official state migratory marine mammal of South Carolina, it is also known as the "black right whale." It grows to 60' in total length and can attain a weight of up to 100 tons. Its heavy body is dark gray to bluish-black with irregular white markings on its underside. Its enormous head makes up 25 percent of its total body length; it lacks a dorsal fin; has wide paddle-shaped pectoral flippers; a deeply notched tail; and it is the only member of the rorqual or Balaenopteridae family with a strongly arched mouth. The top of the head as

North Atlantic right whale.

well as the lower jaw are usually covered in knobby growths called "callosities" that are unique to each individual, and thus are often used in identification. This species is gender dimorphic: the female is larger than the male.

In our region it inhabits the temperate coastal waters of the Atlantic Ocean, migrating north (as far as Newfoundland) in spring, and south (as far as Florida) in the fall. A filter-feeding planktivore, its black baleen reaches 9' in length, a long fibrous broom-like material attached to the upper jaw (and which has replaced regular teeth). Through its baleen plates the whale skims and strains its main food, zooplankton, consuming up to 2½ tons a day.

Like all balaenids, it has paired blowholes (toothed whales have only 1). From these it spouts 2 columns of water which form a v-shape, blasting some 15' into the air.

Though it can hold its breath for up to 30 min. or more, it is a slow swimmer (with a top speed of only 10 mph) and a shallow diver (it seldom goes below 1,000'), spending most of its time near the surface, either alone or in small highly social pods of a dozen or so individuals.

Because it is a baleen whale it does not need (and thus does not have) an echolocation organ (as toothed whales do). It communicates with conspecifics using a variety of low frequency moaning and pulse-like sounds that carry great distances underwater. It shares with its cetacean relatives, however, a love of fun and acrobatics, and is fond of leaping from the water and lobtailing.

Its breeding season takes place in winter. Bulls and cows do not form pair-bonds and will mate with numerous individuals, making this species polygamous. The female breeds only once every 3 yrs. or so. She gestates for about 1 yr., giving birth to a single 20' precocial calf the following winter or spring. It is not known how long it takes for the infant to attain independence. The male does not assist in the rearing of his offspring.

Commercially speaking, its gentle, playful, lethargic behavior, coast-hugging habits, highly oily (and thus valuable) blubber, and easy approachability has long made it the "right" whale. As a result, today, after centuries of over-hunting—combined with the modern hazards of collisions with large ships (vessel strikes), oceanic pollutants, anthropogenic noise, habitat loss, and fishnet entanglements—this cetacid is one of the world's most endangered animals.

A secondary consumer, its natural enemies are sharks and orcas (which feed primarily on the young). The North Atlantic right whale may have once lived up to 100 yrs. or more in the wild; however, due to modern widespread human interference, its life span has been greatly reduced, possibly to as little as 10-20 yrs.

North Atlantic right whale.

NORTHERN BOG LEMMING (*Synaptomys borealis*): This small rodent grows to a total length of about 5", its hind foot measures around ¾", and it can reach a weight of up to 1¼ oz. The rumpled-looking dorsal pelage is brownish to grayish; the ventral pelage is white or silvery gray; the short 1" tail is bicolored (brown on top, tannish-white on the bottom). The coat is coarse; the body is stout and cylindrical; the head is large; the eyes are small and dark; the ears are small and concealed within the fur; the muzzle is snubbed; sensitive vibrissae cover the snout; the limbs are short; the feet are small and delicate.

This microtine cricetid is a riparian nearctic native that primarily inhabits, as its common name indicates, sphagnum bogs; but it is also found in and around temperate tundra, moist savanna, moss-dominant creek banks, spruce forests, alpine fens, meadows, and wet grassland in mountainous regions; prefers settings with logs, stumps, and abundant forest duff, which it uses for hiding cover. Both its common name and its species name point to its range: the mid northern tier of North America, extending over much of Alaska and Canada, and south into a few of the northern states (Washington, Idaho, Montana, Minnesota, and Maine). Lives both on the surface and below ground, fashioning burrow systems, grass runways, and nests to suit its environment and the season. Does not hibernate; active all year. Colonial, it sets up a home range of several thousand square feet, sharing it with a small group of conspecifics.

Lemmings.

Both diurnal and nocturnal, this short-tailed social herbivore feeds on plant matter, such as grasses, sedges, and leafy vegetation, cutting it into 1½" pieces. Its breeding season runs from spring through summer. The polyestrous female can produce up to 3 litters annually, with an average of 3-4 pups each. Communication and perception take place via standard Cricetidae channels: primarily smell and touch, and to a lesser degree, sight and hearing. Its natural enemies are reptiles, birds of prey, wild felids, and wild canids. The life span of the northern bog lemming has not been studied.

NORTHERN BOTTLENOSE WHALE (*Hyperoodon ampullatus*): Also known as the "North Atlantic bottlenose whale," it grows to a total length of 30' or more and can attain a weight of up to 17,000 lb. (nearly 9 tons). It is distinguished by its massively bulbous forehead (the only beaked whale with this feature) and bottle-shaped beak, the two characteristics which give it its species name: *ampullatus* is Latin for "flask."

It has a stout cigar-shaped torso with a sickle-shaped dorsal fin located toward the back end of the body. A single pair of large teeth grow from the front of the

North Atlantic bottlenose whale.

lower jaw of the male. Coloration is varied, from gray to dark bluish-gray, with a lighter underside. This species is gender dimorphic, with the male being larger than the female.

A secondary consumer, this natatorial migratory ziphiid spends its entire life in the cold waters of both the Arctic Ocean and the north Atlantic Ocean (down to New England). Dives to a depth of 5,000', and can hold its breath for 1 hr. Social and curious, it lives in pods of a dozen or so individuals. Using sonar to hunt, its carnivorous diet consists of typical Ziphiidae fare: deep-water squid, fish, and aquatic crustaceans. Known to strand along beaches. The Northern bottlenose whale may live to about 35 yrs. in the wild.

NORTHERN ELEPHANT SEAL (*Mirounga angustirostris*): One of the largest aquatic carnivores in the world, it can reach a total length of almost 22' and a weight of nearly 8,000 lb. Its pelage is countershaded: the dorsum is silvery-gray or uniformly brown, the ventrum is lighter, often yellowish. The male sports a large trunk-like nose that juts out and down, hanging below the lower jaw—hence its common name. The front flippers are smallish with 5 flexible digits and protruding claws; there are no external ears; the rear flippers are webbed and possess 2 lobes (and may be the cause of some purported mermaid sightings). Unlike its cousins the eared seals, whose hind flippers can rotate forward (allowing it to walk quite well on land), the hind flippers of this so-called "earless" seal are formed facing straight backward. Therefore it must drag them awkwardly along when locomoting on land. Despite this, in the water its powerful rear flippers make it an agile swimmer and an excellent diver. Gender dimorphic, the male is larger, more robust, and shaped differently than the female. An example: males sport huge necks composed of thick blubber—protection from injuries (bites and blows) inflicted while sparring with other males.

This littoral marine phocid inhabits the temperate and subtropical coasts of the Pacific Coast, from Alaska south to Baja California. A pelagic species, it is quite at home in open water, and in fact, spends most of its life at sea swimming and diving. For a few months each year, however, it can be found assembled in massive herds on islands and on sandy shorelines (where it is restricted due to its enormous weight) along the west coast of our region. A nearctic migratory animal, it can

Northern elephant seal.

travel as many as 13,000 mi. a year in search of food.

The northern elephant seal is a natatorial carnivore with a voracious piscivorous and molluscivorous appetite, one surrounding fish (ratfish, sharks, hagfish, skates) and cephalopods (squid, octopus). While hunting and feeding, males dive repeatedly to the ocean bottom; females feed in open water and midwater. An individual may forage in this manner continuously for several months, all without sleeping or even resting, making this species both diurnal and nocturnal. It does not drink water, for it receives all of the hydration it needs from its food. It can dive to a depth of 5,000' (nearly 1 mi.) and remain submerged for up to 70 mins., feats made possible, in part, by its ability to temporarily collapse its lungs—the only known mammal able to do so. Its cycle of ceaseless consumption is countered by a lengthy period in which it abstains from eating, which occurs when it hauls out on land in late May to molt. Also a petrivore, just prior to this period it consumes several stones, a rocky addition to its diet that is believed to help it maintain its health while fasting.

Solitary in the water, its breeding season runs from early winter to early spring, at which time individuals form large gregarious colonies on land. Breeding males challenge one another for patches of beach on which to set up a territory. Even after establishing his harem of 50-100 cows, the dominant male, the "beach-master," must constantly defend the females within his territory from rival bulls. Violent battles often ensue, during which males inflate their large probosces, using them to amplify the sound of their roars. Raising their heads to a height of nearly 8', they stare, threaten, menace, and lunge at one another in an attempt to drive their large canines into their opponent's plump necks and thickly cornified (calloused) chests. These brutal contests only end when the weaker one is driven off or both combatants are bloodied and exhausted.

The female gestates for nearly 12 mos., employs delayed implantation, then gives birth to 1 or 2 black, precocial, 65 lb. pups the following year at the rookery site. At this time a large percentage of youngsters are crushed to death by raging 4-ton bulls as they squabble with rivals, oblivious of the youngsters laying on the sand beneath them. For 3-4 wks. following parturition, mothers nurse their neonates with some of the most fat-rich milk ever recorded. At this point, weighing in at

Elephant seal.

around 400 lbs., the now silvery-colored pups are abandoned on the beach by their post-parturient mothers, who return to the sea to mate. Pups must endure a 3-mo. period without adult supervision and protection, while teaching themselves to swim and feed. Only the strongest, most intelligent, and luckiest will survive.

This massive seal transmits and receives information via traditional Phocidae communication and perception channels: sight, touch, hearing, and smell. Vocalizations become all-important during the breeding season, when males snort, roar, and bellow while fighting their male opponents. Its primary natural enemies are orcas and sharks, and many individuals carry toothy scars from attacks (adult females, being smaller, are more susceptible to predation than the massive adult males). Once believed to be extinct (after being relentlessly hunted for its blubbery oil), it has made a full recovery—though its ultimate survival may still be considered tenuous due to continued interference from humans. The northern elephant seal lives an average of 10-12 yrs. in the wild; may live up to 20 yrs. in captivity.

NORTHERN FLYING SQUIRREL (*Glaucomys sabrinus*): This small rodent grows to a total length of 15" and a weight of up to 4 oz. Though it is quite similar to its close cousin the southern flying squirrel, it is larger and more robust and its coat is a darker brown. The extra soft pelage on its dorsum is a grizzled gray-brown; on its ventrum it is light gray. The cheeks are whitish. The body is long, cylindrical, and supple; the head is small and short; the eyes are large and dark—an adaptation suited to nocturnal life;

the ears are large and erect; the nose is snubbed; the limbs are long; the feet are clawed; the 7" tail is long, furry, flattish, and dark-tipped. Its most distinguishing feature is its gliding membrane or patagium: a loose, fleshy, furry fold of skin on each side of the body that is attached between the wrists of the front feet and the ankles of the back feet.

The common name of this unusual night-dweller is rather misleading, for it does not "fly." It glides from tree trunk to tree trunk. This it does by jumping into the air, opening its legs, and spreading its flight fur. Just before reaching its landing site it pulls up, alighting deftly on its target. It can execute flights up to 300' long and also glide in a semicircle. In acquiring the ability to "fly" it sacrificed some terrestrial dexterity, and can be somewhat ungainly on the ground.

This nearctic riparian sciurid is an inhabitant of temperate, mountainous pine forests (and occasionally mixed or hardwood taiga) comprised of an abundance of mature fir, spruce, hemlock, aspen, birch, or maple trees. It prefers cool, moist woodlands with large ground snag, fungal ecosystems, thick shrubbery, decaying logs and stumps, and an ample water supply, such as a marsh, creek, or stream. It occurs in a massive band across the central portion of the northern tier of North America, from Alaska in the west to the Great Lakes region, New England, and Appalachian Mountains in the east; from Canada's Queen Elizabeth Islands, Yukon Territory, and Northwest Territories in the north to California and the southern Rocky Mountains in the south.

Northern flying squirrel.

Only the female is territorial, establishing a small home range averaging 4-35 acres. This species constructs dreys out of twigs and shredded bark set in tree cavities up to 20' off the ground; these are often lined with grass, leaves, pine needles, and fur. It sometimes lives in groups during the winter.

On the herbivorous side of this omnivorous mammal's diet it is a granivore, a frugivore, a nucivore, and a fungivore, while on the carnivorous side it is an insectivore, an invertivore, an ovivore, and an avivore. It has, in fact, the strongest meat-eating proclivities of any of the tree squirrels (besides the southern flying squirrel). Sleeps by day, forages and hunts by night; may cache surplus food.

Because it is nocturnal there is much that we do not know about this scansorial, arboreal, agile, social but largely solitary sciurid. Its breeding season starts in late winter and ends in early spring. After a period of courtship and mating, the doe gestates for around 35 days, giving birth to an average of 3-4 altricial pups in a temporary drey located in a hollow stump or the craw of a tree. After 60 days the young are weaned, becoming completely independent by 3 mos. of age. Lacking paternal instincts, the father does not assist in the rearing of his offspring. As with nearly all other mammal species, the mother *Glaucomys sabrinus* and her young form the nuclear family.

Communicates and perceives via standard Sciuridae channels: sight, smell, touch, and hearing. It is well-known for both its clicking sounds and its soft, bird-like chirping notes. Though it is quite common throughout the northern portions of its range, it is endangered in parts of its southernmost range. It and the southern flying squirrel are the only nocturnal tree squirrels in our region. Benefits the North American ecosystem by dispersing seeds and providing food for its primary natural enemies: birds of prey, mustelids, wild felids, and wild canids. The northern flying squirrel probably lives an average of 2-3 yrs. in the wild; at least twice that long in captivity.

NORTHERN FUR SEAL (*Callorhinus ursinus*): Also known as the "Alaska fur seal," it can attain a total length of 7' and a weight of 600 lb. The male is black, brown, or grayish above, reddish on the underbody, with grizzled coloration around its neck and shoulders. The female is grayish on the dorsum, and reddish-brown on the ventrum. In both genders the fur is dense (providing insulation in cold water), with nearly 350,000 hairs per square inch (in contrast, we humans have approximately 1,500 hairs per square inch on our heads). Its skull is small; the whiskers (which are sensitive to vibrations) are long; the nose is short and conoidal; the eyes are large, dark, and watery; the body is stocky; the tail is small; the flippers are big—with the hind flippers being ¼ the total length of the body (the longest of any member of the Otariidae family).

As with other "eared" seals, its external ears (ear flaps) are visible, and its hind flippers can be rotated forward, facilitating walking and running on land. The neck on both the male and female is extra thick; in the male, however, it is massively enlarged and covered with a short silvery, reddish, or yellowish mane. This species, once known as the "sea bear," is not only gender dimorphic, it is one of the most extremely dimorphic of any mammal, with the male being nearly 400 percent larger than the female.

Northern fur seal.

Both coastal and pelagic, this solitary marine otariid spends 80-90 percent of its life at sea, usually within 100 mi. of shore. During the summer, however, it can be found mostly on the Pribilof Islands in the Bering Sea, the site of its primary breeding grounds. Largely an eastern, western, and northern Pacific Ocean species, its full range extends from the Arctic Ocean south to Baja California. (Outside our region, it is found in Russian and Japanese waters, chiefly the Sea of Okhotsk and the Sea of Japan.) One of the greatest animal migrators, this ocean-going seal is able to travel some 6,000 mi. a year and remain at sea for up to 9 mos. or longer without coming ashore.

Typical of marine seals, this nocturnal carnivore is an opportunistic piscivore and molluscivore with a diet based around fish (rockfish, walleye pollock, hake, capelin, mackerel, anchovy, salmon, sand lance, herring) and saltwater invertebrates (such as squid). Preening and resting on its back at the surface by day, it feeds mostly at night, when many fish species rise up from the depths. Eats small prey whole; tears larger prey into pieces before swallowing. Can swim up to 15 mph, dive to 500', and remain submerged for several minutes.

As noted, its breeding season takes place in summer on the sandy and rocky beaches of islands in the north Pacific. Northern fur seals are guided by natal philopatry, with both genders annually returning to the areas in which they were born. The polygynous bulls arrive early, in May, to set up individual territories, which they defend from rivals. Though normally asocial (at sea the 2 genders seldom gather in groups of more than 2 or 3), the males' goal now is to attract already pregnant cows, who arrive shortly thereafter to give birth. The female gestates for about 1 yr., employs delayed implantation, and produces 1 precocial pup within a day or two after her arrival at the natal rookery. She then comes into estrus and is ready to mate again, and, by her choice of territory, she becomes a member of the harem of that territory's male owner.

After nursing her pup for a few days, the mother leaves it on the beach and swims out to sea to feed, often for up to 2 wks. at a time. Returning to shore, she nurses her youngster again for a few days, then once again heads out to open water to forage. This pattern is repeated for as long as 4 mos., during which time the pup plays and sleeps with other neonates, as it grows and gains weight day by day. Since the father does not aid in the raising of his offspring and the mother offers only the most minimal of care, during this period the pup must essentially fend for itself. After roughly

Northern fur seal.

16 wks., weaning takes place, and the pup, now a juvenile, becomes independent and begins eating solid food. The life cycle of the northern fur seal begins anew.

Information transmission follows normal Pinnipedia communication and perception channels: sight, hearing, touch, and smell. Aquatic and terrestrial vocalizations are varied and include roaring, bellowing, clicking, growling, and bleating. Mothers and their offspring are able to recognize each others' calls for up to 4 yrs., the longest recognition ability presently known.

Its natural enemies are orcas and sharks (which feed on adults) and Steller's sea lions (which feed on pups). This marine mammal can be dangerous if disturbed; is able to move quickly on land and even climb sheer inclines. The male northern fur seal probably lives 10-15 yrs. in the wild; the female may live 20-25 yrs. in the wild.

NORTHERN GRASSHOPPER MOUSE (*Onychomys leucogaster*): This large highly predatory member of the *Onychomys* genus grows to a total length of nearly 8", its hind foot measures 1", and it can reach a weight of up to almost 2 oz. Coloration on the dorsum varies from gray and brown to buff and rusty-yellow; the ventrum is white; the unusual 2½" tail is short, thick, white-tipped, and dichromatic (dark on top, lighter on the bottom). The body is stout; the eyes are large and dark; the ears are large and erect; the muzzle is short; the snout is covered with tactile vibrissae; the limbs are short; the feet are small and delicate.

This solitary cricetid is a nearctic native that inhabits temperate prairies, semiarid savanna, arid deserts, steppes, ravines, chaparral, rangeland, valleys, sageland, sand dunes, shrubland, grassland, and agricultural fields; it is associated with soft sandy soil and high elevations. Its range extends over much of the western half of our region, from central Canada south to northern Mexico, and from California to Iowa; it does not occur east of the Mississippi River. Excavates massive burrow systems (or appropriates the burrows of other underground creatures) that include chambers for various purposes, such as food caching, shelter, protection, sleeping, excretion, and nesting. Innately violent and aggressive, under certain circumstances it is known to kill conspecifics of both genders. Establishes a large home range of several thousand square feet. Does not hibernate; active year-round.

Grasshopper mouse.

A nocturnal obligate carnivore—and a fearsome hunter that prefers foraging on dark nights—over 95 percent of its diet comes from animal matter, mainly insects, which it stalks and kills with the ferocity of a miniature leopard. As its common name denotes, its primary fare consists of grasshoppers; however, it also consumes beetles, caterpillars, moths, scorpions, spiders, and crickets. It also dines on plant material, as well as other small mammals (mainly mice), birds, and lizards; and it is known to engage in cannibalism. Its bite force is more powerful than other types of mice; uses its long fingernails to firmly grasp its living prey, the derivation of its genus name: *Onychomys* means "clawed mouse."

Its breeding season runs year-round, cresting between spring and fall. The iteroparous polyestrous female, who may produce at least 2-3 litters annually, gestates for about 5 wks., giving birth to as many as 8 altricial pups, weaning them within 2 wks. The young are reproductively mature at 3 mos. of age. The father is not paternal and does not contribute to the rearing of his offspring.

Communication and perception take place via standard Cricetidae channels: primarily touch and smell, and to a lesser degree, sight and hearing. Vocalizations include barks, chirps, whistles, and a high frequency canine-like "howl." Its natural enemies are reptiles, birds of prey, wild felids, and wild canids. Beneficial as a soil aerator and recycler, a household pet, a consumer of insect and rodent pests, and as prey for larger carnivores; detrimental as a carrier of deadly diseases and as a possible destroyer of threatened and endangered animals (particularly birds). The northern grasshopper mouse lives an average of 3-4 yrs. in the wild.

NORTHERN IDAHO GROUND SQUIRREL (*Urocitellus brunneus*): This small ground squirrel grows to a total length of about 9", the hind foot or pes measures around 1½", and it can weigh up to 10 oz. Its dorsum is reddish-brown, its ventrum is lighter, usually white, buff, or tan. The front legs and the shoulders are often tawny; the outside of the back legs are brownish cinnamon; the throat is whitish. The body is stout; the head is small; the eyes are dark with whitish rings; the ears are small and erect; the limbs are short; the feet are clawed; the banded brushy 2" tail is reddish brown on top, grizzled along the bottom. Its most conspicuous identifiers are the subtle white spots that dapple its back. Its cryptic, aposematic coloring helps protect it from both aerial and terrestrial predators in its home environment. Gender dimorphism is present: the male is larger than the female.

A nearactic native, this terricolous sciurid inhabits dry open savanna and grassy mountain meadows with thin ground cover, preferably set amidst coniferous forests made up of ponderosa and Douglas pine. It has a very limited range and is found in only 5 counties in western Idaho. Constructs different types of subsidiary burrows for different functions (such as nesting, protection, and hibernacula), with entrances usually located beneath boulders, rocks, stumps, thickets, or fallen timber. Hibernates from midsummer to early spring.

Ground squirrel.

This largely unstudied fossorial sciurid is an omnivore with a diet comprised primarily of plant matter: roots, tubers, stems, bulbs, grass, leaves, flowers, seeds, nuts, and grains; it occasionally preys on insects.

Its polygynandrous breeding season, which includes courting, begins in spring after the bucks and does awake from hibernation. Males fight for access to estrus females; large dominant males mate with the most females. Little is known about the nesting, parturition, and parental behaviors of this rare rodent. The doe digs a natal burrow in which she builds a nest. After a gestation period of about 7 wks., she gives birth to 5 or 6 altricial pups, weaning them shortly thereafter. The male is non-paternal and therefore does not assist in the rearing of his offspring.

Diurnal and solitary (except during the mating season), it communicates and perceives via normal Sciuridae channels: sight, smell, hearing, and touch. Vocalizations include altruistic warning calls to conspecifics. Its main natural enemies are birds of prey and mustelids. It is rapidly declining due to habitat loss. The life span of the fast-living Northern Idaho ground squirrel has not been cataloged, but it appears to be quite short, perhaps 1-2 yrs. in the wild.

NORTHERN POCKET GOPHER (*Thomomys talpoides*): This small member of the genus *Thomomys* grows to a total length of 10", its pes or hind foot measures about 1", and it can attain a weight of up to almost 6 oz. The dorsal pelage runs from yellowish-brown to rusty-brown, and from grayish to grizzled black; the ventral pelage is a lighter shade of that found on the upper body; usually has white patches under the chin; the 3" tail is thick, pinkish, and sparsely furred. The body is stout; the skin is flexible; the fur is soft and short; the head is large; the eyes are small, dark, and beady; the ears have dark postauricular markings and are small, erect, and ovoid; the muzzle is blunt; the snout is covered in tactile vibrissae; the incisors are large, sharp, and protruding; the limbs are short; the heavily clawed feet are large and adapted to digging. Gender dimorphism is present: the male is larger than the female.

This solitary geomyid is a riparian nearctive native that inhabits a myriad of temperate biomes and ecosystems, including valleys, mountains, cropland, prairies, steppe, savanna, gardens, chaparral, cultivated locations, ranchland, taiga, hedgerows, lawns, tundra, meadows, intermontane regions, shrubland, open mixed forests, orchards, and grassland; prefers dense hiding cover, deep, dry pebbly soil, and locations with creeks or streams. It is often found near Mima mounds. Its range covers much of western and central Canada and most of the western U.S., extending into Washington, Oregon, California, Idaho, Montana, North Dakota, South Dakota, Minnesota, Wyoming, Nebraska, Colorado, Utah, Nevada, Arizona, and New Mexico.

Fossorial, as well as highly territorial and aggressive, it excavates massive complex burrow systems up to 500' long, with multiple chambers serving different purposes; ejecta mounds are present near portals; never wanders more than a few yards from the entrance of its subterranean home; plugs portals for protection and climate control; digs subnivean tunnels in winter; viciously guards its home range against conspecifics. Does not hibernate; active year-round.

A sedentary circadian herbivore, the diet of this keystone species consists primarily of seeds, grass, tubers, bulbs, nuts, herbs, grains, forbs, leaves, cacti, stalks, fruit, roots, corms, and agricultural products; it is also a coprophage; hoards and stores surplus food, transporting it in its large external cheek "pockets" or pouches—the source of its common name. Its breeding season takes place in spring. The female gestates for roughly 3 wks., giving birth to an average of 3-4 altricial pups in a leaf-lined maternity nest chamber. Weaning occurs about 5-6 wks. later. The father plays no role in the rearing of his offspring.

Pocket gopher.

Communicates and perceives along routine Geomyidae channels: mainly touch and smell, and to a lesser extent, sight and hearing. Its natural enemies are reptiles, birds of prey, mustelids, wild felids, and wild canids. Besides remaining underground most of the time, its other defenses include running, camouflage, and swimming. Beneficial as a seed disperser, a soil improver, aerator, and recycler, and as food for carnivores; detrimental as an agricultural pest. The northern pocket gopher lives for an average of 1-2 yrs. in the wild; about twice as long in captivity.

NORTHERN PYGMY MOUSE (*Baiomys taylori*): The smallest of North America's many rodent species, it grows to a total length of about 4½" and can attain a weight of up to ⅜ oz. On the dorsum it is brownish-gray with a grizzled silvery wash; the ventrum is whitish to light grayish-silver; the 2" tail is lightly furred and indistinctly bicolored (dark on top, lighter on bottom). Unlike many other muroids, its incisors lack grooves.

This tiny cricetid is a nearctic native that inhabits temperate grassland, prairies, shrubland, desert, oak and pine forests, cactus-rich areas, sageland, cropland, weedy settings, savanna, low-lying marshland, and sandy locations; prefers friable soil, rotting logs, stumps, and abundant debris and hiding cover. Its range is limited to portions of Arizona, New Mexico, and Texas. Semi-fossorial, it often lives in subterranean burrow systems; its globular grassy nests may be found both on the surface and underground; grass runways lead to and from burrow portals. Establishes a home range of 3,000-4,000 sq. ft. Does not hibernate; active all year long.

The diet of this crepuscular and nocturnal omnivore consists of seeds, grasses, grains, cacti, roots, legumes, stalks, and fruit; when available it also consumes insects, gastropods, and other small mammals. Its breeding season runs year-round. The female gestates for about 3 wks., giving birth to an average of 3-4 altricial pups. Communication and perception occur via standard Cricetidae channels: primarily smell and touch, and to a lesser degree, sight and hearing. Its natural enemies are reptiles, birds of prey, wild felids, and wild canids. The northern pygmy mouse probably lives 1-2 yrs. in the wild.

NORTHERN RED-BACKED VOLE (*Myodes rutilus*): Some systematists consider it to be conspecific with the southern red-backed vole; others maintain that they are 2 separate species; thus its taxonomy remains unsettled. Indeed, they are nearly identical and share much of the same ecology and life history. One primary difference, as their common names indicate, is that the northern's range (Alaska and the upper tier of Canada) occurs further north than the southern's. Additionally, the northern also has a shorter thicker tail.

NORTHERN ROCK DEER MOUSE (*Peromyscus nasutus*): Also known as the "northern rock mouse," this large rodent grows to a total length of 10", its hind foot measures around 1", and it can attain a weight of up to 1 oz. or more. The dorsal pelage is brownish-gray; the ventral pelage is white to light gray; the long 5½" tail is furry and bicolored (dark on top, lighter on bottom), and spans more than half the total length of the body. The head is long and tapered; the eyes are large, dark, and protruding; the ears are enormous, erect, hairless, and membranous; the snout is covered in fine tactile vibrissae.

This petraphilic cricetid is a nearctic native that inhabits temperate rock-strewn forests, cliffs, chaparral, evergreen woodland, oakland, scree, boulder fields, canyons, lava mounds, rock walls, talus slopes, and shrubland. Its range covers portions of the states of Wyoming, Utah, Colorado, Arizona, New Mexico, Texas, and Oklahoma, south into northern Mexico. It derives its common name from both its preferred habitat and its excellent rock-climbing abilities.

Deer mouse.

A nocturnal omnivore, its diet consists of seeds, nuts, fruit, and insects. Depending on geography, its breeding season tends to run year-round. The polyestrous female, who may produce at least 2 litters annually, gestates for about 25 days or so, giving birth to approximately 4 pups. Communicates and perceives along standard Cricetidae channels: sight, hearing, touch, and smell. Its natural enemies are reptiles, birds of prey, wild felids, and wild canids. The northern rock mouse probably lives for about 1 yr. in the wild. (Note: Confusion continues over the taxonomic classification of this mouse: some maintain that it is a subspecies of the southern rock deer mouse, while others believe it is a distinct species. I treat it as the latter.)

NORTHERN YELLOW BAT (*Lasiurus intermedius*): This large rare chiropterid grows to a total length of almost 5", its forearm measures a little over 2", its wings span up to 16", and it weighs about ¾ oz. It long soft pelage is golden- to cinnamon-brown on the dorsum (sometimes with black or silver "frosted" hair-tips); it is slightly lighter on the ventrum. Its yellowish tawny fur gives it its common name. It has a high forehead, a short snout, small rounded ears, and a broad curved tragus.

Relatively rare outside Florida, it is found mainly along the Southeast U.S. coast from Virginia to Texas, but it also ranges further south into the Caribbean and Mexico. It shelters mainly inside clumps of Spanish moss, but being a dendrophile, also in trees in both pine and oak forests, as well as the dead fronds of palm trees.

Bat of the Southeastern U.S.

Prefers to be near water; sometimes roosts in buildings. It does not migrate or hibernate, but enters into a state of torpidity during cold spells.

A nocturnal insectivore, it tracks down its prey using echolocation, flitting quietly over open areas, such as fields and forest edges, in the dark. Its consumption of agricultural pests (such as beetles) makes it a friend of the farmer; its consumption of disease-carrying bugs (such as mosquitos) makes it a friend of humanity.

Also known as the "Florida yellow bat," this largely solitary vespertilionid breeds in the fall, with mating probably taking place during flight. The female delays fertilization until spring, bearing up to 4 hairless pups in late spring or early summer. The mother, who has 4 mammary glands, nurses her young in her maternity colony. A secondary consumer, this species' primary natural enemies are reptiles (snakes), birds of prey (owls), and rodents. The life span of the northern yellow bat is unknown, but indications are that it may be somewhat short, perhaps 3-5 yrs. in the wild.

NORWAY RAT (*Rattus norvegicus*): The archetypal rodent, this is the mammal most people think of when the word "rat" is mentioned. Also known variously as the "brown rat," "barn rat," "sewer rat," "house rat," "gray rat," and "common rat," this invasive species grows to a total length of about 20" and a weight of up to 18 oz. It pelage is long and coarse, grayish-brown on the dorsum, grayish-tan on the ventrum. The body is large and muscular; the head is long and robust; the eyes are small, beady, and protruding, providing a wide arc of night vision; the ears are small, thick, naked, and erect; the muzzle is stout; the nose is snubbed and covered in tactile nasal vibrissae; the limbs are short but powerful; the feet are clawed; the long 9" tail is naked, scaly, and used as a counterbalance. This species is gender dimorphic: the male is larger than the female.

Norway rat.

Commonly confused with its cousin the black rat (they are somewhat similar in appearance and both can have brown or black fur), they actually possess many striking differences: the body of the Norway is generally much larger, heavier, and thicker; its head is longer, more robust, and more angular; its eyes and ears are smaller; its tail is shorter in comparison to body length; and its nose is blunt while the black's is pointed.

Its scientific name (*rattus* is Latin for "rat," *norvegicus* is Latin for "Norway") is a misnomer, for it originated in Asia (probably China), arriving in North America in the 18th Century aboard European ships. It one of the world's toughest, most adaptable, and most successful mammals, with the ability to live anywhere on our planet—except the South Pole. In our region it is found in habitats as diverse as arid desert and temperate rain forest, grassland and wetland, chaparral and scrubland, savanna and orchards, old fields and coastline, valleys and mountains. Heavily edificarian, it is a synanthrope that is best known for its association with anthropogenic environments, from isolated rural farms to large metropolitan cities, from sewers to basements, from toolsheds to backyards, from barns to ditches, from garbage dumps to warehouses, from docks to slaughterhouses, from silos to stores. Indeed, it is found everywhere humans live, including (in North America) all 50 states in the U.S., as well as southern and coastal Canada and northern Mexico.

Norway rats are social and colonial, living in an organized hierarchical society built around the leadership of a large alpha male. This fossorial species excavates burrow systems up to 8' long and 4" wide, with multiple entrances, exits, and

Norway rat.

chambers; in these it constructs elaborate nests out of sticks, grass, bark, leaves, and also human byproducts (such as paper, trash, and clothing). Highly territorial, it establishes a home range of about 500 sq. ft., which it scent-marks and defends with great ferocity.

This ubiquitous cosmopolitan murid is a generalist feeder and an opportunistic

forager who, like the raccoon, will eat almost anything it comes across. Though it is most properly called an omnivore, its highly varied diet also makes it (depending on where it lives) part carnivore, herbivore, granivore, folivore, nucivore, lignivore, frugivore, florivore, fungivore, insectivore, vermivore, bulbivore, invertivore, amphibivore, molluscivore, herpivore, avivore, ovivore, carcassivore, mammalivore, piscivore, nectarivore, planktivore, crustaceanivore, and echinodermivore. It can live exclusively on pet food and human garbage if need be and, being cursorial, it possesses the ability to quickly chase down living prey, such as mice and lizards. To a great extent it is this extreme dietary flexibility to which it owes its global success as a mammal.

A number of other unusual traits contribute to its reign as our region's most omnipresent rodent. With no set breeding season it can mate any time of year; and it is not only polygynandrous, but the polyestrous doe is an extra prolific breeder that can have as many as 12 litters annually, with up to 22 altricial pups in each litter, producing 264 young a year. Furthermore, she can mate again only hours after giving birth. Finally, the females, who gestate for about 3 wks., form a sort of matriarchal society in which they nest together and care for one another's offspring, greatly increasing the survival rate of their neonates. Weaning occurs at 1 mo. of age, with dispersal and independence following shortly thereafter. The father, being non-paternal, plays no role in the raising of his young.

This large, intelligent, nocturnal and crepuscular rodent communicates and perceives via the usual Muridae channels: sight, hearing, touch, and smell. Common vocalizations include chattering, squeaking, chittering, chirping, hissing, whistling, and teeth-chomping. Sedentary, its tactile whiskers, finely tuned ears, highly sensitive nose, and phenomenal memory allow it to easily navigate in the dark. Natatorial, esturine, and riparian, it is an adept swimmer that is often found near water, around piers, and on ships, the origin of 2 of its other popular nicknames: "water rat" and "wharf rat." Though terricolous, it is also an expert climber and jumper. Does not hibernate; is active year-round.

Norway rat.

Aggressive, domineering, cunning, adaptable, and seemingly perpetually ravenous, it harms the world ecosystem by damaging crops, contaminating granaries, killing and eating small game birds and livestock, pushing out native species, collapsing human structures (by burrowing under them), threatening endangered animals, causing floods, fires, and electrical outages, and spreading disease (such as bubonic plague) to humans—thereby causing the deaths of untold thousands each year. Overly abundant across its range, it is widely considered one of the most virulent pests the world has ever known. Though quite destructive, it also offers several benefits: it aerates and recycles the soil, it hosts parasites, it consumes human waste, it disperses seeds, it provides food for larger carnivores, and it serves as both a trainable and enjoyable pet and as a useful lab animal for scientific testing. Its natural enemies are reptiles, birds of prey, mustelids, wild felids, and wild canids. The Norway rat lives an average of 1 yr. in the wild (up to 2 yrs. under ideal conditions); it may live twice that long in captivity.

NUTRIA (*Myocastor coypus*): Known in Latin America as the "coypu" (a Native American word), this large semi-aquatic rodent resembles a cross between a beaver and a muskrat. It grows to a total length of 55", its hind foot measures 5", and it can attain a weight of up to 25 lb. The long shaggy guard hair on the upper body is buff-brown; the underfur is yellow-gray; the lower body is a lighter shade of that on the dorsum; the long 17" tail is scaly, thin, round, and sparsely furred. While its common and scientific names are biologically inaccurate, they are metaphorically acceptable: nutria is the Spanish word for "otter"; its binomial name derives from the early Greek words for "mouse" (*mys*) and "beaver" (*castor*), thus meaning "mouse beaver."

This exotic invasive mammal does indeed have the general appearance of a massive muscular mouse—except that its head is much larger (in relation to body size) and its whitish-silvery muzzle is blunt rather than tapered. It is smaller than a beaver but larger than a muskrat. The 3 species can be easily distinguished by their tails: the nutria's tail is round (like a mouse); the muskrat's tail is vertically flattened (like a trout); the beaver's tail is horizontally flattened (like a dolphin).

The nutria's body is stout; its eyes are small and dark; the ears are small, rounded, and erect; the nose is large; the snout is covered with long sensitive silvery whiskers; the incisors are bright orange and protrude from the jaws; the limbs are short and powerful; the front feet are clawed, strong, and prehensile; the 4 inner toes on the clawed hind feet are webbed. The eyes, ears, and nose are located on the upper head, an alligator-like adaptation that allows the nutria to see, breathe, and hear while hiding just below the surface of the water. Gender dimorphism is present, with the male being larger than the female.

Nutria.

This riparian myocastorid is a neotropical native that inhabits wetlands, and more specifically areas with marshes, ponds, and lakes, as well as lethargic streams and shallow river mouths; prefers watery grassy locations near woods; occurs in both freshwater and brackish water. Nonindigenous to our region, in the 1930s it was imported from South America to the U.S. for breeding on Louisiana fur farms, where escapees soon began to populate and spread throughout Dixie. Today, along with its native region (Brazil, Argentina, Chile, Bolivia, and Paraguay), it is found in nearly half of the lower 48 U.S. states as well as Ontario, Canada. Due to human interference (upon which it thrives), it continues to extend its range; it is particularly abundant across the southern and northwestern states.

Sedentary, territorial, and highly social, it lives in colonies of a dozen or more related individuals, consisting of a dominant mature male, several adult females, and their offspring. Excavates burrows in the banks of streams and rivers, on lake shores, and in the sides of dikes and levees; sometimes it appropriates the tunnel systems of other aquatic creatures, such as beavers. Its house is linked to numerous surface runways that lead to and from the entrances. Establishes a home range of roughly 50,000 sq. ft.

This nocturnal natatorial herbivore feasts primarily on aquatic vegetation, consuming the entire plant: roots, stalk, leaves, and bark. It also consumes all manner of green terrestrial vegetation and practices coprophagy as well; often dines, grooms, and rests on 6' wide, circular feeding platforms made of floating plant matter.

In the southern parts of its range its breeding season runs year-round, which includes a vigorous and sometimes violent courtship phase between males and females. The iteroparous polyestrous doe may produce up to 3 litters annually; she gestates for about 4.5 mos., giving birth to as many as 10 precocial pups that can walk and swim at just 1 day old. The mother has 8 mammary

Nutria.

glands set on the side of the body rather than underneath, as in most other mammals. This allows the young to suckle safely at the water's surface. Besides providing nourishing milk, the mother also protects and grooms her neonates, all without aid from the buck (who displays no paternalistic behaviors). They are weaned at approximately 2 mos., becoming reproductively mature as early as 3-4 mos. of age.

Communication and perception occur via normal Myocastoridae channels: touch, smell, hearing, and lastly sight (its weakest sense). Its natural enemies are predatory fish (gar), reptiles (snakes, turtles, alligators), birds of prey (hawks, eagles), and wild canids (fox, coyotes). Has valvular nostrils, allowing them to close when underwater. It can remain beneath the surface for up to 10 mins; as its front legs are much shorter than its hind legs, it walks in a hunched manner. Beneficial as a prey base for carnivores and as a fur provider for humans; detrimental as a soil eroder, destroyer of wild plants, dikes, and wetlands, and as an agricultural pest. Nutria live for about 2-3 yrs. in the wild; longer under optimum conditions; perhaps as long as 10 yrs. in captivity.

Meadow jumping mice.

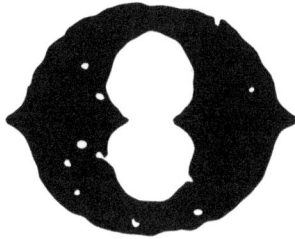

OCELOT (*Leopardus pardalis*): The largest member of its genus, this medium-sized feline grows up to 54" in total length, has a 7" hind foot, stands 20" at the shoulder, and weighs as much as 40 lb. Its tawny-grayish coat is made up of short hair, and is covered with a variety of black and brown markings, including stripes, speckles, spots, whorls, dashes, bars, slashes, clusters, blotches, and rosettes that are bordered in black. The throat, chest, and belly are white or cream; the 15" tail is ringed or barred (with a light underside); the nose is usually pinkish. Black (melanistic) forms have been reported.

Ocelot.

The large yellow-green eyes are round with elliptical pupils. The eyes are encircled in black; two prominent, black, horizontal cheek stripes run back across the face to the neck; a black dashed line runs vertically from each eyebrow over the top of the head. Also known as the "painted leopard," this cat's colors and patterns can vary widely, with each individual having unique facial and body markings. The muzzle is short and blunt; the snout is concave; the limbs are short; the feet are large. Has medium-sized triangular ears; the backs of its ears are black with a white spot in the middle. This species is gender dimorphic: the male is larger than the female.

Its dappled, cryptic coloring serves as ideal camouflage for its habitat: scrubland, chaparral, savanna, brushland, riparian corridors, grassland, disturbed land, mangrove swamp, marshland, coastal wetland, mosaic habitats, agricultural fields, and montane, subtropical, tropical, and high cloud forest, always with thick hiding cover and an abundant rodent population. This widely distributed cat once thrived from Arizona to Louisiana, and as far north as Oklahoma and Arkansas. Its current range, however, runs from southern Texas south into Mexico and Central America. Males establish a home territory of over 35 sq. mi., the borders which they scent-mark to ward off conspecific intruders. Solitary, territorial, nocturnal, and crepuscular, normally it rests in trees, roots, thickets, brushpiles, and various protected cavities during the day. An energetic mammal with high nutritional needs, it is sometimes active during the day, however, particularly when it is overcast.

An opportunistic hunter, it stalks game trails or sits outside the entrances of burrows awaiting the emergence of their occupants. As it will eat almost anything it can catch, its diet is quite varied and includes crustaceans (land crabs), fish, birds, amphibians (frogs), reptiles (tortoises, snakes, lizards), small mammals (mice, rats, pacas, agoutis, rabbits, armadillos, opossums), and sometimes large mammals (anteaters, monkeys, sloths, deer). Like most other mammalian predators, it will take poultry and domestic pets when available. Though it is terricolous, it is also a skilled climber and swimmer that readily hunts in trees and fishes in rivers.

It has no set breeding season, thus mating may take place at any time of the year. The female generally breeds once every 2 yrs. She gestates for around 75 days, giving birth to 1-4 altricial kittens in a carefully hidden den. Weaned at 6 wks., the young attain their independence at around 1 yr., and set off to establish their own territories by 2 yrs. of age. During the pre-dispersal phase the mother feeds, grooms, and protects her kits, and teaches them how to hunt. The polygynous male plays no role in the rearing of his offspring.

Ocelot.

Like all members of the Felidae family, it possesses keen senses, in particular excellent eyesight (including night vision and binocular vision) and smell (which helps it locate prey and find mates over long distances). Vocalizations include meowing, hissing, snarling, growling, yowling, and purring. A secondary consumer, among its natural enemies are birds of prey (eagles), reptiles (snakes), and fellow cat species (mountain lion, jaguar). A natural pest controller, it is beneficial to humans. The ocelot lives up to 10 yrs. in the wild; as long as 20 yrs. in captivity.

OLDFIELD MOUSE (*Peromyscus polionotus*): Also known as the "oldfield deer mouse," this tiny member of the *Peromyscus* genus grows to a total length of 6", its hind foot measures ⅝", and it can reach a weight of ½ oz. Coloration on the dorsum is whitish, fawn, or light cinnamon-gray; the ventrum is white; the short 2½" tail is bicolored (brown on top, white on bottom); the feet are white. It is gender dimorphic, with the female being larger than the male.

Open grassland: oldfield mouse habitat.

As its common name suggests, this terricolous cricetid is a nearctic native that favors open vacant fields; it also inhabits temperate grassland, savanna, shrubland, desert, scrub forests, stump ranches, chaparral, and importantly, dry dunes and beaches—the latter ecosystems where its cryptic sand-colored fur provides ideal camouflage. It is a southeastern species, with a range that covers portions of the states of North Carolina, South Carolina, Mississippi, Alabama, Georgia, Tennessee, and Florida. Fossorial, it excavates large complex burrow systems in friable soil, barricading the shrubbery-concealed portals from within for protection. Territorial, it establishes a small home range which it protects against intruders.

A social, sedentary, crepuscular, nocturnal omnivore, its diet surrounds seeds, nuts, legumes, fruit, grains, herbs, and also insects; hoards and caches surplus food in den larders. Its monogamous breeding season runs year-round. The female gestates for about 3 wks. or so, giving birth to as many as 4 altricial pups in a subterranean maternity chamber; she weans them 3 wks. later. The father is non-paternal, but indirectly assists in the rearing of his offspring by guarding the nest area.

Communicates and perceives via standard Cricetidae channels: sight, touch, smell, and hearing. Its natural enemies are reptiles, birds of prey, mustelids, wild felids, and wild canids. Beneficial as a parasite host, food for carnivores, and a subject of scientific research; detrimental as a household pest. The oldfield mouse lives approximately 1-2 yrs. in the wild.

OLIVE-BACKED POCKET MOUSE (*Perognathus fasciatus*): This small pocket mouse grows to a total length of nearly 6" and can attain a weight of up to ⅜ oz. Its dorsum is olivaceous-gray (the source of the first half of its common name) with a soft cinnamon undercoat and blackish hair tips; the ventrum is whitish or cream; the lateral line is buff; the 2½" tail is long, round, sparsely furred, and pinkish. The body is small and ovoid; the head is large and long (compared to body size) with a convex profile; the eyes are small, dark, and protruding; both ears are small, rounded, and erect, and each one has a yellowish postauricular mark; the tapered muzzle is tipped with sensitive vibrissae; the front legs are short; the hind legs are long, allowing for ricochetal locomotion; the feet are small and have furred soles (providing traction on slippery surfaces); it lacks the bristly posterior spines of some other pocket mice.

This saltatorial terricolous heteromyid is a nearctic native, a riparian, and an edge species that inhabits temperate, dry sandy savanna, floodplains, sageland, semiarid prairie, scrubland, cropland, grassland, sparsely vegetated steppe, shrubland, and chaparral; it is also found along the borders of scrub forests and stump ranches with friable soil and good hiding cover. Its range covers portions of the U.S. states of Wyoming, Montana, Utah, Colorado, North Dakota, South Dakota, and Nebraska, as well as central Canada. Fossorial, it excavates and lives in tunnel systems with multiple compartments and subsidiary burrows, leaving small ejecta mounds of sand near the vegetation-concealed entrance. Summer dens are positioned about 1' underground, while winter dens are located up to 6' below the surface. Probably does not hibernate or estivate, but may enter brief periods of torpidity during extreme heat and cold.

Western grassland: olive-backed pocket mouse habitat.

A sedentary, solitary, crepuscular, nocturnal omnivore, its diet consists of the usual small rodent fare: mainly seeds, but also grains, nuts, fruit, forbs, leaves, and occasionally insects. Caches surplus food in its burrow larders, transporting it in its external cheek pouches or "pockets" (the source of the latter half of its common name). Its polygynous breeding season takes place between spring and summer. The iteroparous polyestrous female may produce 2 litters annually. She gestates for about 30 days, giving birth to as many as 8 altricial pups, probably weaning them at around 1 mo. of age.

Communicates and perceives via standard Heteromyidae channels: sight, touch, smell, and hearing. Its natural enemies are reptiles, birds of prey, mustelids, wild felids, and wild canids. Beneficial as a parasite host and as a prey base for carnivores; detrimental as an agricultural pest and as a carrier of deadly diseases. The olive-backed pocket mouse probably lives 6-12 mos. in the wild; perhaps twice that long in captivity.

ORCA (*Orcinus orca*): The official state marine mammal of Washington (state), its original common name, "killer whale," is misleading, for it is not strictly a "whale" and it is no more a "killer" than any other cetacean (although it *is* a whale killer—the original meaning of the name). Thus this particular name has been retired, replaced by orca, its species name.

The orca is the largest member of the dolphin family, Delphinidae (oceanic dolphins), a group in the suborder Odontoceti, ("toothed whales"), which is part of the order Cetacea (marine

Orca.

mammals). This massive dolphin can grow to 32' in length and weigh 20,000 lb. (10 tons). It has a blunt nose and large eyes. Its stout body is black with a variety of white patches on the underside, running from the throat to the tail; has a narrow oval white blaze near the eyes. Has large paddle-like pectoral flippers and a tall dorsal fin located at the middle of the back. Its jaws are filled with narrow 4"-long teeth. This cetacid is gender dimorphic: the male is larger than the female.

Highly adaptable to climate, it is found in every ocean (inshore and in open ocean), including near the equator. It prefers cooler waters, however, and so is often found in polar seas. In our region it is common along the West Coast of Canada and the U.S.

Along with spyhopping, this carnivorous apex predator uses echolocation to help find its food: mollusks, cephalopods (octopus, squid), fish (including great white sharks), reptiles (sea turtles), sea birds, and other marine mammals (sea otters, seals, sea lions, whales, other dolphins, etc.). Also takes land mammals when available. It will consume up to 500 lb. of fish a day. It is believed that the intensity of its sonar clicks disorients and stuns its aquatic prey. A fearless and pugnacious hunter, it often hunts in cooperative wolf-like packs, driving schools of fish into large compact balls upon which the group feeds. Will temporarily beach itself in pursuit of penguins and seals on land. Known to tip and even wash large waves over ice floes to get at its quarry.

Passes down knowledge and population-specific dialects to its young. With its large brain, complex social life, intricate forms of communication (mainly clicks and whistles), and female-led pods of 100 or more individuals, the orca is arguably the most intelligent animal in the sea. A surprisingly nimble swimmer for its size, this natatorial mammal can travel 100 mi. a day, reach speeds of 35 mph or more, and dive to a depth of 3,500' (over a half-mile).

Killer whales are polygynandrous and have no set breeding season. Despite this, mating seems to crest in the warm months. The iteroparous female is a slow (and also a low) reproducer, bearing young only once every 5-10 yrs. This means that she produces less than a half-dozen calves during her entire reproductive life. She gestates for around 15 mos., giving birth to a single precocial calf the following year. The young are nursed with high protein, fat-rich milk, taught to hunt, and generally socialized during their first year of life, after which they are weaned. These highly social delphinids do not leave their natal pods, even after independence. The male plays no role in the rearing of his offspring.

Orca.

Though fearsome in appearance, quick of movement, and aggressive in behavior, wild orcas are not known to attack humans, and have been known to approach small boats and even swim placidly among bathers. At the top of the food chain, this secondary consumer has no natural enemies. The orca may live 70-80 yrs. in the wild, though 30-50 yrs. is probably more typical.

ORD'S KANGAROO RAT (*Dipodomys ordii*): This common widely distributed rodent grows to a total length of 11" and can attain a weight of up to 3½ oz. or more. Its dorsal pelage is bright rusty-yellow with black-tipped hairs washing over the upper back; the ventral pelage is white or cream; the long 6½" tail is dichromatic (dark on top, lighter on bottom) with a black tufted tip; this species has a prominent white thigh stripe as well as 2 white subauricular patches and 2 white supraocular patches. The body is small and ovoid; the head is large with a convex profile; the eyes are large, dark, and protruding, indicating nocturnality; the ears are small, rounded, and erect; the forelimbs are small; the large powerful hind limbs are designed for saltatorial locomotion, lending it its common name.

This terricolous solitary heteromyid is a nearctic native that inhabits temperate grassland, arid wasteland, plains, savanna, juniper woods, chaparral, cropland, desert, shrubland, and sand dunes, with light vegetation, open ground, and dry conditions. Its range covers portions of the entire western part of our region, from southern Canada in the north to Mexico in the south; from California in the west to Kansas in the east. Fossorial, it excavates burrow systems in slopes, relying on them for shelter, protection, nesting, sleeping, and food storage; plugs entrances to control the subterranean climate in its tunnels; ejecta mounds are often seen near entrances; numerous runways connect portals to the surrounding landscape. Establishes a home range of approximately 500-1,000 sq. ft.

A nocturnal sedentary omnivore, its diet consists of typical Rodentia fare: seeds, grains, fruit, nuts, and insects; its specialized kidneys metabolize water in a highly efficient manner, preventing the need for drinking. Caches surplus food in burrow larders, transporting it in its fur-lined cheek pouches.

Kangaroo rat.

Depending on latitude and food abundance, its breeding season can occur at anytime during the year. The polyestrous female may produce 2 litters annually. She gestates for about 1 mo., giving birth to 1-4 pups in an underground grass-lined natal chamber. Reproductive maturity comes by about 2 mos., making this a highly fecund species.

Communication and perception take place via traditional Heteromyidae channels: sight, hearing, touch, and smell. Vocalizations include squeaking and a bird-like chirp; thumps its hind feet when alarmed. Like many species of rodents it possesses a scent gland on its back (in its case between the shoulders), which it uses to mark territory and communicate with conspecifics. Its natural enemies are reptiles (rattlesnakes), birds of prey (owls), mustelids (weasels, badgers, skunks), and wild canids (coyotes, wolves). It has numerous defensive skills, from the ability to kick sand in the face of its attacker to being able to jump up to 8'. Engages in dust- or sand-bathing. Beneficial as a parasite host and as food for carnivores; detrimental as an agricultural pest; may negatively impact wild vegetation growth. Ord's kangaroo rat probably lives an average of 1-2 yrs. in the wild; perhaps twice that long in captivity.

ORNATE SHREW (*Sorex ornatus*): Its grows to a total length of about 4", its tail is long at 1½", its hind foot measures about ½", and it weighs up to around ⅓ oz. The dorsal pelage is grayish-brown; the ventral pelage is lighter. The eyes are tiny; the external ears are just visible; the body is somewhat stout; the head is short and broad; the muzzle is slightly convex and pointed; and like other shrews, the tactile-whiskered snout is flexible at the tip. Its tail is bicolored: brownish on the dorsum, grayish on the ventrum. It is very similar in appearance to several other shrew species, and is thus difficult to identify in the field (the main differences lie in minor characteristics of their skulls and dentaries).

Ornate shrews.

A riparian soricid, it inhabits brushland, foothills, grassland, coniferous forests, shrubland, valleys, and chaparral, with a preference for palustrine habitats, such as tidal and saltwater marshes, stream banks, rivers, lakes, freshwater swamps, and wetlands generally. Its range extends from central California south into the Baja Peninsula, Mexico. May excavate its own burrows; also appropriates those of other animals.

A carnivore, it is primarily insectivorous, but it is also a molluscivore (slugs, snails), amphibivore (frogs, salamanders), and an all-around invertivore (centipedes). Its breeding season begins in late winter and ends in early fall. The gestation period is unknown but is probably the same as similarly sized shrews: about 3 wks. A secondary consumer, its natural enemies probably include reptiles and birds of prey. The ornate shrew may live 1-1½ yrs. in the wild.

Leaf-nosed bat.

PACIFIC JUMPING MOUSE (*Zapus trinotatus*): Considered by some systematists to be conspecific with the western jumping mouse; others, however, believe the 2 are separate species. Either way, they share an almost identical ecology and life history. One primary difference is that, though their ranges overlap, the Pacific has a much smaller one than the western.

PACIFIC SHREW (*Sorex pacificus*): Though minuscule by human standards, it is the largest brown shrew in its region, growing to a total length of about 6", with a long 3" tail, a hind foot that measures ¾", and a weight of up to ½ oz. Its summer pelage is dark brown; its winter pelage is rusty-brown on the upper body, grizzled buff on the lower body. The head is conoidal; the eyes are tiny; the snout is long and pointed; the flexile nose tip is pinkish brown; the external ears are visible; the limbs and feet are light brown in color; the teeth are a "stained" reddish-brown, particularly at the tips.

Pacific shrew.

This plantigrade mammal is a secondary consumer that inhabits stream banks in moist coniferous forests dominated by microhabitats such as fallen logs and damp leaf litter. An Oregon native, it ranges across the western edge of the Beaver State and south in northwest California, generally hugging the Pacific coast. Constructs its nests in piles of moss, lichen, and leaves, as well as other types of forest floor detritus.

This carnivorous soricid consumes the typical diet of its kind. Utilizing its keen senses of hearing and smell, as well as an aggressive hunting style, it feeds primarily on insect larvae, gastropods, arachnids, chilopods, hymenopterans, annelids, and amphibians. Can run, leap, and dig in pursuit of its prey; known to cache extra food. Being nocturnal, its main natural enemy is the owl, though it is also sometimes preyed upon by amphibians. The Pacific shrew probably lives 1-2 yrs. in the wild.

PALE KANGAROO MOUSE (*Microdipodops pallidus*): This small rodent grows to a total length of about 6½" and can attain a weight of about ⅝ oz. The soft furry coat on the upper body is a pale pinkish fawn color, giving it both its common name and its species name; the lower body pelage is white; the 3½" tail is rat-like, sparsely furred, and lacks both a tuft and the usual black tip. The body is small and ovoid; the head is large and convex; the eyes are dark, large, and protruding; the ears are rounded and erect; the front legs are small and designed for digging and grasping; the hind legs are large, an adaptation for long distance hopping and the source of its common name. The tail provides balance during locomotion while serving as a fat storage compartment: in preparation for inactive periods, the tail grows thicker in the middle, providing energy during lean times.

This sedentary terricolous heteromyid is a nearctic native that inhabits temperate grassy deserts, sand dunes, scrubland, valleys, sageland, dry bottomland, alkali sinks, and open savanna with light vegetation. Its range is restricted to small portions of Nevada and California. Prefers fine sandy soil locations with abundant greasewood and saltbush.

A solitary, nocturnal, and saltatorial omnivore, its diet consists of seeds, nuts, forbs, grains, and insects. Hoards and caches surplus food, transporting it in its fur-lined cheek pouches. Though it lives in arid terrain, it does not need to drink water as its highly efficient kidneys metabolize all of the fluid it needs from the solid food it consumes. Digs 5'-6' long burrows in friable soil, blocking the entrances when inside in order to regulate the temperature and humidity. May estivate and hibernate, but it seems more likely that it enters brief periods of torpor during extreme weather conditions.

Its breeding season seems to take place between spring and fall, with the female giving birth to as many as 6 pups. Gestation, weaning, and independence may be similar to that of other species of kangaroo mice. Communicates and perceives in typical Heteromyidae fashion: primarily touch, smell, and hearing, and to a lesser degree, sight. Its incisors lack grooves;

Nevada scrubland: pale kangaroo mouse habitat.

its back paws are heavily furred, giving greater traction in its sand-dominant habitats. Its natural enemies include reptiles, birds of prey, mustelids, wild felids, and wild canids. It is a proficient jumper, just one of its many defenses against predators. Beneficial as a prey base for carnivores. The pale kangaroo mouse may live an average of 1-2 yrs. in the wild; twice that long in captivity.

PALLID BAT (*Antrozous pallidus*): This large chiropterid grows to a little over 5" in total length, its forearm measures roughly 2½", it weighs around 1¼ oz., and its wings can span 16". Its dorsal pelage is buff, tawny, tan, or cinnamon; the ventral pelage is whitish, a soft pale coloration which gives it its common name: the pallid bat. It has a flatish head, strong muzzle, big eyes, a long pointed tragus, and large ears that reach a length of nearly 1½". The 2" tail is long, but does not extend beyond the interfemoral membrane.

It shelters in trees, rocky crevices, caves, mines, and buildings in arid deserts, grasslands, canyons, and pine and oak forests, always near a reliable water source. It ranges from British Columbia south to Mexico, residing in all of America's West Coast states, east to parts of Idaho, Wyoming, Texas, and Kansas. Highly social, it forms summer colonies of 150 individuals or more, which, atypical for most bat species, contains members of both genders.

This nocturnal insectivore flaps its wings slowly (compared to other bats), possibly an adaptation to being primarily a ground feeder: flying low to the ground, it snatches up its food (beetles, moths, spiders, crickets, grasshoppers) as it drifts past. An excellent crawler, jumper, and climber, it sometimes lands on the ground to glean large prey items (praying mantis, cicadas, scorpions, lizards, mice). Its acutely sensitive ears can pick up the sound of an insect walking across the forest floor from high above. It is

A pallid bat species.

also known to occasionally hawk airborne insects while in flight; it can eat up to half its body weight in pestilent insects each night, making it a mammalian ally of ours.

It mates in the fall, the female delaying fertilization until spring when she and other mothers-to-be form special nursery roosts. After a gestation period of around 60 days, the female typically gives birth to temporarily blind, nearly hairless twin pups, who become independent following 2 mos. of care and feeding. While nursing, the mother carries her altricial offspring aloft during nighttime feeding forays.

Uses numerous different vocalizations to locate and communicate with conspecifics. These calls include clicks, buzzes, rasping sounds, and high-pitched notes. Does not appear to migrate, but hibernates in human structures, rock crevices, canyons, and deep caves during the coldest months. When frightened this vespertilionid has glands on its snout that emit an offensive odor. A secondary consumer, its natural predators are reptiles, birds of prey, and small to medium-sized mammals, including wild felids and wild canids. The pallid bat lives for about 10 yrs. in the wild; longer in captivity.

PALMER'S CHIPMUNK (*Neotamias palmeri*): Once thought to be a subspecies of the Uinta chipmunk, due to recent genetic studies it is now recognized as a separate species. While it shares much of the same ecology and life history as its close cousin, there are several differences, among them: it has 3 dark dorsal stripes (the Uinta has 3-5), it is smaller, growing to a total length of only about 8" (the Uinta grows to 10"), and its geographical range is restricted solely to Nevada (the Unita's range extends over 8 western states).

PANAMINT CHIPMUNK (*Neotamias panamintinus*): This colorful rodent grows to a total length of almost 9", its hind foot measures about 1", and it can attain a weight of up to 2½ oz. Its pelage is pale rusty-brown on the dorsum, cream-white or grayish-buff on the ventrum. Its alternating dark and gray-white dorsal and lateral stripes are somewhat indistinct; median stripe is brownish; facial and malar stripes mimic those on the back; sides are brownish-cinnamon; its head (which is less curved than other chipmunks), outer back legs, and posterior are gray, grayish, or grizzled gray; eye stripe is dark brown; the bushy 4" tail is covered in grizzled black, gray, and brown fur. Its warm weather coat is more vivid than its cold weather coat. Its cryptic coloration allows it to blend in with its woodland environment where the scattering shadows of sticks, grass, bushes, and branches predominate. Gender dimorphism is present: the female is larger than the male.

Chipmunks.

This medium-sized sedentary sciurid is a nearctic native that inhabits temperate montane coniferous forests, and also brushland and chaparral—or similar biomes with logs, stumps, and forest detritus. Petraphilic, it is especially fond of rocky settings, such as scree, cliff faces, boulder fields, outcrops, ledges, and talus slopes. Known particularly for its presence in the Panamint Mountains of Death Valley (from which it derives both its common name and its species name), its range is narrow, extending only over portions of southeast California and southwest Nevada. It dens in fallen trees, tree cavities, rock crevices, and sometimes underground. Since it stores food, it probably hibernates during extreme cold. Its home range is unknown, but is likely similar to that of others of its kind (perhaps around 1,000 sq. ft.). Though social, it is also territorial and aggressive and will readily defend its turf.

A terricolous diurnal omnivore, it partakes of the usual chipmunk fare: seeds, grains, flowers, forbs, leaves, woody plant parts, lichens, fruit, nuts, insects, and carrion. Is usually able to procure all of its water needs from the food it consumes. As noted, it hoards surplus edibles. Its breeding season occurs during the spring, with the doe gestating about 5 wks. She bears an average litter of 4-6 altricial pups in an underground natal nest chamber, weaning them around 5 wks. later. The buck does not contribute to the rearing of his offspring.

Communication and perception take place via normal Sciuridae channels: sight, touch, smell, and hearing. Vocalizations are identical or similar to others of its kind. Its natural enemies are reptiles, birds of prey, mustelids, wild felids, and wild canids. Benefits its ecosystem by hosting parasites, dispersing seeds, and providing food for carnivores. The Panamint chipmunk probably lives an average of 2-4 yrs. in the wild.

PANAMINT KANGAROO RAT (*Dipodomys panamintinus*): Named for the Panamint Range, a group of low mountains located within Death Valley National Park, this medium-sized heteromyid grows to a total length of about 13" and a weight of up to 2 oz. It shares a nearly identical ecology and life history with its close cousin the Stephens' kangaroo rat. Outside of genetics (where there are a few differences), the main dissimilarities concern size and range: The Panamint is slightly larger than Stephens', while its range is quite narrow, occupying only a small area in western Nevada and eastern and central California.

PIÑON MOUSE (*Peromyscus truei*): Also known as the "piñon deer mouse," this large rodent grows to a total length of 9", its hind foot measures about 1", and it can reach a weight of around 1 oz. Coloration on the dorsum runs from rusty-brown to brownish-gray; the ventrum is creamy gray; the 4½" tail is bicolored (dark above, light below), lightly furred, and tufted. Its cryptic coat colors vary with region and habitat. The body is thin, cylindrical, and covered in long, soft satiny fur; the head is long and tapered; the eyes are large, dark, and protruding, denoting nocturnality; the ears are large, membranous, and erect; the conical muzzle is tipped with dark sensitive vibrissae; the hind limbs are large; the feet are small and delicate.

This terricolous sedentary cricetid is a riparian nearctic native that inhabits temperate, rugged, piñon-juniper areas, hence its common name; it also occurs in dry savanna, desert, chaparral, cliffs, scrubland, sand dunes, mixed forests, mountains, palustrine environments, ranchland, shrubland, sageland, and canyons; has

Piñon mouse.

a preference for rocky settings, snag, cacti, logs, hillsides, and hiding cover. A western species, its range covers portions of Oregon, Idaho, Wyoming, California, Nevada, Utah, Colorado, Oklahoma, Arizona, New Mexico, and Texas, running south into Mexico. It dens in rock and tree crevices; scansorial, it is an adept climber and spends much of its time in trees; territorial, it establishes a large home range of several thousand square feet, vigorously defending it against intruders.

An omnivore with a granivorous, frugivorous, folivorous, fungivorous, and nucivorous appetite, it is also a part-time insectivore and arthropodivore. A scatter hoarder, it caches surplus food in shallow pits that it digs throughout its home range; highly efficient kidneys allow it to go for long periods without drinking water. Leaves little mounds of seed and nut shells around tree bases, and on logs, boulders, stones, and stumps.

Its breeding season runs from spring to fall. The polyestrous female may produce 3 litters annually; she gestates for about 1 mo. or so, and gives birth to an average of 2-4 altricial pups in a leaf-lined tree nest or a grass-lined rock crevice chamber, weaning them around 1 mo. later. Both males and females become reproductively mature by approximately 2 mos. of age, making this one of the most prolific of all mouse species. The male does not appear to aid in the rearing of his offspring.

Communication and perception occur via normal Cricetidae channels: sight, touch, hearing, and smell. Its natural enemies are reptiles, birds of prey, wild felids, and wild canids. Beneficial as a seed disperser, a parasite host, and as food for carnivores; detrimental as a carrier of dangerous diseases (such as the hantavirus and Lyme disease). The piñon mouse lives an average of 6-12 mos. in the wild; perhaps twice that long in captivity.

PIUTE GROUND SQUIRREL (*Urocitellus mollis*): Once believed to be a subspecies of the similar Townsend's ground squirrel, due to recent genetic studies it is now recognized as a separate species (it has 38 chromosomes, while Townsend's has 36 chromosomes). Morphologically, however, it is identical to Townsend's ground squirrel, with the same ecology and life history (including coloration, size, weight, habits, habitat, reproductive life, diet, and longevity). One dissimilarity: the Piute has a different geographical range, one covering portions of Oregon, California, Idaho, Nevada, and Utah.

PLAINS HARVEST MOUSE (*Reithrodontomys montanus*): This small rodent grows to a total length of nearly 6", its hind foot measures about ⅝", and it can attain a weight of up to ⅜ oz. The upper body is buffy with a gray-black wash and an indistinct dark dorsal stripe; the lower body is white; the short 2½" tail is thin, round, lightly furred, and dichromatic (dark on top, lighter on bottom). The head and muzzle are tapered; the eyes are large, dark, and protruding; the ears are large, thin, rounded, and erect; the ears often have light postauricular spots; its incisors are grooved.

Harvest mouse.

This endothermic cricetid is a nearctic native that inhabits dry prairies, savanna, cropland, scrubland, desert, woods, rangeland, chaparral, and well-drained grassy areas in general. As its common name indicates, its range extends over the Great Plains of America's heartland, covering portions of the states of Montana, Wyoming, North Dakota, South Dakota, Colorado, Nebraska, New Mexico, Arizona, Kansas, Oklahoma, Missouri, Arkansas, and Texas, running south into northern Mexico. It may nest underground, under rocks, or under logs. Its species name, *montanus* ("mountain"), is somewhat inaccurate as it is primarily a creature of level open land.

A nocturnal omnivore, its diet is made up of standard Rodentia fare: seeds, nuts, grass, grains, flowers, fruit, and insects. Caches surplus food. Its breeding season runs year-round. The polyestrous female may produce 2 litters annually; she gestates for about 3 wks., gives birth to an average of 4-6 altricial pups, and weans them about 2 wks. later. Communicates and perceives via standard Cricetidae channels: primarily smell and touch, and to a lesser extent, sight and hearing. Its natural enemies are reptiles, birds of prey, wild felids, and wild canids. Beneficial as a seed disperser and as a prey base for carnivores. The plains harvest mouse probably lives 6-12 mos. in the wild.

PLAINS POCKET GOPHER (*Geomys bursarius*): This medium-sized prairie rodent grows to a total length of about 14", its hind foot measures 1¾", and it can reach a weight of up to 12 oz. The pelage coloration on the dorsum ranges from tan and buff to grayish-brown and black; the ventrum is a lighter shade of the dorsal color; the longish 4" tail is round, lightly furred, and highly tactile. Its upper incisors are grooved, always visible, and grow continually throughout its life; its lips close behind the teeth.

The fur is soft and satiny; the cylindrical body and massive head are robust; the head is flattish in order to expedite digging and life below ground; the neck is thick and short; the jaw is strongly built; the eyes and ears are small; sensitive vibrissae on the snout aid navigation in the dark; the limbs are large and muscular; the front feet are large and heavily clawed, designed for soil excavation. It is mildly gender dimorphic, with the male being slightly larger than the female.

Plains pocket gopher.

This fossorial geomyid is a nearctic native that inhabits savanna, mesquite prairies, yards, cropland, mixed woodland, scrubland, cultivated pastures, roadsides, and moist grassland; prefers sandy, loamy, friable soil. Its range extends over much of the Great Plains, from North Dakota and Minnesota in the north to Texas and Louisiana in the south, from Colorado and Wyoming in the west to Illinois and Indiana in the east.

It lives in informal colonies, but because it is solitary it remains separate from conspecifics. Excavates large complex burrow systems with multiple chambers and passageways; these may reach a depth of 6' and can be located by the many fan-shaped ejecta mounds in the area—some as large as 1' in height and 2' in width; excess dirt is pushed out of its tunnels using its chest, often leaving "gopher cores" or soil castings at the surface. Establishes a home range of some 5,000 sq. ft.; territorial, it viciously protects its turf against intruders. Does not hibernate; active year-round.

A crepuscular nocturnal herbivore, its diet consists primarily of subterranean fare, such as roots and tubers; will also dine on surface plants when available; transports food in its external fur-lined pouches or "pockets," the source of its common name; obtains most of its water requirements from the solid foods it consumes. Its breeding season runs through the spring. The female gestates for 3-4 wks., giving birth to an average of 3-4 pups in a globular, grass-lined underground nest; here she nurses, grooms, and protects her young until weaning and independence arrive some 60 days later.

Communicates and perceives via standard Geomyidae channels: mainly touch and smell, with reduced sight and hearing. Its natural enemies are reptiles, birds of prey, mustelids, wild felids, and wild canids. Beneficial as an ecosystem engineer that aerates and recycles soil and as food for predators; detrimental as an agricultural pest and a lawn pest. The plains pocket gopher lives an average of about 2 yrs. in the wild.

PLAINS POCKET MOUSE (*Perognathus flavescens*): This silky furred rodent grows to a total length of about 5" and can attain a weight of ⅜ oz. Its coloration on the dorsum is buff with a gray or black wash; the ventrum is white; the 2½" tail is round with sparse pinkish-gray hairs; the lateral stripe is distinct; each ear often bears a yellowish postauricular spot. The body is ovoid; the eyes are dark and large; the ears are rounded and erect; the incisors are grooved.

This sedentary terricolous heteromyid is a nearctic native that inhabits temperate open landscapes, as well as grassland, sand dunes, rocky brushland, sandy deserts, plains, mixed woodland, and savanna. Has a preference for soft friable soil, meager vegetation, and scrubby hillsides. As its common name denotes, its range encompasses much of the Great Plains, extending from North Dakota in the north to northern Mexico in the south, from Utah in the west to Missouri in the east.

Solitary and fossorial, each individual excavates a small vertical burrow system with multiple tunnels and compartments, where it sleeps, shelters, and nests. For protection and humidity control, the entrances—which are only about ½" in diameter and hidden beneath vegetation—are sealed off when the animal is at home. This species enters a period of torpidity during adverse weather; may estivate and hibernate under extreme circumstances. Establishes a home range of 3,000-4,000 sq. ft.

A nocturnal omnivore, it consumes seeds, nuts, grass, legumes, grain, agricultural products, leaves, fruit, insects, and non-insect arthropods. Under normal conditions it does not need to drink water, most of which it can obtain from the solid foods it consumes. Caches surplus edibles, transporting them in its fur-lined cheek pouches or "pockets," the source of its common name.

Its breeding season takes place between spring and summer. The iteroparous polyestrous female may produce up to 3 litters annually. She gestates for between 3 and 4 wks., giving birth to an average of 3-4 altricial pups in a subterranean maternity chamber; weaning occurs within 3 wks. Communication and perception run along normal Heteromyidae channels: sight, touch, smell, and hearing. Its natural enemies are reptiles, birds of prey, mustelids, and wild canids.

Dry grassland: plains pocket mouse habitat.

Beneficial as a seed disperser, a soil aerator and recycler, and as food for carnivores; detrimental as an agricultural pest. The plains pocket mouse lives for an average of 6-12 mos. in the wild; perhaps twice that long in captivity. (Note: Once considered conspecific with the Apache pocket mouse, the larger Apache is now regarded by many systematists to be a subspecies of the plains, and has been classified with the trinomial *Perognathus flavescens apache*.)

POCKETED FREE-TAILED BAT (*Nyctinomops femorosaccus*): Sometimes placed in the genus *Tadarida*, this molossid grows to a total length of nearly 5", its forearm measures 2", and it weighs about ½ oz. Receives its common name from a pocket-like fold of skin on the underside of the interfemoral membrane. Its fur over the entire body is dark gray-brown with whitish roots. Its ears are connected at their base.

A petraphile, it shelters in rocky cliffs, crevices, and slopes, and also caves and buildings, near good-sized water corridors. It inhabits semiarid desert, scrubland, and oak and pine forests across the southwestern U.S., from southern California, east to Texas, south into Mexico. This chiropterid forms small colonies of 50-100 individuals.

At her nursery roost the female bears a single pup in early summer. This highly vocal, medium-sized insectivore is a fast, powerful nocturnal hunter that uses echolocation to detect moths, beetles, ants, flies, crickets, froghoppers, lacewings, and other agricultural pests. A secondary consumer, its primary natural predators include reptiles and birds of prey. Currently little else is known about the pocketed free-tailed bat.

POLAR BEAR (*Ursus maritimus*): Also known as the "sea bear" or "ice bear," it stands 4' at the shoulder, grows to 11' in total length, and can reach 1,200 lb.—though weights of up to 2,200 lb. have been recorded (making it the second heaviest land carnivore in the world after the brown bear). Though the coat of this archetypal symbol of the arctic North is typically white, it can range in color from yellowish to tan, and from gray to brownish. Despite its outward appearance, its skin is black (which helps to absorb sunlight and thus heat) while its hairs lack pigment (which aids in camouflage in its nearly all-white habitats). Hollow and transparent, each hair is filled with both air and light-scattering particles that cause its fur to luminesce in the sun. This phenomenon, combined with the millions of small light-refracting salt particles that get stuck in its fur (from swimming in the sea), make its coat appear white when light strikes it. Its hollow-

Polar bear.

haired pelage, which also gives it added buoyancy in the water, yellows with age.

This large heavyset ursid has a small head, a black nose pad, black lips, short rounded ears, and a long neck. Has a convex snout like its cousin the brown bear (from which it may have originally evolved), but lacks its prominent shoulder hump and long claws. Its extra-sharp teeth supply seizing and puncturing power, useful for hunting and defense. A plantigrade, all four of its paws are large, fur-covered, and snowshoe-like (providing grip on rocks, ice, and snow, as well as insulation against the cold). Its short, well-curved claws are usually black, and furnish not only extra traction, but prey-grasping power as well. Its wide front paws are designed for paddling. An idea of the scale of this creature comes from the adult male's hind foot, which measures up to 13" long by 9" wide. Its massive internal volume means that it has a small surface area-to-volume ratio, which reduces heat loss, helping it maintain the correct body temperature. This species is gender dimorphic, with the male being much larger than, and over twice as heavy as, the female.

Polar bear.

Though it is a terricolous mammal, one with legs and feet designed for walking on land, as its scientific name indicates (*Ursus maritimus* means "maritime bear"), it is also a marine (though not an aquatic) mammal—one with a preference for sea ice. It inhabits circumpolar regions and biomes, in particular, the environs of the Arctic Ocean. It thrives in, on, and around intertidal areas, tundra, rocky shoreline, mud and salt flats, sand bars, oceanic cliffs, subarctic grassland, and sandy beaches. In cold months it lives on pack ice; in warm months, when the pack ice thaws and recedes, it takes to land, roaming along the coastline, island-hopping, and foraging around polynyas. A chionophile, it is found only in the Northern Hemisphere. In our region the polar bear lives strictly in the far north: Alaska, northern Canada, and Arctic islands, straying only as far south as fast ice grows. It occasionally enters northern New England. Establishes a home range, but does not appear to be particularly territorial.

In a region where it can be light or dark for months at a time, this natatorial mammal is both diurnal and nocturnal. Thus, while it is most active during the day, it may also forage and roam about at night. It is active at all times of the year. Neither the male or the female hibernate in the traditional sense. However, after mating, the female dens up in a snow shelter for 4-8 mos. (from fall to spring), at which time she gives birth. She does not eat, drink, urinate, or defecate during this state of torpor: beforehand she feeds abundantly, then alters her metabolism in order to live off the stored fat while she "sleeps." In spring, when the cubs have reached a weight of 20-30 lb., the family group leaves the den.

With its protein-rich meat and calorie-rich blubber, the primary prey of this fearsome boreal meat-eater—the most carnivorous of the North American bears—is the ringed seal. Also omnivorous, the polar bear will consume algae, kelp, grasses, leaves, berries, mushrooms, fish, birds and their eggs, crabs, shellfish, rodents, reindeer, other seal species, walruses, whales, and carrion. It sometimes resorts to cannibalism. Synanthropic, it also eats human garbage. Spends up to 50 percent of its time foraging. Uses 2 primary types of hunting: still-hunting (waiting for its prey to emerge from its shelter) and stalk-and-pounce hunting (creeping up to its prey and leaping on it). It does not normally

Polar bear.

hoard or cache surplus meat; will share food with conspecifics.

With a long slender neck for poking its head into seal ice holes, a heightened olfactory sense, extra large teeth, stealthy hunting techniques, a white waterproof coat that camouflages it while on the hunt, and the ability to swim (up to 6 mph) for hours, dive to depths of 15', and hold its breath underwater for 2 min., the polar bear is well designed for stalking and killing. It can run 25 mph for short distances. Its sense of smell is so sharp it can catch the scent of an animal hiding as deep as 3' under the snow and as far as 1 mi. away. Its narrow torpedo-like skull cuts through the water, making it easier to swim quickly and dive efficiently. It swims only with its broad front paws, which are shaped like large paddles (its motionless hind legs drag behind). It can swim for at least 9 consecutive days and a distance of over 400 mi. without coming ashore. Has been observed swimming out at sea 200 mi. from land.

Its mating season lasts around 3 mos., running from early to late spring. Males fight violently over access to estrus females. Breeding boars and sows form temporary pair-bonds at this time. The female employs delayed implantation, with a total gestation period of up to 8 mos. long. The following year she gives birth to 1-4 extremely altricial cubs in the late fall or early winter. Parturition takes place in a maternity snow den that she excavates on land, always near the ocean (when necessary females will also dig dens in the snow on pack ice or in the soil beneath the permafrost). The entirely helpless young remain inside the den for 2-3 mos., growing quickly on their mother's fat-rich milk. After emerging, they stay with her for another 2-3 yrs. During this time the sow feeds and protects them, while teaching them various survival skills—including how to hunt. Fiercely protective of her babies, she will fearlessly attack and drive off males that are much larger than herself. The female polar bear has one of the lowest reproductive rates of any mammal, producing only five litters throughout her entire lifetime. The polygynous boar departs after mating, and does not contribute to the rearing of his offspring.

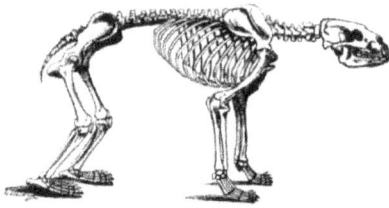

Polar bear skeleton.

It communicates and perceives using traditional Ursidae channels: sight, touch, smell, and hearing. Vocalizations include chuffing, roaring, snorting, hissing, and growling. This nomadic, solitary, arctic native is considered an umbrella species. Has been known to interbreed with its relative the brown bear (grizzly bear), producing a fertile hybrid called a "pizzly." A secondary consumer and an apex predator, it has few natural enemies outside the wolf (which preys on its young) and conspecifics. On average the polar bear lives about 15 yrs. in the wild; a very small percentage may reach 20-25; in captivity it has been known to live to 45 yrs. of age.

PRAIRIE VOLE (*Microtus ochrogaster*): This small microtine grows to a total length of around 7", its hind foot measures nearly 7/8", and it can attain a weight of up to about 2½ oz. The pelage is not only rough, it is longer than in some other vole species. While many variant colorations can occur, its upper body is typically sienna mixed with yellowish and blackish hairs; the lower body is light brown, from which it derives its Greek species name (*ochrogaster* means "brown-bellied"); its ears are small, giving it its Greek genus name (*Microtus* means "little ear"); the short 1½" tail is bicolored (dark on the dorsum, light on the ventrum). Its incisors lack grooves.

This terricolous sedentary cricetid is a nearctic native that inhabits temperate dry grassland, weed-fields, chaparral, prairies, savanna, thickets, meadows, shrubland, woodlands, and fallow pastureland. As its common name suggests, its range centers around the prairielands of the Great Plains, extending from south central Canada in the north to Texas in the south, from Colorado in the west to West Virginia in the east.

Social and usually monogamous, it lives in mated pairs or small colonies, depending on food supply and the season. Constructs winding grass tunnel systems where it shelters and nests; also semi-fossorial, it may excavate shallow burrows in friable soil. Populations seem to cycle up and down every few years. This species is natally philopatric; establishes a home range of about 5,000 sq. ft. or so.

Prairie vole.

A diurnal, crepuscular, and nocturnal omnivore, it consumes typical rodent fare: seeds, leaves, grains, fruit, nuts, wood, tubers, bark, grass, forbs, leaves, stalks, roots, and insects. Hoards and caches surplus food in underground larders. Its breeding season runs year-round. The iteroparous polyestrous female may produce 2 litters annually. She gestates for about 3 wks., gives birth to 2-4 altricial pups in a grass-lined nest, and weans them approximately 3 wks. later. Both the male and the female become reproductively mature at around 1 mo. of age. The father indirectly aids in the rearing of his offspring by guarding the maternity site.

Communication and perception take place via traditional Cricetidae channels: sight, hearing, touch, and smell. Its natural enemies are reptiles, birds of prey, mustelids, wild felids, and wild canids. Beneficial as food for carnivores; detrimental as an agricultural pest. The prairie vole lives an average of 6-10 mos.; twice as long in captivity.

PRONGHORN (*Antilocapra americana*): The only species in its family, *Antilocapridae* ("antelope goat"), the pronghorn is neither an antelope or a goat, but the single remaining member of a prehistoric group that is most closely related to the giraffe.

Also known as the "American antelope" and the "prongbuck," it grows up to 5' in total length, stands 3½' tall at the shoulder, and weighs as much as 140 lb. A medium-sized antilocaprid with a barrel-shaped body and long legs, its dorsum is tan to reddish brown, while its ventrum, cheeks, lower jaw, rump, chest, and sides are white. Two white blazes form over the throat and it has a black nose and a black eye-band. Bucks have black neck patches and a distinctive black band running from under the eyes to the throat. Both genders sport a short erectile mane, as well as a bright white rump patch (also erectile), which they flash when startled or in danger.

Its most distinguishing characteristic is its set of black "horns" (the source of the second half of its common name), which possess characteristics of both true horns and true antlers, but which are neither. The male's bow-shaped horns grow to 20" in length and curve inward and back at the top, ending in sharp hooked tips. Additionally, a thick, hooked, pointed prong protrudes midway up on the front of each horn (the source of the first half of its common name). Unlike true antlers, its horns are encased in keratin (the same material out of which our fingernails are composed); and unlike true horns, the pronghorn sheds and regrows them every year. The female may or may not grow horns; when she does, they average about 3-4" in length, are straighter than the male's, and lack a prong. Gender dimorphism is evident: the male is larger, heavier, and more distinctly colored than the female.

Pronghorn.

This widespread artiodactyl, territorial ungulate, and nearctic native inhabits open prairie, grassland, chaparral, treeless plains, brushland, scrubland, sagebrush areas, rangeland, and desert with clean reliable water sources. Its range covers southern

Pronghorn herd.

Canada, south into the Rocky Mountain states of the American West, and further south into northern Mexico. Its home range varies from 5 to 15 sq. mi., depending on the season and terrain. Social and gregarious, throughout the year pronghorn herds change size and mixture, depending on numerous factors. In general, females form matriarchal groups made up of around a dozen mature does and fawns, while bucks form bachelor herds comprised of young males and yearlings. Mature males split off in order to establish their own territories in time for the rut.

The pronghorn is cursorial; that is, it is specifically designed for running. It is so perfectly adapted to this form of locomotion that it is the fastest land animal in the Western Hemisphere: it can bound some 20', run at 30 mph for up to 15 mi., and easily reach a top speed of 70 mph—which it can maintain for 4 min. There was a price for this biological perfection, however, for the adjustments Nature has made to its bones, muscles, and joints to allow it to run at these speeds created deficiencies in other areas. For example, it cannot jump natural or human-made barriers (such as fences) as deer do.

Like many prey animals, it has protruding eyes, which allow a wide arc of vision. In the pronghorn's case, its large motion-detecting eyes see 320 degrees around, and can pick up slight movements from as far as 4 mi. away. An excellent swimmer, its coat (with hollow outer hairs) lays flat in winter to maintain heat, and fluffs out in summer to allow heat to escape.

An herbivorous ruminant, grazer, and browser, its summer food consists of grass, shrubs, forbs, and cacti; in winter it switches over mainly to sagebrush. Its 4-chambered stomach is well-suited to digesting the roughage that makes up so much of its diet. If it consumes enough moisture-laden green vegetation, it does not need to drink. Because it feeds on deleterious plants (such as nonnative vegetation), it is of great benefit to us and is thus considered a mutualist species.

Its breeding season is dependent on latitude and is thus highly variable. Over its full range, however, the rut runs from midsummer to early fall. Bucks scent-mark their territory using ground scrapes into which they urinate; they then paw the soil into a pungent muddy mixture, creating a territorial warning smell to other bucks. The male competes with other males over rights to females, the winners forming small harems (of up to 20 does) that each one defends until the mating season ends. This competitive jostling can lead to violent male-on-

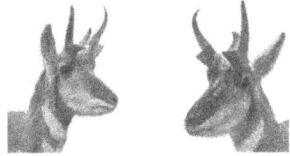

Two male pronghorns.

male encounters that involve agonistic body language, pushing and shoving, threatening vocalizations, and stabbing contests (that may result in serious wounds and even death).

The doe breeds once a year, gestates for around 8 mos., and gives birth to 1-2 precocial fawns in the spring. Though the young are weaned at around 3 wks., they do not become independent for at least a year or more. Bucks are non-paternal and therefore do not contribute to the raising of their offspring.

It communicates and perceives through normal Antilocapridae channels: eyesight, sound, body language, and scent. The latter is expedited using a variety of glands scattered over the body. Also known as the "pronghorn antelope," the natural enemies of this primary consumer are birds of prey (such as golden eagles), wild felids (such as mountain lions), and wild canids (such as wolves). Nearly driven to extinction by 1900, under judicious management today this icon of the American prairie is making a comeback. The pronghorn lives on average 8-10 yrs. in the wild; it may live up to 15 yrs. in captivity.

PUMA: See Mountain lion.

PYGMY RABBIT (*Brachylagus idahoensis*): Growing to a total length of only 9-12", and weighing up to 1 lb., this is North America's smallest leporid. Its dorsum is buff-gray, its ventrum is lighter, the back of the neck and the outside of the legs are cinnamon colored. It has short hind legs; furred hind feet; black and white whiskers; a small tail; and short, rounded, furry ears.

It inhabits mainly sageland, but also grassland, mountainous areas, valleys, slopes, shrubland, plains, steppe, and plateaus. Its historic range covers sections of Washington, Oregon, Idaho, Montana, Wyoming, Utah, Nevada, and California. Does not seem to migrate. Unlike most other rabbits in our region it usually excavates its own burrows, which run to about 3' deep and have multiple entrances.

An herbivore, this diminutive lagomorph dines primarily on sagebrush, but also on grass and forbs. Breeds in the spring. On average the polyestrous female gives birth to 4-6 kits. She may have up to 3 litters each mating season. A primary consumer, its natural enemies include birds of prey, weasels, badgers, foxes, bobcats, and coyotes. The pygmy rabbit lives from 3-5 yrs. in the wild.

Shrubland slope: pygmy rabbit habitat.

PYGMY SHREW (*Sorex hoyi*): Its body is a mere 2" long. Including its 1½" tail, it has a total length of 3½". Its hind foot measures about ½" and it weighs just ⅛ oz. This makes it not only the smallest shrew in North America, but also the smallest mammal in our region. Indeed, it is one of the tiniest mammals in the world, weighing less than a single sheet of paper. (The only mammal that is smaller, by mass, is *Suncus etruscus*, the Etruscan shrew of Europe, North Africa, and Asia.) Also known as the "American pygmy shrew," its dorsal pelage is a grizzled grayish-brown; the pelage on the venter is silvery-gray. The body is stout; the head is cone-shaped; the eyes are minuscule and nearly covered with fur; the ears are nearly fur-hidden; the nose is convex, long, and pointed, with a flexile tip; the limbs and feet are tiny, pinkish, and finely formed.

A riparian, it inhabits a diversity of biomes and ecosystems, including boreal marshes, damp meadows, floodplains, timberland, foothills, wet mixed-grass prairies, freshwater pools, open fields, brushland, fen, grassland, forbland, bogs, and the mesic areas of deep forests dominated by moist soil, sphagnum moss, detritus, and rotting logs. A nearctic soricid, it ranges across the northern tier of North America, from the Pacific Coast to the Atlantic Coast, covering Alaska and Canada, south into the Rockies, and east into the Midwestern states and Appalachia. Active all year-round, it is relatively rare (and nearly impossible to see) wherever it is found.

Shrew skull, top and bottom.

A solitary, stealthy, and ferocious hunter, it is a carnivore whose diet includes dipterans, coleopterans, lepidopterans, amphibians, annelids, hymenopterans, and arachnids. It may feed on carrion as well. Diurnal, crepuscular, and nocturnal, this fossorial mammal is an expert excavator who spends much of its time foraging in leaf litter and damp earth in search of its quarry.

Detailed information about the mating system of the pygmy shrew is lacking. Its breeding season occurs during the summer months. The female gestates for about 3 wks., giving birth to 3-7 young in a well-concealed nest. She grooms, protects, and nurses them until they are weaned. The role of the male in the care of his offspring is not

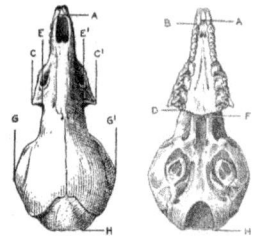

known.

This shrew species communicates and perceives using sight, touch, hearing, and smell. It vocalizations are typical for a member of the Soricidae family: squeaks, whistles, purrs, peeps, hisses, chattering, and twitters. Scent glands produce a pungent odor that may act as both an attractant (to conspecifics) and a repellant (to predators). A secondary consumer, its natural enemies are fish, reptiles, birds of prey, and small mammals. The pygmy shrew may live around 12-16 mos. in the wild.

PYGMY SPERM WHALE (*Kogia breviceps*): Also called the "lesser sperm whale" and the "short-headed sperm whale," this small, compact cetacean grows to a length of 13', weighing in at around 900 lb. (a little less than half a ton). It has a small dorsal fin set back on the spine and a slender lower jaw with less than a dozen teeth on each side. There is some variation in coloration, but it is generally grayish-blue to bluish-gray on the dorsum, with a lighter pinkish-tan ventrum. Its small head (described by its species name, *breviceps*, Latin for "small head") combines traits of its namesake the sperm whale,

Pygmy sperm whale.

as well as its primary enemy the shark: the former gives it a somewhat blunt whale-like appearance, the latter a somewhat predatory shark-like appearance.

Though an air-breathing mammal, behind its eyes there is a whitish crescent-shaped arch on the skin, a false gill aperture, probably an evolutionary adaptation: by impersonating a shark, the pygmy intimidates real sharks, helping to prevent predatory attacks. Jutting sharp teeth along the tiny lower jaw complete the masquerade. This toothed whale has another unusual defensive strategy, this one similar to that used by the squid: when alarmed an intestinal sac emits up to 1 quart of rust-colored liquid that may act as a "smokescreen" while it darts away to safety. Like a number of other cetacids, the female is larger than the male, thus it is gender dimorphic.

The blowhole of this diurnal physeterid is not located in the center, but like its namesake the sperm whale, it is set slightly to the left, with the right side, or right nostril, being devoted to sound-production. This design, as well as the accompanying organs inside, causes the head to be slightly asymmetrical in shape. Also, like its larger cousin, the pygmy possesses a special organ (located in its head) that is filled with a translucent waxy oil called "spermaceti." The exact function of the spermaceti organ is not known, but it almost certainly aids in echolocation, buoyancy, or navigation. The pygmy's big brain is rich in magnetite crystals, which may serve to help it navigate using the earth's magnetic field.

Though it ranges primarily throughout the temperate and tropical coastal waters of the world, in North America this cosmopolitan mammal inhabits both the Pacific Northwest and the Atlantic coastline north to Canada. Using echolocation, it hunts squid, octopus, shrimp, and fish, diving to a depth of up to 1,000' and staying underwater for nearly 1 hr. As with many other whale species, the little that is known about this rare, seldom-seen mammal comes mainly from stranded individuals. Even its family name has yet to be agreed upon: is it Physeteridae or Kogiidae?

Pygmy sperm whale.

The iteroparous females give birth to a single 4' precocial calf that weighs about 110 lb. A slow swimmer, it sometimes floats motionlessly (and unnoticed) like a log at the surface (a behavior known, naturally, as "logging"), making it vulnerable to vessel strikes. A secondary consumer, along with great white sharks, its other main natural enemy is the orca. This natatorial kogid is sometimes confused with its smaller relative, the dwarf sperm whale. The solitary, shy, reclusive pygmy sperm whale may live as long as 25 yrs. in the wild.

Wolverine.

RACCOON (*Procyon lotor*): The official state furbearer of Oklahoma and the official state wild animal of Tennessee, this well-known mammalian "bandit" grows to an average 12" at the shoulder, it is 38" in total length, it has a 5" hind foot, and it typically weighs about 25 lb. (though it can reach 50 lb.). Also known variously as the "North American raccoon," "northern raccoon," "common raccoon," or simply "coon," its gray-brown fur is grizzled and tinged with yellow and cinnamon. Its ventral fur is lighter, usually cream or grayish. Depending on the individual and its geographic location, its dense pelage can vary widely in markings and coloration. It has a broad face; small ears; a tapered slightly concave snout; a compact body; a bushy ringed 16" tail; and shortish legs ending in 4 highly sensitive, extremely dexterous 5-toed paws. It may be instantly identified by its familiar black facial mask (which aids in reducing light glare), bordered above by two white eyebrow stripes. Though a medium-sized mammal, this procyonid is the largest member of its family. Gender dimorphism is present, with the male being larger than the female.

Raccoon.

Both a nearctic and neotropical native, as a riparian mammal it is usually found near water, such as along shorelines, creeks, rivers, estuaries, swamps, ponds, lakes, flood plains, wetlands, and littoral zones in general. Though it prefers damp forested areas, it easily adapts to suburban and even urban environments; it can also handle extremes in both biomes and weather, and is found in areas ranging from boreal bogs and grassland to tropical rainforests and savanna. Lives across most of the U.S. and southern Canada, though it is rare or absent in the Rocky Mountain states. Occurs as far south as South America. Dens in hollow trees, as well as crevices, attics, tunnels, caves, sheds, sewers, houses, barns, and the abandoned burrows of other animals—almost anywhere in its range, in fact, that offers suitable protection, water, and food. Solitary, crepuscular, nocturnal, and sedentary, this native North American establishes a 10-mi. home range, though it is not particularly territorial.

Its five-toed footprints can resemble tiny bare human hands; with them it can open doors of all kinds, making a nuisance of itself. Like bears, Sasquatch, and humans, it is a plantigrade mammal—that is, it walks on the soles of its feet, rather than on its toes as digitigrade mammals do (e.g., bobcats and coyotes).

An opportunistic omnivore, its diet is made up of plants, roots, fruits, tubers, seeds, grains, nuts, insects, small mammals, birds and their eggs, carrion, and crops. In water it will take invertebrates, amphibians, fish, and turtles and their eggs. A synanthrope, it readily scavenges campsites and human dumping grounds, knocking over garbage cans to get at their contents. Prefers examining and washing its food in water before eating, hence its species name *lotor*: the "washer."

Does not hibernate during cold weather, but will go into a state of temporary dormancy under adverse winter conditions, living off stored body fat. Though it may appear harmless and ungainly, even comical, the stocky raccoon is a cunning, nimble, and formidable fighter that can run 15 mph, climb trees, swim, bite, and tear. It is capable of jumping to the ground from a height of 30' or more, and can easily defend itself against animals many times its size.

Its polygynandrous mating season occurs in early spring, with males traveling great distances to find a mate. Breeding females and males develop temporary pair-bonds that dissolve soon afterward. The sow gestates for about 2 mos., and bears 1 litter a year in late spring. Her 1-7 altricial kits are blind, weigh just 2 oz. at birth, and sport little black face masks. They are weaned by 2½ mos., after which they leave the maternal den. Independence occurs at around 1 yr. of age—though family members usually den near one another even after dispersal. The boar does not contribute to the rearing of his offspring.

Raccoon.

Communication and perception take place via established Procyonidae channels: sight, hearing, smell, and especially touch. One of Nature's more vocal creatures, it makes over 100 different noises and vocalizations, including: snarls, screams, purrs, hisses, whimpers, screeches, grumbles, mews, growls, chitters, cries, snorts, barks, and whistles. A secondary consumer, among its natural enemies are reptiles, birds of prey, wild felids, and wild canids. A well-known crop-destroyer and transmittable disease-carrier, the intelligent, mischievous, inquisitive, and sometimes destructive, raccoon lives an average of 2-3 yrs. in the wild, rarely more than 5-10 yrs.; may reach 20 yrs. of age in captivity.

Rafinesque's big-eared bat.

RAFINESQUE'S BIG-EARED BAT (*Corynorhinus rafinesquii*): Its grows to a total length of nearly 4½", its forearm measures about 2", its wingspan is around 12", and it weighs ⅜ oz. Employs countershading: the fur on its dorsum is gray-brown, the fur on its ventrum is brown with white tips. Its tragus is tall and narrow and there are 2 fleshy glands on the nose, which has elongated nostrils. Its most prominent trait are its large ears, which grow to just about 1½", nearly half the length of its body. When awake and alert, particularly during flight, it leans its erect antennae-like ears forward for better reception. While at rest, however, this vespertilionid coils them up like the horns of a ram (helping to conserve body heat and water), giving it the nickname the "ram-eared bat."

It is often confused with its cousin, Townsend's big-eared bat. But the latter is usually lighter in color and has a buff underside, while Rafinesque's big-eared bat has grizzled belly fur.

A riparian species, when roosting it seeks out hollow trees and snag in oak and pine forests near water (wetlands, bogs, rivers, swamps, ponds), but will also shelter in caves, wells, mines, bridges, and buildings—depending on the season, latitude, and local population. From fall to late winter it goes into a shallow hibernation, sustaining itself on fat reserves in special cold weather roosts known as hibernacula. It ranges over much of the American Southeast as well as the southern extremes of the Midwest. This nocturnal acrobatic flyer (it can both hover and take flight from nearly any position) is an invertivore that climbs into the air after sundown, using echolocation to detect its prey (mainly moths, but also flies, ants, beetles, and bees), using gleaning and hawking methods to catch it.

The troglophilic male and female mate just prior to autumn hibernation, after which, so one might assume, the female would begin gestating. But this means that her young would be born in the middle of winter, a hazardous time for any mammalian neonate. Thus, she delays fertilization until spring when conditions are more advantageous for both mother and infant. Sitting upright in the nursery roost, the mother gives birth to a single altricial pup in early summer, catching it in her interfemoral membrane. She carefully raises it (even carrying it in flight) until it is mature enough to fly and feed on its own—which takes roughly 5-6 wks. A secondary consumer, its natural enemies include reptiles and small mammals. Rafinesque's big-eared bat lives 5-10 yrs. in the wild.

RED BAT (*Lasiurus borealis*): This medium-sized tree bat grows to a total length of 5", its forearm measures a little under 2", the wings span up to 13", and it weighs ½ oz. Its most distinguishing characteristic is its color: the male's long soft coat is usually bright red, the female's is usually a duller brick-red. Both genders bear a white patch on their shoulders, while the coat hairs are white-tipped, giving it a grizzled or "frosted" look. This coloration makes for excellent "dead leaf" camouflage as its hangs motionless in the tree foliage by one foot, shrouded in its hair-covered tail membrane which it wraps around itself for warmth (it is from this trait that it derives its genus name: *Lasiurus* means "hairy tail"). The ears and the tragus are small and rounded; the wings are long and narrow; the uropatagium is furred on the upper side, an adaptation to low temperatures; its skull is small and robust; the muzzle is short and blunt. Because the pelage of the male and female are different colors (a rarity among North America mammals), this vespertilionid is gender dimorphic.

A nearctic chiropterid, it prefers rural forested regions where it shelters in trees and vegetation, such as bushes, hedgerows, Spanish moss, windbreaks, clearings, orchards, and forest edges. Unusual for a bat, it will sometimes roost in the open. Its species name, *borealis* ("northern") identifies it as a North American mammal; in its case, one that ranges over a large part of our region, from Canada through the U.S. and south into Mexico and beyond (into the Caribbean). Mostly solitary, it migrates south in the fall, north in spring, at which time it hibernates in microhabitats, such as tree cavities and ground leaf litter, to escape the cold.

Red bat.

This hardy fast-flying insectivore acts as a natural pest controller. Leaving its roost shortly after dark, it drops silently from its perch (usually located in dense leafy trees) into the night air, and, using echolocation, it quickly swoops down on its prey: beetles, flies, leafhoppers, ants, planthoppers, spittlebugs, and moths, among many other types of agricultural nuisances. Uses both hawking and gleaning techniques as it hunts; can often be found flitting about street lights in search of airborne insects.

The red bat mates once a year in the fall, but the female delays fertilization until spring, gestating for about 85 days, giving birth to 1-4 altricial pups (that is, they are temporarily blind and hairless, and thus completely dependent on their mother for survival). The female is quite unique in having 4 mammary glands (like female humans, most female bats have only 2). She nurses, grooms, and protects her offspring for around 5 wks., after which they are weaned and ready for independence. Also unlike many other types of bats, the female does not carry her young during flight. A secondary consumer, its natural enemies include birds of prey and small mammals. The life span of the red bat is unknown.

RED FOX (*Vulpes vulpes*): The official state land mammal of Mississippi, it is the largest member of the fox family, standing up to 18" at the shoulder. Its total length is a little over 3'; its hind foot measures 7", and it can attain a weight of 30 lb. Its dorsal coat is cinnamon-red in color; its underside is white, cream, or grayish. The chin and chest are white to buff. Upper legs are a dark shade of rust; lower legs and paws are black. A variety of markings and color phases are common. The archetypal fox, it has a small head; forward-facing eyes with elliptical pupils; pointed ears with cream fronts and black backs; a long narrow muzzle; a dark brown nose; and a long, bushy, 16" white- or black-tipped tail with black along the venter. The male is larger than the female, thus it is gender dimorphic.

An edge species, it inhabits a wide diversity of biomes and ecosystems, such as fields, meadows, scrubland, woods, brushland, desert, chaparral, farmland, mixed forest, tundra, rangeland, stream bottoms, montane forest, savanna, ranchland, prairie, dune, forest edges, pastureland, woodlots, ecotones, and mosaic habitat in general. Also a synanthropic species, it thrives in anthropogenic areas or human-altered landscapes, including suburbs and even densely populated inner cities. It is commonly found, for example, in or around backyards, golf courses, ditches, gardens, school campuses, high rises, garbage dumps, subdivisions, cemeteries, hedgerows, and parks and other large, cultivated green spaces.

Red fox.

Though it is the most widely distributed carnivore in the world, debate continues as to whether it is native to North America. One school of thought maintains that it is native only to the northwest, and that its origins elsewhere in our region are unknown. Another believes it to be entirely nonnative, asserting that it began as an Old World species (the European red fox) that was introduced to North America (mainly along the East Coast) around 1750 from England. Wherever and however it arose, due to its adaptability, versatility, and high rate of reproduction, today it thrives from Alaska to Central America, and is found in nearly every U.S. state and most of Canada.

Both a primary cavity excavator and a secondary cavity excavator, it prefers loose soil and often digs its own earthen den (usually on slopes, stream banks, or in rock piles), but it will appropriate the burrows of other animals (rabbit, woodchuck, badger). Sedentary, dens are reused repeatedly and may contain numerous passages, tunnels, entrances, and escape exits. This medium-sized, catlike canine requires a nearby water source and good hiding cover, the latter which it uses for shelter, protection, and travel corridors. Depending on available prey and water, it establishes a lifelong home range of up to 50 sq. mi., which it defends against conspecifics.

Red fox.

More an omnivore than a carnivore, its winter diet consists of rodents (voles, mice, squirrels, gophers, woodchucks), shrews, opossums, rabbits, raccoons, porcupines, moles, weasels, mink, muskrats, birds and their eggs, and carrion. In summer it eats primarily both vegetation (grasses, fruit, nuts, grains, vegetables) and invertebrates (insects and crayfish), as well as reptiles. Will prey on poultry, domestic pets, and human refuse when available. Often employs a sit-and-wait hunting technique, leaping suddenly on its unsuspecting prey. Hoards and caches surplus food under ground or snow.

Nocturnal, crepuscular, reclusive, cunning, and wary, it is not often seen. Unlike its cousin the wolf, it is solitary and therefore does not live in packs. Can run 30 mph, jump 6' high fences, and swim fast-flowing rivers. It rarely dens up in winter, but instead sleeps out on open ground encircled by its bushy tail for warmth. While sleeping thus, it may allow itself to become buried in snow—an added means of heat retention as well as protection against predators.

Depending on where it lives within its range, its breeding season generally runs from winter to spring and entails monogamous, polygynous, and polyandrous matings. However, a temporary pair-bond is eventually established between the dog (male) and the vixen (female). Males compete violently with one another over estrus females. The female gestates for about 7 wks., giving birth to an average of 7-8 altricial pups in a maternity den, where, for their first month, they eat predigested food regurgitated by the mother. She may move her babies to different dens several times during the first month (to avoid predation). Though the pups begin exploring outside the den by 1 mo. of age, they are not weaned until they are around 2 mos. old.

Red fox (juvenile).

Dispersal takes place at 4 mos. The father participates in the rearing of his young by bringing live food to the entrance of the natal den. (Sometimes adult male siblings and related females also assist the new *vulpes* family.) After the pups become independent the mother and father also disperse, reverting to a life of solitude until the following winter, when the mating season begins again.

It communicates and perceives via normal Canidae channels: sight, touch, hearing, and smell, the latter which entails copious scent-marking using urine, dung, and a variety of oil-secreting glands located over the body (rump, paws, face). One of the most vocal of the canids, it makes over 2 dozen sounds, from barking, howling, yipping, whining, and growling, to coughing, screaming, squealing, huffing, and gekkering. Though it can carry and transmit rabies, it consumes large quantities of pestilent rodents, rabbits, and birds (pigeons), and is thus beneficial to us. A secondary consumer, its natural enemies include birds of prey, wild felids (lynxes, mountain lions), other wild canids (coyotes, wolves), and ursids. The red fox lives about 3-4 yrs. in the wild; up to 10 yrs. in captivity.

RED SQUIRREL (*Tamiasciurus hudsonicus*): The smallest tree squirrel in the range it occupies, it grows to a total length of about 15", its hind foot measures around 2", and it can attain a weight of nearly 9 oz. Though coloration can vary greatly with the population, season, and latitude, generally the dorsum is a grizzled brownish-gray with a bright cinnamon-red wash, the source of its common name; the ventrum is white, cream, buff, or light gray; in summer a distinct black lateral line separates the upper and lower body; the 6" tail is bushy and rust-colored, and entirely bordered with a black band that has a white edge; the eye rings are white; the paws are cinnamon colored; the ears are large, erect, and tufted in winter.

This solitary sedentary sciurid is a riparian nearctic native that inhabits mountain taiga; or more descriptively, all types of cool northern forests: coniferous, deciduous, and mixed, from wilderness areas to urban environments. It is also found in palustrine landscapes, preferring those with downed trees, deadfall, and snag.

Red squirrel.

From its predilection for evergreens it receives its popular nickname the "pine squirrel." Its range is extensive, covering most of the upper tier of North America, from Alaska in the west to Quebec in the east, from Manitoba in the north to New Mexico in the south (this part of its range includes much of the Rocky Mountain corridor); it is also found in parts of the upper Midwest, New England, and the Southeast. Nests in tree cavities, underground dens, and logs, utilizing grass and bark as construction material. Scansorial, it is an excellent climber; arboreal, it is an agile tree acrobat, both abilities which it uses to great effect when fleeing from predators. Highly territorial, it establishes a large home range of several thousand square feet, which it scent-marks and aggressively defends against intruders.

A sedentary, diurnal, and crepuscular omnivore, its diet consists of standard rodent fare: seeds, grains, flowers, fungi, buds, fruit, nuts, bark, stems, wood, tree sap, insects, eggs, birds, and other small mammals. Debarks and girdles trees in search of cambium, often causing abnormal growth as the trees mature; will steal food from fellow squirrels when the opportunity presents itself. A preferred food is the seed of the pine cone; thus a sign of its presence in an area are kitchen middens comprised of both discarded pine cone scales and stored surplus food—the latter which it caches for lean periods. A scatter hoarder as well as a larder hoarder, it also buries edibles in small holes in the ground (another sign that it is nearby).

Its polygynandrous breeding season occurs once, sometimes twice, a year (late spring and late summer), depending on geography. Mature males and females engage in energetic mating chases, with dominant bucks acquiring the majority of does. The female gestates for 1 mo. or so, giving birth to an average of 4 altricial pups in a leafy tree nest or a grass-lined subterranean chamber. Weaning takes place by 2 mos. of age, after which the young disperse to embark on a life of independence, beginning the red squirrel life cycle over again. The father is not paternal and therefore does not contribute to the rearing of his offspring.

Red squirrel.

Communication and perception run along normal Sciuridae channels: sight, touch, smell, and hearing. Vocalizations include growling, chattering, chirping, trilling, and screeching. Its natural enemies are those common to most small rodents: snakes, hawks, owls, eagles, mink, fishers, martens, weasels, lynxes, mountain lions, foxes, coyotes, and wolves. Beneficial as a seed disperser, a parasite host, and as a prey base for carnivores; detrimental as a tree destroyer and a household pest. Dangerous if cornered. The red squirrel lives for about 3-4 yrs. in the wild; perhaps twice that long under optimal conditions.

RED-TAILED CHIPMUNK (*Neotamias ruficaudus*): This bright but darkly colored medium-sized chipmunk grows to a total length of 10", its hind foot measures nearly 2", and it can weigh up to 2 oz. Its upper body is grayish-brown, the lower body is tawny. The sides, shoulders, neck, and upper front legs are cinnamon-orange; top of the head, upper back legs, and posterior are grayish; the 4½" tail is brushy and bicolored: reddish rust above, dark red below. Its 5 dark dorsal stripes run from black to brownish, its 4 light dorsal stripes run from buff to whitish; facial stripes mimic the back stripes; the eye stripe is brownish; there is 1 white stripe above and below each eye. The pelage is lighter in winter. Its genus name, *Neotamias*, means "new dispenser" (a reference to its biological function as a seed disperser); its species name, *ruficaudus*, means "red-tail." Gender dimorphism is present: the female is larger than the male.

This sprightly sciurid is a nearctic native that inhabits rugged, mesic, subalpine coniferous forests made up of dense stands of fir, cedar, and spruce pines, among others.

Chipmunk.

It is also fond of boulder fields, blueberry meadows, and woody edgeland with diverse shrubbery, thick understory, blowdown, abundant snag, stumps, logs, deadfall, and disturbed habitat, such as fire burns and clear cuts. Both terrestrial and semi-arboreal, it is comfortable on the ground and in trees; it is also slightly riparian and fossorial. Its range extends over the northern Rocky Mountains, from Canada (British Columbia and Alberta) in the north to Washington, Idaho, and Montana in the south. Nests above ground (in rock crevices, trees, bushes) and also underground (burrows). Estivates during hot weather; enters light hibernation or short episodes of torpor during cold weather, occasionally rising to feed from its larder. Establishes a home range of about 500 sq. ft., agonistically defending it against conspecifics when necessary.

Diurnal and sedentary, what little is known of its diet reveals that it is a generalist omnivore, consuming the usual chipmunk fare of seeds, grasses, fruit, forbs, flowers, shrubs, foliage, leaves, bulbs, fungi, woody vegetation, insects, birds' eggs, and chicks. A scatter hoarder, it caches surplus food both in its house and also in ground holes which it excavates throughout its territory, transporting it in its cheek pouches.

Its breeding season occurs in spring after snowmelt begins and hibernation ends. The doe gestates for about 1 mo., giving birth to a single litter of about 6 altricial pups in the maternal nest chamber. The mother may move her neonates to a different nest site periodically to discourage predation. As with most other members of this family, it is probable that the male does not aid in the rearing of his offspring.

Communication and perception run via standard Sciuridae channels: sight, smell, touch, and hearing. Vocalizations include chipping, trilling, chucking, barking, and whistling. Engages in dirt-bathing. Its natural enemies are most likely those of other sciurids: reptiles, birds of prey, mustelids, wild felids, and wild canids. The red-tailed chipmunk lives an average of 2-4 yrs. in the wild; twice that long in captivity.

RED TREE VOLE (*Arborimus longicaudus*): Due to its habit of living in treetops over 100' off the ground, this tree vole is rarely seen and thus its ecology and life history are incompletely understood. It grows to a total length of about 8", its hind foot measures up to $7/8$", and it can attain a weight of nearly 2 oz. The dorsal pelage is cinnamon or tannish-red (lending it its common name) with a light black wash over the back; the ventral pelage is whitish; the lengthy $3\frac{1}{2}$" tail is dark and furry. The coat is long, dense, and silky; the eyes are small; the ears are erect, membranous, and partially obscured by its coat. Gender dimorphism is present: the female is larger than the male.

This sedentary somewhat terricolous cricetid is a nearctic native that inhabits temperate, mesic, fog-enshrouded forests dominated by Douglas fir trees—and sometimes by spruce, redwood, and hemlock trees as well. Its range is small and restricted to the Pacific Northwest, where it extends over the moist coastal regions of Oregon and California. This

Mature western hemlock and Sitka spruce forest: red tree vole habitat.

species is quite arboreal (the female more than the male) and spends most of its time in trees; here it finds shelter, protection, mates, food, and nesting sites. However, it is also fossorial (the male more than the female), and will readily reside in an underground burrow system when necessary. Establishes a home range of probably not more than 1,000 sq. ft.

A solitary nocturnal herbivore, it feeds almost exclusively on the pine needles of the Douglas fir. And though it consumes other foods, such as stems, leaves, twigs, and bark, it is unable to survive long without Douglas fir needles, which also help fulfil its water requirements. Its breeding season generally runs year-round, cresting in spring and fall. The female gestates for an average of 40 days, giving birth to several altricial pups which she weans a month later. As in most other mammalian species, the female is the sole parent, providing sustenance, grooming, and protection until the young become independent.

Red tree vole.

Communication and perception take place via standard Cricetidae channels: sight, hearing, touch, and smell. Its natural enemies are birds of prey, mustelids, procyonids, and wild felids. Its dendrocentric lifestyle gives it its genus name (*Arborimus*), while its long tail gives it its species name (*longicaudus*). Its specialized diet makes the unadaptable red tree vole quite vulnerable to extreme environmental changes that can result from, for example, logging, fire, and habitat fragmentation. Beneficial as food for carnivores (in particular the endangered spotted owl). The red tree vole probably lives for about 6-12 mos. in the wild; perhaps twice as long under ideal conditions.

Red wolf.

RED WOLF (*Canis rufus*): Stands 20" at the shoulders, grows to 5' in total length, has a 10" hind foot, and weighs up to 90 lb. This little known species is named for the rust-colored cast of its grayish coat. Dorsally it is grizzled with gray and black hairs from head to tail tip. The cheeks, throat, underside, inner legs, and paws are cream. The top of its muzzle, its ear backs, and its upper outer legs are cinnamon; the 15" tail is often black-tipped. It molts once a year in the summer. As with other wild canids, depending on the individual, time of year, and geographic location, variations are seen in both fur length and pelage coloration. The latter runs from white to blonde, brown to red, and gray to blackish. Midway in size between a coyote and a gray wolf, it has a wide head, long muzzle, large pointed ears, long rangy limbs, and big paws. Gender dimorphism is present: the male is larger than the female.

This sedentary terricolous dog is a nearctic native that inhabits temperate wetlands, brushland, forests, prairies, bayous, hilly country, intermontane regions, marshes, mountains, intermontane regions, coastal plains, and swamps. Though it once ranged across most of the eastern and southeastern U.S. (from Texas to Florida, from Illinois to New Jersey, and from Pennsylvania to Georgia), by 1980 it was declared extinct in the wild. The remaining genetically pure individuals were captured and bred in captivity, after which their descendants were placed in a reintroduction program. Today, having been extirpated from nearly 99 percent of its original territory, the red wolf is found only in southwestern Louisiana, southeastern Texas, and North Carolina and Tennessee (the latter 2 states where it has been reintroduced). It dens in a wide variety of microhabitats, such as dense vegetation, culverts, downed trees, under stumps, or in the burrows of other animals, which it enlarges to suit its needs. Scent-marks its 50 sq. mi. home range, ferociously defending it against other canines.

This lean canid is an opportunistic carnivore whose diet includes fruit, insects, birds, mice, rats, nutria, muskrats, rabbits, raccoons, wild boar, deer, and carrion; will take livestock and domestic pets when available. Usually hunts alone, but being social it will forage in packs when necessary (such as when hunting large prey). May travel up to 2 dozen mi. a day seeking food.

Its breeding season runs from midwinter to early spring. Dogs (males) and bitches (females) are monogamous and mate for life. Only the pack's dominant male and female breed each year, however. The alpha female gestates for around 2 mos., giving birth to an average of 4-5 altricial pups in late spring in a den excavated in a bank, brushpile, hollow log, rocky outcrop, or the side of an irrigation ditch. The alpha male helps his mate rear their young, and both are often aided by the rest of the closely related pack. By age 3 the young disperse to establish their own territories and find mates.

The ecology and life history of this crepuscular, nocturnal, secretive, shy mammal have yet to be studied in detail, but it probably shares many of the same traits and habits possessed by its cousin the gray wolf. The red communicates and perceives through regular Canidae channels: touch, sight, smell, and hearing. Vocalizations include yaps, growls, barks, and howls. A secondary consumer, as an apex predator the adult red wolf is not part of the diet of any animal—though the young may be hunted by birds of prey and wild felids. Its only real natural enemies are other wild canids, such as coyotes and gray wolves, all who are in constant competition with one another for food and territory.

Red wolf pack.

Its coyote-like appearance, lanky build, and sandy-grizzled coloring, along with its coyote-like DNA, have prompted theories that it is not a pure wolf species, but rather a gray wolf/coyote hybrid (it may be, however, that these commonalities merely indicate that wolves and coyotes share a common canine ancestor). Beneficial due to its consumption of pestilent animals, such as rats, nutria, and raccoons; it also helps control populations of ungulates, like deer. At one time on the brink of extinction, today conservation efforts continue in a campaign to preserve this still highly endangered species. (Unfortunately, it readily breeds with the coyote, making the loss of red wolf genes through natural hybridization a real threat to its survival.) The life span of the red wolf in the wild is 4-5 yrs.; it lives to 12 yrs. or more in captivity.

RIBBON SEAL (*Histriophoca fasciata*): This medium-sized pinniped, the only living member in the genus *Histriophoca*, grows to a total length of 6' and a weight of up to 200 lb. The male is reddish-brown, dark brown, or black; the female is lighter in color, usually grayish. Both genders sport 4 large white or cream-colored, ribbon-like bands over their bodies (though the female's rings are not as bright): one around the neck, one around each of the 2 front flippers, and one around the tail. From these striking, unique ringed markings this seal gets its vernacular or common name.

The body is fusiform, slenderish, and long; the head is small and short; the snout is snubbed; the neck is long; the eyes are large, dark, and watery; the nostrils are large; the flippers are smallish, with the front pair heavily clawed (for gripping the ice) and the rear pair pointing straight backward (making locomoting on land somewhat ponderous). Gender dimorphism is present, with the male being slightly larger and more dynamically colored than the female (whose markings look "faded" in comparison).

A pelagic polar phocid, this singular "earless" seal inhabits the cold, open marine waters and drifting pack ice of the Arctic and northern Pacific Oceans, with a range that extends from Alaska south to (on rare occasions) Washington and California—and also west to Asia and Russia. Rests, molts, and pups on thick ice floes. Being solitary it does not congregate in groups. Rarely seen since it spends most, if not all, of its life at sea far from shore.

Carnivorous, diurnal, and migratory, this natatorial mammal specializes in a piscivorous and molluscivorous diet that centers around cod, capelin, eelpout, pollock, sculpin, squid, octopus, and shrimp. Uses its sensitive vibrissae and a type of echolocation to help navigate and sense its prey in cold dark waters. An excellent swimmer and a powerful diver, it can go down to a depth of 2,000' and stay submerged for as long as 30 mins. while hunting.

Like many other true seals, it is polygynous with a breeding season that takes place from late spring to early summer, the period when the pack ice is just beginning to melt and break apart. Territorial bulls defend their patches against male rivals, while vocalizing to draw the attention of cows. The female gestates for about 11 mos., employs delayed implantation, and gives birth to a single, lanugo-covered pup at the ice rookery the following spring. Embryonic diapause, the

Ribbon seal.

postponement of implantation (in the case of the ribbon seal, lasting 2-4 mos.), allows pups to be born when there is a maximum amount of ice surface available. After 4 wks. on a diet of high protein, fat-rich milk, the pup is weaned. Wearing a new bluish-gray coat on its upper body, the now independent 60 lb. youngster begins to forage and hunt on its own using techniques taught to it by its mother. The father plays no role in the rearing of his offspring.

When locomoting on land it moves in a undulating, snake-like fashion, the only seal known to do so. Communication and perception are typical for members of the Phocidae family: sight, smell, hearing, and touch. Its primary natural enemies are orcas, polar bears, sharks, and walruses. The ribbon seal lives 20 yrs. or more in the wild.

RICHARDSON'S GROUND SQUIRREL (*Urocitellus richardsonii*): This prairie dog-like ground squirrel grows to a total length of 15", its pes measures nearly 2", and it can reach a weight of 17 oz. Countershaded, the pelage on its upper body is a plain, dull, tawny-gray interspersed with black-tipped hairs and indistinct off-white flecking; the underbody is whitish to tan; the 4" tail is thin, darkish on the dorsum, tan on the ventrum, and edged in brown. Its coat is short, thick, and lacks conspicuous markings such as stripes and spots. The throat, neck, and outer legs may be tinged with cinnamon. The body is robust; the head is small and the profile is convex; the eyes are dark and ringed in buff; the ears are small, rounded, and erect; the nose is snubbed; the limbs are short; the digits are clawed. This species is gender dimorphic, with the male being larger than the female.

Fossorial, its preferred habitat is temperate savanna with soft sandy or gravelly soil for ease of digging. This includes grassland, hedgerows, sagebrushland, cultivated areas, meadows, prairie, farmland, cropland, gentle hills, and previously plowed fields. It is thus an anthropogenic mammal that is comfortable living alongside us and our artificially manipulated environments. This is essentially a Northwestern-Midwestern squirrel, with a somewhat limited range that extends from central Canada (Alberta, Manitoba, Saskatchewan) in the north to Montana, North Dakota, South Dakota, Minnesota, and Iowa in the south. It once occurred further south and west into Idaho, Nevada, Colorado, and Wyoming, the latter for which it was nicknamed the "Wyoming ground squirrel." Its range, however, has been on the decline for decades.

Richardson's ground squirrel.

This omnivorous diurnal sciurid feeds primarily on seeds, grasses, grains, leaves, stems, legumes, flowers, and nuts, as well as on insects and carrion. It has cannibalistic tendencies and may occasionally consume the corpses of its fellows. Hoards and caches surplus food, which it transports to its burrow in large cheek pouches. It earned the nickname "picket pin" due to its habit of sitting upright in order to better surveil its territory for signs of danger. Establishes a home range of about 1,000 sq. ft. in which it excavates a complex and extensive burrow system comprised of numerous entrances, exits, and chambers.

This terricolous rodent spends the vast majority of its life resting, eating, and sleeping underground in its extensive burrow system. Estivates during hot weather. After the hibernation period (early fall to early spring, depending on latitude), bucks and does emerge from their hibernacula, inaugurating their polygynandrous breeding season. The doe gestates for around 3 wks. or so, giving birth to 5-10 altricial pups in a round, grass-lined, natal nest chamber located below ground. She weans her young a month later, after which they attain their independence. With a kinship system that is matricentric in character, young females tend to remain within the philopatric matriarchal group, while young males go off to begin their largely solitary lives, socializing with conspecifics mainly only during the mating season and inclement weather—when they become semicolonial. The father plays no role in the upbringing of his offspring.

Communicates and perceives via traditional small mammalian channels: sight, hearing, touch, and smell (emits a powerful oil from scent glands located on its hindquarters). Uses a myriad of vocalizations, including chips, whistles, churrs, chirps, and various high-pitched alarm calls that fall outside the range of human hearing (ultrasounds). Another one of its monikers, "flickertail," derives from its habit of flicking its tail while vocalizing—an appellation that became a nickname for one of the states in its range: North Dakota, the "Flickertail State."

Though often referred to as a "gopher," it is not a true gopher (which belongs to the Geomyidae family), it is a member of the Sciuridae family. Additionally, Richardson's ground squirrel is often confused with the black-tailed prairie dog. Though there are many differences, the most apparent is that the black-tailed is almost twice the size of Richardson's.

This rodent's main natural enemies are reptiles, birds of prey, mustelids, wild canids, and ursids. Considered a pest due to its destruction of crops and for hosting the bacteria that transmits bubonic plague (thus, touching it, or even breathing its aerosol, is potentially hazardous). The male Richardson's ground squirrel lives an average of 3-4 yrs. in the wild; the female probably lives 4-5 yrs. in the wild; in captivity an individual can attain an age of 7 yrs.

RINGED SEAL (*Pusa hispida*): The smallest of the aquatic carnivores, this plump pinniped grows to a total length of around 5½' and a weight of up to 225 lb. Its base pelage coloration is highly varied, but on the dorsum it is generally silverish, bluish-gray, or brownish-black, while on the ventrum it is lighter, usually whitish, yellowish, or gray. Irregular silvery-whitish rosettes or rings cover the upper body and sometimes the lower body as well—the source of its common name. There may also be spots, streaks, splotches, lines, and marbling over the torso. There is often black around the eyes, sometimes profuse enough to form a "mask."

The body is stout but fusiform; the head is small; the eyes are large; the snout is snubbed; the vibrissae are light in color; the neck is thick; the tail is short; the smallish front flippers are heavily clawed (an adaptation for digging in snow drifts and carving through ice up to 6' thick or more). Overall, the face has a rather cat-like appearance. An "earless" or true seal, unlike its otariid cousins, its hind flippers are fixed backward, making locomoting on land somewhat laborious. Gender dimorphism is present, with the male being larger than the female.

Ringed seal.

This wary, jar-shaped, solitary, ice-affiliated phocid inhabits the cold open waters of polar seas, but is also found in lakes. Both pelagic and coastal, it spends nearly its entire life offshore, hauling out on large thick ice floes, where it socializes, rests, breeds, pups, winters, molts, and forms herds and colonies. Also associated with sheltered rocky coasts, bays, islands, and inlets, it prefers using air holes in the ice for entering and leaving the water, and also for avoiding predators. Though it is a holarctic species that normally haunts the circumpolar regions of the Northern Hemisphere, it has been found as far south as the American Northwest and the American Southeast. Establishes a home range of approximately 5-20 sq. mi., though it is somewhat migratory and may travel up to 1,000 mi. in summer when the ice begins to break up. The most common of all the Arctic seals, it consistently ranges further north than any other known mammal.

A carnivorous piscivore, this generalist feeder consumes mainly fish (cod, smelt, herring, burbot, whitefish, vendace, roach, perch, ruffe, sculpin), but will also eat marine invertebrates, including octopus and squid, when available. Can dive to a depth of 300' and remain underwater for up to 20 min.

Its breeding season occurs during the spring, with fasting males and females overseeing territories both above and below the ice. The polygynous bulls, who emit a powerful female-attracting odor that smells similar to gasoline, spar over receptive cows. The female gestates for 11 mos., employs delayed implantation (which lasts for about 3½ mos.), giving birth to 1 precocial, wooly, lanugo-covered, 9 lb. pup the following year in either a natural snow cave or a specially constructed subnivean den. The latter can be 2' high and 10' long, and is built on landfast ice or pressure ridges over a well-maintained breathing hole. The multi-chambered design of this camouflaged cavity-lair helps insulate and protect the baby from freezing temperatures, cold winds, and hungry predators. The youngster drinks milk for nearly 2 mos., the longest nursing period

Polar sea: ringed seal habitat.

of any of the true or "earless" seals (Phocidae). The father does not contribute to the raising of his offspring.

Sometimes confused with spotted seals and harbor seals, this semiaquatic marine chionophile communicates and perceives mainly via sight, touch, and hearing. Acoustic sounds and cues—all which seem to aid in socializing, avoiding predation, and navigation—include snorting, pulsing, groaning, bleating, whining, croaking, grunting, chugging, chirping, and roaring. Its natural enemies are polar bears, sharks, and orcas (which feed on adults and pups) and arctic foxes and gulls (which feed on pups). The ringed seal probably lives 15-20 yrs. in the wild, though in some cases it may attain an age of 40 yrs.

RINGTAIL (*Bassariscus astutus*): The official state mammal of Arizona, it stands about 6" at the shoulder, is 32" in total length, has a 3" long hind foot, and weighs about 2 lb. The base coat on its upper body is tan to cinnamon, with light grizzling (black-tipped hairs) covering much of the dorsum—excellent camouflage for its preferred habitat. Its ventrum is cream to buff in color. This unique procyonid appears to be a combination of different creatures: its head is canid-like; its ears are vespertilionid-like; its body is felid-like; and its long, bushy, black and white-banded, 17" tail, from which it derives its common name, is raccoon-like. Its limbs are short. Does not have a face mask, but its large, dark round eyes are set in thin black circles, which are, in turn, surrounded by broad white or cream rings.

Though it is a member of the raccoon and coati family, it has partially retractable claws, licks its paws and fur, and is, overall, quite cat-like in appearance and habits, to which it owes its other common names: "civet cat," "miner's cat," and "ring-tailed cat." It also has fox-like traits, which are reflected in its scientific name: translated from the Graeco-Latin, it means "shrewd little fox."

Ringtail.

A petraphilic nearctic native, it inhabits temperate rock piles, talus slopes, deserts, mines, sand dunes, canyons, chaparral, scrubland, caves, mountains, and montane forests, always near water. Though a riparian that is found as far north as Oregon, its primary range covers the drier southwestern part of our region, from Arizona, New Mexico, and Texas, south into Mexico, and from California and Nevada east into Utah, Colorado, Oklahoma, and Kansas. It dens in hollow trees, rock crevices, mine shafts, attics, and the burrows of other animals. Uses urine to scent-mark its home range.

Quick and agile, this lithe procyonid is an excellent jumper, leaper, and climber. Using its long fluffy tail for extra balance, it can easily scale canyon walls, rock faces, trees, and cacti. It possesses a trait known as "hind foot reversal," making it one of the few mammals that can descend vertically headfirst. This it does by rotating its hind feet 180 degrees, giving it additional grip on the way down. It can also climb cliffs by hopping quickly back and forth from wall to wall, while it can "chimney stem" small crevices by placing all four of its feet against one side and its back against the opposite side.

Its ears rotate independently, aiding it in hunting and avoiding predators, while its white face-mask functions as a reflector, directing the dim light of the moon and stars into its eyes, enhancing night vision. Its stiff vibrissae serve as sensory antennae as it moves about in the darkness; its strong sharp teeth allow it to break open hard nuts; its long tongue is designed for licking juices and nectar from fruits and flowers. In sweltering arid areas, where water is scarce, the ringtail can concentrate its urine, enabling it to slow down its body's water loss. Secretes a disagreeable odor from its rump glands when in danger, which drives off most intruders.

Solitary, sedentary, nocturnal, and crepuscular, the diet of this carnivorous omnivore includes grains, fruit, nectar, seeds, nuts, invertebrates, amphibians, reptiles, birds, small mammals, and carrion. Its breeding season runs from midwinter to midspring. The female gestates for 6-7 wks., giving birth to 2-4 altricial kits in a concealed maternity nest in late spring to early summer. The young are weaned by 3 mos., hunt on their own at 4 mos., and are independent by 6-8 mos. The male contributes to the rearing of his offspring by bringing food to the den and by engaging in play with them as they mature.

Ringtail.

Possessing keen physical senses, it communicates and perceives via biological channels typical of the Procyonidae family: sight, touch, hearing, and smell. Its many vocalizations include growls, squeaks, howls, chitter, mews, barks, chatter, chirps, screams, clicks, whimpers, and the whistle-grunt. A secondary consumer, its natural enemies are birds of prey, bobcats, coyotes, and its cousin the raccoon. As a natural pest controller it is of benefit to humans. The ringtail lives 8-9 yrs. in the wild; up to 15 yrs. in captivity.

RISSO'S DOLPHIN (*Grampus griseus*): It grows to 14' in total length and weighs up to 1,200 lb. Has a large squarish head, a stout body, long narrow pointed flippers, and a tall killer whale-like dorsal fin set at the center of the body. Lacks a beak, but instead has a very blunt snout, with a prominent furrow running centrally from its tip up to the forehead. The juvenile may be brown or gray with a lighter underside, growing overall

244 ~ NORTH AMERICA'S AMAZING MAMMALS

lighter with age. Some adults are nearly completely white. Numerous scars often cover the body, perhaps resulting from sparring with other dolphins or from battles with squid. It has up to 14 teeth in the lower jaw (only), with which it grasps its prey, crustaceans, cephalopods, and fish, swallowing everything whole.

This delphinid is mainly solitary, but it also travels in small pods. Cooperates with conspecifics while searching for food. Uses echolocation to communicate, hunt, and navigate. Can dive to ½ mi. and hold its breath for up to 30 min. Newborn calves are precocial, about 5' long, and weigh around 45 lb.

This natatorial pelagic carnivore inhabits temperate, subtropical, and tropical waters throughout our region, from Alaska to the Gulf of Mexico, and from New England south. Travels in large pods (including superpods containing many thousands of individuals), often associating and traveling with other cetacean species. Some populations migrate. Playful and gregarious, this secondary consumer readily engages in tail-slapping (the surface of the water), breaching, and head-bobbing; also communicates (and hunts) using a wide array of sounds, such as clicks, whistles, grunts, and squeals. Risso's dolphin may live as long as 35 yrs. in the wild. The habits of this cetacid make it difficult to study, thus little else is known.

Risso's dolphin.

ROCK POCKET MOUSE (*Chaetodipus intermedius*): This small, coarse-haired pocket mouse grows to a total length of about 7" and can attain a weight of up to ⅝ oz. Its upper body is buff with dull dark grayish-tipped hairs; the lower body is cream; the grayish 4" tail is bicolored (dark above, lighter below) and has both a crest and a tufted tip. It may have an orangish lateral strip; its rump spines are not prominent. Gender dimorphism is present: the male is larger and heavier than the female.

This petraphilic heteromyid is a nearctic native that favors dry rocky locations, sand dunes, cliffs, scree, lava flows, canyons, talus, boulder fields, grassland, and deserts. A purely southwestern species, its range extends over portions of the U.S. states of Arizona, Utah, New Mexico, and Texas, and south into northern Mexico. Excavates minuscule tunnels under rocks, where it spends much of its time resting, grooming, and feeding. Does not hibernate but may enter torpor during extremely cold weather.

A nocturnal omnivore, its seed-dominant diet, along with an occasional insect, is typical of its genus. Caches food in burrow larders, transporting it in its fur-lined cheek pouches. Its breeding season runs from early spring to midsummer. The female gives birth to an average of 4 pups in a subterranean nest chamber. This rodent communicates and perceives along normal Heteromyidae channels: touch and smell, and to a lesser degree sight and hearing. Its natural enemies are reptiles, birds of prey, wild felids, and wild canids. The rock pocket mouse lives for 2-3 yrs. in the wild.

Boulder field: rock pocket mouse habitat.

ROCK SQUIRREL (*Otospermophilus variegatus*): The largest ground squirrel in its range, it grows to a total length of about 20" and a weight of up to 30 oz. Though pelage coloration varies with the individual and the population, generally the dorsum is blackish, but lightly washed in wavy mottled grayish-brown that turns rust-colored toward the rump; its ventrum is whitish, buff, pinkish, or tan. Its nearly 10" tail, the shaggiest of the ground squirrels, is mottled with rust, gray, black, and white hairs, and edged in white. The body is stout; the head is large with a rounded profile; the eyes are large, dark, and ringed in pinkish-tan; the ears are large and erect; the nose is snubbed; the limbs are short; the paws are clawed.

A nearactic riparian native, as its vernacular name implies, the preferred habitat of this petraphile is bare rock settings, under and around which it can forage, hunt, burrow, den, hide, rest, hoard food, hibernate, mate, nest, and give birth. Such habitats include open temperate landscapes with soft well-drained soil and abundant rocks: boulder fields, rock outcrops, cliffs, stone heaps, chaparral, canyons, rocky hillsides, and talus and scree slopes. Needs good hiding cover with some vegetation, woody refuse, and a stable water source; thus it is often also found in grassland, scrubland, and savanna with arroyos and scattered trees. Will occupy human-made structures as well, including stone walls, bridges, and buildings.

Rock squirrel.

The rock squirrel is an inhabitant of the American Southwest with a range that extends over portions of California, Idaho, Nevada, Utah, Colorado, Oklahoma, Arizona, New Mexico, and Texas. It is also found in parts of northern and central Mexico. Scent glands are used to mark territories and home ranges. It is slightly colonial, with loose female groups overseen by an alpha male.

A typical omnivorous sciurid, it consumes a wide variety of vegetable and animal matter: seeds, roots, leaves, stalks, grasses, tubers, grains, nuts, bark, fruit, annelids, insects, birds' eggs, birds, and small mammals. Hoards and stores excess edibles, transporting large portions in its massive cheek pouches.

Dislikes extreme temperatures and estivates during hot weather. During cold weather northern populations hibernate for brief periods, emerging in time for the breeding season, which begins in May. Bucks violently compete with other males over estrus does, who gestate for about 1 mo., then give birth to 4-8 altricial pups in an underground maternal nest chamber. The young are weaned at 8 wks. of age, the males eventually dispersing into the wild, the females remaining with their mothers. The non-paternal fathers play no role in the rearing of their offspring.

This crepuscular, diurnal rodent communicates and perceives via sight, smell, touch, and hearing. Vocalizations include chits, barks, growls, whistles, chucks, trills, squeals, and teeth-chattering. Its primary natural enemies are reptiles (rattlesnakes, bullsnakes), birds of prey, procyonids, mephitids, wild felids, and wild canids. It is immune to rattlesnake venom. An agile runner, this ground squirrel can also climb trees and bushes with ease; it is often seen sitting atop rocks and boulders. Beneficial as a seed disperser, as a host for parasites, and as a prey base for carnivores. However, it is also regarded as a crop destroyer, one that can carry the flea-borne bacteria that causes both sylvatic plague and Rocky Mountain spotted fever. The rock squirrel probably lives 2-3 yrs. in the wild.

ROCK VOLE (*Microtus chrotorrhinus*): This rarely seen medium-sized rodent grows to a total length of about 7", its hind foot measures around 7⁄8", and it can attain a weight of 1½ oz. Its upper pelage is longish and grayish-brownish-silverish in color; the lower pelage is grayish; the 2" tail is lightly bicolored (dark above, lighter below). It can be distinguished from other voles by its bright orange facial fur and yellowish nose (from whence it derives its nickname, the "yellownose vole"). Its scientific name means "little ear" (*Microtus*) and "colored nose" (*chrotorrhinus*). It is gender dimorphic, with the male being slightly larger than the female.

This boreal petraphilic cricetid is a nearctic native that inhabits cool, moist, mossy biomes, ecotones, and ecosystems, such as mixed forests, grassland, clear cuts, and foothills, particularly—as its common name denotes—mountainous areas with talus, rocks, and boulders. It prefers an elevated location with adequate ground cover, rotting logs, ferns, and a clean stable water source (creeks, streams, ponds). Its range extends over the northeastern part of North America, covering portions of Canada (Ontario, Quebec, Labrador, New Brunswick, Nova Scotia) and the states of Maine, Vermont, New Hampshire, New York, Pennsylvania, West Virginia, Virginia, Tennessee, North Carolina, and South Carolina.

Lives in small subterranean colonies connected by numerous shallow tunnels and surface runways; often dens and nests under fallen logs, rocks, and forest duff; does not hibernate; active year-round; establishes a home range of a few thousand square feet.

Forest stream biome: rock vole habitat.

A gentle diurnal (and somewhat nocturnal) omnivore, most of its diet is comprised of green vegetation (moss, fungi, stems, forbs, grass, sedges, leaves) and fruit, but it also consumes insects. Chops up vegetable matter, storing the cuttings in its underground larders. Its breeding season runs from spring to fall. The polyestrous female may produce up to 3 litters annually. She gestates for about 3 wks., giving birth to an average of 3-4 pups in an underground moss- and grass-lined nest chamber.

It communicates and perceives via well-established Cricetidae channels: touch and smell, and to a lesser degree sight and hearing. Its natural enemies are reptiles, birds of prey, other rodents, mustelids, wild felids, and wild canids. Beneficial as a soil recycler and as food for carnivores. The rock vole lives an average of 6-12 mos. in the wild; twice that long in captivity.

ROUGH-TOOTHED DOLPHIN (*Steno bredanensis*): Named for its ridged teeth (which aid in grasping prey), it grows to a total length of nearly 9' and can weigh up to 350 lb. Its dorsum is dark gray to purple-black; the ventrum is whitish-pink. There is some mottling over the body. This little delphinid has a small sloping forehead, long narrow beak, large flippers, and long slender body, giving it a somewhat prehistoric even dinosaur-like appearance. This cetacid is gender dimorphic: the male is larger than the female.

It prefers warm tropical zones, and in our region is found in the temperate waters of the Atlantic Ocean. Playful, but less active and energetic than other members of its family, it is gregarious, travels in pods of up to 100 individuals or more, and also associates with other cetacean species. Little is known about its mating system or reproductive life.

Like some other members of the Delphinidae family, this secondary consumer will help hold injured members of its pod at the surface so they can breath. It can dive to 500' or more and hold its breath for at least 20 min. Echolocates and hunts in small groups, feeding on typical cetacean fare: fish, cephalopods, and aquatic crustaceans. Known to strand. The rough-toothed dolphin may live as long as 50 yrs. in the wild, though 20-30 yrs. is probably more typical.

Rough-toothed dolphins.

ROUND-TAILED GROUND SQUIRREL (*Xerospermophilus tereticaudus*): This wiry little ground squirrel grows to a total length of about 11", its hind foot measures 1½", and it can attain a weight of 7 oz. Its pelage is plain rusty-gray-brown on the dorsum, whitish-gray on the ventrum. The top of the head is darkish, the cheeks are a light buff. It lacks markings of any kind and has a drab coat that acts as concealing coloration; one that helps it blend in with its sunbaked monochromatic surroundings. The head is small with a rounded profile; the eyes are large, dark, and ringed in tan; the ears are small and rounded; the limbs are short; the back feet are long; the digits are clawed. Its 4" tail is thin and round, giving it its common name. It possesses characteristics of other members of the Rodentia order, such as the chipmunk, prairie dog, gopher, and rat—though on close inspection it would be difficult to confuse it with any of these.

The round-tailed inhabits temperate arid ecosystems and flat, low-humidity biomes with extreme temperature changes, such as scrubland, playa, chaparral, salt flats, xeric

Round-tailed ground squirrel.

shrubland, alkali sink communities, and deserts, such as the Mohave and the Sonoran; it is especially fond of gravelly soil, and more particularly sandy soil, so it is also found in and around mesquite bush, creosote scrub, and rolling sand dunes. It has a small distribution, with a limited range that extends over portions of the southwestern U.S. states (California, Nevada, and Arizona) and parts of northern Mexico. Crepuscular and partly diurnal, this common rodent rests or estivates in its burrow during the day to escape the heat, coming out in the morning and evening to forage. Semicolonial, it lives in groups, but uses its own separate tunnel systems with entrances that are well hidden beneath vegetation. Spends most of its life underground.

This energetic fossorial sciurid is an omnivore that concentrates its diet around seeds, roots, grains, stems, flowers, leaves, tubers, nuts, insects, small mammals, and carrion. Derives most of its water from the succulent foods it consumes.

While some believe it hibernates during the cold months, it is more likely that it enters a state of intermittent torpor. This lasts from early fall to midwinter (depending on latitude), when it emerges in time for its polygynandrous breeding season. Does gestate for around 1 mo., giving birth to an average of 6 altricial pups. They are weaned at 5 wks. of age, dispersing several months later. Parenting is left solely to the female, with the male playing no role in the rearing of his offspring.

Communication and perception channels are normal for the Sciuridae family: sight, smell, hearing, and touch. It scent-marks its home range. Vocalizations include peeping, squeaking, and whistling. Its main natural enemies are birds of prey, mephitids, wild felids, and wild canids. Like all ground squirrels it benefits the world ecosystem by aerating and recycling the soil, hosting numerous parasites, and providing food for carnivores. The round-tailed ground squirrel probably lives for about 4-5 yrs. in the wild; perhaps twice that long in captivity.

ROUND-TAILED MUSKRAT (*Neofiber alleni*): This rodent can grow to a total length of 20" (though 15" is more usual), its webbed hind foot measures about 2", and it can attain a weight of up to 13 oz. Its thick rich fur is dark brown to black on the dorsum, light brown or buff on the ventrum. Its body is stocky; the eyes and ears are small; the 6½" tail is black, scaly, and sparsely furred. Though its genus name means "new beaver," it is not a beaver—though as fellow members of the Rodentia order, the 2 species are related. Gender dimorphism is present: the male is slightly larger than the female.

This aquatic mammal is often confused with its cousin the muskrat. Not only is it smaller, however, its common name reveals one of the major differences between the

2: the round-tailed's tail is round (like a mouse) while the muskrat's tail is vertically flattened (like a salmon). The round-tailed is also sometimes misidentified as a nutria, and indeed the nutria also has a round tail. The primary difference here is in size: at only 15"-20" long, the round-tailed is less than half the length of the 55" long nutria. Lastly, the round-tailed is also sometimes confused with the common rat, and is thus called the "water rat" (or "Florida water rat") in some areas. But though they can be similar in size, a rat's snout is more pointed and its body is more cylindrical than the round-tailed—which has a bluntish nose and a stout body.

Round-tailed muskrat.

This shy, elusive, mainly nocturnal (but also diurnal and crepuscular) cricetid is a nearctic native that inhabits shallow, open marshy areas, including saltwater, freshwater, and brackish wetlands, swamps, lakes, ponds, seepage slopes, wet fields, bogs, and river deltas, particularly those with soft, peaty, or sandy bottoms. Threatened throughout most of its range, it is strictly a southeastern species, occupying only portions of the states of Georgia and Florida.

It weaves both feeding platforms and dome-shaped houses out of grass cuttings, locating them on floating mats of decaying aquatic plants (such as sphagnum moss). The latter measure roughly 24" in width and 15" in height; each domicile contains a number of entrances and exits, both above and below water, as well as a grass-lined nest chamber. Highly social, this species prefers living in a colony with conspecifics, often sharing its home with others; in wet fields it constructs grassy runways, as well as complex tunnel systems beneath the mud; establishes a home range of several thousand square feet.

A nonterritorial herbivore, the diet of this small muskrat is comprised almost solely of aquatic vegetation, consuming the sprouts, roots, stems, stalks, and seeds of a variety of plants such as arrowhead, water lilies, and pickerelweed. Leaves behind cuttings of cattails and other types of aquatic plants, including grasses and sedges. Its breeding season runs year-round, with the polyestrous female able to produce up to 6 litters (of about 3-4 pups per litter) annually. Gestation lasts about 1 mo., while weaning occurs at around 3 wks.

It communicates and perceives along normal aquatic Cricetidae channels: mainly smell and touch, with reduced sight and hearing. Its natural enemies are reptiles (snakes, alligators), birds of prey (hawks, owls), and wild felids (bobcats). Beneficial as a provider of homes for other burrowing creatures and as food for carnivores; detrimental as an agricultural pest (that can wreak havoc in sugarcane fields). The round-tailed muskrat probably only lives 1-2 yrs. in the wild; perhaps 3 yrs. under optimum conditions.

SAGEBRUSH VOLE (*Lemmiscus curtatus*): This small vole grows to a total length of about 6", its hind foot measures around ⅝", and it can attain a weight of up to 1⅜ oz. Coloration on the upper body is light gray; the lower body is silvery-white; the short 1" tail is furry and dichromatic (dark above, lighter below). The body is stout and cylindrical; the fur is thick and long; the eyes are small; the ears are small and rounded; the snout is tapered and covered in tactile vibrissae; the feet are light-colored (usually white or silvery gray).

This solitary crepuscular cricetid is a nearctic native that inhabits temperate chaparral, shrubland, brushland, grassy hills, rangeland, dry prairies, canyons, grassland, scrubland, and, as its common name denotes, sagebrush-dominant locations. A purely Western rodent, it is found in southwestern Canada and in portions of the states of Washington, Oregon, Idaho, California, Nevada, Montana, Wyoming, Utah, Colorado, North Dakota, and South Dakota. Gregarious, social, and semicolonial, it excavates short shallow tunnels with multiple portals, usually located under grass clumps; numerous surface runways connect burrow systems, where related adults sometimes live in communal family groups. Does not hibernate; active throughout the year.

Vole.

A motile herbivore, its diet consists primarily of grasses, plants, roots, leaves, twigs, flowers, stalks, and bark. Produces plant clippings and grass cuttings, which it leaves strewn around its burrows. Unlike many other rodents it does not cache food. Its semi-monogamous breeding season runs year-round. The iteroparous female of this fecund polyestrous species may produce 3-4 litters annually. She gestates for 3-4 wks., giving birth to an average of 4 altricial pups in an underground grass-lined nest chamber; weaning occurs at 3 wks. of age. The male displays paternal behaviors and often participates in the rearing of his offspring.

Communication and perception take place via standard Cricetidae channels: smell and touch, with reduced sight and hearing. Its natural enemies are reptiles, birds of prey, mustelids, wild felids, and wild canids. Beneficial as a prey base for carnivores. The life span of the sagebrush vole is unknown.

SALT-MARSH HARVEST MOUSE (*Reithrodontomys raviventris*): This small rodent grows to a total length of around 6½", its hind foot measures ¾", and it can attain a weight of ½ oz. The dorsal pelage is dark brown turning to cinnamon-buff on the sides; the ventral pelage is pinkish-white or reddish; the long 3½" tail is slightly bicolored (dark above, lighter below); the eyes are small and dark; the ears are darkish.

This terricolous sedentary cricetid is a nearctic native that inhabits the temperate wetland ecosystems indicated by its common name: tidal salt marshes and brackish bogs.

Has a fondness for pickleweed; prefers thick hiding cover; moves to elevated ground during high tides. Its small range is limited to the area around San Francisco Bay, California. It dens above ground, constructing globular, grass-lined surface nests inside clusters of dense vegetation. Though it is not specifically natatorial, it has an adaptation that suits its watery life: a semi-water resistant coat that helps keep it afloat while swimming.

A gentle nocturnal omnivore, its diet centers around aquatic plants, but it also consumes seeds, grains, leaves, wood, nuts, and insects; possesses the unusual ability to drink saltwater. Its breeding season runs from spring to fall. The female probably gestates for about 3 wks., giving birth to 2-4 altricial pups, which she weans sometime within the following month or so.

Communicates and perceives via standard Cricetidae channels: sight, hearing, smell, and touch. Its natural enemies include reptiles,

Harvest mice and nest.

ardeids, birds of prey, mustelids, wild felids, wild canids, and other rodents. The shy and endangered salt-marsh harvest mouse probably lives an average of 6-10 mos. in the wild.

SAN DIEGO POCKET MOUSE (*Chaetodipus fallax*): This medium-sized pocket mouse grows to a total length of about 8" and can attain a weight of up to ¾ oz. Its pelage is countershaded and cryptic; coloration on the dorsum is dark brown; on the ventrum it is white; the upper and lower body colors are divided by a yellow-gold lateral stripe; the long 4½" tail is crested and bicolored (dark above, lighter below); it has large eyes as well as prominent black rump spines and white hip spines.

This sedentary terricolous heteromyid is a nearctic native that inhabits temperate open locations, such as arid desert, grassland, shrubland, chaparral, sand dunes, savanna, and weedy forests; has a preference for sparse vegetation and sandy rocky areas. As its name indicates, its range is situated primarily around San Diego, California, though it extends as far south as Mexico's Baja Peninsula. Fossorial, it excavates large burrow systems where it spends much of its time resting, sheltering, sleeping, and eating. Does not hibernate; active all year long. Territorial, it establishes a home range of 2,000-3,000 sq. ft, which it defends against conspecifics.

A solitary nocturnal omnivore, its diet is typical of many rodents: seeds, nuts, grains, stalks, leaves, bark, wood, and insects. Caches surplus food in underground larders, transporting it in its fur-lined cheek pouches. Able to procure most of its water needs from the solid food it consumes. Its breeding season runs year-round, with activity climaxing in late spring and early summer. The polyestrous female, who can produce as many as 3 litters annually, gestates for 3-4 wks., giving birth to an average of 3-4 pups in a subterranean, grass-lined maternity nest. Here she nurses, protects, and grooms them until they are weaned. The male does not appear to aid in the rearing of his offspring.

Communication and perception occur via standard Heteromyidae channels: sight, smell, touch, and hearing. Its natural enemies are reptiles, birds of prey, mustelids, wild felids, and wild canids. An agile, fast-moving, saltatorial mammal, its prodigious speed and unpredictable hopping style offer protection against predators. Beneficial as a soil aerator and recycler, seed distributor, parasite host, and food for carnivores. The San Diego pocket mouse lives for an average of 1 yr. in the wild; at least twice that long under optimum conditions.

SAN JOAQUIN POCKET MOUSE (*Perognathus inornatus*): This medium-sized pocket mouse grows to a total length of about 6½" and can attain a weight of nearly ½ oz. The soft upper body pelage is pinkish-orangish-tan with black-tipped hairs, giving it a grizzled appearance; the lower body is creamy; the long 3" tail is essentially unicolored (tannish) and tufted; the large eyes may be indistinctly encircled by cinnamon spots;

Weedy field: San Joaquin pocket mouse habitat.

each ear has a whitish subauricular spot; the vibrissae are somewhat short. Unlike many other pocket mice, it lacks rump and hip spines. Its genus name, *Perognathus*, derives from the Greek words *pera* ("pouch") and *gnathus* ("jaw")—a reference to its fur-lined cheek "pockets" or pouches; its species name, *inornatus* ("inornate") refers to its lack of conspicuous markings. Gender dimorphism is present, with the male being larger than the female.

This sedentary terricolous heteromyid is a nearctic native that inhabits temperate, dry, open spaces, such as grassland, brushland, shrubland, alkali sinks, sand dunes, savanna, washes, hillsides, weedy fields, scrubland, and desert, preferably with friable soil. Its range is revealed by its common name: the central western region of California, extending over portions of the San Joaquin Valley, the Mojave Desert, the Tulare Basin, and the Sierra Nevada Mountains. It enters torpor to conserve energy and to escape extreme weather. Fossorial, its constructs a burrow system for protection, shelter, and nesting; entrances are concealed below shrubbery.

A nocturnal omnivore, its diet consists of grains, grass, leaves, nuts, shrubs, seeds, forbs, worms, and insects; hoards and caches surplus food, carrying it to its underground larder in its large cheek pouches. Its breeding season stretches from spring to summer. The iteroparous polyestrous female, who may produce up to 2 litters annually, bears about 4 altricial pups in a subterranean nest chamber.

Communicates and perceives according to standard Heteromyidae channels: sight, smell, hearing, and touch. Its natural enemies are reptiles, birds of prey, wild felids, and wild canids. Engages in sunbathing. Beneficial as a seed distributor, parasite host, soil aerator and recycler, and as a prey base for carnivores. The San Joaquin pocket mouse probably lives 1-2 yrs. in the wild; perhaps twice that long under optimum conditions.

SASQUATCH (*Homo sapiens cognatus*): Males can grow to a height of 10' (or more) and may weigh as much as 1,000 lb. (½ ton); females are thought to reach a height of about 8' and weigh up to 600 lb. Thus this species is gender dimorphic. (Revealingly, this gender body-size dimorphism is similar to that found in *Homo erectus*, chimpanzees, and modern humans, all which range up to 20 percent.) Its coat is highly variable in color and can run from white or gray (rare) to brown or black (common). However, its pelage is very often reddish-brown, an ideal cryptic coloration for its primary habitat: mosaic terrain, one with good hiding cover, easily accessible travel and dispersal corridors, and abundant water and food sources, chiefly in or near densely forested, mountainous regions.

This robustly designed, principally nocturnal biped is completely hair-covered, except for its palms, foot soles, and parts of its face (around the

Frame 352 from the famous Patterson-Gimlin film, showing an alleged female Sasquatch. Despite ongoing attempts, this 1967 home movie has never been scientifically disproven.

eyes and nose); it also lacks a beard and its head and facial hair is short. Its common identifying characteristics (reported by eyewitnesses unknown to each other and often living thousands of miles and sometimes decades apart) describe a creature with both human and apelike traits: a small head (in relation to body size); a sloping forehead (also described as having "no forehead"); a pronounced brow ridge; lack of a neck (often described as an extremely thick "bull" neck); a flat face; thin lips; large dark sunken eyes (that produce eyeshine at night); no whites in the eyes (that is, the eyes possess dark sclera); a covered humanlike nose (though flattish and much broader); humanlike ears (though proportionally smaller than ours); a long upper lip; wide mouth; high and extremely broad shoulders; muscular chest and back; massive biceps; thick arms with large "cupped" hands that hang nearly to the knees; hips nearly as wide as the shoulders; grayish to blackish skin; and, as noted, long thick hair (not fur), up to 1' in length.

The scientific belief that we share our planet with wild, living, humanlike creatures is not new. In the 1700s, these four primitive "manlike apes" ("anthropomorpha") were proposed by renowned Swedish zoologist Carl Linnaeus, the Father of Taxonomy.

Known for thousands of years by native peoples as the "hairy wild man of the forest," it has a smooth compliant gait (also probably used by early bipedal hominids); a slightly stooped forward walk; whitish foot soles; a strong stench; and a distinct sagittal crest that gives it a "coned head" appearance (this feature is also found not only in gorillas, but also in various species of Australopithecines, an extinct type of early hominid). Besides being bipedal, it is also quadrupedal, as it has been seen walking and running on all fours.

As in humans, there is wide variability in physical appearance, particularly in the face—explaining some of the regional variations among different populations that have been reported. Importantly, these same traits continue to be remarked upon by eyewitnesses across the U.S. and Canada to this day. (Note that many of the characteristics listed above are plainly observable in the famous 1967 Patterson-Gimlin film, which shows a purported female Sasquatch at Bluff Creek in northern California.)

Despite its seeming predilection for cool, coniferous, montane habitat, Sasquatch has been seen in nearly every type of biome and ecosystem, from arid desert to tropical jungle. And yet, though it has been recorded in all 50 U.S. states, in most Canadian provinces, and across much of Mexico, its primary North American range seems to lie in the Pacific Northwest (where a majority of sightings still take place). Wary and reclusive, it often uses waterways, ravines, and gallery forests for concealment, water, and short-distance movements. It inhabits remote unexplored regions as well as urban areas heavily populated by people. For instance, overt signs of its presence (described below) have been found overlooking and even abutting backyards, busy highways, city parks, and shopping centers.

Considered more human than ape (that is, it is a true wild man and wild woman), nonetheless our "uncivilized" cousin displays behavior that is convergent with both *Homo sapiens* and nonhuman primates: it is exceedingly intelligent, inquisitive, powerful, cunning, quick, agile, shy, playful, and nonaggressive—unless provoked or cornered. Moves with a smooth fluid motion, passing quickly through even impenetrable brush and woods with ease and grace. It has been observed *running* up steep (sometimes rocky) hills, something nearly impossible for most humans.

This relict hominoid seems to be mildly territorial and somewhat nomadic, perhaps migrating short distances to follow game animals as the seasons change. Extrapolating from numerous firsthand field reports, its social structure appears to be patriarchal, built around a monogamous nuclear family overseen by the father. These familial units may occasionally congregate together in small temporary troops or clans that are supervised by an alpha male. Some individuals seem to be solitary, and, as in humans (and as in numerous other mammal species, in fact), there may be a small percentage of hermit-like rogue males (and possibly females).

Sasquatch displays great curiosity toward people and will study and observe us for hours at a time if able to do so; it seems to have a particular fascination for human women and children. It will engage in gifting and trading under certain circumstances, for example, leaving eye-catching stones, bones, sticks, and feathers where sympathetic humans are likely to find them (on stumps, boulders, tree branches, etc.). It is also known to approach and enter campsites, tents, yards, and even homes (if entry can be made quietly and easily). It has also been reported to rock cars, trucks, vans, and trailers back and forth, slap the sides of houses (with an open palm), and knock (usually 3 times) on doors before vanishing into the night.

It does not wear clothing, build houses, or make tools (though there is circumstantial evidence that it may use unworked rocks, bones, and sticks as tools). And while it does not make or use fire, it seems to be aware of it, for it constructs what I call "pseudo-campfires" (a simple pile of sticks, or more typically, a small crude circle of rocks, sometimes with dead branches placed oddly and randomly inside of it). This behavior could be interpreted as a concealment strategy: the pseudo-campfire appears to be nothing more than the result of humans in the area, helping mask Sasquatch's presence.

Sasquatch sign, indicating that it is or has been occupying an area, is abundant and easily seen, even to the untrained eye. It creates and scatters, for example, peculiar items throughout its habitat that are incongruent with their surroundings. Seemingly an ardent petraphile, these include curiously placed stones, freshly broken rocks, piles of shiny minerals (like quartz, which it seems particularly drawn to), and huge boulders (that are too heavy for humans to lift but have obviously been transported from elsewhere).

It is well-known for its strange tree "art," tree markings, and tree structures, such as teepee-like "huts"; twisted and broken branches; wrenched tree trunks; gently bent branches; and arched trees. It also leaves behind "weavings" of bark, sticks, and branches (sometimes these are made from sizeable trees); large covered ground "nests" produced from carefully interwoven branches, sticks, and boughs (often spacious enough for an enormous animal to crawl into); crude fence-like barriers (seemingly meant to impede access to an area or funnel prey for ambushing); and, most predominately, "X"s: symmetrically crossed sticks or logs (sometimes small, sometimes gigantic) that may serve as warning signs, territorial markers, navigation markers, trail markers, boundary markers, or entrance and exit symbols.

The gibbon, like us, is a member of the Hominoidea superfamily. This humanlike ape shares a number of traits with Sasquatch, including an "overpowering and deafening cry," monogamous relationships, and both bipedal and quadrupedal locomotion.

Hunters and hikers report coming across trees (broken off at ground level or pulled from the soil) that have been flipped upside down and rammed deeply into hard ground crown first, their massive root balls pointing skyward. (Occasionally these are also hung from living trees.) Gargantuan logs are sometimes "suspended" horizontally (and perfectly leveled) in trees (often quite high off the ground) through the clever use of small tree branches, as well as wedging and weaving.

Sasquatch is known to "decorate" these structures with snag, stones, human refuse, and animal parts, hanging or placing such things as hides, legs, antlers, skulls, and

various bones, on branches in an obviously intentional manner. It also creates stick-breaks, small tree-snaps, and large tree-breaks, and leans sticks, deadfall, and large poles (debarked trees) against logs, boulders, and other trees—again, usually in a peculiar but recognizably artistic fashion.

The enormity and weight of some of the timber used in these structures (many weigh several tons), as well as the complexity of many of the designs, would suggest that more than one individual was involved in the making of these weird, out-of-place objects. On occasion an outdoorsman will discover purposefully stacked rocks (cairns) and humanoid-shaped stick structures (perhaps a type of Sasquatch symbolism). The meaning and purpose of these bizarre objects and extraordinary structures can only be guessed at, for most of them are located in out-of-the-way, hard-to-reach places, and nearly all are both quite primitive and fantastically constructed. (For a discussion on the theories surrounding the creation of these objects and structures, see Appendix E.)

Known to mimic other creatures, such as birds (most notably owls), elk, coyotes, and even humans, Sasquatch seems to have a rudimentary language and perhaps a primitive alphabet: not only does it speak in a strange guttural chatter (recorded on numerous occasions), but it also engages in rock-clacking, tree-knocking, and the construction of ground glyphs (letter-like symbols made from sticks) which it leaves in both conspicuous and inconspicuous places. It seems particularly preoccupied with the number 3, as many of its glyphs and tree structures have 3 obvious parts, some resembling 3-pronged spears, others appearing like the capital letter "A." As noted, tree knocks often come in series of threes as well.

Our planet brims with as yet unknown humanlike, human-looking primates. For instance, the African Sanje mangabey (a close cousin of the white-cheeked mangabey pictured here) was only discovered and recognized by mainstream science in 1980. Yet this Old World monkey species was known to native African peoples for thousands of years prior. Many people believe that the same will turn out to be true of Sasquatch, whose reality was accepted by Native Americans eons ago.

Vocalizations include many typical primate sounds: grunts, growls, howls, screams, wails, whoops, mouth pops, and whistles, the latter sound which has been memorialized in Native American art (ceremonial masks, totem poles, ancient pictographs) as the "whistling" apelike folk figures Bukwas (Wild Man of the Woods) and Dzunkwa (Wild Woman of the Woods). Comprehensive acoustical analysis of Sasquatch calls has shown that they are not made by wolves, coyotes, mountain lions, deer, elk, or any other known animal, and that, just as importantly, humans do not have the biological apparatus to either produce or reproduce such sounds, some which turn out to be, as we will see momentarily, far below human hearing.

The orangutan is the closest living relative of Gigantopithecus: a 10' tall, 1,000 lb. member of the Hominidae family that allegedly went extinct 100,000 years ago.

Sasquatch possesses the oft reported abilities to quickly walk up steep rocky hillsides without using its hands, calmly stride across double-lane roads in 3 or 4 steps, and rise from sitting on the ground to a full standing position without using its arms—and in one single fluid motion. It can form a power grip, permitting it (like chimpanzees, gorillas, and humans) to accurately hurl projectiles (for example, pine cones, sticks, pebbles, rocks, and even boulders and logs). It seems to throw such objects as a means of diverting or drawing attention, or to drive away unwanted visitors. It may live in underground cave systems or lava tubes (not uncommon in the northwest U.S.), helping to explain its ability to quickly disappear from view. Juveniles are known to climb, perch, swing, play, and leap about in trees.

In order to survive in the many different ecosystems it inhabits, this giant omnivore is probably an opportunistic feeder. Although we cannot be sure what its diet fully entails, like other large primates it is likely that it eats some if not all of the following: roots, leaves, stems, tubers, shoots, vines, buds, flowers, soil, bark, forbs, herbs, honey, seeds, nuts, berries, insects, and eggs.

Also a skilled carnivorous hunter and tracker, this ambush predator can sit or stand perfectly motionless for hours, as well as trail, capture, and kill all manner of game. It has been observed, for instance, carrying dead rabbits in its hands, as well as dead deer (indications are that this is its favorite food), sheep, and other types of mammals draped over its shoulders. Hunting methods seem

L-R: skeletons of a gibbon, orangutan, chimpanzee, gorilla, and human (not drawn to scale). Sasquatch may possess characteristics of all five body types.

to include running down its prey, lying in wait (ambush), and pitching heavy objects (rocks). Since it does not use fire it must consume meat raw, as chimpanzees do—and as our own ancestors did before the discovery of fire.

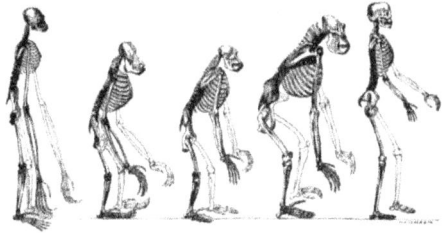

A related sign of its presence are large animal bones that have been broken or twisted in half (to allow access to the marrow), and which lack teeth marks, claw marks, knife marks, or hatchet marks, ruling out both known animals as well as human beings. This leads objective researchers to the conclusion that the creator of these wrenched and smashed bones must have opposable-like thumbs and muscle strength far beyond that of the average person.

Sasquatch, a classic ridge-walker, is a master of surveillance: a carefully positioned "scout" or "lookout" scopes out the surroundings from elevations such as slopes, cliffs, and hilltops. Thus he is usually aware of our presence long before we are aware of his. If unwanted human interlopers enter its home range, it will silently and unobtrusively observe, then "escort" them out, paralleling them off trail (just out of sight), sometimes all the way back to their vehicles—without being noticed.

If this strategy does not succeed it may resort to branch-breaking or tree-snapping; or a tactic far more intimidating, infrasonic sounds: low-pitched, inaudible sound frequencies. Since they can penetrate solid objects (such as trees), these long distance-traveling sound waves—which fall below the level of human hearing (generally about 20 Hz.)—can, some maintain, lead to a host of psychophysical effects, from anxiety, exhaustion, sadness, panic, hallucinations, and stress, to tinnitus, nausea, headaches, chills, disorientation, and even paralysis. (Tigers, for example, use infrasound to stun and paralyze their prey, attract mates, and scare off intruders.)

At this juncture in its interaction with unwanted human visitors, Sasquatch may also foot-stomp, or emit deep low bellows, angry snarls, and hair-raising primate-like cries and shrieks that cannot be made by any known animal. All are meant to inspire fear and dread in their human listeners, and ultimately a speedy retreat out of the area—tactics which, according to those who have experienced them, are quite effective.

When it wants to remain unseen, Sasquatch will tread softly or walk on logs to avoid leaving footprints. When it disregards these precautions, however, it leaves inline tracks with toes pointed straight forward and massive walking stride lengths of up to 72" (the average human walking stride length is about 28"). The footprints themselves are humanlike, plantigrade, and "living." In other words, its prints present sole-walking feet

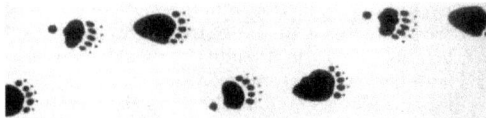

Black bear track: easily distinguishable from Sasquatch track.

that bend and grip, complete with splayed toes (showing that the subject has never worn shoes), an angled (or sometimes square) toe row, hair impressions, friction ridges (also known as dermal ridges or dermal papillae), and pressure cracks, along with overt evidence of a midtarsal break and a pad under the hallux (big toe). Its prints are sometimes confused with bear tracks, which can be similar in appearance. However, it is not difficult to differentiate between the two: bears leave claw marks; Sasquatch (a primate with toenails like ourselves) does not. (See the American black bear for more on this topic.)

In particular, the flow pattern of Sasquatch friction ridges (which run up and down the sides of the feet) are unique, and thus strongly evidential, for they do not match those of apes (which flow diagonally) or humans (which flow side to side). Additionally, its footprints are deeply impressed into the soil (showing tremendous weight), with some measuring 24" long and 10" wide. Sasquatch sometimes also leaves behind massive handprints (complete with detailed fingerprints), knuckleprints, and nearly complete body prints (in mud, grass, and pine bough nests). Importantly, nearly all of the characteristics laid out above are consistent no matter when or where Bigfoot trace evidence is found, even when its discovery is separated by over a century and the width of our continent.

Such wild animal sign, *taken all together*, is obviously not faked, and indeed cannot be faked. For one thing, a number of Sasquatch trackways have been found that run on for many miles. Not only that, they are laid down in some of the world's most difficult, most unforgiving, impenetrable terrain. Third, many are only discovered accidently, usually by lost hikers, rangers, loggers, and hunters. Though these few facts

Typical human, side-to-side walking gait, with feet directed slightly outward.

alone thoroughly destroy the hoax theory, skeptics have as yet failed to offer a rational explanation for how and why such sign commonly appears in unmapped, uninhabited, inaccessible locations—where a person would not normally ever see it. Thus, we are only left with the obvious explanation.

Understandably, mainstream scientists want hard evidence. Plenty of circumstantial and even direct evidence exists, but almost none are willing to examine it. Those who have are compiling a wealth of valuable data on this enigmatic North American mammal—all without waiting for a body to be found. This is being done using a new scientific methodology called metagenomics, which focuses on its own category of physical evidence, one known as "environmental DNA," or eDNA.

Human footprints adjusted to show Sasquatch's inline walking gait, with feet pointing straight forward.

As a mammal (or any other living organism) travels through its habitat, it automatically sheds thousands of cells, microscopic traces of bodily substances that are cast off in the form of DNA. This exuviated genetic material can be discharged through the skin, saliva, urine, mucus, blood, fur, or dung. As it falls, it is deposited, for example, on leaves, bark, grass, sediment, snow, or in water (streams, creeks, ponds). After collecting these materials from nature, scientists extract and purify any eDNA that is found, and examine it in an attempt to identify what animal species produced it.

Recently, analysis of samples of purported

Sasquatch eDNA were collected from the field containing its tissue, saliva, blood, and—not *fur* but—naturally tapering *hair* (showing that it has never been cut). This genetic material reveals the owner to be a hybrid creature (half-human, half-unknown nonhuman primate) that appeared on the scene around 15,000 yrs. ago (crossing from Asia into North America over the Bering land bridge during the Pleistocene Epoch), but which descends from ancestral hominids over 1 million yrs. old.

Unfairly, while anecdotal data is sometimes accepted in place of hands-on field studies for other elusive mammals, it is usually rejected outright when it comes to Sasquatch. Yet despite this unscientific attitude (for real science is an open-ended search for truth—whatever it turns out to be), North America's "Yeti" continues to be seen, heard, smelled, tracked, photographed, and video-recorded by untold thousands, including military personnel, police, doctors, and scientists from many different disciplines. (Note that the majority of these sightings are never reported for fear of ridicule.)

The male gorilla is another hominid that has much in common with Sasquatch, including superior strength, opposable thumbs, and a sagittal crest.

Though subjective, the sheer weight of these eyewitness reports, many which include Class A visual sightings as well as audio and video recordings (often made by trained and experienced individuals), is highly compelling—to those who care to impartially study them for themselves. To this avalanche of information must be added a mountain of real-world objective evidence: plaster castes of hundreds of large humanlike footprints and strange inline trackways that a number of forward-thinking ichnologists, primatologists, anthropologists, and paleoanthropologists have pronounced to be irrefutable evidence of an unrecognized, upright-walking, North American hominid.

While hoaxing occurs on occasion (baggy costumes, stiff gaits, shallow trackways, "dead" footprints, and high human foreheads quickly expose the fraudsters), it is a fact that those who are most skeptical are also generally those who refuse to review the physical, trace, and photographic evidence, spend time researching in the field, or examine the voluminous records of eyewitness accounts that have been collected over the years, not only from North America, but from around the world.

Reinforcing the credibility of recent scientific findings are Native Americans, such as the Hupa, who call Sasquatch "Omah." Having accepted its existence for millennia, they have carefully preserved it in their lore, each tribe, people, and region giving it their own individual name—creating a data bank of hundreds of monikers that span centuries and millions of square miles. Early European Americans too have been chronicling evidence of what they called the "wild man" for hundreds of years, all of this long before the modern hoax theory arose. New England's Academy of Applied Science at Boston, Massachusetts, was once interested enough that it supported a 5-yr. search for the mysterious cryptid.

Top: *Homo erectus* skull (circa 1 million years old). Bottom: skull of *Homo sapiens* (modern human). *H. erectus* has several commonalities with Sasquatch: a flat face, a sloping forehead, powerful jaws, and possibly a covered but flattish nose.

The reality of this giant upright-walking mammal, also popularly known by its Chinook name, "Skookum," was once openly admitted, accepted, and officially recognized by the U.S. government. In the latter half of the 20th Century, the U.S. Army, *backed by the Pentagon*, listed it as a real living

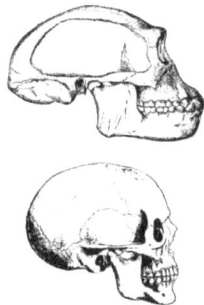

animal in its 1975 edition of the *Washington Environmental Atlas*, devoting one-third of a page to it (most animals received only one-sixth of a page). Included in the work because of "overwhelming evidence," the Bigfoot entry was published under the heading, "Some Important Wildlife of Washington," noting that "FBI laboratories" had performed detailed examinations of "alleged Sasquatch hair samples" and determined that "no such hair exists on any human or presently-known animal for which such data are available."

Some scientists surmise that this mysterious evolutionary package may be, in fact, a descendant of Gigantopithecus (*Gigantopithecus blacki*), a prehistoric hominid (belonging to the Hominidae family and the Pongo subfamily) that went "extinct" around 100,000 yrs. ago. It is believed that this enormous ape stood about 10' tall and weighed at least 1,000 lb., the same measurements as a typical male Sasquatch. If they are related, or perhaps even the same creature, it means that "Giganto"—a forest-dwelling native of Asia (and like Sasquatch, probably both bipedal and quadrupedal)—did not die out, but instead survived into the Upper Paleolithic period, at which time it, or its descendants, crossed over to North America via the Bering land bridge.

If correct, we might be able to glean valuable information about Sasquatch from the Bornean orangutan (*Pongo pygmaeus*), a red-haired hominid who happens to be Gigantopithecus' closest living relative. A connection between Sasquatch and Giganto would also help explain the many ancient Bigfoot legends, myths, and artistic renderings that were created by Native Americans, *an Asian people who also traveled east over the Bering land bridge*, long prior to the arrival of Europeans in North America. It is certainly of interest that the Malaysian people gave this particular pongid the name orangutan: in the Malay language it means "wild" (*hutan*) "person" (*orang*).

Due to its secretive nature, fear of man, probable burial of its scat, tracks, and dead, its wide array of stealth tactics, its military-like control of its wilderness domain, its great speed and legerity, and its vast prehistoric knowledge of the woods and mountains, naturally—like the wolverine, mountain lion, and other large but secretive mammals—it is seldom seen by humans. This phenomenon is augmented by numerous clever evasion tactics it utilizes: strategical use of its camouflaging semi-transparent fur; freezing in place; shadowing people (moving when they move, stopping when they stop); evacuating the area quickly (and usually noiselessly); tree-hiding; tree-peeking; and branch-waving, a technique meant to simulate blowing wind (another behavior it shares with several nonhuman primates).

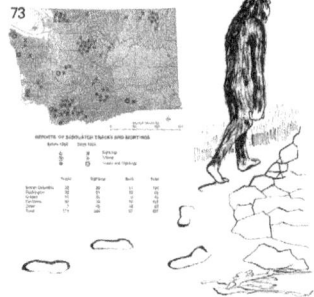

SASQUATCH

73

Part of the section dedicated to Sasquatch, from the 1975 *Washington Environmental Atlas*, published by the U.S. Army Corps of Engineers—with the support of the Pentagon.

Estimates of its population range from a few thousand (in the Pacific Northwest alone) to millions worldwide. If it has a low population density, this mammal would be quite rare, also helping to explain why it is not often observed, and also why no skeletons have been found (even the most experienced wilderness authorities have never come across, for example, a black bear skeleton). On the other hand, its evasiveness may simply indicate our inferior skills and its superior skills when it comes to knowledge of the natural world. In any event, the 2017 recognition of a new species of great ape, the 4½' tall, 200 lb. Tapanuli orangutan (*Pongo tapanuliensis*), proves that our planet continues to hide large primates, even in the technologically advanced 21st Century. (Large nonprimate mammals continue to be discovered as well, such as the wood bison, which was found in Canada in 1957.)

Being an apex predator, this log-walking hominid has no natural enemies. It may have an average life span that lies midway between chimpanzees and Neanderthals (40 yrs.) and modern humans (80 yrs.); that is, roughly 60 yrs. (my estimate).

At the moment there are no, and there cannot be any, Bigfoot experts. One day this will change. Until then, much of what we think we know about Sasquatch is necessarily based on eyewitness accounts, speculation, opinion, theory, and educated guesswork. This is also true, however, of a number of recognized living species. The book on this puzzling giant anthropoid remains open.

SEA OTTER (*Enhydra lutris*): In our region this mammal has 2 recognized subspecies: 1) the northern or Alaska sea otter (*Enhydra lutris kenyoni*) lives from Alaska south to Oregon; 2) the southern or California sea otter (*Enhydra lutris nereis*) lives off the coast of California. (A third subspecies, the Asian sea otter (*Enhydra lutris lutris*), lives outside North America.) Though the northern sea otter is slightly larger than its California cousin, the southern sea otter, both share nearly identical traits and habits. Thus, they are presented here under the same entry, and as a single entity: the sea otter (*Enhydra lutris*), also sometimes called the "North American sea otter."

It can grow up to 6' in length, has an 8" hind foot, its tail is 14" long, and it can attain a weight of 100 lb. Its velvety pelage is cinnamon to dark brown, the head and neck are tan, yellow-gray, blonde, or brown. The entire coat may be grizzled with gray or blonde-tipped hair. Its head is broad; the muzzle is short; its nose is large and dark; its eyes are small and dark; its ears are small and rounded; its neck is thick. Its front legs are short, with fused paws containing retractable claws. Its back legs, which are longer than the forelimbs, sport large flat paws with webbing between the toes.

Like many other riparian mustelids, it has long stiff whiskers, which it uses as sensory antennae in the dimly lit, nearshore marine ecosystems and biomes it inhabits. These include shallow rocky coastlines, subtidal zones, estuaries, and barrier reefs, with soft, sandy, muddy, or silty bottoms; it is particularly partial to marine shorelines with abundant shellfish and thick kelp forests, the latter which it uses for protection, resting, birthing, foraging, hunting, and feeding.

A nearctic native, its northern range (inhabited by *Enhydra lutris kenyoni*) covers the Aleutian islands, southern Alaska, British Columbia, Washington, and Oregon; its

Sea otters.

southern range (inhabited by *Enhydra lutris nereis*) extends from central to southern California. Males establish 1 sq. mi. home ranges in the water along the coast. They defend these littoral territories against other males, particularly when females in estrus are in the vicinity. Sedentary and neritic, sea otters do not hibernate, are active all year, and occupy the same area for most of their lives.

Mainly diurnal, but also crepuscular, this opportunistic carnivore is an avid molluscivore and a piscivore. Its diet includes bivalves (such as mussels, clams, and abalone), cephalopods (octopus, squid), echinoderms (sea urchins, starfish, sea cucumbers), gastropods (snails), chitons, limpets, marine worms, and crustaceans (lobsters, crabs), as well as dozens of other saltwater species, including fish and sometimes sea birds. A petraphile, it is one of the few nonprimate mammals that uses stone tools: while foraging on the ocean floor it carries a small stone (usually in a pouch-like flap of skin that extends from under each forearm across the chest) that it uses to pry animals, like mollusks, from rocks. After bringing its prey to the surface, the sea otter floats on its back, and lays the rock on its chest. Then, holding its quarry in its dexterous, hand-like forepaws, it vigorously taps it on the stone until it cracks open, permitting access to the meat inside. It then rolls over in the water to rinse its food, or to drop the rock when it is no longer needed. Stores excess food in its chest "pockets."

Shares with other members of its scientific family both a sense of playfulness and a high metabolic rate; to sustain itself it must eat nearly 25 percent of its own body weight each day. Uses kelp as a bed to rest in, gathering in "rafts" (groups of 12 or more individuals) for safety and socializing; wraps itself in kelp fronds to keep from drifting away. While rafting in the kelp canopy it will often clean and groom itself, spreading water-resistant oils through its fur that help with insulation and buoyancy.

Though one of the smallest of our marine mammals, it is the heaviest member of the weasel family. Its scientific name is Graeco-Latin for "in-water otter," and it is indeed one of the world's most aquatically oriented mammals: the sea otter is born at sea, eats at sea, sleeps at sea, mates at sea, nurses at sea, and plays at sea, only coming ashore during severe weather. Using its flattened rudder-like tail, webbed feet, and flipper-like hindpaws, it can dive to a depth of 330' and remain submerged for 5 min. Though it lacks heat-preserving blubber, it makes up for this by wearing the densest fur of any animal, with a coat containing up to 1 million hairs per sq. in. (we have a mere 100,000 hairs covering our entire head, or about 1,500 hairs per sq. in.).

Sea otter feeding.

It has no set breeding season. The polygynous boars mate with any receptive sow that is in estrus, and at any time of year. The female employs delayed implantation, gestating for an average of 8 mos., giving birth to 1 pup at sea (twins are rare)—or sometimes on land or an ice floe. Unlike the young of most other mammals, the North American sea otter baby is both precocial and altricial, for it is born with its eyes open, newly emerging teeth, and a full coat of buoyant fur that keeps it afloat; yet, being a semiaquatic mammal living in the harsh environment of the ocean, it is completely dependent on its mother for the first 6 mos. of life. Only then is it weaned and ready for independence. As the mother floats on her back in the water, the pup plays and naps on her chest while she grooms and nurses it (with warm, high protein, fat-rich milk). During the pre-weaning stage she also protects it from danger and teaches it how to dive and hunt. The male is non-paternal and therefore plays no role in the rearing of his offspring.

Communication and perception take place through normal Mustelidae channels: sight, touch, hearing, and especially smell, which is used to identify individuals (each which has its own unique scent). Vocalizations include: whistling, growling, whining, squeaking, snarling, screaming, whimpering, cooing, grunting, squealing, and hissing.

A secondary consumer, its natural enemies in the sea are sea lions, sharks, and orcas; birds of prey take pups at the surface; on land it is preyed upon by bears, coyotes, and bobcats. Among its defense strategies are flight, concealment, and climbing onto land. A keystone species, it helps maintain the balance of kelp ecosystems by consuming kelp-destroying species, such as the sea urchin (whose main predator, the starfish, is dying out along the Pacific Coast). The male (North American) sea otter lives an average of 10-15 yrs. in the wild; the female lives about 15-20 yrs. in the wild.

SEI WHALE (*Balaenoptera borealis*): What might be thought of as a smaller version of the fin whale, this rorqual is a member of the baleen whale family. It grows to a total length of 65' and weighs some 20 tons. Its upper body is dark or bluish gray, its sides are a mottled gray, its underside is white. Unlike the fin, it has uniform coloring over the left and right sides of the bottom jaw. Dark gray baleen or whalebone, 3' long, curtain-like bristles in its upper jaw or rostrum, has replaced what were originally teeth. Has some 60 throat grooves on the underside, which expand and retract while feeding. It can spout up to 14' in the air through its double blowholes. This cetacid is gender dimorphic: the female is larger than the male.

Its common name is the English form of *seihval*, a Norwegian word meaning "coalfish" (*sei*) and "whale" (*hval*), which was given to this particular cetacean due to its habit of following coalfish in search of food. The sei whale has a flat v-shaped head; a long slender body; a dorsal fin located on the back half of the spine; short pectoral flippers; and a large centrally notched tail fluke with pointed tips.

Sei whale.

A pelagic species, this secondary consumer inhabits all temperate and subtropical oceans, usually avoiding the coldest waters of the polar regions and the warmest waters of the tropical regions. Highly social and migratory, it normally forms small pods of about 5-10 individuals; however, it has been known to gather in superpods of many thousands.

Both a piscivore and a planktivore, its carnivorous diet consists mainly of small crustaceans and fish, which it skims—as it swims along on its side—into its enormous maw, where it is filtered by baleen as the seawater is expelled. It eats up to 2 tons of food a day. Able to swim up to 30 mph, it is one of the fastest whales in the world. It is not a deep diver, nor can it hold its breath for more than a few minutes. However, since most of its food is near the surface, it does not need to be able to dive to great depths.

Data on the social and reproductive life of this massive natatorial balaenopterid is scarce. The female gestates for around 1 yr., giving birth to a single 15' calf in the fall. The sei whale may live to around 75 yrs. in the wild. Little else is known.

SEMINOLE BAT (*Lasiurus seminolus*): This medium-sized chiropterid grows to a little under 4½" in total length, its forearm measures just under 2", its wingspan is about 12", and it weighs nearly ½ oz. Its ears are short and rounded; the tragus is short; the wrists and underarms are furred; and the interfemoral membrane is furred on the dorsal side. Due to its size and coloring it is sometimes mistaken for the red bat. Though it does indeed have the silver-tipped pelage of the red bat (giving both a "frosted" appearance), the Seminole bat is a darker mahogany brown, lending it its alternate common name: the "mahogany bat." Ventrally it is lighter, and sometimes the chest is buff or white. It shelters, usually alone or in small groups, inside clusters of Spanish moss as well as under tree bark. Like most bat species it prefers open ground beneath its roosting area, allowing it to drop down and take flight without hindrance. This microchiropteran is gender dimorphic: the female is larger than the male.

Spanish moss-covered trees in Louisiana: Seminole bat habitat.

It is found primarily in mixed forests, with a range covering the same territory as the early Seminole Indians: the American Southeast and more specifically the Gulf Coast states—hence, its common name. Crepuscular and nocturnal, this insectivorous vespertilionid leaves its roost around sunset, using echolocation to track down its insect prey (beetles, wasps, flies). It hunts over ground and water, as well as around tree tops and street lights. A fast and efficient flyer, it consumes large amounts of agricultural pests, making it a friend of the farmer.

Its reproductive life has not been sufficiently studied. The female probably practices delayed fertilization. She gestates for about 85 days, bearing 1-4 altricial pups in late spring or early summer. This species migrates but does not hibernate since much of the area it occupies remains temperate throughout the winter. If the weather gets too cold it may go into a temporary state of torpor. A secondary consumer, its natural predators include falcons, blue jays, snakes, and opossums. The life span of the Seminole bat is not known.

SHADOW CHIPMUNK: (*Neotamias senex*): Also known as "Allen's chipmunk," and once believed to be a subspecies of the similar Townsend's chipmunk, based on recent genetic studies it has been accorded full species status. Nonetheless, it shares much of the same ecology and life history as its close cousin—though the range of the shadow occurs further south (Oregon, California, and Nevada) than Townsend's (Washington and Oregon).

SHORT-BEAKED COMMON DOLPHIN (*Delphinus delphis*): One of two species of common dolphin (the other is the long-beaked common dolphin), it grows to a total length of almost 9' and can weigh up to 300 lb. Its coloration is striking, for it is the only dolphin that has a blackish upper body, whitish undersides, and a curved lighter-colored pattern wrapping around the sides (which gives it its alternate name, "short-beaked saddleback dolphin").

This common natatorial carnivore ranges along the Atlantic Coast from Canada to Florida, and along the Pacific Coast from Washington to Mexico; also found in the Caribbean and the Gulf of Mexico. Some 120 small teeth interlock inside its jaws, useful for catching and grasping prey. This highly social delphinid travels in gregarious pods of from a few hundred individuals to superpods comprised of over 100,000.

Like others of its kind it is altruistic, and will push sick or injured pod members to the surface to breathe, holding them up with its flippers. A nimble

Common dolphin.

acrobat and a fast swimmer, it can reach speeds of nearly 40 mph and dive to a depth of over ¼ of a mile. Will bow-ride large moving objects in the sea, from whales to boats. Feeds on aquatic crustaceans, cephalopods, and fish. A secondary consumer, its natural enemies are sharks and orcas. Has been known to mass strand. The short-beaked common dolphin lives to an estimated 35 yrs. in the wild.

SHORT-FINNED PILOT WHALE (*Globicephala macrorhynchus*): Despite its common name, this toothed cetacean is not a whale (though it possesses whale-like behaviors), it is a dolphin. Another one of its common names, "blackfish," is even less accurate, for it is not always black and it is not a fish. Its scientific name, *Globicephala macrorhynchus*, gives a more factual description, translating as "globe head, large snout."

This highly social delphinid grows to 20' in total length and a weight of 6,500 lb. or more. Its long robust body is gray to black with light gray markings on the throat and belly; has a light-colored "saddle" over its mid-back; there is often a whitish "slash" above or behind the eyes. Distinguished not only for its square bulbous forehead and blunt beak, but also for its post-anal keel and large sharply hooked, wide-based dorsal fin, set slightly forward on the back. Easily confused with its cousin the long-finned pilot whale (with whom its range overlaps), as its common name suggests, the short-finned has smaller, less-sharply curved flippers, which only grow to one-sixth its body length. This cetacid is gender dimorphic: the male is larger than the female.

This natatorial nomadic dolphin prefers open subtropical and tropical waters, but can also be found in temperate coastal regions, and has been reported from the North Atlantic Ocean to the Gulf of Mexico. Gregarious, it forms close emotional bonds with

Pilot whales (stranded).

fellow pod members, often traveling in large matricentric groups of hundreds of individuals. Its many interesting behaviors include logging (laying motionless at the surface).

Also known as a "pothead" (due to its large roundish head), like other members of the Delphinidae family it possesses a complex form of what might be considered language, communicating with conspecifics using a broad range of sounds, from whistles and clicks, to squeals and snorts. A largely nocturnal carnivore, it hunts its prey, cephalopods and fish, by echolocation: a series of sonar beams which emanate from organs inside its large forehead. It can consume up to 100 lb. of food a day.

This polygynandrous cetacean has no set breeding season. The cow gestates for an average of 14 mos., giving birth to a single 6' long, 130 lb. precocial calf. The young are usually weaned at around 2 yrs. of age, though some continue to nurse until they are 5 yrs. old. Females, who experience menopause as they age, will act as "wet nurses," readily caring for and breastfeeding each others' young. The bull does not participate in the rearing of his offspring.

This high-speed dolphin is known to lobtail (slap the water surface with its tail), porpoise (clear the water when fast-swimming), and spyhop (lift its head out of the water to surveil its surroundings). Mass strandings are not uncommon. A secondary consumer, its natural enemies are large shark species and orcas. The male short-finned pilot whale may live 40-50 yrs. in the wild; the female may attain an age of 55-65 yrs. in the wild.

SHORT-TAILED SHREW (*Blarina brevicauda*): Though it grows to a total length of 5", its tail is only 1" long, giving this shrew both its common name and its species name (*brevicauda* means "short tail"). Its hind foot measures up to ¾", and it can attain a weight of 1 oz. It has a cylindrical body covered by a soft fur coat that is almost uniformly silvery gray in color, above and below. The head is conoidal; the eyes are tiny; the nose and feet are pinkish; the muzzle is somewhat short and rounded for a shrew. Gender dimorphism is present, with the male being larger than the female.

Also known as the "Northern short-tailed shrew," this habitat generalist can be found in both wet and dry regions, woods and fields, marshes and chaparral, farmland and uncultivated meadows, wild places and urban areas. Solitary, diurnal, nocturnal, very active, and extremely fossorial, its powerful mole-like front feet make it an expert climber and digger, one that spends most of its life underground. Thus, it prefers moist friable soil, usually with good hiding cover, downed trees, rocky locations, and abundant leaf litter under which it can nest, tunnel, hide, and feed. Its range extends from central and eastern Canada south into the American South and Midwest, and as far west as the prairies of the northern Great Plains.

Short-tailed shrew.

Its omnivorous diet includes a wide assortment of edibles, ranging from fungus, nuts, seeds, earthworms, snails, and centipedes, to beetles, reptiles, amphibians, rodents, and birds. Resorts to cannibalism when necessary. Hoards and caches extra food. This large husky soricid is one of the fiercest hunters in the animal kingdom: furtively chasing down its quarry, it leaps upon it, inflicting a series of rapid violent bites about the face and neck. Its poisonous saliva stuns and paralyzes its victims, allowing it to consume its prey while it is still alive (and at its freshest and most nutritious). Energetic with a high metabolism, it can eat up to 3 times its own weight every 24 hrs.

Despite the fact that it is one of our most common animals, as with many other mammals in our region, in particular shrews, its mating habits have not been adequately studied and are therefore shrouded in mystery. Its breeding season seems to run year-round, though it crests between spring and fall. The female may bear up to 3 litters a year. She gestates for about 3 wks., gives birth to an average of 6 altricial young, and weans them at about 3 wks. of age. The male's role in the raising of its offspring is not known.

Having relatively inferior vision, its sense of touch and smell are highly developed. Not only is its flexile snout covered in tactile whiskers (which serve as sensory antennae), but it also uses ultrasonic clicking sounds (similar to bat echolocation) to help it navigate through its often dark domain. A secondary consumer, its natural enemies include fish, reptiles, birds of prey, mustelids, and wild canids. Marks its territory and defends itself by exuding a noxious odor from its scent glands. It is important to humans as a pest controller. The short-tailed shrew may live 2-3 yrs. in the wild.

SILKY POCKET MOUSE (*Perognathus flavus*): This little mammal is one of our region's smallest mice, one of our most diminutive rodents, and the smallest member of its family (Heteromyidae). It grows to a total length of about 4½", the hind foot measures about ½", and it can attain a weight of up to ⅜ oz. Named for its soft satiny fur, on the dorsum its coat is buff with black-tipped hairs washing over the back; on the ventrum its pelage is white to tan; its 2½" tail is tufted and slightly dichromatic (dark above, lighter below). The head and eyes are large; the feet are small and pinkish-white; the ears are small, rounded, and erect; surrounding the top, back, and bottom of each ear is a large, curved, whitish-yellowish-tan fur patch that mimics the shape of the ear.

This little motile heteromyid is a nearctic native that inhabits temperate, often harsh, open prairies, chaparral, grassland, slopes, deserts, plains, bottomland, piñon-juniper forests, savanna, shrubland, rangeland, valleys, sand dunes, mesas, and intermontane locations generally, preferably with gravel-dominant ground, rotting logs, sandy soil, rocks, surface duff, and dispersed vegetation. A purely western mammal, its range cover portions of the states of Wyoming, South Dakota, Utah, Colorado, Nebraska, Kansas, Arizona, New Mexico, Oklahoma, and Texas, extending south into central Mexico.

Excavates tiny burrows, expertly concealing their entrances and exits beneath shrubbery, grass clumps, rocks, and cacti; each burrow contains multiple compartments: a den chamber, a food storage chamber, a nesting chamber, and a toilet chamber. Portals are closed off when the owner is at home—useful for controlling temperature and humidity and for protection against predators. It will appropriate the abandoned burrows of other fossorial animals. Though it is active year-round, it enters brief states of torpidity during extremely cold weather.

Silky pocket mouse.

A sedentary, crepuscular, nocturnal omnivore, its diet consists of seeds, grass, fruit, forbs, various plants, and insects. Does not normally drink water, which it is able to metabolize from the vegetation it consumes. Stores surplus food in its burrow larders, carrying it home in its fur-lined cheek "pockets" or pouches. Its breeding season runs from spring to fall. The female gestates for about 1 mo., giving birth to an average of 3-4 pups in a grass-lined underground natal nest.

Communicates and perceives via standard Heteromyidae channels: smell and touch, and to a lesser degree sight and hearing. Its natural enemies are reptiles, birds of prey, wild felids, and wild canids. Beneficial as food for carnivores. The silky pocket mouse probably lives 1-2 yrs. in the wild.

Silver-haired bat.

SILVER-HAIRED BAT (*Lasionycteris noctivagans*): This medium-sized bat grows to nearly 4½" in total length, its forearm measures almost 2", its wingspan is around 12", and it weighs ½ oz. Its dorsum is blackish with frost- or white-tipped fur, giving it its common name. The ears are rounded, hairless, and short; the nose is slightly upturned. It has more fur than many other bats (particularly on its interfemoral membrane), and both the ears and wing membranes are often dark brown or black. A nocturnal flying mammal, its species name, *noctivagans*, translates as "evening vagabond" or "night wanderer."

Abundant in the northern parts of our region, this chiropterid mainly inhabits old growth deciduous and coniferous forests across its range, from Alaska through Canada and most of the U.S. south into Mexico. In the U.S., however, it is not usually found in either the southernmost states or the Southwest states. Largely solitary, it shelters in trees, snag, blowdown, deadfall, under bark, in rotting trees, in caves, bird nests, buildings, and in squirrel and woodpecker holes. A hardy riparian species, it prefers undisturbed habitat near established water sources, such as creeks, ponds, estuaries, reservoirs, rivers, and lakes. Migrates north in summer, south in winter, hibernating through the coldest months in hibernacula, typically trees, wood piles, rock crevices, cliffs, and human-made structures. Establishes a home range of up to 12 sq. mi.

This slow-flying but agile vespertilionid leaves its roost around sundown, using echolocation to track down its favorite prey: moths. Utilizing both hawking and gleaning hunting methods, it eats nearly a dozen other types of insects (ants, midges, mosquitoes, flies, beetles, leafhoppers), making it highly beneficial to us. The female

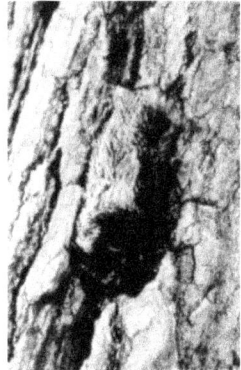

Silver-haired bat.

delays fertilization, gestates for about 55 days, then gives birth to 1-2 altricial pups in early summer. The natural enemies of this secondary consumer are birds of prey (mainly hawks and owls) and mammals (from skunks and raccoons to foxes and wildcats). The silver-haired bat lives 8-10 yrs. (or more) in the wild.

SINGING VOLE (*Microtus miurus*): Also known as the "Alaska vole," this small rodent grows to a total length of about 6", its hind foot measures ¾", and it can attain a weight of up to 2 oz. or so. On the dorsum the pelage is grayish-buff with grizzled black-tipped hairs washing over the back; the ventrum is tan or pale gray; the short nearly 2" tail is furry. The body is ovoid; the eyes and ears are small; the feet are pinkish and heavily clawed.

This subterranean cricetid is both a holarctic and a nearctic native that inhabits subalpine and alpine tundra, particularly locations with willow trees, birch trees, and a fresh water source. A far northwestern species, its range covers Alaska and Canada's Yukon and Northwest Territories. Fossorial, it excavates burrow systems (leaving behind little ejecta mounds) with surface runways that fan out from the entrance; will appropriate the abandoned tunnels of other burrowing animals.

Social and colonial, its breeding season occurs during the summer, with females giving birth to an average of 8 pups in an underground nest chamber. A diurnal and nocturnal vegetarian or herbivore, it has a strong dietary preference for lupines, but it also consumes weeds, twigs, leaves, tubers, and sedges. Semi-scansorial, it will climb into shrubbery and short trees in search of edibles. Caches tough fibrous foods in its burrow larders; cuts up lighter green vegetation, forming (sometimes large) stacks of clippings around tree bases throughout its small home range.

Communication and perception run via standard *Microtus* channels: touch and smell, with reduced sight and hearing. Its thin, piercing, trill-like call gives it its common name. Its natural enemies most likely include some of those that feed on its more southern cousins: birds of prey, mustelids, wild felids, and wild canids. The precise life span of the singing vole is unknown.

SISKIYOU CHIPMUNK: (*Neotamias siskiyou*): Once believed to be a subspecies of the similar Townsend's chipmunk, based on recent genetic studies it has been accorded the distinction of being a separate species. Nonetheless, it shares much of the same ecology and life history as its close cousin—though the range of the Siskiyou occurs further south (Oregon and California) than Townsend's (Washington and Oregon).

SMOKY SHREW (*Sorex fumeus*): It grows to a total length of 5", its tail is 2" long, its hind foot measures ½", and it weighs up to ⅓ oz. In summer its dorsal coat is brown, its ventral coat is buff; in winter its dorsal coat is a sooty gray, its ventral coat is light gray. The pelage of some individuals may be slightly grizzled. From these particular colorations it receives its common name: "smoky shrew." The body is stout and cylindrical; the head is flatish and conoidal; the eyes are tiny; the external ears are small but visible; the snout is tapered; the long tail is bicolored (brown on the dorsum, yellowish on the ventrum); its teeth tips are often "stained" dark brown.

Shrew.

A riparian, it inhabits damp coniferous and deciduous forests, swamps, woodland edges, fields, wetlands, and stream banks at high elevations that are dominated by rotting wood, snag, hollow trees, deadfall, and copious amounts of leaf litter—which it uses for nesting, protection, tunneling, and food. It ranges across the northeastern part of our region, from southeastern Canada through the northeastern U.S., south to Tennessee and Georgia. Its large burrow systems are often located under stumps, roots, rocks, leaf mold, and river banks. Its tiny burrow entrances are a mere ½" in diameter.

This nearctic soricid is a fossorial omnivore whose diet includes fungi, chilopods, lepidopterans, coleopterans, annelids, isopods, gastropods, arachnids, and amphibians. Diurnal and nocturnal, with its high metabolism it must eat nearly constantly, and thus it actively forages and hunts both day and night throughout the year. Despite the coldest winter weather, it does not hibernate.

There is no set breeding season, but it generally crests between spring and fall. As is typical for many short-lived species, it may produce up to 3 litters per year. The polyestrous female gestates for about 3 wks., with the average litter size being 4. The altricial young are weaned by about 4 wks. The maternal nest, built with forest floor detritus and discarded fur, may be constructed above or below ground. The male does not appear to help in the raising of his offspring.

It communicates and perceives via normal Soricidae channels: sight, smell, touch, and sound. Like some other shrews, it may use echolocation to navigate and hunt. Vocalizations include squeaking, purring, whistling, chattering, and twittering. A rapacious insectivore, it is of great benefit to us. A secondary consumer, its natural enemies are reptiles, birds of prey, mustelids, didelphids, wild felids, and wild canids. The smoky shrew probably lives no more than 1½ yrs. in the wild.

SNOWSHOE HARE (*Lepus americanus*): Also known as the "varying hare," this secretive leporid grows to a total length of nearly 21", its 3" ears are black-tipped and ringed, and it weighs up to 3 lb. Its common name derives from its massive, fur-covered hind feet, which can grow to nearly 6" long. Its summer coat is dark brown, and grizzled with white and black hairs. In winter its pelage turns white, often with brown mottling. These color changes, known as molting, are photoperiodic in nature; that is, they are dictated by the amount of light that falls each day. In the case of the snowshoe hare, the shorter the length of day the more white its coat becomes; the longer the length of day the more brown its coat becomes.

This abundant well-known lagomorph shelters in grassy indentations (known as "forms"), hollowed out logs, or the abandoned burrows of other animals. It inhabits fields, thickets, boggy areas, low brushy forest, piney uplands, young coniferous stands, and northern riparian woodland, from Alaska and Canada south into the upper tier of the lower 48 U.S. states and down into the Sierra Mountains, the Rocky Mountains, and Appalachia.

Elusive and shy, this solitary mammal is both crepuscular and nocturnal, readily engages in dust-bathing, and spends much of the daytime napping and grooming. Although classed as an herbivore that feeds mainly on grasses and fruit as well as on tree buds, twigs, needles, and bark, this coprophage is also mildly carnivorous, for when the opportunity arises it will eat carrion—including the remains of its own kind.

Its breeding season begins in spring and ends in summer. The polyestrous doe may bear up to 4 litters annually, and because she possesses two uteruses, she can conceive a second litter while still pregnant with the first (a phenomenon known as "superfetation"). She gestates for about 5 wks., giving birth to 1-6 precocial leverets, which she weans in 2-4 wks. The non-paternal buck plays no role in the raising of his offspring.

It communicates using various vocalizations, from hissing and snorting to squealing and foot-thumping. A primary consumer, its natural enemies are crows, ravens, owls, hawks, eagles, squirrels, weasels, fishers, mink, martens, foxes, wolves, coyotes, wolverines, lynxes, bobcats, mountain lions, and bears. To evade predators it uses its cryptic coat to camouflage itself, freezing motionless in thickets until the danger has passed; or it will run in large circles at up to 30 mph, zigzagging as well as leaping distances of 12' (to break its scent trail). As a last resort it will leap into a body of water and swim, if one is nearby.

Its population rises and falls precipitously every decade or so, the reasons for which are still not completely understood. The snowshoe hare may live as long as 5 yrs. in the wild, but an average of 1-2 yrs. is more likely.

Snowshoe hare.

SONOMA CHIPMUNK (*Neotamias sonomae*): Named after Sonoma County in California's wine region (where it was probably first identified), this large chipmunk grows to a total length of nearly 11", its hind foot measures around 1½", and it can attain a weight of up to about 4½ oz. Its soft thick pelage is dark rusty-brown on the dorsum, tawny-cream on the ventrum. It has 5 indistinct dark dorsal stripes that alternate with 4 indistinct light dorsal stripes; facial stripes mimic the back stripes, but are more vivid. Has less gray than most other chipmunks; each ear bears a black subauricular mark; its bushy 5" dichromatic tail is dark above, reddish-brown below, and edged with yellowish-brown fur. Its summer coat is lighter in color than its winter coat.

This timid wary sciurid is a sedentary nearctic native that inhabits temperate, open coniferous forest and grassland dominated by brush, duff, snag, deadfall, and fallen timber. Riparian by nature, it is also found along creeks, streams, and rivers, and in dense savanna, sageland, rocky settings, and hardwood stands. Its range is limited to one small section of the western U.S.: northwestern California. It builds its house in either stumps, logs, and rock crevices, or underground. Does not seem to hibernate, but dislikes extreme weather and may enter brief periods of torpidity when the temperature drops.

What little is known of its omnivorous diet suggests that it is similar to others of its kind: seeds, grains, fruits, forbs, stalks, bulbs, flowers, leaves, foliage, woody material, herbs, fungi, birds' eggs, chicks, and small mammals (mice). Forages both on the ground and in bushes and trees. Hoards and caches surplus food, transporting it to various larders in its cheek pouches.

This solitary, diurnal, paludal, small stripped squirrel breeds once each spring. Bucks fight fellow rivals for access to estrus females. Does gestate for 1 mo. and give birth to a litter of about 4 altricial pups in the natal nest. Weaning probably takes place within 3-4 wks., after which the young disperse. The non-paternal buck does not participate in the raising of his offspring.

Communication and perception run along normal Sciuridae channels: sight, smell, touch, and hearing. It is known for its high-pitched bird-like alarm call. Its natural enemies are no doubt the

Chipmunk.

same as other rodents in its genus: reptiles, birds of prey, mustelids, wild felids, and wild canids. Benefits its ecosystem by dispersing seeds and spores and by providing food for carnivores. The life span of the Sonoma chipmunk is unknown, but it may be assumed that it lives approximately 2-4 yrs. in the wild; perhaps longer under ideal conditions.

SOUTHEASTERN MYOTIS (*Myotis austroriparius*): This smallish chiropterid grows to a total length of almost 4", its forearm measures nearly 2", its wingspan is around 9", its tail is almost 2" long, and it weighs ⅜ oz. Its dull, short, wooly fur is brown on the upper body, tan or grizzled gray-brown on the underbody. It inhabits caves, hollow trees, buildings, bridges, and culverts, ranging across parts of the southeastern U.S. north to the southern Midwest. Gender dimorphism is present: the female is larger than the male.

As its species name suggests (*austroriparius* translates as "southern river," or more loosely, an inhabitant of "watery southern regions"), it is a riparian that prefers areas such as forested wetlands, orchards, snag, and hollow trees with an abundant water supply nearby, such as a stream, river, pond, floodplain, or lake. Its ideal roost is a cave with a water source inside it. (Unfortunately, this means that many pups who fall from the walls and ceiling perish from drowning.) An edificarian species, it is also found in urban and suburban environments that include woodlands, buildings, storm sewers, boat houses, bridges, and culverts.

Southeastern myotis.

An insectivore, this common vespertilionid drops from its perch after sunset and heads for water, where, using echolocation and the trawling foraging method, it swoops down near the surface in search of its prey: primarily moths, mosquitos, and beetles—making it our benefactor. Forms cave colonies of up to 100,000 bats packed so tightly together that there may be as many as 150 individuals per sq. ft. Will share its roost with other bat species.

The female gives birth to 1-3 altricial pups in the spring, each one weighing only 1 gm. A secondary consumer, its primary natural predators are large carnivorous insects, birds of prey, snakes, and small carnivorous mammals, such as opossums. The southeastern myotis may live as long as 10 yrs. in the wild; over 20 yrs. in captivity.

SOUTHEASTERN POCKET GOPHER (*Geomys pinetis*): It grows to a total length of about 13" and can attain a weight of up to nearly 14 oz. Its cryptic pelage coloration correlates with the color of the soil it lives in, and so runs from blonde and cinnamon to gray and dark brown; the ventral area is generally tannish or grayish; the short 3½" tail is sparsely furred. The body is cylindrical and muscular; the upper incisors are grooved; the eyes and ears are small; the front paws are large and heavily clawed—all adaptations for its fossorial lifestyle. Gender dimorphism is present: the male is larger than the female.

This sedentary solitary geomyid is a nearctic native that inhabits temperate open forests, meadows, grassland, pastures, farmland, roadsides, scrub woodland, cropland, hedgerows, ditches, and well-drained sandy, hilly areas generally—both in wilderness locations and in urban settings. Strictly a mammal of the southeastern U.S., its range extends over portions of the states of Alabama, Georgia, and Florida. Individuals dig complex burrow systems up to 200' long, where they spend most of their time; ejecta mounds are left around portals; highly territorial, it violently guards its home range against conspecifics. Does not hibernate; active all year.

A crepuscular nocturnal herbivore, it feeds almost exclusively on subterranean plant matter (tubers, roots, bulbs), but will also consume grasses, stalks, rhizomes, forbs, bark, twigs, sedges, succulents, and leaves; caches surplus food in its burrow larders, carrying it in its cheek "pockets," the source of its common name. Its breeding season runs year-

Rolling farmland: southeastern pocket gopher habitat.

round, with maximum reproductive activity occurring in early spring and late summer. The polyestrous female may produce 2 litters annually; she gestates for around 1½ mos., gives birth to 1 or 2 altricial pups, and weans them approximately 4 wks. later.

It communicates and perceives via standard channels inherent to its genus: hearing, touch, smell, and to a lesser degree, sight. Its natural enemies include birds of prey, mustelids, and wild canids. A commensal rodent, its massive tunnel systems provide homes for countless species, from insects to other mammals. It is also beneficial as a soil aerator and recycler, as a parasite host, and as food for carnivores; but it is detrimental as a lawn, park, golf course, and agricultural pest. The average life span of the southeastern pocket gopher is probably 1-2 yrs. in the wild.

SOUTHEASTERN SHREW (*Sorex longirostris*): Also known as "Bachman's shrew" (after its discovery by the American naturalist John Bachman in the early 19th Century), this minuscule mammal grows to a total length of around 4", its tail is about 1½" long, its hind foot measures ½", and it weighs up to ⅛ oz. Its pelage on the dorsum is a rusty-brown color; on the ventrum it is fulvous. Its body is stout and cylindrical; its head is cone-shaped; its small external ears are visible; its eyes are tiny; the muzzle is tapered, whiskered, and convex; the pinkish nose has a flexible tip; its tail is long; the limbs and feet are tiny; the tips of its teeth are brownish.

A secondary consumer, this nearctic riparian inhabits palustrine biomes and ecosystems, including moist fields, brushland, mixed forest, shrubland, grassland, woodlands, riversides, bogs, chaparral, fen, and swamp and bog edges. Prefers herbaceous wetland with soft soil, dense detritus, snag, mossy logs, deadfall, blowdown, and rotting trees. It ranges across the southeastern U.S. and through the lower Midwest, from Florida west to Louisiana, from Missouri east to Maryland.

A fossorial omnivore, its diet includes vegetable matter, gastropods, coleopterans, lepidopterans (larval stage), orthopterans, arachnids, and chilopods. Almost nothing is known about its mating system. The polyestrous female probably gestates for about 3 wks. and bears up to 3 litters a year, with an average of about 3-4 young per litter. Mostly nocturnal, its main natural enemy is the owl. The southeastern shrew may live 12-14 mos. in the wild.

SOUTHERN BOG LEMMING (*Synaptomys cooperi*): This small rodent grows to a total length of about 6", its hind foot measures around ¾", and it can attain a weight of up to 1¾ oz. Its dorsal pelage is quite variable, running from tan and buff to brownish-gray and grayish-brown; its ventral fur is silvery-gray; the short ¾" tail is bicolored (dark above, lighter below).

Lemming.

This terricolous sedentary cricetid is a riparian nearctic native that inhabits temperate grassland, meadows, fen, mixed woodlands, shrubland, fields, clearings, savanna, chaparral, and forests; it depends on locations with good hiding cover, preferably with downed trees, deadfall, snag, and friable soil. Despite its common name, it is not always found around bogs, marshes, and freshwater wetlands. A midwestern to northeastern species, its range extends over mid and eastern Canada, and in the U.S. from South Dakota in the west to New Jersey in the east, from Michigan in the north to Georgia in the south. Constructs surface runways under ground duff and shrubbery, as well as an elaborate shallow underground tunnel system with numerous multipurpose chambers. Social and semicolonial, it sometimes lives in conspecific groups, often sharing its 20,000 sq. ft. home range with other rodent species. Does not hibernate; active all year.

A crepuscular and nocturnal omnivore, its diet consists of standard *Synaptomys* fare: seeds, grass, clover, moss, fungi, fruit, tubers, roots, algae, sedges, bark, leaves, stems, wood, insects and mollusks. Produces piles of 2"-3" grass cuttings, which it leaves strewn about its small home range. Its breeding season runs year-round. The polyestrous female, who can bear up to 3 litters annually, gestates for 3-4 wks., and gives birth to an average of 4 altricial pups in a concealed, bolus-like, subterranean or surface nest, weaning them 3 wks. later.

Communication and perception take place via normal Cricetidae channels: sight, smell, hearing, and touch. Its natural enemies are reptiles, birds of prey, mustelids, wild felids, and wild canids. Like some other lemmings species its population may vacillate wildly from year to year. Beneficial as a soil aerator and recycler, as a parasite host, and as food for predators. The southern bog lemming probably lives 6-12 mos. in the wild.

SOUTHERN FLYING SQUIRREL (*Glaucomys volans*): The smallest of our region's tree squirrels, it grows to a total length of 10", its pes measures about 1½", and it can attain a weight of up to 3 oz. Its coat is thick, soft, glossy, and velvety. Strongly countershaded, on the dorsum it is a gray-brown or fulvous; on the ventrum it is a creamy white. Its body is small, slender and limber; its head is small and convex from top to snout; the eyes are large, dark, and ringed in black fur; the ears are large and erect; the limbs are long; the feet are clawed; the brushy 4" tail is grayish above, whitish below.

A black-edged patagium (a loose membrane of furred skin on each side of the body) is attached to the wrists of the manus and to the ankles of the pedes, which, when stretched out, enable it to leap into the air and glide (not "fly") from tree trunk to tree trunk—hence its common name. While in the air its flat tail serves as both a rudder and a stabilizer, helping it navigate its flight, which can span some 250' from an elevated takeoff point to a low landing point. As it approaches its target, it moves its body into an upright position, the now forward-facing patagium and downward hanging tail helping slow its descent before

Southern flying squirrel.

gently alighting on all 4 feet. In gaining its remarkable flight abilities it sacrificed terrestrial dexterity, and is thus a somewhat awkward walker. The southern flying squirrel should not be confused with its close relative the northern flying squirrel, which is generally larger, heavier, and darker.

Nocturnal, arboreal, and gregarious, this tiny riparian rodent inhabits temperate montane woodlands with a preference for hardwood forests. These should contain such deciduous trees as maple, beech, poplar, hickory, and oak, preferably accompanied by abundant snag, hollow trees, deadfall, and a readily accessible water supply. This squirrel is found less commonly in coniferous forests. It usually constructs its own leafy drey in the craw of a tree; however, as a secondary cavity user it is quite happy to take over the abandoned nests of primary cavity excavators, such as woodpeckers.

Its range is massive, extending over some 1,000,000 sq. mi., from Canada in the north to Central America in the south. This includes the entire eastern half of the U.S., from Minnesota and Texas in the western portion to New England and Florida in the eastern portion. This species is communal, particularly during cold weather, when it will share nests with conspecifics. Does not hibernate, and is active year-round; though it will enter a brief state of torpor under extremely cold conditions.

This silky-coated sciurid is an omnivore with herbivorous and especially strong carnivorous habits that also make it a lignivore, granivore, nucivore, frugivore, florivore, fungivore, invertivore, insectivore, ovivore, avivore, and carcassivore. Scavenges and hoards surplus food. Along with the northern flying squirrel, it shares the strongest predisposition toward meat-eating of any of the tree squirrels.

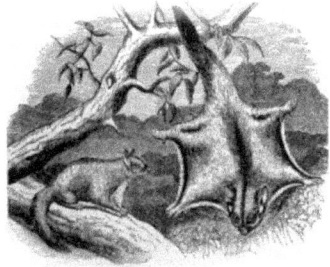

Flying squirrel in flight.

As with many other nocturnal mammals, the mating habits of this species have not yet been studied in detail. Its polygynandrous breeding season occurs twice a year, typically in winter and summer, depending on latitude. The doe gestates for nearly 6 wks., bearing 2-4 altricial pups in a leaf-lined, twig and bark tree nest. Weaning takes place at around 2 mos. The buck does not display paternal behaviors and thus plays no role in the parenting of his offspring.

Communication and perception run along typical Sciuridae channels: sight, smell, touch, and hearing. Vocalizations include bird-like chirps, trills, chits, and squeaks. The latter is a type of high-pitched sound that may actually be a form of echolocation, helping it maneuver and hunt in the dark. It is beneficial as a seed disperser, a consumer of pestilent insects, a parasite host, and as a prey for larger carnivores. However, it is the only known animal capable of carrying the bacteria that causes typhus fever, hence contact with it (and its nests) should be avoided. It and the northern flying squirrel are the only nocturnal tree squirrels in our region. Its natural enemies are reptiles, birds of prey, mustelids, procyonids, and wild felids. The southern flying squirrel lives an average of 4 yrs. in the wild; at least twice that long in captivity.

SOUTHERN GRASSHOPPER MOUSE (*Onychomys torridus*): Also known as the "scorpion mouse," this stout little rodent grows to a total length of around 6", its hind foot measures ¾", and it can attain a weight of up to nearly 1 oz. Its dorsal pelage is reddish-buff to goldish-gray; the ventral pelage is white to pinkish-brownish-orange; the thick 2½" tail is white-tipped; the eyes and ears are large.

This motile cricetid is a nearctic native that inhabits temperate, low-lying, arid shrubland, grassland, desert, rangeland, sand dunes, valleys, scrubland, chaparral, and savanna, preferably with widely dispersed yucca shrub, huisache trees, creosote bush, mesquite trees, and cacti (particularly cholla). Strictly a species of the American West

A grasshopper mouse.

and Southwest, its range extends over portions of the states of Nevada, California, Utah, Arizona, New Mexico, and Texas, as well as south into northern Mexico. Semi-fossorial, it excavates its own burrows, or takes over those of other animals. Though its home range—covering approximately 350,000 sq. ft.—is quite large for such a small rodent, massive space is required to sustain its largely carnivorous lifestyle. Solitary and extremely territorial, it actively defends it turf against conspecifics. Does not hibernate; active year-round.

A nocturnal omnivore, its diet consists primarily of insects (such as ants, beetles, and grasshoppers—the latter creature from which it derives its common name), arthropods (such as spiders and scorpions), and small mammals (such as voles and other mice). In addition, it consumes seeds, fruits, grains, vegetables, nuts, and green matter generally, and, being a cannibal, it also devours its own kind (if food is scarce, it will even eat its own male or female mate). Though tiny, this pugnacious mouse is an adept and savage wolf-like little hunter that carefully stalks its mammalian prey, then quickly dispatches it with a vicious bite to the head or neck. Its breeding season runs throughout the year, with the usual peaks in spring and summer. Males fight over estrus females, who can produce as many as 5-6 litters annually. The iteroparous polyestrous female gestates for about 1 mo. and gives birth to an average of 3 altricial pups in a subterranean natal chamber.

An endothermic onychomid, it communicates and perceives along channels typical of its genus: sight, touch, smell, and hearing. When alarmed it rears up on its back legs and gives a tiny, yet long, loud wolf-like howl, a piercing high-toned call for which it is well-known. Its natural enemies are probably reptiles, birds of prey, wild felids, and wild canids. Beneficial as a consumer of insects and other pests, and as food for carnivores; detrimental as an agricultural pest. The southern grasshopper mouse lives an average of 2-3 yrs. in the wild.

SOUTHERN PLAINS WOODRAT (*Neotoma micropus*): This medium-sized rodent grows to a total length of 15", its hind foot measures nearly 2", and it can attain a weight of up to 11 oz. The coloration on its upper body is pale to bright gunmetal gray, or occasionally tannish-sienna with black hair tips; the lower body is light gray; the tail is bicolored, gray above, whitish below. Throat, breast, and feet are white; the pelage is soft and thick. Its body is compact and ovoid; the head is small with a convex profile; the eyes are small, beady, and protruding; its ears are rather large, pinkish, and erect; the muzzle is robust; the nose is pinched, snubbed, and covered in sensitive vibrissae; the limbs are short; the feet are clawed; the 7" tail is short, thick, and lightly furred. Gender dimorphism is present: the male is slightly larger than the female.

This endothermic cricetid is a nearctic native that inhabits semiarid grassland, scree, level desert, plains, bottomland, thorny thickets, dry shrubland, rock talus, dunes, rocky lowlands, mesquite and creosote scrubland, chaparral, savanna, cliffs, valleys, and brushy hillsides; has a preference for landscapes that contain one or more of the following: cactus plants, hollow trees, rocky outcroppings, snag, ground detritus, hiding cover, deadfall, blowdown, and rotting logs. Its range is limited to the American Midwest and Southwest, from Kansas, Oklahoma, and Colorado in the north to Texas and New Mexico in the south. It also occurs in parts of northern Mexico.

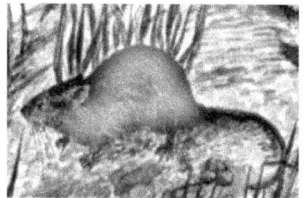

Southern plains woodrat.

Partly fossorial, its home, which it constructs above or below ground (depending on soil density and availability of building material), is made of branches, sticks, cactus thorns, leaves, dung, grass, and sometimes human refuse. Usually built near or beneath cacti, the structure contains numerous chambers and escape holes, as well as well-worn pathways that fan out from its many entrances. Succeeding generations often use the same house, adding on to it over a span of many thousands of years, with the result that some can reach a height of 6' or more. Establishes a home territory of about 1,000 sq. ft., which it scent-marks and defends aggressively against conspecifics.

This nocturnal terricolous herbivore consumes mainly cactus-related items, such as cactus seeds, pulp, fruit, and leaves. This is supplemented with a diet of roots, beans, grains, tubers, pods, nuts, and woody plants and material. Hoards and caches food in a storage compartment in its den. It is able to derive 100 percent of its water requirements from the succulent foods it eats.

Its polygynandrous breeding season occurs in the spring, with the bucks and does engaging in elaborate courtship rituals. The female gestates for around 30 days or so, giving birth to as many as 4 altricial pups, which she weans at 1 mo. of age—the start of their life of independence. It seems likely that the male does not assist in the rearing of his offspring.

This solitary mammal is a primary consumer that communicates and perceives via regular Cricetidae channels: sight, smell, hearing, and touch. Vocalizations include squeaking, squealing, and chattering. It is known to engage in hind foot-drumming. Benefits the natural environment by aerating and recycling the soil, hosting parasites, dispersing seeds, creating homes for other animals, and by providing food for carnivores. Harmful to humans (and other animals) by acting as a carrier of various deadly diseases. Its natural enemies are reptiles, birds of prey, mustelids, procyonids, wild felids, and wild canids. The average life span of the southern plains woodrat in the wild is probably about 6-8 mos.; twice that long in captivity.

SOUTHERN POCKET GOPHER (*Thomomys umbrinus*): While some view this rodent as conspecific with Botta's pocket gopher and Townsend's pocket gopher, others maintain that all 3 are separate species. Thus, though these 3 rodents share similar ecologies and life histories, their taxonomy remains unsettled.

Southern red-backed vole.

SOUTHERN RED-BACKED VOLE (*Myodes gapperi*): Also known as "Gapper's red-backed vole," it grows to a total length of about 6½", its hind foot measures approximately ¾", and it can attain a weight of up to 1¾ oz. Coloration on the upper body is bright reddish-cinnamon (giving it its common name), which is interspersed with tan, buff, and gray hairs, as well as black-tipped hairs; lateral coloration tends toward yellowish-gray; the underbody is buffy-white; the short 2" tail is thin and bicolored. The body is ovoid; the eyes are small; the ears are large, rounded, and erect; the feet are small, pinkish, and delicate.

This sedentary terricolous cricetid is a muskeg-loving nearctic native that inhabits mesic, boreal, mixed forests, swamps, fen, stone piles, taiga, sedgeland, damp mossy tundra, peatland, and bogs, with plenty of hiding cover, fallen trees, stumps, rocks, clumps of vegetation, deadfall, and exposed tree roots. Its range occupies a large swath of North America, spanning the entire southern to middle tier of Canada, as well as Alaska, New England, the upper Midwest, Appalachia, the Northwest, and the Rocky Mountain states. Dens and nests under brush, rocks, moss, and decaying wood; uses both natural surface and subterranean runways to travel about its home range; will appropriate the underground burrow systems of other animals when available. Solitary and territorial, it actively defends its turf against intruders. Does not hibernate; active year-round.

A diurnal, crepuscular, and nocturnal omnivore, its diet consists of typical *Myodes* fare: seeds, grains, nuts, bulbs, shoots, fruits, stalks, flowers, lichens, roots, leaves, tubers, bark, wood, fungi, and insects. Leaves grass cuttings around its home; hoards and caches surplus food. Its breeding season runs from winter to fall. The iteroparous polyestrous female, who may produce several litters annually, gestates for about 3 wks., gives birth to an average of 5 altricial pups in a concealed, grass-lined, globular nest, and weans them 3 wks. later.

Communication and perception fall along normal vole channels: touch, smell, hearing, and sight. Its natural enemies are reptiles, birds of prey, mustelids, wild felids, wild canids, and ursids. Arboreal and saltatorial, it often escapes predators by scampering up rocks and vegetation. Beneficial as a seed and spore disperser and as a prey base for carnivores. The southern red-backed vole lives for about 1 yr. in the wild; slightly longer under optimum conditions.

SOUTHERN SHORT-TAILED SHREW (*Blarina carolinensis*): The smallest of the short-tailed shrews, it grows to a total length of about 4½", its tail is 1⅛" long, its hind foot measures ¾", and it weighs up to ¾ oz. Its upper body is gray, its lower body is light gray. The body is robust; the skull is cone-shaped; the eyes are beady; the neck is thick; the external ears are just visible; the snout is convex, pointed, and covered in vibrissae; the nose is pinkish and flexible at the tip; the limbs are short; the feet are five-toed with sharp claws designed for digging. Its tail is somewhat stubby for a shrew, hence its common name.

This tiny mammal is attracted to swampy areas and moist woodlands characterized by soft friable soil and a generous amount of leaf litter. It is also found, however, in dry farmland, open meadows, and disturbed habitats. As its moniker indicates, it is found mainly in the southeastern U.S., though its range can extend into the central Midwestern states.

Shrew.

An omnivorous soricid, its diet is comprised mainly of chilopods, gastropods, arachnids, annelids, amphibians, and small mammals (mice); but it also consumes some vegetable matter (fungi, fruit). Its venomous saliva enters the wounds of its prey via biting, paralyzing then killing it. Caches surplus food. It is nocturnal and somewhat social, living most of its life hidden away in subterranean darkness, occasionally sharing its burrows with conspecifics.

Its breeding season takes place between spring and fall. The female gestates an average of 25 days, giving birth to about 4 altricial young in a well-hidden nest lined with leaves and grass. Over the next few weeks she nurses, grooms, cares for, and protects her infants, weaning them at 2 or 3 wks. of age. The male does not assist in the rearing of his offspring.

Communicates with conspecifics mainly through touch and smell, and to a lesser degree sight and hearing. A secondary consumer, its natural enemies are reptiles, birds of prey, mustelids, mephitids, and wild canids. The southern short-tailed shrew probably lives 10-14 mos. in the wild; up to 24 mos. in captivity.

SOUTHERN YELLOW BAT (*Lasiurus ega*): This large solitary chiropterid grows to a total length of 5", its forearm measures nearly 2", and it weighs ¾ oz. Its soft silky fur is a golden-buff to tawny-yellow, lending it its common name. There is sometimes a blackish cast to the pelage. It has large ears; a broad curved tragus; a keeled calcar; and a furred tail membrane. It has 30 hard sharp teeth, which it uses to crush and kill its prey.

It shelters mainly in tree holes and forest foliage (in particular it enjoys dead palm fronds still attached to their trees), but it can also be found in human-made structures. It ranges over the American Southwest, from California and Arizona to New Mexico and Texas, south into Mexico and beyond.

The female, who has 4 mammary glands, practices delayed fertilization. She gestates for about 85 days, giving birth to 1-4 pups in the late spring or early summer in a maternity colony that may contain tens of thousands of individuals. This secondary consumer is a nocturnal insectivore that echolocates to find its food, using both hawking and gleaning techniques. Feeding requirements unknown. Predators unknown. Life span unknown. Status unknown.

SOUTHWESTERN MYOTIS (*Myotis auriculus*): Also known as the "southwestern bat," it is 3½" in total length, its tail and its forearms each measure nearly 2", it has a wingspan of 11", and it weighs ¼ oz. Its calcar is un-keeled; it has a slender pointed tragus; long ears; and a large skull with a sagittal crest. The fur on its dorsum is dull cinnamon brown to grizzled brown; the ventrum is grizzled gray or buff. It lacks gender dimorphism.

Southwestern myotis.

This nearctic petraphilic vespertilionid inhabits a myriad of biomes and environments: rocky cliffs, mines, caves, and buildings in arid desert, scrubland, chaparral, oak forests, and mesquite regions of the southwestern U.S. (mainly Arizona and New Mexico), south into Mexico. In particular it is attracted to ponderosa pine forests near water. Is also found in urban areas.

This secondary consumer uses echolocation to find its prey. A nocturnal insectivore, it feeds on large quantities of flies and other protein-rich insects, making it our ally. The female gives birth to a single ½" altricial pup in the summer, grooming and protecting it, and supplying it with vitamin-rich milk until it is mature enough to fly and feed on its own. Probably hibernates and migrates. The southwestern myotis may live for around 3 yrs. in the wild. Due to lack of study little else has been learned about this chiropterid species.

SPERM WHALE (*Physeter macrocephalus*): The official state animal of Connecticut, it is the largest of the toothed whales, growing to a total length of nearly 70' and weighing up to 100,000 lb. (50 tons) or more. Its heart alone weighs nearly 300 lb. The sperm whale's coloring ranges from light gray to black. Unique among whales, it has a massive squarish head that makes up one-third of its entire body length, and which contains the largest eyes of any toothed whale and the largest brain of any animal in the world. It has small front flippers (used for steering) and an enormous fluked tail (used for propulsion), which moves up and down instead of back and forth (as in most fish). Its dorsal fin is not plainly visible, but seems to have been subsumed into the body, appearing as a small rounded hump on the back.

Sperm whale.

Given the species name *catodon* by some systematists, its single blowhole is part of a highly modified nasal complex that began (like the human nose) with two identical air-breathing nostrils, but which through time evolved into two different functioning halves: one side of the blowhole (the left) remains devoted to breathing, the other (the right) to sound-production (this, the right "nostril," even contains an anatomical structure that acts like human vocal cords). No doubt these are adaptations to echolocation. Thus the sperm whale's single air-breathing blowhole is located on the left of center, while the less prominent sound-producing half is on the right. This "offset" design has produced a slight asymmetry in the skull of the sperm whale—as well as in other toothed whales. (Note that unlike toothed whales, baleen whales, which use sounds for communication rather than for feeding, do not possess this apparatus, and so have two fully functioning blowholes.)

Sperm whale.

The sperm whale displays the most extreme gender dimorphism of any cetacid, with the male being substantially larger than the female. Migratory and gregarious, it ranges all the oceans, preferring sub-Arctic and sub-Antarctic waters. It takes its common name from a barrel-shaped structure in its head called the "spermaceti organ," which produces a semitransparent waxy oil. The precise function of this substance is unknown, but it may aid the whale in echolocation, navigation, and buoyancy. Early Americans used spermaceti (meaning "whale seed") for a variety of purposes, including lamp oil, medicines, candles, water-proofing, and machine lubricants.

This shy carnivorous physeterid feeds on squid, octopus, rays, and fish (including sharks), consuming nearly 2,000 lb. (one ton) of food a day. The "clicks" that it makes while echolocating are the loudest sounds ever recorded in the animal kingdom. Probably used to stun, paralyze, or even kill its prey, they can reach a level of over 200 decibels (a thunderclap is about 120 decibels). They are also used to communicate underwater in a highly sophisticated language that we have yet to decode. Its clicks are so loud that they can be heard by other sperm whales thousands of miles away, even on the other side of the world.

Pods of sperm whales often hunt together in pack-like fashion to corral small fish. The deepest diving mammal on earth, it will descend nearly two miles into the dark depths and stay down for as long as 1½ hrs. in search of its favorite food: the 45' giant squid—with which it wages violent deep-sea

Lower jaw of the sperm whale.

struggles (sperm whales often carry large dinner-plate size scars on their bodies from the giant squid's suckers). This whale can swim nearly 30 mph, and maintain this speed for nearly 60 min.

Its small, long, slender lower jaw is filled with several dozen peg-like teeth, which fit neatly into hollow sockets in the toothless upper jaw. Since most of its food is swallowed whole, the function of these teeth is not fully understood, for they do not appear to be used for grasping, hunting, or feeding. (They may now be a useless evolutionary vestige of its prehistoric origins as a carnivorous land mammal.)

The breeding season of this polygamous cetacean takes place mainly in spring, at which time males engage in violent competition over females. The cow, who only reproduces once every 4 or 5 yrs., gestates for about 15 mos., giving birth to a single precocial calf that is weaned at around 2 yrs. of age. Females form protective pods with one another and their young, often choosing to remain in warm waters. Sometimes the usually solitary males also form pods, venturing into colder waters to the north and south.

Sperm whale.

Also known as the "cachalot," this secondary consumer's main predators are orcas, false killer whales, and pilot whales. (Sharks sometimes attack calves.) Unlike a number of other whale species, however, it fights back against it attackers. It periodically excretes ambergris, a rare, waxy intestinal substance that is used in the making of perfumes and medicinal drugs. The focal character in Herman Melville's Victorian novel *Moby Dick*, the sperm whale lives to about 70 yrs. in the wild.

SPINY POCKET MOUSE (*Chaetodipus spinatus*): It grows to a total length of nearly 9" and can attain a weight of up to ¾ oz. The longish dorsal fur is a dull tannish-yellow with black-tipped hairs; the ventral fur is white to tan; the long 5" tail is dichromatic (dark on top, lighter on bottom) and crested at the terminal end. This rodent lacks an obvious lateral line; possesses 2 small whitish subauricular patches; spines cover the rump area and grow sporadically up to the neck.

Southwestern mesa: spiny pocket mouse habitat.

This coarse-haired heteromyid is a nearctic native that inhabits arid hot deserts and talus slopes at low elevations; also found on rocky hillsides, boulder fields, rugged foot hills, river washes, scree, and mesas, with scattered scrub and overall dry conditions; has a preference for mesquite- and tamarisk-rich locations with gravelly soil. Its range is limited to the American Southwest, covering portions of the states of Nevada and California, as well as Mexico's Baja Peninsula. Semi-fossorial, it spends much of the daytime in its underground burrow system, coming out to forage at night. Probably does not hibernate, but may enter brief periods of torpidity during extreme conditions.

A nocturnal herbivore, its diet consists mainly of seeds, grasses, and plant stalks. Caches surplus food in its burrow larders, transporting it in the cheek pockets from which it derives its common name. Almost nothing is known about its reproductive activities, but they are likely to be similar to other pocket mice in its range. Communication and perception channels are typical of its genus: sight, hearing, smell, and touch. Its natural enemies are similar to those of other members of the Heteromyidae family: reptiles, birds of prey, mustelids, wild felids, and wild canids. The life span of the spiny pocket mouse is unknown, but it probably lives 1-2 yrs. in the wild.

SPOTTED BAT (*Euderma maculatum*): This rare flying chiropterid grows to 4½" in total length and its forearm measures 2". Its coloration is striking: the fur on the upper body is black with two large white spots on the shoulders and a large white patch on the hindquarter, giving the appearance of a skull. From this it gets its alternate common name: the "death's head bat." Even more distinctive are its two massive ears, which, at 2" in height (almost half the length of its body) look like a second pair of miniature wings. Normally, these are held erect, particularly during hunting (when they are directed forward). When at rest, however, it curls them backward. Along with its remarkable body coloration, the face of this secondary consumer is black while the wings and translucent ears are buff or pinkish.

Spotted bat.

A riparian, paludal, petraphilic vespertilionid of the North American West, it roosts in rocky cliffs, cracks, and crevices. It inhabits a variety of biomes and environments, including arid deserts, caves, marshy areas, waterholes, canyons, forests, flat grassland, scree, buildings, mountainous terrain, aspen woods, rock talus, chaparral, and watery corridors bounded by piñon-juniper woodland. Its range covers Canada south to Mexico—with some density in the American Southwest.

Little is definitively known about this rapid flying microchiropteran. Possibly solitary, it is a nocturnal insectivore, feeding primarily on moths, making it our friend. The female gives birth to a single pup in late spring or early summer. Communicates using a variety of sounds, from clicks, ticks, and chirps, to squeaks, hisses, and teeth-clicking. The spotted bat probably lives from 2-6 yrs. in the wild.

SPOTTED GROUND SQUIRREL (*Xerospermophilus spilosoma*): This diminutive rodent—in our region the only ground squirrel with spots—grows to a total length of 10", its pes or hind foot measures 1½", and it can attain a weight of up to 4½ oz. On the dorsum the color of the pelage is smoky gray to cinnamon-brown; on the ventrum it is cream white. The short 3½" tail is round, thinly furred, dichromatic (grayish brown on top, tawny on the bottom), with a black tip. Its most conspicuous identifier is what gives it its common name: a scattered pattern of light white squarish spots over its back. Found only on the upper body, they run from the shoulders back to the rump. (Coloration and markings can vary with geography and the individual.) The body is slender and cylindrical; the head is small and rounded; the eyes are large and dark; the ears are small and erect; the nose is snubbed; the limbs are shortish; the feet are clawed. It molts twice annually: once in spring and again in the fall. It is sometimes confused with the Mexican ground squirrel and the thirteen-lined ground squirrel, both, however, which possess rows of dorsal spots rather than randomly distributed ones (as in the spotted ground squirrel).

Spotted ground squirrel.

The drab coloration of this shy and agile nearctic sciurid provides ideal camouflage for the arid and semiarid biomes and ecosystems it inhabits: temperate grassland, desert, savanna, sandhills, shrubland, dunescapes, mesas, pine woodlands, plains, chaparral, rangeland, parkland, and scrubland with dry, sandy, loamy, or gravelly soil (for ease of digging). A high desert mammal, its range covers various parts of the American West and Southwest, extending from South Dakota in the north to central Mexico in the south; from Colorado in the west to Oklahoma in the east, as well as sections of all the states in between (Texas, Kansas, Utah, Wyoming, Nebraska, Arizona, New Mexico). Its subterranean home is nearly 20" underground. The entrance to its burrow is often camouflaged beneath shrubbery or rocky outcroppings. It estivates on hot days; during inclement weather it remains in its den resting, feeding, and grooming itself.

A terricolous omnivore, its diet is comprised mainly of herbivorous foods, such as seeds, roots, grass shoots, stems, tubers, plants, flowers, cacti, wood, grains, and nuts. The carnivorous side of its diet includes insects, reptiles, and fellow members of the Rodentia order.

Both diurnal and crepuscular, it enters into hibernation in late summer, which lasts until spring, with the bucks emerging first, followed by the does, who gestate for 3-4 wks. and bear 5-10 pups in a natal nest chamber underground. (The females of southern populations may produce 2 litters a year.) Weaning takes place by 3 wks. of age.

This fossorial species communicates and perceives through normal Sciuridae channels: sight, hearing, smell, and touch. It is well-known for its hind foot-stomping alarm display, meant to both scare off burrow intruders and alert conspecifics of danger. Engages in sunbathing and sand-bathing. Benefits the ecosystem by eating pests, aerating and recycling the soil, dispersing seeds, controlling the growth of vegetation, and providing food for carnivores. Its main natural enemies are reptiles and birds of prey. The spotted ground squirrel probably lives from 2-4 yrs. in the wild.

STAR-NOSED MOLE (*Condylura cristata*): It grows to a total length of 8", its tail is about 3" long, its hind foot measures around 1", and it can attain a weight of nearly 3 oz. Its dorsal pelage is dark brown to black, its ventral pelage is light brown to light gray. It has the typically thick cylindrical body of a mole, with short coarse fur, tiny eyes, a scaly tail, and massive five-toed front paws with enormous curved claws adapted for excavation. Its most distinctive feature, however, is the physical trait from which its common name derives: a naked star-shaped nose comprised of 22 pink, moveable, highly tactile "tentacles." Eleven of these organs surround the nostril on the left side of the nose tip, while the other 11 surround the nostril on the right side.

The only surviving member of its genus (*Condylura*), it is a riparian species that inhabits damp fields, wet forests, peatland, swamps, lakes, marshy areas, lawns, ponds, streams, and wetlands generally. Its range covers the central northeastern territory of our region, extending from North Dakota to Quebec, Canada, into parts of the Midwest states (Indiana, Ohio), south to coastal

Star-nosed mole.

Virginia and Georgia. A social animal, it may establish a home range of up to 50 sq. ft., space which it sometimes colonizes and shares with conspecifics. Its 1½" burrow entrances are enveloped by ejecta mounds.

This fossorial talpid is a strict carnivore with a hardy vermivorous appetite. Its diet primarily surrounds earthworms, aquatic worms, and leeches, but it will also consume insects (mainly various fly species), beetles, mollusks, crustaceans, and fish when available. Natatorial, it is an excellent swimmer and diver, and is equally adept at capturing its food on land, in water, and even under ice. Uses it sensitive nose feelers to locate prey, then folds the appendages out of the way while feeding. Its tunnel systems, which leave characteristic "mole-hills" at the ground surface, can be quite complex and include a nesting chamber lined with shredded vegetation.

Diurnal, crepuscular, and nocturnal, this active eulipotyphlid's breeding season runs through the spring. The female gestates for around 1½ months, giving birth to 2-6 altricial young, each with a tiny nose-star. The nest is constructed in a hummock or other well-drained mounds of earth. Weaning takes place at around 1 mo. of age, after which the offspring become independent.

Closeup of star-nosed mole's snout.

It communicates and perceives via normal Talpidae channels: sight, hearing, and smell. However, its greatest sense is touch, for it possesses one of the world's most sensitive sensorial organs: its 22-tentacle nose-star can detect information at a microscopic level. A secondary consumer, its natural enemies are fish, amphibians, birds of prey, mephitids, mustelids, wild felids, and wild canids. Benefits the world ecosystem by eating numerous pests, by providing food for other animals, and by aerating and recycling the soil. The life span of the star-nosed mole is probably 1-2 yrs. in the wild.

STELLER SEA LION (*Eumetopias jubatus*): One of the world's largest pinnipeds, and the biggest of the eared or walking seals, it grows to a total length of nearly 11' and a weight of up to 2,500 lb. It is named after Georg W. Steller, a German naturalist who described it in the mid 1700s as the "lion of the sea" due to its lion-like roar and feline-like yellowish irises. Pelage coloration in the male is blondish-yellow to golden-brown on the upper body, reddish-brown on the lower body. The female is brown on the dorsum and the ventrum. Both bulls and cows are somewhat darker on their abdomens and chests. Unlike many other types of seals, its thick coarse coat does not darken when wet; however, like other members of its family it sheds its fur once a year.

The head is large and uniquely shaped with a low forehead; the face is otter-like; the nose is snubbed; the neck is thick; the body is robust; and the blackish flippers are long, with the hind pair capable of rotating forward, facilitating locomotion on land. An excellent

Steller sea lion.

climber, it will sometimes scramble up onto high cliffs and rock shelves. Heavy layers of blubber insulate it from the cold; its sensitive facial whiskers or vibrissae are able to pick up vibrations in the water, helping it hunt and navigate; the massive front flippers are used for propulsion, while the powerful hind flippers are used for steering. Gender dimorphic, the male—who sports a dense mantle of hair over its neck, shoulders, and chest—is much larger than the female.

Also known as the "northern sea lion," this bulky marine otariid is a native of the cold waters of the Pacific Ocean, where it is found along temperate, undisturbed, sometimes boulder-strewn coasts on rocky, sandy, or gravel beaches, from California and Alaska to Russia and Japan. Known to swim up rivers. Also associated with rocky reefs, it returns to familiar haul-out sites for resting and traditional rookeries for whelping. It is often solitary at sea but will sometimes form groups and "raft" at the surface.

Gregarious, highly vocal, and endangered, this saltwater carnivore is an opportunistic feeder with a piscivorous and molluscivorous diet that includes around 100 different species of fish, such as cod, sandlance, flounder, pollock, salmon, sole, sandfish, rockfish, blackfish, capelin, hake, greenling, and mackerel, as well as cephalopods (squid, octopus), arthropods (crabs), and gastropods (clams). Likes to feed within 20 mi. of shore. Also a petrivore, it swallows pebbles, and even stones up to 4" in diameter, for reasons which are not yet fully understood. A nocturnal hunter, while foraging it can dive to a depth of nearly 1,400' and remain submerged for as long as 2 min. Caches surplus food.

Steller sea lion (juvenile).

Its polygynous reproductive season takes place during the summer, at which time dominant males establish breeding territories, compete for estrus females, and gather together, mate with, and eventually oversee as many as several dozen cows. (Bachelors and non-estrus females form their own colonies away from the main breeding herds.) The female gestates for 12 mos. and employs delayed implantation, giving birth to a single, 3' long, 45 lb. pup on her island rookery the following summer. Within 2 wks. the mating process begins again. Pups are weaned at around 1 yr. of age but may stay with their mothers for as long as 3 yrs. Until they are fully grown, pups endure a harsh struggle for survival, facing death from storms, disease, cold, occasionally violent adults, predation, rejection, abandonment, and compression of the body (being crushed by adults). Dominant males protect their harems but otherwise do not directly assist in the raising of their young.

The only living member of its genus, it communicates and perceives via standard Otariidae channels: sight, touch, hearing, and smell. Its usual vocalizations are growling, bellowing, bleating, and low-frequency roaring, the latter for which it is commonly known. This species is slowly declining throughout most of its range, for like all marine mammals it is under constant threat from anthropogenic influences, such as human-induced climate change, disturbance (by watercraft, landcraft, and aircraft), reduction of prey by fisheries, net entanglement, oil spills, toxic industrial runoff, human predation, loss of habitat, vessel strikes, and habituation to humans, to name only a few. Its natural enemies are sharks and orcas. On average the Steller sea lion lives for about 25 yrs. in the wild.

STEPHENS' KANGAROO RAT (*Dipodomys stephensi*): This medium-sized, highly endangered heteromyid grows to a total length of about 13" and a weight of up to 2½ oz. Shares a nearly identical ecology and life history with its cousins the Dulzura kangaroo rat and the agile kangaroo rat. Outside of genetics (where there are several differences), the main dissimilarity concerns their ranges: Stephens' range is quite narrow, only occupying a small area in southwest California (mainly in and around San Diego County).

STEPHENS' WOODRAT (*Neotoma stephensi*): Very similar to, though smaller than, the bushy-tailed woodrat, this gray furry-tailed species shares much of the same range (in its case, Utah, Arizona, and New Mexico), ecology, and life history as its close cousin.

Bushy-tailed woodrat.

STRIPED DOLPHIN (*Stenella coeruleoalba*): This small gregarious delphinid, also known as the "blue-white dolphin," grows to a total length of 8' and weighs up to 330 lb. It has a sleek, hydrodynamic body, a tall dorsal fin set in the middle of the back, narrow pectoral flippers with black stripes, and a long beak containing 44-50 teeth. It is slate blue on the dorsum, light blue or whitish on the ventrum, with backward sweeping, flaring blue and white stripes, hence its common name.

   This pelagic carnivore inhabits both temperate and tropical waters around the world, including the Atlantic and Pacific Oceans. One of the more vocal of the cetacids, it communicates mainly via whistles and clicks. Highly active, this secondary consumer travels in large groups of 500-1,000 individuals, performing playful acrobatics (leaps, flips, twists, bow-riding) as it speeds through the water. An opportunistic feeder, its diet includes squid, octopus, aquatic crustaceans, and fish. The striped dolphin lives to an estimated 50 yrs. in the wild.

Striped dolphin.

STRIPED SKUNK (*Mephitis mephitis*): Grows to 32" in total length, has a 3" hind foot, and weighs up to 15 lb. With its two familiar broad white stripes running over a black body from head to rump, and a bushy 15" striped tail, this is the archetypal North American skunk. Its conspicuous markings serve as "warning flags" to predators. Though normally placid, if its warning is disregarded it raises its tail, stomps its forefeet, and sprays a sulfuric musk up to 15' (the fetid smell can travel over a mile) that burns the throat and nose and can cause temporary blindness. Its family name Mephitidae, meaning "harmful odor," is well earned.

   The head and ears are small; the face is tapered; the muzzle is slightly concave; the eyes are small and dark; the legs are short. It has a single vertical forehead stripe that runs from the nose, back up between the eyes, to the top of the head. Pelage coloration, markings, patterns, and even body size, can vary greatly depending on the individual and geography. This species is gender dimorphic: the male is 15 percent larger than the female.

   Inhabits a wide range of biomes and ecosystems: forest, grassland, seashore, chaparral, farmland, plains, savanna, woodland, rangeland, desert, fields, scrubland, and suburban and urban areas. As a riparian it prefers habitats with a reliable fresh water source, such as a stream. A native North American, its nearctic range extends from Canada to Mexico, and includes all of America's lower 48 states. Dens in hollow logs, rock piles, dense vegetation, or in any suitable crevice; a secondary cavity user, it will also den underground, sometimes in the abandoned burrows of other mammals. Will occasionally den communally with conspecifics. A plantigrade, its home range is roughly 1 sq. mi.

   This opportunistic omnivore is a dedicated insectivore, feeding on a host of invertebrates

Striped skunk.

such as worms, grubs, crickets, grasshoppers, and beetles. Its also eats seeds, grains, fruit, vegetables, nuts, crayfish, amphibians, reptiles, birds, eggs, small mammals, and carrion. Will consume pet food and human garbage as well. Appears to be immune to the venom of various creatures, such as bees, scorpions, and snakes, which make up part of its diet.

With its massive and powerful front claws, this solitary, nocturnal, crepuscular mephitid is an expert digger and rooter that can run 10 mph. It does not hibernate, but an individual living in a cold climate enters a state of torpor during extreme winter weather, which slows down its metabolism, preserving energy.

Its breeding season takes place from midwinter to midspring. The doe employs delayed implantation, assuring that the birth of her kits occurs during the warmer months when food is abundant. She gestates for around 2½ mos., giving birth to 2-8 altricial young. Weaning takes place at about 2 mos., with independence and dispersal occurring by 1 year of age. The polygynous buck does not contribute to the rearing of his offspring, making the female the sole nurturer, guardian, and teacher.

It communicates and perceives using customary Mephitidae channels: touch, hearing, smell, and sight. Though usually silent, it can produce a host of vocalizations, including snarling, whimpering, growling, teeth-clicking, screeching, cooing, screaming, squealing, twittering, hissing, and churring. A secondary consumer, its natural enemies are foxes, bears, coyotes, mountain lions, badgers, and especially great horned owls. (If sprayed, predators will remember the aposematic skunk's bold black and white markings and avoid it in future.) Beneficial as a consumer of large quantities of insects and other pests; detrimental as a carrier of deadly diseases. The striped skunk lives an average of 5 yrs. in the wild; it may live up to 10 yrs. in captivity.

Striped skunk.

SWAMP RABBIT (*Sylvilagus aquaticus*): Also known as the "cane-cutter rabbit," it is the largest member of the cottontail family, growing to a total length of 21" and reaching a weight of 6 lb. Its dorsal pelage is cinnamon-brown with white and black tips, giving its short dense fur an overall grizzled appearance. The ventral coat is tan to white. The ears are short for a cottontail; the eyes are encircled with rust-colored fur; the feet are cinnamon-colored; the thin tail is cottony-white on the underside—one of the reasons it is classed with the family of cottontails.

It shelters in logs, thickets, and the burrows of other animals. As its common and scientific names indicate, this leporid is a riparian, that, in its case, prefers swampland. For the swamp rabbit this type of biome extends to canebreaks, marshes, bottomland, flood plains, rivers, and watery areas generally. It ranges throughout the American Southeast and Midwest, from Indiana to Georgia, and west through Tennessee and onto Texas, Oklahoma, and Missouri.

Both crepuscular and nocturnal, this herbivorous coprophage is also a folivore, a granivore, and a lignivore that feeds mainly on sedges and grasses; it also consumes woody plants and tree parts (bark, leaves, branches, stems, twigs, seeds).

Depending on the region, its breeding season generally starts in midwinter and ends in midsummer. The doe gestates for about 35 days, giving birth to 1-6 slightly altricial kittens in a shallow grassy nest hidden underground or inside a log. Polyestrous, she may produce as many as 6 litters a year. The neonates are weaned and become

Swamp rabbit.

independent after 2 wks. or so. The buck does not contribute to infant care.

The natural enemies of this primary consumer are birds of prey (owls, hawks), reptiles (snakes, alligators), wild felids (cougars), and wild canids (foxes, coyotes). When evading predators it employs the usual cottontail defense strategies: freezing and running in a zigzag pattern. It has 2 others that most rabbits do not, however: the ability to swim, as well as submerge itself just under the water's surface with only its nose showing. The swamp rabbit probably lives for 1-2 yrs. in the wild; perhaps up to 8 yrs. in captivity.

SWIFT FOX (*Vulpes velox*): The smallest canid in our region, it stands 1' tall at the shoulder, is 30" in total length, and weighs up to 6 lb. Employing countershading, its upper coat is buff-colored, with varying tones of grizzled gray, brown, cinnamon, and yellow, while its inner ears, throat, belly, and inside legs are cream to white. Its outer legs are cinnamon; its bushy 11" tail is black-tipped; the face is often grizzled; the muzzle may have a black swash on each side. Pelage color, patterns, and markings vary across its range, but all serve as worthy camouflage for the dry ecosystems it inhabits. Its large ears dissipate heat, important for regulating body temperature in arid regions. This species is gender dimorphic, with the male being slightly larger than the female.

Fox family.

It inhabits America's western grasslands, mesas, pastures, and prairies, from southern Canada south into the Great Plains states (largely east of the Rocky Mountains). Prefers areas rich in prairie dog towns. Highly endangered, small isolated populations may be found as far south as New Mexico. It is not as wary (of humans) as other canid species, which has helped lead to its extirpation in much of its historic range. Shy and graceful, it spends most of its day in underground burrows, which it excavates in loose or sandy soil. Daytime denning helps it keep cool, reducing water loss. Its year-round dens have multiple entrances and exits; establishes a home range of about 1 sq. mi.

Though as an omnivore it will consume seeds, nuts, grasses, grains, and fruit, its preferred diet is one that includes insects, fish, birds, eggs, amphibians, reptiles, mice, rats, squirrels, rabbits, prairie dogs, and carrion. During the cold months it caches surplus food beneath the snow. Nearctic, nocturnal, solitary, evasive, secretive, vocal, and nonterritorial, its common name derives from its agility and speed (nearly 40 mph), which it uses to great effect when hunting and fleeing from predators. Its thick undercoat helps protect it from the heat during the day and from the cold at night.

Monestrous and monogamous, its breeding season takes place in winter. The vixen gestates for about 2 mos., giving birth to 1-5 altricial pups in the spring in a subterranean maternity den. They are weaned by 7 wks. and independent by 6 mos. The dog (male) plays no role in the rearing of his offspring.

Debate continues as to whether the swift fox and the kit fox (with which the former is often confused) are the same species, 2 distinct but closely related species, or 2 subspecies of *Vulpes velox*. (I treat them as similar but separate species.) Though scientific tests yield contradictory results on this question, the swift fox is bigger, has a broader muzzle, eyes that are more widely set apart, larger ears, and a longer tail. The two inhabit different ranges as well.

Communicates and perceives via standard Canidae channels: touch and smell, and to a lesser degree sight and hearing. Vocalizations include growls, howls, whines, and purring. A secondary consumer, its natural enemies are birds of prey and coyotes. The swift fox lives an average of 4 yrs. in the wild; at least twice that long in captivity.

Fisher.

T

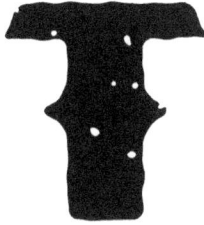

TAIGA VOLE (*Microtus xanthognathus*): Also known as the "yellow-cheeked vole," this large vole grows to a total length of 9", its hind foot measures ¾", and it can attain a weight of up to 6 oz. Its dorsal pelage is coarse and drab grayish-brown; the ventral pelage is whitish, silvery, or grayish; the long 2" tail is dichromatic (dark above, lighter below); the nose is rust-colored and covered in sensitive vibrissae. The body is ovoid; the eyes are small, dark, and beady; the ears are medium-sized, rounded, and erect; the upper incisors are grooved; the limbs are short; the paws are small and delicate.

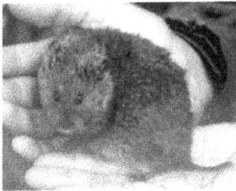

Taiga vole.

As its common name indicates, this terricolous sedentary cricetid is a nearctic native that inhabits temperate spruce taiga, often located near wetland and tundra; requires creeping rootstalk, damp friable soil, and dense mossy ground cover. Its largely palustrine range is limited to the extreme northern and central-northern portions of our region, extending from Alaska, the Yukon Territory, and the Northwest Territory, eastward through Alberta, Saskatchewan, and Manitoba.

Fossorial, it excavates shallow, multi-entrance burrow systems beneath ground detritus; these contain multiple chambers used for various activities, such as resting, nesting, sleeping, and food storage; it creates numerous long runways in moss and other types of forest debris; leaves massive ejecta mounds near portals. Both solitary and colonial (depending on various conditions), during cold weather, for instance, it may form small informal colonies. Territorial, it establishes a modest home range which it scent-marks and defends against outsiders.

A crepuscular herbivore, its diet consists primarily of standard Cricetidae fare: grasses, seeds, lichens, tubers, horsetails, fruit, grains, nuts, fungi, leaves, and roots. Hoards and caches surplus food; the vast majority of its diet is made up of rhizomes. Its breeding season occurs between spring and fall. The polyestrous female probably gestates for 3-4 wks., gives birth to an average of 7-8 altricial pups in a well-concealed, subterranean grass-lined nest, and weans them 2-3 wks. later.

Taiga vole (juvenile).

Communication and perception run along standard *Microtus* channels: hearing and smell, with reduced sight and touch. A highly loquacious rodent, its vocalizations include a myriad of altruistic warning calls, such as chirping, that aid in the survival of the colony. Its natural enemies are birds of prey, mustelids, wild felids, wild canids, and ursids. Beneficial as a seed and spore disperser and as a prey base for carnivores. The average life span of the taiga vole in the wild is about 1 yr.; longer under optimum conditions.

TAWNY-BELLIED COTTON RAT (*Sigmodon fulviventer*): This vole-like rodent is the largest of the cotton rats, growing to a total length of about 11" and a weight of up to 8 oz. On the dorsum its coloration is a grizzled or speckled mixture of black, white, gray, and brown fur; the ventrum is tawny or fulvous, giving it its common name and its species name; the 4" tail is scaly but lightly covered in grayish-black fur. The body is small, compact, and cylindrical; the head is small, broad, and short, with a convex profile; the eyes are medium-sized, dark, and protruding; the ears are small, rounded, and erect; the muzzle is short and robust; the nose is snubbed and covered in tactile vibrissae; the limbs are short; the feet are small, delicate, and clawed. Its genus name, *Sigmodon*, means "s-tooth," a reference to the s-shape pattern on the crowns of its molars.

Cotton rat.

This large cricetid is a nearctic native that inhabits tall grasslands, dense marshy areas (with cattails), and thick savanna, but it is also found in juniper and oak woodlands, yucca and mesquite landscapes, weedy shrubland, level desert, dry plains, chaparral, sand dunes, and along fencerows. Fossorial, it prefers good hiding cover and soft soils. Its range is restricted to the American Southwest, extending from Arizona, New Mexico, and Texas, south into central Mexico. It excavates burrows beneath vegetation; tunnels through grass; nests are made from woven grass and leaves. Avoids livestock fields. Does not hibernate; active all year long.

A nocturnal omnivore, it has a well-rounded diet consisting of seeds, grasses, tubers, roots, leaves, nuts, grains, flower parts, crops, worms, insects, small reptiles, birds' eggs, and birds. Does not hoard or cache food. Its breeding season seems to run year-round. The doe gestates for about 1 mo., giving birth to 4-10 precocial pups, which are weaned and independent by 2 wks. of age. The father is non-paternal and takes no part in the raising of his offspring.

Sigmodon molars (upper on left, lower on right) showing "s"-shaped crowns.

It communicates and perceives along typical Cricetidae lines: sight, hearing, smell, and touch. Its natural enemies are reptiles, birds of prey, and wild canids. It is an agile runner and swimmer. It aids in the equilibrium of its ecosystem by hosting parasites and providing food for carnivores. Benefits humans as a subject of scientific research, but harms us as a crop destroyer. The tawny-bellied cotton rat lives, at most, for perhaps 3-4 mos. in the wild; less under strenuous conditions.

TEXAS ANTELOPE SQUIRREL (*Ammospermophilus interpres*): This chipmunk-like ground squirrel grows to a total length of about 8½" and a weight of around 4 oz. Its ecology and life history are nearly identical to its close relation the white-tailed ground squirrel, except that it is smaller and occupies a more limited range: Texas, New Mexico, and parts of northern Mexico. It also possesses a gray tail-tip with two dark bands, characteristics lacking in the white-tailed.

TEXAS KANGAROO RAT (*Dipodomys elator*): This rare 4-toed rodent grows to a length of nearly 14", its hind foot measures about 1½", and it can attain a weight of up to 3½ oz. Its upper body is golden brown with a blackish sheen; its lower body is white; the long 8" tail is thick, whip-like, and bicolored, with a dark stripe on top, and a light stripe below; unusual for a kangaroo rat, its tail has a white tuft or "banner" at the tip. There is an indistinct face-mask; each eye has a white supraocular mark and each ear has a white subauricular mark (these sometimes connect); a white thigh stripe runs laterally over the upper leg.

The body is compact and ovoid; its head is large (in comparison to its body) with a convex profile; the eyes are large, dark, and protruding, offering excellent wide-arched night vision; the ears are small, naked, and erect; the muzzle is short; the nose is pinched, snubbed, ringed in black, and covered in sensitive vibrissae; the manus are short, while the pedes are long, powerful, and kangaroo-like, providing excellent speed and jumping abilities that lend it its common name. Gender dimorphism is present: the male is larger than the female.

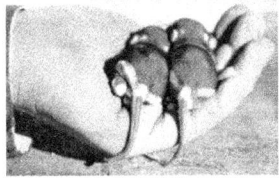

Kangaroo rat babies (2 wks. old).

This solitary heteromyid is a nearctic native that inhabits temperate mesquite scrubland, chaparral, prickly pear regions, short herbaceous grassland, and patchy woodland, usually with sparse woody vegetation and semi-firm but malleable soil (such as loam or clay). It is often found along well-drained roadways, crop field edges, and fencerows, preferably with some canopy and deadfall. Its range, which has been shrinking annually, is now quite narrow, covering only north-central Texas. Fossorial, it excavates large burrow and tunnel systems 3' deep under and around vegetation; these include numerous entrances, exits, and escape holes; here it eats, sleeps, nests, and hides from predators. Asocial, it is highly antagonistic toward conspecifics, except during mating; does not hibernate but is active all year; extremely nocturnal, it dislikes even moonlight; establishes a small home range of about 1,000 sq. ft.

An opportunistic omnivore, its diet centers around seeds, forbs, nuts, stems, grains, grasses, leaves, and domestic crops (such as wheat, sorghum, and oats); supplements these foods with an occasional insect. Hoards and caches surplus edibles in its burrow. Like others of its kind, it has 2 external fur-lined cheek pouches that it uses to transport food.

Its breeding season runs year-round. The doe probably gestates for about 1 mo., and gives birth to 2-4 altricial pups in an underground natal nest. She weans them 3 wks. later. It is not known if the buck participates in the raising of his young; since most male rodents are non-paternal, however, it seems unlikely.

Communicates and perceives along traditional Heteromyidae channels: sight, smell, hearing, and touch. It is an excellent runner and, being saltatorial, it is an adept leaper; it also engages in dust-bathing. Its natural enemies are reptiles, birds of prey, mustelids, wild felids, and wild canids. Benefits its ecosystem by aerating and recycling the soil, hosting parasites, dispersing seeds, and providing food for carnivores. Like many other aspects of the ecology and life history of the Texas kangaroo rat, its life span has not been studied, but it probably lives for about 1-2 yrs. in the wild; twice that long in captivity.

TEXAS MOUSE (*Peromyscus attwateri*): Also known as the "Texas deer mouse," it grows to a total length of around 8½", its large hind foot measures about 1", and it reaches a weight of up to 1¼ oz. On the dorsum its pelage is a drab, sooty brownish-gray; lateral coloration is rusty-brown; on the ventrum it is whitish; its long 4½" tail, which it uses as a counterbalance while running and jumping, is dichromatic, furry, and tufted at the tip. The eyes and ears are large; the ankles are dark; the feet are pinkish-white.

This solitary, medium-sized petraphile is a sedentary cricetid and nearactic native that inhabits temperate scrubland, rock outcroppings, chaparral, oak woodland, juniper forest, rock bluffs, talus slopes, cliffs, meadows, scree, rock piles, cedar glade, shrubland, and rocky situations generally. Its range is limited to the central

Talus slope: Texas mouse habitat.

Southwest, covering portions of the states of Texas, Oklahoma, Kansas, Missouri, and Arkansas. Dens and nests in rock crevices; establishes a home range of about 20,000 sq. ft.

A crepuscular and nocturnal omnivore, its diet consists of typical deer mouse fare: seeds, grains, nuts, fruit, leaves, flowers, stems, and insects (mainly crickets, beetles, grasshoppers). Its breeding season runs year-round, reaching its zenith between fall and winter. The polyestrous female may produce 2-3 litters annually. She gestates for 3-4 wks., bears an average of 3 altricial pups in a grass-lined nest, and weans them at 1 mo. of age. The male does not appear to aid in the rearing of his offspring.

Communication and perception occur via normal Cricetidae channels: smell and touch, and to a lesser degree sight and hearing. Its natural enemies are reptiles, birds of prey, mustelids, wild felids, and wild canids. Scansorial and semi-arboreal, it uses its proficient tree- and cliff-climbing abilities as a defense against predators. Beneficial as a seed disperser and as food for carnivores; detrimental as a carrier of deadly diseases. The Texas mouse lives an average of 3-6 mos. in the wild; longer under optimum conditions.

TEXAS POCKET GOPHER

TEXAS POCKET GOPHER (*Geomys personatus*): This large sturdy rodent grows to a total length of almost 13", its hind foot measures about 1½", and it can attain a weight of up to nearly 1 lb. Has soft, sooty, drab gray pelage on the dorsum; whitish-gray pelage on the ventrum; the long 4½" tail is lightly furred. As with most other pocket gophers the body is stout, the head is large, the snout is blunt and robust, the nose is covered in sensitive vibrissae, the eyes and ears are small, the limbs are muscular, and the feet are heavily clawed. The skull has a sagittal crest, and the upper incisors are grooved.

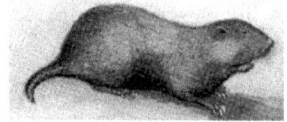

Pocket gopher.

This fossorial geomyid is a nearctic native that inhabits temperate grassland, river bottoms, and coastal areas, preferring soft, deep, friable, sandy soil. Its range is limited to the southernmost tip of Texas and a small portion of northeastern Mexico. An adept excavator, it constructs large elaborate burrow systems that can reach 100' in length. A strict herbivore, its diet consists mainly of roots, grasses, and other subterranean plant matter, which it pulls down into its tunnel from below the ground's surface. Its reproductive behaviors have not been adequately studied, but they may be similar to other members of the *Geomys* genus.

Communicates and perceives via standard pocket gopher channels: touch and smell, with reduced sight and hearing. Its natural enemies are probably reptiles, birds of prey, wild felids, and wild canids. Beneficial as a soil aerator and recycler, a habitat influencer, and as a prey base for carnivores; detrimental as a soil eroder and an agricultural pest. More taxonomic research is needed on this species and how it is related to numerous other geomyids—who seem to share similar ecologies and life histories. The average life span of the Texas pocket gopher is likely around 1-2 yrs. in the wild.

THIRTEEN-LINED GROUND SQUIRREL

THIRTEEN-LINED GROUND SQUIRREL (*Ictidomys tridecemlineatus*): Our only striped ground squirrel, this small rodent grows to a length of about 12", its hind foot measures nearly 2", and it can attain a weight of up to 10 oz. The base color of its pelage is tawny-tan; it is slightly darker on top; the abdomen is lighter. A cinnamon-rust wash sometimes appears around the cheeks, outer limbs, and sides. It takes its common name from its most conspicuous markings, 13 dorsal stripes: 6 narrow, plain, horizontal, tan stripes separated by 7 wide, white spot-filled, dark brown stripes, each which runs from the neck to the rump. (Note: there can be more or less than 13 stripes, and they can sometimes be in reverse order of the configuration described above.)

Also known as the "striped gopher," in the 19th Century its feline-like markings inspired one of its other common names at the time: the "leopard seed-lover." The body is smallish, supple, and slender; the head is small and gopher-like; the eyes are medium-sized, dark, and ringed in white; the ears are tiny and rounded; the limbs are short; the digits are clawed; the 5" tail is round, thinly furred, brushy, and tipped with orangey-buff hair.

A nearctic native, its preferred habitat is temperate, short (and sometimes long), well-drained grassland, whatever its form: prairie, brushy pastures, yards, rangeland, golf courses, roadside fields, wood edges, meadows, lawns, fencerows, farms, and cemeteries. Needs sandy or loamy soil for excavating. Will inhabit any mowed or cleared (treeless) land within its range, which occupies the central portion of North America, extending from Canada (Manitoba, Saskatchewan, and Alberta) in the north to Arizona, New Mexico, and Texas in the south; from Montana, North Dakota, and South Dakota in the west to the American Midwest (Illinois, Iowa, Michigan, Indiana, and Ohio). Digs intricate 20' long burrow systems with multiple entrances and exits, which it conceals under thickets or fence posts and defends against intruders; establishes a home range of about 2 acres. Diurnal, it is most active at midday, particularly during warm weather. It will sometimes appropriate the abandoned tunnels of fellow mammals such as prairie dogs and pocket gophers.

Thirteen-lined ground squirrel.

This omnivorous sciurid has a diet that is typical of its kind: roots, stalks, tubers, flower heads, grasses, herbs, leaves, seeds, nuts, grains, vegetables, fruits, agricultural crops (wheat, oats, corn), insects (ants, grasshoppers, crickets, caterpillars, beetles), birds' eggs, small birds (chickens), small mammals (shrews, mice), and carrion. Transports its victuals in large cheek pouches; hoards and caches surplus food in underground storage chambers. Though mainly solitary it may become semi-gregarious during the cold months, when it will hibernate with conspecifics in a small squirrel colony known as a "scurry."

Terricolous and fossorial, it builds up fat reserves for winter hibernation, which begins in midfall; it curls up into a tight furry sphere and slows its breathing and heartbeat in order to conserve energy. Emerges from its hibernacula in spring (males first, then females), at which time its breeding season begins. After gestating for about 1 mo. the doe gives birth to as many as 10 altricial pups in her natal nest, weaning them at 6 wks. of age. The vast majority of young become food for predators before reaching adulthood. The father does not assist in the rearing of the offspring he has sired.

Communicates and perceives through its senses of vision, hearing, touch, and smell—the latter especially useful for transmitting information via the aromas it discharges from its facial scent glands. Its primary natural enemies are reptiles, birds of prey, and medium-sized carnivores. Escapes predators by standing guard at its burrow entrance and diving below ground at the first sign of danger. For safety it usually remains near a specially designed escape burrow, sitting bolt upright with its nose pointed upward, a position that helps optimize surveillance of the surrounding area. When disturbed or frightened it emits a soft, altruistic, bird-like alarm whistle that alerts other squirrels nearby; it can dash nearly 10 mph along intricate runways that it has hidden beneath dense, grassy hiding cover; runs with its tail pointing backward horizontally.

Thirteen-lined ground squirrel.

Because of its predilection for roadside environments, it is often seen scurrying across highways, with the result that many thousands are killed by vehicle strikes each year. It benefits the ecosystem by aerating and recycling the soil, consuming vermin, dispersing seeds, acting as a host for numerous endoparasites and ectoparasites, providing subterranean homes for other animals, and serving as food for large carnivores. Some farmers consider it a pest due to its affinity for cultivated produce. The thirteen-lined ground squirrel probably only lives 2-3 yrs. in the wild.

Townsend's big-eared bat.

TOWNSEND'S BIG-EARED BAT (*Corynorhinus townsendii*): This medium-sized chiropterid grows from 3½" to a little over 4" in total length, its forearm measures just under 2", its wingspan is nearly 13", and on average it weighs around ½ oz. Its dorsal color ranges from gray to brown, the ventrum is light brown. It has broad wings; a short muzzle; expanded nostrils; and two large glands on the nose. It is named for its massive ears, which can attain a length of 1½" (nearly half its body length). They are so large that their bases connect at the middle of the forehead. During flight the ears rotate forward, then swivel in circles while echolocating; when it rests the bat curls its ears into coils, giving them the appearance of small ram's horns; hence its alternate common name: the "ram-eared bat."

It lives in desert and scrubland, as well as pine, aspen, and oak forests, from Canada to Mexico, with dense populations in the Western U.S. states. (Several subspecies occur in other parts of the U.S., and reside in different habitats.) It shelters in rocky cliffs, mines, caves, and sometimes human-made structures.

The female gestates for around 75 days, giving birth to a single temporarily blind and naked pup at her maternity roost in late spring or early summer (the baby's eyes open and its fur begins to grow within a few days). At this time the males remain solitary or form bachelor colonies.

This highly sensitive vespertilionid migrates short distances, enters periodic torpor during the summer (perhaps to conserve body fluids), and hibernates during the winter (to conserve fat). A nocturnal insectivore, it feeds at night, using its biological sonar to detect its quarry. An agile flyer, it can catch a moth in mid-flight (known as "hawking") or pick a beetle off a tree limb while hovering (known as "gleaning"). Its large consumption of pestilent insects (flies, beetles, wasps) makes it our ally. It will sometimes drink by skimming the surface of various bodies of water. A secondary consumer, its natural predators include snakes, hawks, owls, rats, raccoons, and wild cats. One of the longer-lived Chiroptera species, the life span of Townsend's big-eared bat in the wild is at least 10 yrs; twice that long under optimum conditions.

Big-eared bats.

TOWNSEND'S CHIPMUNK (*Neotamias townsendii*): One of the largest chipmunks in our region, it is named after the 19th-Century American naturalist John Kirk Townsend. This cautious primary consumer grows to a total length of nearly 15", its hind foot measures almost 2", and it can attain a weight of up to 4 oz. Darker than most of its kind, its pelage is a dusky rust-brown on the upper body, creamy white on the lower body; both its dark and light dorsal stripes lack sharpness, as do the facial stripes; the 6" tail is long, bushy, darkly grizzled on top, cinnamon below, and edged in tannish fur. Like most other chipmunks its winter coat is duller and darker than its summer coat. Its dark cryptic coloring allows it to easily blend into the shadow-filled world it inhabits. Gender dimorphism is present, with the female being larger than the male.

This riparian terrestrial sciurid is a nearctic native that inhabits temperate, mesic deciduous and coniferous forests with plenty of shrubbery, duff, and logs. It is also found in humid grassland, rocky settings, clear-cuts, and moist open ground with deadfall and good hiding cover. Its range covers the coastal and western zones of Washington and Oregon and the southwestern portion of British Columbia, making it the only chipmunk that lives along the seacoast in these areas.

Sedentary, terricolous, and slightly fossorial, it constructs its home underground in 30' long burrow systems, which it uses for protection, sleeping, resting, hoarding, staying warm in winter, and staying cool in summer. It is fond of nesting in talus piles and slopes. Being semi-arboreal it is also adept at tree-climbing. Estivates during hot weather; in severe winter conditions it may employ torpor for brief periods. Solitary and territorial, it establishes a home range of several thousand square feet, which it defends against conspecifics.

A wary diurnal omnivore with a strong predilection for fungi, it feeds primarily on seeds, grains, roots, nuts, fruit, lichens, tubers, and grasses. These foods are supplemented with insects, and probably birds' eggs and chicks. Stores surplus edibles, which it transports in its cheek pouches.

Townsend's chipmunk.

It breeds in spring after emerging from hibernation. The doe gestates for about 1 mo., giving birth to an average of 3-4 altricial pups in a subterranean maternal nest chamber. Weaning comes at about 7 wks. of age, independence a few weeks later. The buck does not contribute to the rearing of his offspring.

Communication and perception run along traditional Sciuridae lines: sight, touch, smell, and hearing. Its natural enemies are reptiles, birds of prey, mustelids, wild felids, and wild canids. Beneficial as a seed and spore disperser, a host for parasites, and as a prey base for carnivores. On average Townsend's chipmunk probably lives from 2-4 yrs. in the wild; twice that long in captivity.

TOWNSEND'S GROUND SQUIRREL (*Urocitellus townsendii*): It grows to a length of about 11", its pes measures 1½", and it can reach a weight of 12 oz. Its upper body is an unmarked, drab, light gray with a tinge of pinkish coloration; its lower body is whitish, tan, or fulvous; the short 2½" tail is thin and a rusty-fulvous color with whitish edging. The cheeks and back legs are sometimes lightly washed in cinnamon-colored fur. The plain, unmottled body is somewhat stout; the head is small; the eyes are small, dark, and ringed in buff; the ears are smallish, rounded, and erect; the limbs are short; the feet are clawed. This species is gender dimorphic: the male is larger than the female.

A terricolous sciurid, it inhabits temperate arid ecosystems in south-central Washington. Prefers dry desert, dunes, chaparral, sageland, grassland, ranchland, discarded farmland, scrubland, hedgerows, old fields, shrubland, and well-drained cropland, from wilderness to suburban areas. It is often found around canal and railroad embankments. Not especially social, it lives in groups, but with each individual excavating its own burrow system within the larger colony. Tunnels are generally 5'-6' underground and can extend up to 50' or more.

Townsend's ground squirrel.

An endothermic omnivore, its diet consists mainly of tubers, forbs, grasses, seeds, shrubbery, leaves, stems, roots, and woody material (bark, cambium, wood), occasionally supplemented with insects. Climbs vegetation while foraging; hoards surplus food.

Diurnal and fossorial, it spends much of the year estivating and hibernating (from early summer to late winter, depending on the latitude), awakening in time for its polygynous spring breeding season. The doe gestates for about 3 wks., gives birth to as many as 12 altricial pups in an underground natal nest, and weans them at about 1 mo. of age (though they do not become independent until they are about 1 yr. old). The buck is non-paternal and does not contribute to the rearing of his offspring.

Communication and perception take place via the usual Sciuridae channels: sight, hearing, smell, and touch. A variety of calls are used for various functions, from defense and mating, to family life and warnings. Its natural enemies include reptiles, birds of prey, mustelids, and wild canids. Though farmers often consider it vermin (since it can damage cultivated crops), this rare rodent benefits the world ecosystem in many ways, from aerating and recycling the soil and dispersing seeds, to acting as a host for endoparasites and ectoparasites, providing subterranean homes for other animals, and by serving as food for both carnivores and humans. This species is in decline. The Townsend's ground squirrel lives for about 3-4 yrs. in the wild.

TOWNSEND'S MOLE (*Scapanus townsendii*): Named after 19th-Century American naturalist John Kirk Townsend (the first to describe it), this eulipotyphlid grows to a total length of 9", its tail is 2" long, its hind foot measures about 1", and it weighs up to 5 oz. This makes it the largest mole in our region. Its pelage is gray to grizzled black in color. Its massive body is cylindrical, and like nearly all of its kind, its eyes are tiny and nearly nonfunctional; the nostrils are set atop the nose; the ears are hidden under its coat; its short velvety fur is highly flexile (allowing it to easily move forward and backward in tight quarters); its tail is short and nearly hairless; and its enormous, spade-like front feet, which turn palms-out while digging, are tipped with huge curved claws—all adaptations to its life as a fossorial mammal. Gender dimorphism is present, with the male being larger than the female.

Floodplain: Townsend's mole habitat.

A nocturnal riparian talpid, it inhabits flowered meadows, prairies, grassy fields, pastures, hedgerows, floodplains, cropland, suburban lawns, shrubland, orchards, and mixed alpine forest, preferring locations with deep, soft, damp soil. Its range is narrow, occurring only along a short stretch of the American West Coast, from southwestern British Columbia, south through Washington and Oregon, and continuing into northwest California. Mostly solitary, it establishes an average home range of about ½ acre, which it violently defends against unwanted guests.

Active all year long, as with many other members of the insectivorous Soricomorpha order, it is a rapacious vermivore, whose diet is primarily made up of earthworms. Besides annelids, however, it also consumes vegetable matter (including subterranean farm crops), gastropods, chilopods, and small mammals. This motile endotherm spends most of its life underground, where it constructs temporary, shallow foraging burrows, as well as deeper, permanent, complex tunnel systems, some of them reaching a depth of 10'. Here it spends its life foraging, hunting, resting, and searching for mates.

Its polygynandrous mating season occurs during the month of February. The female gestates for 1-1½ mos., bearing an average of 3 altricial young in a spherical nest chamber in spring. She weans them at around 4 wks. of age, after which they disperse, leaving the natal tunnels to set up their own territories. The male does not assist in the rearing of his offspring.

Communicates and perceives via normal Talpidae channels: hearing, smell, and, to a lesser extent, sight (its vestigial eyes see objects as light and dark shapes). Like most subterranean mammals, its strongest physical sense is touch, which enables it to read its environment through its highly sensitive whiskers. It greatly benefits the world ecosystem: 1) as a carnivore it is a secondary consumer that feeds on animals we consider pests; 2) its body supports numerous parasitic communities; 3) an ecosystem engineer, its burrows help aerate, recycle, and drain the soil; 4) its underground tunnel systems provide homes for other animals.

Its natural enemies are birds of prey, mustelids, and wild canids. The life span of the Townsend's mole is not known, but it probably lives 2-4 yrs. in the wild.

TOWNSEND'S POCKET GOPHER (*Thomomys townsendii*): Some believe that this large rodent is conspecific with Botta's pocket gopher and the Southern pocket gopher, while others maintain that all 3 are separate species. Thus its taxonomy remains confused and unsettled. Townsend's range is limited to parts of Oregon, California, Nevada, and Montana.

TOWNSEND'S VOLE (*Microtus townsendii*): One of our region's largest voles, it grows to a total length of around 9½" and it can attain a weight of nearly 4 oz. Its upper body is a cryptically colored drab brown to gunmetal grayish-black; its lower body is dull whitish-gray; the long 3" tail is dark gray and bicolored. Its body is robust; its fur is short and coarse; the eyes are dark and beady; the ears are large, wide, erect, and rounded; the snout is covered in tactile vibrissae; the feet and claws are dark. The female is slightly larger than the male, indicating gender dimorphism.

This motile sedentary cricetid is a riparian nearctic native than inhabits temperate dense grassland, damp forest, dry savanna, sedgeland, seepage slopes, and moist meadows, preferring settings with an abundance of stumps, logs, deadfall, blowdown, and ground litter. It is particularly

Seepage slope: Townsend's vole habitat.

fond of wetlands and can often be found near freshwater and saltwater marshes, and along both alpine and subalpine creeks, streams, and rivers. Its range is limited to the West Coast of the U.S. (Washington, Oregon, California) and southwestern Canada (British Columbia).

Strongly fossorial and natatorial, it is an excellent excavator and swimmer that constructs extensive burrow systems, often inside hummocks; when these are near a body of water, it locates its tunnel portals under the water's surface for protection from predators; surface runways link the den site with surrounding forage areas. Though normally social, it becomes territorial during the mating season. Does not hibernate, active year-round, but enters brief periods of torpidity when necessary.

A diurnal, crepuscular, nocturnal herbivore, its diet is made up of typical *Microtus* fare: seeds, grass, tubers, forbs, grains, rushes, leaves, roots, fruit, sedges, nuts, and bulbs. Hoards and caches surplus food; leaves grass cuttings around its home. Its breeding season runs from spring through fall, with males competing for access to estrus females. The polyestrous female can produce 4 litters or more a year. She gestates for approximately 3 wks., gives birth to an average of 3-4 altricial pups in a well-concealed nest (depending on the season, either underground or above ground), and weans them about 2 wks. later. The non-paternal male sometimes inadvertently aids in the rearing of his offspring by providing territorial protection of the nesting area.

Communication and perception occur via standard Cricetidae channels: sight, hearing, smell, and touch. Its natural enemies are reptiles, birds of prey, mustelids, wild felids, and wild canids. Beneficial as a seed disperser, parasite host, soil aerator and recycler, commensal homebuilder, and food for carnivores; detrimental as a lignivore that may injure young trees. Townsend's vole lives an average of 1-2 yrs. in the wild.

TROWBRIDGE'S SHREW (*Sorex trowbridgii*): This soricid grows to a total length of about 5", its tail is 2½" long, its hind foot measures ½", and it weighs up to ⅓ oz. Its pelage on the dorsal surface runs from gray to brown; on the ventral surface it is light gray to light brown. The body is stout; the head is conoidal; the eyes are tiny; the external ears are visible; the muzzle is convex and covered in vibrissae; the nose is tapered and has a pinkish, highly flexible tip; the tail is rat-like and mostly hairless; the limbs are short; the feet are delicate.

It inhabits mature, coniferous-dominant montane forests and mixed woodlands with dense leaf litter and loose well-drained soil. It is also found in chaparral and foothills. Its range covers southwestern Canada, south into Washington, Oregon, and California. A secondary consumer, this insectivore and granivore forages in leaf mold and in its tunnel system, feeding on seeds, chilopods, arachnids, and annelids, as well as a wide variety of insects.

Shrew.

The breeding season of this short-lived forest-dweller takes place from midwinter to early spring. The polyestrous female bears 2-3 litters a year, giving birth to an average of 3 altricial young. The male seems to play no role in the rearing of his offspring. It communicates via typical Soricidae channels: sight, touch, hearing, and smell. Its primary natural enemies are ambystomatids (Pacific giant salamanders), birds of prey (owls), and mustelids (weasels). Trowbridge's shrew probably lives about 12-18 mos. in the wild.

TRUE'S BEAKED WHALE (*Mesoplodon mirus*): Named after American biologist Frederick William True (who identified it in the early 1900s), this mesoplodont has a thickset body, grows to 18' in total length, and weighs up to 3,200 lb. The adult male, or bull, grows two small teeth at the front of its jaw. Perfectly camouflaged for its watery environment, it employs countershading. Thus, like most whales its upper body is dark in color (in its case a bluish- or brownish-gray) while its underside is paler (often a light gray, light blue, or white). Dappled dark spots are sometimes present. It has a narrow short snout or "beak"; a rounded forehead; small flippers; and a little, triangular, hooked dorsal fin located toward the terminal end of the body. In keeping with other beaked whale species, this ziphiid is gender dimorphic, with the female being generally larger and heavier than the male.

True's beaked whale.

A secondary consumer, its diet is comprised of squid and probably fish. Prefers deep water, at least 3,000'. Uses echolocation when feeding, diving, and navigating. It occasionally strands; will sometimes breach. As a mammal, the mother, or cow, suckles her calf, providing vitamin-rich milk until it can fend for itself. Considered quite rare, True's beaked whale is an inhabitant of the Atlantic Ocean, and has been reported from Canada south to the Caribbean Sea. Seldom seen, little else is currently known.

TUNDRA VOLE (*Microtus oeconomus*): Also called the "root vole," this medium-sized rodent grows to a total length of nearly 9", its hind foot measures about 1", and it can attain a weight of up to almost 3 oz. Depending on latitude and ecosystem, the dorsum is usually drab gray to dull yellowish-brown, often with a wash of grizzled black-tipped hairs over the back; lateral coloration may be tawny, buff, rusty, or cinnamon; the ventrum is typically cream to white; the short 2" tail is bicolored (dark above, lighter below). The eyes are small, dark, and beady; the ears are short and small. Gender dimorphism is present: the male is larger than the female.

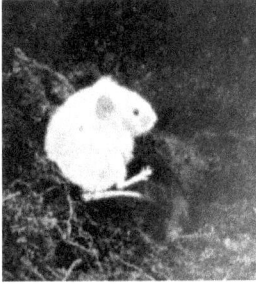

Vole.

This sedentary motile cricetid is a palearctic and nearctic native that inhabits cool mesic forests, bogs, dense grassland, steppe-meadows, muskeg, arctic tundra, marshes, taiga, sedgeland, and mixed montane woodland, preferably near a creek, stream, pond, or lake interspersed with stumps, brushpiles, deadfall, exposed tree roots, blowdown, ground snag, logs, and good hiding cover. Its common name suggests its range: the far northern tier of northwestern North America, covering portions of Alaska and Canada (British Columbia, the Yukon and Northwest Territories, and Nunavut). This makes the tundra vole the most northern member of its genus.

Being terricolous but semi-fossorial, it dwells largely above ground, living in shallow burrows that it excavates beneath soil litter and forest duff; these are connected by hidden surface runways that help protect it from predators. Normally social, it becomes increasingly territorial as food supplies begin to dwindle. Does not hibernate; active year-round.

A diurnal, crepuscular, nocturnal herbivore, it feeds mainly on sedges, roots, and tubers, but it also consumes seeds, stalks, lichens, shrubbery, mosses, flowers, grasses, herbs, leaves, bark, and wood. Hoards and caches surplus edibles, in particular rhizomes. Its breeding season runs from spring to fall. The iteroparous polyestrous female can produce several litters annually. She gestates for about 3 wks., gives birth to an average of 6 or 7 altricial pups in a concealed sedge-lined nest, and weans them 2-3 wks. later. Along with nursing, the female also appears to be the sole provider of grooming and protection of the young.

Communication and perception occur via standard *Microtus* channels: smell and hearing, with reduced sight and touch. Its natural enemies are birds of prey, mustelids, and wild canids. Beneficial as a parasite host, food for carnivores, and a subject of scientific research; detrimental as a farm and forestry pest and as a carrier of deadly diseases. The tundra vole probably lives 6-12 mos. in the wild; slightly longer under optimum conditions.

Gray wolf.

U

UINTA CHIPMUNK (*Neotamias umbrinus*): This medium-sized dendrophile grows to a total length of about 10" and can attain a weight of up to 3 oz. Like most of its kind it is rusty-brown on the upper body, whitish on the lower body. Unlike them, however, its 3-5 dark dorsal stripes are generally brown (rather than black) while its 4 light dorsal stripes tend to be white (instead of whitish or gray). The head crest and the haunches are usually gray; facial stripes mimic the back stripes. Its bushy 4½" tail is grizzled on the dorsum, buff on the ventrum, edged in white, and black-tipped. Its summer coat is brighter than its winter coat. Gender dimorphic, the female is larger than the male.

This solitary brownish sciurid is a nearctic native that inhabits temperate montane pine forests, craggy slopes, mixed woodlands, and open ground dominated by abundant logs, vegetation, and rocks. Its range is restricted to several western states, including portions of California, Nevada, Colorado, Idaho, Montana, Wyoming, Arizona, and Utah. Unusual for a chipmunk, though it is terrestrial, it is primarily arboreal, making it an adept tree-climber that likes to den and nest in trees. It is also known, however, to reside in burrows, which it conceals by excavating them under bushes and stones. At the onset of cold weather (normally October, depending on geography) it enters a state of torpidity from which it frequently awakes in order to feed. Establishes a home range of several thousand square feet, which it violently defends against conspecifics.

Montane forest and rugged slopes: Uinta chipmunk habitat.

A sedentary diurnal omnivore, its diet consists of seeds, fruit, grains, nuts, leaves, flowers, stems, grasses, buds, fungi, pollen, insects, and birds' eggs. Builds up bodily fat reserves in the fall. Hoards and stores surplus food in subterranean larders, which, as noted, it relies on during winter months. Transports edibles in its cheek pouches.

After emerging from hibernation in spring or early summer, this pert little rodent begins its polygynandrous breeding season. The doe produces 1 litter a year, gestating for about 1 mo. and giving birth to 3-5 altricial pups in a natal tree nest or an underground maternal chamber. Weaning occurs at around 60 days, with dispersal and independence occurring shortly thereafter.

Communication and perception take place via standard Sciuridae channels: sight, smell, hearing, and touch. Vocalizations include trademark chipmunk sounds: chips, trills, chucks, and squeals; when running it maintains its tail in a horizontal position. Its natural enemies are reptiles, birds of prey, mustelids, wild felids, and wild canids.

Benefits its ecosystem by aerating and recycling the soil, hosting parasites, dispersing seeds, and providing food for carnivores. The average life span of a wild Uinta chipmunk is probably about 2-3 yrs.; perhaps as many as 4-5 yrs. under optimal conditions.

UINTA GROUND SQUIRREL (*Urocitellus armatus*): This large ground squirrel grows to a total length of 12", its hind foot measures 2", and it weighs up to 15 oz. Its pelage is a lightly maculated grayish-brown on the upper body, fulvous on the abdomen, light buff-gray on the flanks. The top of the nose, the upper back, and the outer limbs may be cinnamon-colored; the 3½" dichromatic tail is blackish-brown on top, grayish below, with buff edging. The body is stout; the head is large with a convex profile; the ears are small, rounded, and erect; the eyes are large, dark, and ringed in tan; the nose is snubbed; the limbs are short; the digits are clawed.

The preferred habitat of this terricolous nearctic sciurid is sageland, meadow, savanna, and suburban lawns—that is, temperate grassy areas with soft soil and moist conditions. Semi-riparian, it can be found from low to high elevations, as long as there is abundant succulent plant material and a reliable water source. Its distribution is quite limited, with a range that extends only over parts of Idaho, Montana, Wyoming, and Utah.

A fossorial omnivore, its diet includes seeds, leaves, grains, nuts, insects, arthropods (pillbugs, ticks, centipedes, mites, spiders, scorpions), and small mammals. An excellent swimmer, it readily enters water to forage for aquatic plants. May consume cultivated crops, causing serious damage to farms. Hoards and caches surplus food.

Ground squirrel.

This diurnal rodent estivates during hot weather and hibernates during cold weather. The latter period runs from winter to spring, after which it emerges to start its breeding season. Dominant bucks and does are the ones mostly likely to mate. The doe gestates for around 30 days, giving birth to an average of 4 altricial pups in a natal chamber below ground. After weaning (about 1 mo. later), young females tend to stay with the matricentric group while young males disperse in search of other scurries, eventually establishing their own territories.

Communication and perception take place via regular Sciuridae channels: sight, hearing, touch, and smell. Scent-marking is particularly important. Vocalizations include squeals, chirps, chits, churrs, and chips. Spends at least 75 percent of its life underground, thus its primary natural enemies are subterranean carnivores, such as the badger. The life span of a wild Uinta ground squirrel is unknown, but it is probably similar to others of its kind (3-4 yrs.).

UNDERWOOD'S MASTIFF BAT (*Eumops underwoodi*): This large chiropterid grows to a total length of nearly 7", its forearm measures almost 3", its wings span up to 22", and it can attain a weight of 1½ oz. This makes it the second biggest of the dog-faced bats (after its cousin, the western mastiff or western bonneted bat). Like all free-tailed bats, its tail extends out past the tail membrane, in its case growing to a length of 2⅜". Dichromatic, the fur on its dorsum is dark gray-brown; on its ventrum it is lighter. Its enormous 1⅜" long ears are connected at their bases and extend forward over the forehead, giving it its alternate common name: "Underwood's bonneted bat."

A mastiff bat species.

This secondary consumer shelters in dry trees, hollow trees, and palm fronds throughout its range: southwestern U.S., south into Mexico and Central America. In common with other molossids, it has long narrow wings that are adapted to fast, powerful, straight-line flight. It is a nocturnal echolocator, foraging late into the night for insects (moths, beetles, grasshoppers), emitting loud calls while on the hunt. Breeding, gestation, and parturition are probably similar if not identical to the western mastiff. Its life span is not known. Underwood's mastiff bat has been little studied.

UTAH PRAIRIE DOG (*Cynomys parvidens*): The smallest of our prairie dogs, this endangered Western ground squirrel grows to a total length of nearly 15" and can attain a weight of up to 2¾ lb. The dorsal pelage is reddish-brown; lateral coloration is cinnamon-brown interspersed with black-tipped hairs; the ventrum is rusty-clay-brown; the brushy 2" tail is white-tipped; has dark cheek patches as well as dark supraocular patches.

This motile endothermic sciurid is a nearctic native that inhabits temperate flat grassland, plains, prairie, and savanna; prefers dry friable soil and scattered low vegetation. As its common name indicates, its range is limited to the state of Utah. Lives in "towns" or colonies made up of individual clan units, each led by an alpha male who oversees several females and their combined young; excavates massive burrow systems that are continually surveilled by "guard dogs" or sentinels. Hibernates (or enters torpidity) during harsh winter weather.

Prairie dogs at their burrow.

A diurnal omnivore, it consumes seeds, grasses, stalks, flowers, leaves, farm crops, insects, and bovine dung. Its breeding season begins in spring after emergence from hibernation, with the female gestating for around 1 mo. and giving birth to as many as 10 altricial pups in an underground nest chamber. Weaning takes place at 6-7 wks. of age.

Communication and perception are typical of the genus *Cynomys*: smell and touch, with reduced sight and hearing. Its natural enemies are reptiles, birds of prey, procyonids, mustelids, wild felids, and wild canids. Beneficial as food for carnivores; detrimental as an agricultural pest and as a carrier of deadly diseases, such as the plague (which can be transmitted by touching the carrier or by breathing its aerosol). Some consider this rodent to be a subspecies of the white-tailed prairie dog. The life span of the Utah prairie dog is unknown.

Mountain goat.

V

VAGRANT SHREW (*Sorex vagrans*): Also known as the "wandering shrew," it grows to a total length of nearly 5", its tail is about 1½" long, its hind foot measures around ⅝", and it weighs up to ⅜ oz. Its summer pelage on the dorsum is reddish-brown or brownish-gray, on the ventrum white to grayish-buff; its winter pelage is gray to black over the whole body. The body is stout and cylindrical; the head is flattish; the external ears are visible; the muzzle is convex and covered with vibrissae; the nose is tapered with a flexible tip; the tail is monocolored and mouse-like; the limbs are tiny and delicate.

Somewhat riparian, it inhabits a wide variety of biomes and ecosystems: mixed mesic forests, pond fringes, prairie, sagebrush areas, disturbed timberland, transitional areas, open woodlands, fen, swamps, alpine tundra, savanna, shrubland, damp meadows, ravines, foothills, forbland, draws, steppe, mountainous regions, and parkland. Prefers areas with soft acidic soil and a fresh water source, such as a river. It ranges over the Pacific Northwest, from southwestern Canada (including Vancouver Island)

Shrew.

south through parts of Washington, Idaho, Montana, Wyoming, Utah, Nevada, and California. A solitary creature, it aggressively defends its baseball diamond-sized home range against intruders.

Like other shrews, this medium-sized omnivorous insectivore has a high metabolic rate, forcing it to feed almost constantly. Eating more than its own body weight each day, it consumes seeds, fungi, insect larvae, earthworms, carrion, and even its own kind. Also, like many other soricids, its uses echolocation to navigate in its subterranean world, emitting clicking noises as its travels through its dark burrows and under leaf litter.

Its breeding season probably falls between midwinter and midsummer. The polyestrous female can bear up to 3 litters a year. She gestates for about 3 wks., giving birth to 3-7 altricial young in a domed, grass-lined nest often constructed under logs or stumps. The young are weaned by 2-3 wks. of age. Its natural predators are birds of prey (such as owls) and wild felids (such as bobcats). The vagrant shrew—named not for any nomadic inclination, but for the fact that its spends most of its time away from its den in search of food—probably lives about 1 yr. in the wild.

VIRGINIA OPOSSUM (*Didelphis virginiana*): Though there are some 250 species of marsupials in the world (such as the kangaroo and the koala), North America has only 1: the Virginia opossum. The official state marsupial of North Carolina, this primitive, nonplacental, house cat-sized mammal descends from ancestors that lived 70 million yrs. ago. Also known as the "North American opossum," it grows to a total length of 40", its tail can reach 21" long, its hind foot measures 3", and it weighs up to 14 lb. The dorsal fur coloration ranges from grizzled black, gray, and white to grizzled brown, reddish, and white; the ventrum can be lighter, darker, or the same color as the upper body. The face, throat, and muzzle are white; the ears are black; the nose is pinkish. There is often a vertical black forehead stripe running from between the ears down to just above the eyes. Wide variations in coloration can occur depending on geographic location.

Virginia opossum.

The body is stout and heavyset; the head is triangular; the rounded ears are large and leaf-like; the muzzle is tapered and covered in highly tactile vibrissae; the limbs are short; and its hind feet are similar to tiny human hands, complete with clawless, opposable big toes known as halluces. Its long, scaly, prehensile tail serves as a fifth appendage, and is used to grasp and hold objects (making it very useful when climbing trees). Gender dimorphism is present, with the male being slightly larger than the female in some populations.

This husky didelphid is an adaptable tropical native that inhabits a myriad of biomes and ecosystems, including scrubland, thickets, open woodland, farmlands, brushland, wasteland, and mixed temperate and tropical forests. Somewhat riparian, it gravitates to areas with dependable water sources, from bogs and marshes to rivers and ponds.

An edificarian, it easily accommodates itself to wilderness, agricultural, suburban, and urban areas, denning under logs, rock piles, stumps, and buildings, as well as in the abandoned burrows of other animals.

Though originally a South American animal, today its range covers most of the eastern section of North America, from southern Canada south to Central America. This includes the Southern and Midwestern states and parts of the Northeastern states. Introduced populations thrive along the West Coast as well. As a warm-weather species, it is not designed for extreme cold and is therefore susceptible to frostbite on its hairless ears, toes, and tail (some of which may be missing as a result). Despite this physical limitation, the already massive range of this synanthropic species continues to expand across our region each year; in particular it is growing along with the spreading human population with which it is often associated.

Nocturnal and solitary, the Virginia opossum is an opportunistic omnivore that, like the raccoon, will eat nearly anything edible. This makes its diet quite varied, encompassing nearly the full gamut of wild foods that are available to a land mammal: seeds, plants, leaves, nuts, fruit, grains, gastropods, annelids, insects, reptiles, amphibians, eggs, nestlings, birds, other mammals, carrion, pet food, poultry, and human refuse. Though it consumes the poisonous pit viper (a snake that is deadly to humans), it is immune to its venom. Its 50 sharp teeth aid it in hunting, feeding, and defense.

Virginia opossum (juvenile).

The Virginia opossum's polygynous mating season runs roughly from winter to fall (depending on location), with jacks competing over access to breeding jills. The polyestrous jill can produce up to 3 litters a year. She gestates for about 2 wks., giving birth to as many as 25 altricial young, each about the size of a kidney bean. These minuscule embryonic neonates must make the perilous journey from under the mother's tail to her abdominal pouch, where, if they succeed, they will find warmth, protection, and 13 milk-producing mammary glands. Here the survivors remain for about 2 mos., then, leaving their furry pocket-like home, they begin eating solid food and riding on their mother's back, for at least another 5 wks. or so. At around 3 mos. of age the young become independent. The father does not assist in the rearing of his offspring.

Communication among Virginia opossums runs along typical Didelphimorphia channels: sight, touch, hearing, and particularly smell (both the male and female have powerful scent glands that emit musk-like odors, which are used to identify individuals, locate mates, and drive off rivals and predators). Vocalizations include chirping, clicking, and lip-smacking.

Mother opossum and young.

Its natural enemies are reptiles, birds of prey, procyonids, wild felids, and wild canids. It defends itself by growling, screeching, hissing, bearing its teeth, attacking, running, hiding, urinating, swimming, climbing, and, as noted, excreting a noxious substance from its rump glands. It may also "play dead," an involuntary catatonic state induced by shock: lasting from a few mins. to 6 hrs., the opossum falls to the ground, becomes stiff, begins to drool, and its breathing slows down. This death-feigning behavior (from which we derive the term "playing possum") usually discourages predators who do not eat carrion.

Economically, it plays a dual role: it provides us with sport, meat, pelts, and medicines, but it is also injures people, carries diseases, and feeds on farm crops and animals. The Virginia opossum lives about 1½ yrs. in the wild; 3-8 yrs. in captivity.

Caribou.

WAGNER'S BONNETED BAT (*Eumops glaucinus*): Also known as "Wagner's mastiff bat," the "Florida mastiff bat," and the "Florida bonneted bat," this rare flying chiropterid grows to a total length of just over 5", its forearm measures nearly 3", and it weighs up to 1½ oz. Medium-sized, its upper fur is gray to black, the underside is lighter. The head is broad and somewhat flattish; its ears are rounded; the tragus is square-tipped and wide; the antitragus is oval and well-developed. Possesses a small gorilla-like sagittal crest. Being a molossid its tail projects beyond the tail membrane. Its plagiopatagium is connected to the back of its foot; it is capable of producing a strong musk-like odor (perhaps used in mating or defense).

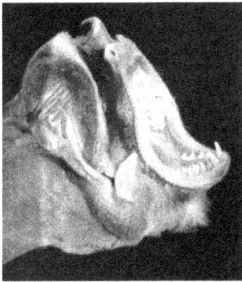

A mastiff bat species.

It inhabits forest, scrubland, desert, and cities in tropical and subtropical environments. Its range covers Florida, Cuba, Mexico, and regions south into Central and South America. Probably does not hibernate. Nocturnal and insectivorous, its long narrow wings permit quick powerful flight as it echolocates in the darkness in pursuit of its prey (moths, beetles, flies, grasshoppers). The female gives birth to 1 altricial pup. A secondary consumer, its natural predators are birds of prey. Due to pesticides it is disappearing from much of its former range. The life span of Wagner's bonneted bat is unknown.

WALRUS (*Odobenus rosmarus*): Known in Old English as the "*morse*," its modern common name, walrus, is a Dutch word deriving from the Scandinavian *hvalros*, meaning "whale horse." One of the most gigantic of the pinnipeds, there are 3 subspecies: the Pacific walrus, *Odobenus rosmarus divergens* (the largest); the Atlantic walrus, *Odobenus rosmarus rosmarus* (the second largest); and the Laptev walrus, *Odobenus rosmarus laptevi* (the smallest). (Unless noted, all 3 are described here under one generalized heading.)

Walrus.

This marine behemoth grows to a total length of almost 12' and a weight of some 3,000 lb. Its thick skin appears nearly naked, but is covered in short, cinnamon tinted hair; the skin itself runs from light gray and tannish to yellowish and reddish-brown in color. The face is sometimes framed in blackish skin. Nearly every feature of this mammal is massive: the rounded head; the broad short muzzle; the neck; the body; and the flippers. Its eyes are small and its tail is short, however. Its adipose tissue ("blubber"), which acts as insulation against the cold as well as protection against predators, is from 2"-10" thick—depending on where it is located on the body.

The walrus' most distinctive feature, unique among marine mammals, is its pair of long tusks, which continue to grow throughout its life. Found on both the adult male and the adult female (they are much smaller in the latter), they are not true tusks, but actually elongated upper canine teeth. Those of the male Pacific walrus can reach a length of 40" (over 3') and a weight of 12 lb. each. These impressive pieces of ivory dentition serve multiple purposes, such as dominance displays, weapons against rivals, and defense against predators. Additionally, they function as "ice picks" (useful for digging through up to 8" of ice) and "grapnels" (used to haul the walrus' enormous bulk out of the water and onto the ice).

Walruses.

Its large hands have nails or claws on all 5 digits. These paddle-like appendages (its front flippers) are used for steering, while its strong webbed feet (its rear flippers) propel it through the water. Though it is "earless" (that is, it does not have external ear flaps), like its "eared" seal cousins it can turn its hind flippers forward under its body, facilitating terrestrial locomotion. Despite this, it is somewhat ungainly on land due to the fact that the bones above both its elbows and its knees are enfolded within its body, which restricts its movement. It maintains open breathing holes by knocking new ice away with its head. Gender dimorphic, the male is larger, heavier, more robustly built, and thicker skinned than the female.

This cold weather loving odobenid is a native of the Arctic coasts, where it spends most of its time on and around ice shelves and the edges of pack ice, resting, sunbathing, and foraging. It avoids deep water; instead, it is found in shallow marine waters (about 55' deep or less) near ice (preferred by females) rimmed by rocky and sandy beaches (preferred by males). In the eastern part of our region it ranges over the eastern Atlantic Ocean (from Nunavut, Quebec, and Prince Edward Island, to Labrador and New Brunswick, Canada); in the western portion it ranges from Russia to Alaska.

This huge molluscivore consumes all manner of small benthic invertebrates. Moving slowly along the murky bottom of the sea, it searches for its prey in the sand and muck using its mouth and short quill-like but highly sensitive vibrissae. Once located, it either crushes the shell of the mollusk with its 18 teeth or it pries it open with its tongue. It then forcefully sucks out the innards, swallows them whole, then disposes of the shell fragments. True to its carnivorous status, it sometimes feeds on larger game, like seabirds and other marine mammals, such as seals and even small whales. In pursuit of its prey it can swim 15 mph, dive to a depth of 300', and can stay submerged for up to 30 min. In order to control its body temperature while deep-diving, blood flow is redirected away from the skin to the internal organs, temporarily turning the skin pinkish. It can eat up to 100 lb. a day; hoards and caches surplus food.

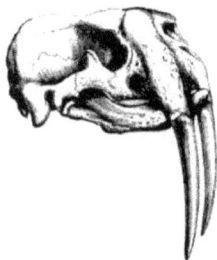

Walrus skull.

Males and females live in separate hierarchal herds for much of the year, joining together primarily during feeding and migration. The two genders also mix during the polygynous breeding season, which runs from midwinter to early spring, forming large blended herds of as many as 2,000 individuals or more. A descendant of a prehistoric terrestrial carnivore, this natatorial mammal remains closely linked to terra firma, and thus must return to shore to breed. Dominant bulls emit underwater mating calls to draw the attention of cows. At the same time they violently challenge other males for mating rights, which involves loud bellowing and stabbing one another with their tusks. The male with the largest tusks and the most strength, power, size, and endurance is the most attractive to breeding cows—and the one who most often comes away the victor. Thick folds of skin, as well as "bosses" (large bumps) on the tough hide around the neck, chest, and shoulders, help protect males from serious injury. On occasion, however, these bloody contests can lead to broken tusks, serious bodily trauma, and death.

After mating (which takes place in the water), the female gestates for nearly 1½ yrs., with about ⅓ of that time dedicated to delayed implantation. Advantageously, this 4-5 mo. interval allows the young to be born in the spring, as the weather is warming up. Cows give birth to a single, precocial, 125 lb., 3½' calf once every 2 yrs. or so. Though it can swim shortly after birth, the baby often prefers riding on its mother's back. Lacking tusks and the adult's bulk, for the first 2-3 yrs. it is completely dependent on her for milk and protection, after which it is weaned and becomes autonomous. The father plays no role in the affairs of the offspring he sires. The walrus nuclear family unit, as with most other mammals, is the mother and her young.

Walruses.

Communication and perception occur via standard Odobenidae channels: sight, hearing, touch, and smell. Above-water and below-water vocalizations include bellowing, barking, grunting, thumping, whistling, trumpeting, and the emission of bell-like tones. It sleeps "standing up" in the water, buoyed at the surface by 2 air sacs located in its neck. Social and gregarious, it assists sick and wounded fellow herd members, even protecting one another against their natural enemies: orcas and polar bears. The walrus may live 30-40 yrs. in the wild, though 20-25 yrs. is probably more usual.

WASHINGTON GROUND SQUIRREL (*Urocitellus washingtoni*): This small mammal grows to a total length of 10", its pes or hind foot measures 1½", and it can attain a weight of up to 10 oz. The pelage on the upper body is grayish-buff with overt white flecks or spots scattered across the back from neck to rump; the lower body is tannish-gray-white; dorsum and ventrum are separated by a horizontal line running laterally along the sides; the short 3" tail is grayish-brown with a dark tip. The nose and back legs may be washed in cinnamon. The body is stout; the head is large with a rounded profile; the ears are small, rounded, and erect; the eyes are large, dark, and ringed in tan; the nose is snubbed; the limbs are short; the toes are clawed. It is the only ground squirrel in its range with speckled markings; possesses some prairie dog characteristics. This species is gender dimorphic, with the male being larger than the female.

A nearactic native, it inhabits temperate open grassland, agricultural fields, chaparral, savanna, hills, shrub-steppe, and sageland, preferably with deep, sandy, silty, or loamy soil. Its distribution is small, with an ever-shrinking range that now extends only from southern Washington into northern Oregon. Excavates extensive burrow systems where it feeds, rests, estivates, hibernates, gives birth, and rears its young. Unlike many other species of ground squirrels, its burrow opening is not concealed, but is plainly visible with trackways spreading outward from the entrance. It can often be seen standing upright at its entryway, where it nervously guards its small home range.

Ground squirrels.

An herbivorous sciurid, it feeds mainly on various species of grass, which it supplements with stems, nuts, grains, leaves, and grains, as well as cultivated crops; it has a keen taste for plantain, mallow, and alfalfa.

Diurnal and colonial, it estivates during the driest, hottest months, then enters hibernation as the weather begins to turn cold, living off fat it builds up while awake and active—which is typically only from about late winter to early summer. Thus it spends nearly 70 percent of its life underground in varying states of torpor or sleep. Following the end of hibernation in midwinter, bucks emerge from their dens first, followed by the does, at which time the males begin aggressively competing for mating rights. The female's gestation time is unknown, but it probably lasts for about 1 mo. or so. An average of 8 altricial pups are born in an underground nest chamber, with weaning following around 30 days later. The father does not provide any care in the raising of his offspring.

This fossorial rodent communicates and perceives via traditional Sciuridae channels: sight, smell, touch, and hearing. Its natural enemies are reptiles, birds of prey, mustelids, and wild canids, and most specifically other burrowing animals, such as the badger and the burrowing owl. Benefits the world ecosystem by aerating and recycling the soil, hosting various parasites, and providing food for carnivores. It is considered a pest by farmers, however, due to its consumption of cultivated crops. This species is on the decline. The life span of the Washington ground squirrel is not known, but it probably lives 3-5 yrs. in the wild.

WATER SHREW (*Sorex palustris*): Also known as the "American water shrew," this large soricid grows up to 6" long, its 3" tail makes up half its total body length, its hind foot measures ¾", and it can attain a weight of ⅝ oz. The dorsal coat is grizzled iridescent gray to black, the ventral coat is light gray to white. The body is stout but mouse-like; the head is long and flattish; the eyes are tiny; the neck is thick; the muzzle is tapered, convex, and covered in tactile whiskers; the nose is pointed with a flexile tip; the dichromatic tail is quite long; the limbs are short; the small forefeet and the wide hind feet both have 5 toes. This species is gender dimorphic: the male is larger than the female.

Water shrews.

A nearctic native, as both its common name and its scientific name indicate (its species name, *palustris*, comes from the Latin word for "marsh"), this riparian mammal inhabits watery environments, preferring damp boreal forests with good hiding cover, thick understories, rocky overhangs, mossy logs, exposed root systems, and decaying trees, as well as seeps, vernal pools, streams, creeks, bogs, marshes, rivers, ponds, and lakes. Once found as far south as Arizona, today it seems to be restricted to a range extending from Alaska east to Nova Scotia, south into the Rockies and east to the Appalachians. It is, then, primarily a species of the montane forest regions of the northern tier of North America. Fossorial, its excavation technique is somewhat mole-like: dirt is pulled out with the front feet and pushed backward with the rear feet. Solitary, it does not easily tolerate the presence of conspecifics: disputes and even violent, injury-inflicting battles are not uncommon among both genders.

A member of the genus *Sorex* (the long-tailed shrews), this voracious secondary consumer feeds primarily on aquatic insects, such as ephemerids and plecopterids and their larvae and nymphs. However, being omnivorous, it also consumes fungi, vegetable matter, annelids, gastropods, amphibians, and small fish. With its high metabolic rate it must feed almost continuously, and will perish if it goes more than 3-4 hrs. without eating. Thus it is diurnal, crepuscular, and nocturnal.

Natatorial and semiaquatic, it is ideally suited for a carnivorous life spent mostly in, around, and under the water. It is an excellent swimmer and diver, whose soft coat is water resistant: its fur traps air bubbles, which not only helps it maintain body heat, but makes it buoyant, permitting it to easily bob back up to the surface when necessary. All 4 of its feet are trimmed with stiff fibrillae (tiny hairs) that aid in swimming, increase the surface area of the feet, and trap air, allowing it to run along the surface of the water.

Its mating season extends from winter to fall. The iteroparous polyestrous female, who may bear 2-3 litters in 1 yr., gestates for about 21 days, giving birth to an average of 5 altricial young in a nest lined with shredded vegetation—and usually constructed under a stump, log, or pile of forest detritus. The male appears to play no role in the rearing of his offspring.

This species communicates and perceives via normal Soricidae channels: touch and sight, but it is most attuned to its senses of hearing (it echolocates) and smell (it emits a feculent odor from its scent glands). Aids the world ecosystem by preying on pests, providing food for other animals, and serving as a host for various parasites. Its natural enemies are fish (mainly bass, trout, and pike), reptiles, birds of prey, and mustelids. The water shrew may live to about 1½ yrs. in the wild.

WATER VOLE (*Microtus richardsoni*): Also known as the "North American water vole," this large rodent is our region's biggest vole. It grows to a total length of 10" or more, its large hind foot measures about 1", and it can attain a weight of up to nearly 2 oz. On the dorsum its long fur is brownish-gray to cinnamon-brown with black-tipped hairs washing over the back; lateral coloration is slightly lighter; the ventrum is silvery-white; the long nearly 4" tail is bicolored (dark above, light below). The body is stout; the eyes are small and beady; the ears are large and erect.

As its common name indicates, this semiaquatic cricetid is a nearctic native that inhabits the banks and edges of subalpine and alpine creeks, streams, rivers, ponds, marshes, and lakes, but it is also found in moist taiga, grassland, meadows, and forests; it prefers habitat with fallen timber, rotting stumps, wet wood, thick hiding cover, deadfall, blowdown, and clean, fast-moving, glacial watercourses. Its range extends unevenly over portions of the

Water vole.

American West (Washington, Oregon, Montana, Idaho, Wyoming, Utah) and southwestern Canada (British Columbia, Alberta). Does not hibernate; active year-round.

A semi-social riparian, it forms small colonies near water. Fossorial, it excavates 5" wide burrows under willow trees or in the banks of creeks and streams; it positions its burrow entrances near or along the water's edge; leaves ejecta mounds near portals; damp, grassy, surface runways connect the den to the water. Natatorial, it is a skilled swimmer, an ability it uses to escape predation. Territorial, it establishes a home range of around 10,000 sq. ft., scent-marking its boundaries as a discouragement to intruders.

A nocturnal omnivore, its diet consists of typical arvicoline fare: seeds, grass, fruit, flowers, forbs, herbs, twigs, sedges, willow tree parts, and green leafy plant matter generally; also consumes insects when available. Caches surplus food for lean periods. Its breeding season runs from spring through summer. The female, who may produce up to 2 litters annually, gestates for approximately 3 wks., giving birth to an average of 4 altricial pups in a subterranean grass-lined nest.

Communication and perception fall along standard *Microtus* channels: smell and touch, with reduced sight and hearing. Its natural enemies are reptiles, birds of prey, wild felids, and wild canids. Beneficial as a seed disperser, parasite host, and as food for carnivores; detrimental as a forestry and agricultural pest. The water vole probably lives 1 yr. or less in the wild.

WESTERN BONNETED BAT (*Eumops perotis*): Also known as the "greater bonneted bat," it is the largest molossid in North America. It grows to a total length of nearly 8", its forearm measures about 3", its wingspan is nearly 2', and it weighs just over 2 oz. It is sometimes called the "western mastiff bat" due to its head, which can bear a general resemblance to a small mastiff dog. Its soft pelage is dark gray-brown with whitish roots on the upper body; the lower body is paler. Easily identified by its large, rounded, 1¾" long ears, which are joined over the forehead, and which, hat-like, project strongly forward over the face, giving it its common name. It has large feet and its long narrow wings allow fast powerful flight; possesses a pouch-like gland on its throat that produces an unpleasant odor (this organ is more developed in the male than in the female).

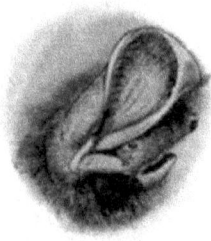

Bonneted bat.

Shelters in cliffs, canyons, rocky places, slabs, crevices, tunnels, and buildings over 30' tall. Due to its large size, wing design, and overall physical build, it cannot take flight from the ground, and therefore needs large bodies of water to drink from while airborne. It also requires 10'-15' of clear and open space below its roost crevice (usually a vertical slit a few inches wide) so that it can drop downward before taking flight (and return safely as well). This cliff-dweller can be found in a wide variety of habitats, from arid desert, floodplain, and scrubland to coastal grassland, montane meadow, and semiarid forest. It ranges over much of the American Southwest, from California east to Texas, and south to South America; does not migrate.

An excellent climber and a strong, nimble, straight-flying chiropterid, it leaves its roost after dark, squeaking and chattering loudly as it echolocates in search of its quarry: primarily moths, but also crickets, ants, bees, grasshoppers, leafbugs, wasps, cicadas, and dragonflies. During foraging it soars high into the night sky and sometimes travels great distances, searching out thermal air currents which carry insects aloft. It hunts, using both hawking and gleaning techniques, for most of the evening, returning to its roost near dawn. In consuming large quantities of insect pests it is a friend of humanity.

This secondary consumer breeds in spring. The female gestates for around 85 days, giving birth to 1-2 pups in late spring or early summer. The males and females remain together all year-round (forming small colonies of 10-100 individuals) instead of dividing into separate roosts during parturition (as many other bat species do). The life span of the western bonneted bat has not been adequately studied.

WESTERN GRAY SQUIRREL (*Sciurus griseus*): Also known as the "California gray squirrel," it is the largest gray squirrel and the largest native tree squirrel in its range, growing to a total length of 24", with a pes that measures 4" and a weight of up to 34 oz. Countershaded, its entire upper body is slate gray, intermingled with white-tipped hair; its ventral fur is white or light gray; its long 12" tail is bushy and grizzled grayish-silver with whitish edging. Occasionally there is a cinnamon wash on the face, throat, shoulders, flanks, tail, or front and back legs.

The body is robust; the head is well proportioned with a convex profile; the ears have rusty-colored backs and are large, tuftless, and erect; the eyes are large, dark, and ringed in tan; the nose is snubbed; the limbs are short; the digits are clawed. It often carries its tail in an "s" shape over its back, providing shade from the summer sun, warmth during cold weather, and camouflage in its gray-brown woodland environment. Its tail also serves a multitude of arboreal functions, such as providing balance and stability while leaping from tree branch to tree branch. It molts twice a year: in spring and in fall.

This crepuscular diurnal tree squirrel is a nearctic native that inhabits coniferous forests and mixed woodlands along North America's West Coast, from Washington and Oregon in the north to Baja California in the south, and east into parts of Nevada. Mildly riparian, it prefers open habitat with hollow trees, standing snag, and a fresh water source.

An endothermic omnivore, it consumes fungi, seeds, nuts, sap, bark, and fruit, occasionally supplementing its diet with insects and bird's eggs. It is particularly fond of acorns and the seeds of pine cones. It hoards extra food, burying it around its small home territory. During the cold lean months it uses its keen sense of smell to relocate these underground caches, one of the many keys to its survival. To keep cool, in summer it builds large moss- or lichen-lined tree dreys made from sticks, leaves, twigs, grass, and shredded bark; to keep warm, in winter it lives in leaf-lined tree crevices and holes. A secondary cavity user, it will sometimes appropriate the abandoned holes of woodpeckers.

Tree squirrel.

Active all year-round (though it remains in its tree den during inclement weather), this scansorial non-hibernating sciurid begins its breeding season in spring (depending on latitude). The doe gestates for around 6 wks., then gives birth to as many as 5 altricial pups in early summer in a fur-lined brood nest. This enormous bolus-style drey can be up to 3' in diameter, and is visible from long distances as a black mass perched high up in a tree. Fiercely defending her natal chamber from conspecifics, the mother weans her young by 3 mos. of age (though they usually remain in the nest for several more months).

A shy plantigrade that dislikes the presence of humans, it communicates and perceives via traditional Sciuridae channels: sight, smell, hearing, and touch. It is well-known for its barks, foot-stamping, and tail-flicking—the latter habit for which it has been nicknamed the "banner-tail squirrel." Its natural enemies are birds of prey, procyonids, wild felids, and wild canids. Has a positive impact on the ecosystem by dispersing seeds (that is, tree-planting) and by providing food for various carnivores; it is considered a pest by nut farmers, however. In the wild the western gray squirrel lives an average of 6-7 yrs.

WESTERN HARVEST MOUSE (*Reithrodontomys megalotis*): This small rodent grows to a total length of about 6", its hind foot measures ¾" long, and it can attain a weight of up to ¾ oz. Its upper body is brown with black flecking; it may have an indistinct brown dorsal stripe and chest patch; lateral coloration is rusty-yellow; the ventrum is white; the 3½" tail is lightly furred and dichromatic (dark above, lighter below); the coat is coarse and short; the eyes are large and dark; the ears are large, rounded, and erect; the upper incisors are grooved; the feet are whitish, small, and delicate.

This common, solitary, sedentary cricetid is a riparian nearctic native that inhabits open grassland, desert, cropland, chaparral, prairies, scrubland, mesa, sand dunes, hedgerows, wetlands, steppe, dry weed beds, abandoned fields, fallow pastures,

Western harvest mouse.

shrubland, pine-oak woodlands, meadows, estuarine biomes, and montane forests; it is found in both wilderness and urban ecosystems; requires thick ground cover and fresh water. Its massive range covers much of the American Midwest, Great Plains states, Rocky Mountain states, and the West Coast, running from southwestern Canada to central Mexico, and from California to Indiana (it is not found on the eastern side of the U.S. or Canada). It dens under shrubbery, rocks, logs, and ground litter; may create surface runways, but usually appropriates them from other animals (such as voles); its nests can be located above ground (in a bush), on the ground (beneath grass), or underground (in a small burrow). Establishes a home range of about 35,000 sq. ft.

A nocturnal opportunistic omnivore, the diet of this keystone species consists of seeds, nuts, grains, flowers, grain, fruit, and insects. An adept climber, it forages on the ground and also above ground in vegetation; caches surplus food in subterranean larders. Its breeding season runs all year, with the fecund polyestrous female capable of bearing several litters annually. She gestates for 3-4 wks., gives birth to an average of 4-5 altricial pups in a grass-lined globular nest, and weans them within 1 mo. or so. The father does not aid in the rearing of his young.

Communicates and perceives along channels standard to its genus: sight, smell, hearing, and touch. Its natural enemies are numerous and include arachnids (scorpions), reptiles (snakes), birds of prey (owls, hawks, falcons), mustelids (weasels, skunks), wild felids (bobcats, lynxes), wild canids (fox, coyotes, wolves), and other rodents (shrews, squirrels). Beneficial as a seed disperser and as food for carnivores. The western harvest mouse lives for an average of 6-12 mos. in the wild.

WESTERN JUMPING MOUSE (*Zapus princeps*): It grows to a total length of nearly 10½", its hind foot measures around 1", and it can attain a weight of up to 1 oz. It sports a broad, buff-grayish, grizzled dorsal band; lateral coloration is golden- to reddish-yellow; a distinct lateral line separates the dorsum from the sides; the abdomen is yellowish-white; the long 6" tail is bicolored (dark on top, lighter below); the medium-sized ears are narrow, erect, and lightly bordered in white.

This sedentary motile dipodid is a riparian nearctic native that inhabits a wide range of elevated temperate biomes and ecosytems, including forest, damp meadows, sedgeland, bushland, bog, tall grassland, and montane fields; prefers thick green hiding cover located near bodies of water, such as creeks, rivers, pools, and lakes. Its range covers much of the western and northwestern U.S. (Alaska to New Mexico) as well as southwestern and central Canada (Yukon Territory to Manitoba). It is not found in the eastern half of North America. It excavates its own burrow systems, but may also take over both the tunnels and surface runways of other animals. Enters hibernation in fall, living off stored fat reserves; emerges in late spring; may estivate during extremely hot weather.

Western jumping mouse.

A common nocturnal omnivore, its diet consists of standard *Zapus* fare: seeds, fruit, fungi, leaves, herbs, grasses, and arthropods. Will climb up into shrubbery while foraging. Its breeding season begins following the end of hibernation, and runs from late spring to midsummer. The iteroparous polyestrous female can produce up to 3 litters annually. She gestates for about 3 wks., gives birth to an average of 5 altricial pups in a spherical underground or surface nest chamber, and weans them at about 1 mo. of age. The non-paternal male does not participate in the rearing of his offspring.

Communication and perception run along typical Dipodidae channels: smell and touch, while sight and hearing are slightly reduced. Its natural enemies are birds of prey, mustelids, procyonids, wild felids, and wild canids. When disturbed or threatened it engages in tail-drumming; other defenses include performing 5' jumps using its large hind legs (from whence it derives its common name); swimming or climbing to safety; and freezing in place. Beneficial as a seed disperser, pest controller (of animals that negatively impact humans), and as a prey base for carnivores. The western jumping mouse probably lives for an average of 1-2 yrs. in the wild; longer under optimum conditions.

WESTERN PIPISTRELLE (*Pipistrellus hesperus*): Also known as the "canyon bat," with a total body length of less than 3½", a forearm less than 1½" long, a wingspan of as little as 7", and a body mass as small as ⅛ oz. (about the same weight as a single sheet of paper), it is the smallest bat in North America. Its fur color varies from yellowish and gray to reddish and brown on the dorsum; its ventrum is lighter, from tan to whitish. Its calcar is keeled; its tragus is short and blunt. Outstanding physical characteristics include blackish ears, nose, wings, uropatagium, feet, and face mask, producing a striking contrast against its buff, grizzled, or gold-sheened pelage. It is gender dimorphic: the female is larger than the male.

Pipistrelle bats.

Its scientific name matches its common name: *Pipistrellus hesperus* means "bat, western" (that is, "western bat"). This vespertilionid can be found in caves, scrubland, arid desert, loose rock, mines, buildings, forest, canyons, and rocky outcroppings, crevices, and cliffs near water sources. Its range extends from Mexico north across much of the southwestern U.S. and up into Washington. It hibernates during the cold months.

This slow, weak, erratic flyer—which is often mistaken for a large moth—is one of the few diurnal chiropterids in our region, moving through the air at 5 mph or less. It sallies forth from its roost to feed during early morning or late afternoon hours (sometimes at noon), using echolocation to catch a myriad of insect pests, such as moths, beetles, ants, wasps, flies, and mosquitos, as it gently flutters up and down just a few feet off the ground or along cliff faces. It drinks water while in flight, lightly skimming the surface of rivers and lakes. Avoids strong winds due to its small size and lack of wing power.

The female gestates for 35-45 days, usually giving birth to 2 altricial pups in late spring to early summer. A secondary consumer, its natural enemies include owls and probably other typical bat predators (reptiles and small mammals). The western pipistrelle lives to about 12 yrs. in the wild.

WESTERN POCKET GOPHER (*Thomomys mazama*): Once thought to be a subspecies of the mountain pocket gopher, it has now been awarded separate species status. Nonetheless, it continues to possess much of the same ecology and life history as its close cousin. The main difference is their slightly overlapping ranges: the mountain's covers portions of California and Nevada, while the western's range extends over parts of the states of Washington, Oregon, and California.

Western pocket gopher.

WESTERN SPOTTED SKUNK (*Spilogale gracilis*): Grows to nearly 2' in total length and can attain a weight of up to 2 lb. Closely related to the eastern spotted skunk (the 2 were once thought to be conspecific), like its cousin, the western is small, nimble, and covered in fine, dense, silky black fur, though it has more irregular white markings. Particularly identifying are 3 longitudinal stripes on each side of the forebody and 3 vertical stripes running up and down the rump area. It has a stout anatomy; a broad head; a conspicuous white forehead patch; a tapered face; a slightly concave muzzle; small rounded ears; short legs; and small paws with 5 toes each. The tail is brushy and white-tipped, with white fur along the ventral back half and black fur along the dorsal front half. It is gender dimorphic, with the male being larger than the female.

This nearctic palustrine mephitid inhabits grassland, farmland, shrubland, woodland, desert, conifer stands, mountains, talus, savanna, fields, canyons, orchards, wetland, scree, intermontane regions, forest, chaparral, inland cliffs, cropland, rangeland, and rocky areas. It ranges (as its common name denotes) across the western side of North America, from southern Canada south to Mexico and Central America, covering most of the western U.S. states in between. It dens in loose soil, hollow logs, fallen timber, snag, attics, blowdown, brushpiles, deadfall, canyon walls, hillsides, and beneath buildings. It usually remains hidden during daylight hours; establishes a home range of up to 4 sq. mi.

A carnivore, frugivore, insectivore, and invertivore, it shares the same omnivorous dietary preferences as the eastern spotted skunk: roots, fruit, insects, eggs, reptiles, and small mammals (mice, voles, rabbits). It has long recurved claws on its front paws that not only aid it in excavating burrows, but also in digging up subterranean food. Kills live prey with quick savage bites to the neck.

Its breeding season takes place in early fall. Unlike the eastern, the western doe employs delayed implantation, allowing her to bear young when food and weather conditions are most favorable for their survival. She gestates for around 7 mos., giving birth to 2-6 altricial kits in the spring. When this small matriarchal family is foraging they engage in a behavior known as "caravanning," in which the young closely follow their mother single file. The buck plays no role in the rearing of his offspring.

Western spotted skunk.

Synanthropic, secretive, curious, sedentary, and nocturnal, it does not hibernate, but enters into a state of torpor during severe winter weather, conserving body fat. Communicates and perceives like other members of the Mephitidae family: primarily through touch and smell, and to a lesser degree sight and hearing. Its vocalizations include screeches, grunts, hisses, and barks.

To defend itself, its 2 scent glands produce a smelly sulphuric oil, which it shoots from fleshy tips located under its tail. These 2 "nipples" can be aimed accurately to a distance of up to 4 yd. Sometimes it sprays its victims while "hand-standing" on its front paws, with its hindquarters and tail in the air. A secondary consumer, its natural enemies are birds of prey (golden eagles, great horned owls), wild felids, and wild canids. Beneficial as a domestic pet and also as a pest controller (it consumes large quantities of insects and rodents); detrimental as a carrier of deadly diseases and as a home, yard, and farm pest. The western spotted skunk lives to about 3 yrs. in the wild; up to 10 yrs. in captivity.

WEST INDIAN MANATEE (*Trichechus manatus*): The official state marine mammal of Alabama and Florida, this massive gentle sirenian grows to a length of 15' and a weight of 3,500 lb. or more (though 1,500 lb. is more usual). Also known as the "sea cow," its portly but seal-shaped body is covered in thinly distributed, inconspicuous, sensory hair; its thick skin is dull grayish to brownish-black in color; its dorsum is often darker than the ventrum. It may appear to have a green "coat" due to the growth of algae on its back. A descendant of a prehistoric, wading, 4-legged land mammal (one that also gave rise to the elephant), the manatee has 2 small front flippers (arms or front legs) but no rear flippers (legs). When it took to the sea its hind legs were no longer needed and were subsumed into the body, replaced by a horizontally flat, paddle-shaped tail that propels it through the water with an up-and-down motion (compared to the side-to-side motion used by most fish).

West Indian manatee.

The body is bulky; the head is large and oval-shaped; the face is wrinkled; the eyes are small, dark, and somewhat dull; the flexile nose is blunt and somewhat pig-like (though much broader than a porcid) and covered in stiff sensitive whiskers; the nostrils are valvular (for closing underwater); the lips are thick and bristly; the upper lip has a deep cleft; it lacks external ears; the front flippers have 3-4 nails. Though an air-

breather, it is perfectly
adapted to aquatic life, with
lungs aligned side-by-side,
heavy marrowless ribs, and a
broad spatulate tail, all
which help it maintain
balance, orientation, and

Manatee skeleton.

stability in the water. The front flippers are used for steering, massaging its flanks, and removing unwanted detritus from its teeth. Gender dimorphism is evident: the female is larger than the male.

A native of the Atlantic Ocean, this coastal trichechid inhabits the warm shallow waters of the southeastern U.S., from Virginia through Florida (where it is most abundant), to the Gulf of Mexico and the Caribbean, south to Brazil. Moves freely in and out of saltwater, freshwater, and brackish water; congregates in slow-moving rivers, canals, lakes, estuaries, tributaries, lagoons, streams, marine bays, and along sandy shorelines; will travel over 100 mi. inland; prefers calm, quiet waters less than 20' deep. Due to its low energy diet, low metabolic rate, and lack of insulating blubber (most of its inner body is devoted to a massive digestive system), it particularly enjoys gathering in warm-water springs as well as the tepid waters discharged by power plants. It seldom ventures into either deep offshore waters or water with a temperature below 70 degrees.

An opportunistic herbivore, this sirenid feeds primarily on coarse sea grasses that it yanks from sea and river bottoms. It roots about in the sand and muck using its flippers to swim, pivot, "walk" along the sea floor, and dig up its food; supplements its diet with water hyacinth, mangrove leaves, acorns, algae, and hydrilla. On occasion invertebrates and small fish are inadvertently swallowed while it is eating. Its prehensile lips possess the ability to grasp and manipulate aquatic plants, facilitating feeding. Due to the low protein value of its diet it must eat for up to 8 hrs. a day, consuming as much as 200 lb. of plant mass, or 10-15 percent of its body weight, every 24 hrs. Due to the abrasive nature of the vegetation it eats (as well as the sand it accidentally ingests), as in elephants, its teeth wear down and must be constantly replaced, a process that continues throughout its life (old teeth move forward and fall out, superseded by new ones that grow from behind). To aid in processing the large amount of roughage it consumes, it has evolved into what is known as a "hindgut digester" with an intestine 100' long (by comparison, a human's small and large intestines together average about 28' long).

Despite its generally slow movements and rotund appearance, it is quite agile and can swim quickly and gracefully, attaining an upper speed of 20 mph—though it normally swims along at 5 mph or less. It can remain underwater for up to 20 mins., though 3-4 mins. is more usual; may sleep "standing up" in the water just below the surface, or it may rest on the bottom, occasionally surfacing to take a gulp of air. This diurnal and nocturnal cousin of the elephant is solitary, and therefore nonterritorial. While lacking a social structure, it does form temporary mating herds or aggregations when mature females are in estrus, and also during cold weather and times of food surplus. It is curious and approachable by nature, as well as affectionate and playful, and will often

West Indian manatee,

greet its fellows with a "kiss" on the lips and a flipper "hug" before rollicking in the water together. Some populations migrate south with the approach of winter.

This docile member of the Trichechidae family is a slow reproducer with no set breeding season, though spring and summer are favored times for mating. The cow is courted by the mating herd's mature bulls. She becomes pregnant only once every 2-3 yrs., gestates for up to 14 mos., and gives birth to a single 65 lb., 4' calf underwater, helping it to the surface to breathe. The calf can begin consuming solid food (aquatic vegetation) within a month after birth. However, it is not fully weaned until it is 1-2 yrs. of age. The calf nurses underwater from the mother's 2 mammary glands, located behind her front flippers. As with nearly all mammal species, the mother-offspring relationship forms the one and only true nuclear family among West Indian manatees. Fathers play no role in the rearing of their young.

Communicates and perceives via channels that are normal for members of the Sirenia order: sight, sound, touch, and smell. Vocalizations include squeals, grunts, squeaks, screams, and chirps. As a regional heterotherm it has the capacity to adjust the blood flow in specific regions of its anatomy, which helps it regulate its overall body temperature. It does not appear to have any natural enemies. Nearly all injuries and premature (unnatural) deaths are anthropogenic in nature: boat strikes, swallowing garbage and fishing line and hooks, net and trap-line entanglements, pollution, drowning in canal locks, habitat loss, hunting, etc.

It is highly beneficial to us as an object of ecotourism and as an aquatic weed controller (helping to keep boating channels clear of overabundant plant life). Vulnerable and endangered with a population that is steadily declining each year, this mild-mannered marine mammal is protected by numerous laws. The West Indian manatee may live 40-50 yrs. in the wild; up to 70 yrs. in captivity.

WEST INDIAN MONK SEAL (*Monachus tropicalis*): Also known as the "Caribbean monk seal" or "sea wolf," the last confirmed sighting was in 1952 (near Jamaica). In 1967 it was placed on the endangered list, and in 2008, after a meticulous but unsuccessful search for surviving individuals, the U.S. declared the species extinct. Because it disappeared before in-depth scientific studies could be done, its ecology and life history are incomplete.

This warm-water phocid grew to a total length of about 8' and weighed up to 600 lb. Its pelage on the dorsum was gray to brown in coloration; on the ventrum it was lighter in color. Like other true seals it was somewhat ungainly on land. A coastal pinniped, it inhabited sandy beaches in warm subtropical waters, with a range that extended across the Caribbean Sea and the Gulf of Mexico, making it the only known pinniped to have inhabited this particular area. (At one time it was found as far north as South Carolina.) Its diet, reproductive behavior, and communication and perception channels were likely similar to other members of the Phocidae family. It may have had a life span of 20-25 yrs. in the wild.

West Indian monk seal.

First reported by Christopher Columbus in 1494, due to its passivity and approachability this tropical marine carnivore was easily killed (for its fur, meat, and oil). By the early 20th Century the species may have already been incapable of sustaining itself, and the last known individuals soon perished.

Species thought to be extinct sometimes reappear and rebound, and, indeed, in recent years some claim to have seen it in and around the Caribbean. Mainstream scientists believe that these sightings are misidentifications and that the West Indian monk seal is "gone for good." Yet, many mammals that were once declared "extinct"—like the New Guinea highland wild dog, Nelson shrew, mountain pygmy possum, Chacoan peccary, Laotian rock rat, Cuban solenodon, Philippine naked-backed fruit bat, and monito del monte—have "risen from the dead" Lazarus-like, and proven them wrong. The West Indian monk seal may prove science wrong yet again.

WHITE-ANKLED MOUSE (*Peromyscus pectoralis*): Also known as the "white-ankled deer mouse," this medium-sized mouse grows to about 9" in length, its hind foot measures about ¾", and it can attain a weight of up to 1½ oz. On the dorsum it is tannish-gray; the lateral coloration is buff-gray; the ventrum is white; the long 4½" tail is sparsely furred, ringed, and dichromatic; the dark beady eyes and erect ears are somewhat large for its body size. The ankles are white or whitish, giving it its common name. In the field it is sometimes difficult to distinguish from other similar species.

A member of the *Peromyscus* genus.

This motile cricetid is a nearctic native that inhabits rough mountain terrain, oak forests, arid desert, grassland, foothills, rocky fields, bushland, and scrub woodland. Its range is limited to the American Southwest, covering portions of Arizona, New Mexico, Texas, Oklahoma, and northern Mexico. It dens and nests under rocks, dense vegetation, and logs.

An omnivore, its diet consists of typical Southwestern *Peromyscus* fare: seeds, grains, fruits, nuts, cacti, and insects. Little is known of its reproductive activities, but they are probably similar to other Western members of its genus—as are its communication and perception channels, natural enemies, and life span.

WHITE-BEAKED DOLPHIN (*Lagenorhynchus albirostris*): Grows to 10' in total length and weighs up to 700 lb. The upper body is gray to black, the lower body is white to grayish; sides may have white-gray flared markings; back half of the body will often have a whitish-gray "saddle" on top. In some cases the body is mottled. As its name indicates, it has a white beak—though in some cases it is gray. It has a somewhat stubby body; a short snout; long pointed flippers; and a hooked dorsal fin set midway on the back.

It inhabits the cold deep waters of the Atlantic, but sometimes comes inshore during the warmer months. Its diet is typical of the Delphinidae family: aquatic crustaceans, cephalopods, and fish, which it

White-beaked dolphin.

hunts (sometimes in cooperative packs) using a sophisticated form of biological sonar called echolocation. Sometimes it congregates and travels in superpods of over 1,000 individuals. It can dive a half-mile deep and swim up to 30 mph, and will feed alongside whales and other dolphin species.

Like most of its kind, this lively delphinid is frolicsome, curious, and fun-loving, and readily takes to riding the bow waves of boats and large whales, breaching, and playing with various objects, such as seaweed, shells, and flotsam. Vocalizations include clicks, whistles, and squeals. A secondary consumer, its natural predators are sharks, orcas, and polar bears. The white-beaked dolphin lives 40-50 yrs. in the wild.

WHITE-EARED POCKET MOUSE (*Perognathus alticola*): This small mouse, a close cousin of the great basin pocket mouse, grows to a total length of about 7", and earns its common name from its whitish-furred ears. It is light brown on the dorsum, with a grayish dorsal wash flowing over the back; its lateral coloration is buff-white; the ventrum is whitish; the long tail is bicolored (dark on top, lighter below) and tufted at the terminus. Gender dimorphism is present, with the male being larger than the female.

This endothermic heteromyid is a nearctic native that inhabits temperate, open, elevated grassland, overgrown fields, shrubland, montane woodland, steppe, rangeland, coniferous forest, scrubland, chaparral, sageland, and arid piñon-juniper communities;

it generally occurs in association with friable soil, ferns, sagebrush, Joshua trees, and ponderosa pine. Endemic to California, its minuscule range is limited to small portions of the counties of San Bernardino, Ventura, Kern, and Los Angeles. It dens and nests underground; establishes a small home range of about 1,500 sq. ft.

A nocturnal herbivore, its diet consists primarily of seeds, herbs, and grains; caches surplus food in underground larders, transporting it in its cheek "pockets" or pouches—from which it derives the latter half of its common name. Little is known about its reproductive activities. Communication and perception channels are probably similar to other members of its genus. Its natural enemies include reptiles, birds of prey, mustelids, wild felids, and wild canids. The life span of the white-eared pocket mouse is not known.

White-footed mouse.

WHITE-FOOTED MOUSE (*Peromyscus leucopus*): Also known as the "wood mouse" and the "white-footed deer mouse," this small mouse grows to a total length of around 8", its hind foot measures about ¾", and it can attain a weight of up to 1½ oz. Pelage coloration is variable, but the dorsum is generally cinnamon-brown to grizzled white, tan, and gray; lateral coloring is orangish to grayish; the ventrum and feet are white, giving it its common name; the long 3½" tail is lightly furred and slightly dichromatic (dark above, light below); many individuals have a dark narrow dorsal stripe. The eyes are large, dark, beady, and rounded, enhancing night vision while permitting a wide field of vision; the ½" ears are large, rounded, and erect; the snout is covered in tactile vibrissae (more adaptations to a life in the dark). In the field it may be difficult to distinguish this rodent from other members of the *Peromyscus* genus.

A terricolous sedentary cricetid, this highly adaptable nearctic native inhabits temperate dry woodland, chaparral, warm mixed forest, farmland, brushland, roadsides, and desert scrubland; it is found in both wilderness and urban areas. An edge species, it thrives in and around ecotones, or any setting where one habitat sharply abuts another; prefers locations with dense hiding cover, fallen trees, stumps, deadfall, and brushpiles. Its massive range extends from Arizona, Wyoming, and Colorado in the West to New England in the East, from Saskatchewan, Quebec, and Montana in the North to Texas, Alabama, and Georgia in the South; it is found as far south as southern Mexico (it does not occur along the East Coast from Virginia to Florida nor in any of the far Western states).

It does not hibernate (except perhaps in the northernmost part of its range); active year-round; may enter brief periods of torpidity during extreme weather conditions. Territorial and solitary, it establishes a home range of about 40,000 sq. ft. Will appropriate the dens, nests, and burrows of other animals when possible; during cold weather it may become semicolonial, grouping together with conspecifics for warmth and safety.

A common, crepuscular, nocturnal omnivore, its diet consists of the usual Cricetidae fare: seeds, grains, leaves, nuts, fruits, flowers, bark, stalks, wood, fungi, and insects; it hoards and caches surplus food; occasionally forages in trees. Its breeding season fluctuates with geography, generally peaking between spring and fall. The iteroparous polyestrous female may produce 3-4 litters annually. She gestates for 3-4 wks., gives birth to an average of 4-5 altricial pups,

White-footed mouse.

and weans them at 3 wks. of age. While the father may remain with the mother after mating, he evacuates the nest site before birthing begins, and thus plays no role in the rearing of his offspring. As with most other mammal species, the nuclear family of this rodent is the mother and her young.

Communicates and perceives along standard *Peromyscus* channels: sight, smell, hearing, and touch. Its natural enemies are reptiles, birds of prey, mustelids, wild felids, and wild canids. Both semi-natatorial and semi-scansorial, it can climb and swim to escape predators when necessary. When alarmed it engages in foot-drumming with its front paws. Beneficial as a seed and spore disperser, a pest controller (consuming creatures we consider vermin), and as a prey base for carnivores; detrimental as a forestry pest, house pest, and carrier of deadly diseases (such as the hantavirus and Lyme disease). The life span of the white-footed mouse is about 6-12 mos. in the wild; twice that long in captivity.

WHITE-FOOTED VOLE (*Arborimus albipes*): Little is known about this small rodent, the most uncommon of our region's voles. It grows to a total length of about 7", its hind foot measures ¾", and it can attain a weight of up to 1 oz. The long dorsal pelage is darkish brown; the soft ventral pelage is brownish-gray, washed over with grizzled black-tipped hairs; the thin 3" tail is lightly furred and dichromatic (dark above, light below). The eyes are small and dark; the ears are rounded and hairless; and the front paws have well-developed (though tiny) claws—all adaptations to a fossorial way of life. Its common name and its species name derive from one of its more noticeable features: its white-topped feet.

This motile riparian cricetid is a nearctic native that inhabits temperate redwood forests, often in the vicinity of alder stands, rotting logs, and languid creeks. Its palustrine range is limited to the northwest coast of America, covering portions of the states of Oregon and California.

Vole.

A nocturnal terricolous herbivore, its vegetarian diet consists of grass, moss, herbs, roots, leaves, and pollen, and green plant material generally. A tiny tree vole, it gets its genus name from its habit of foraging in trees (though it will search for food on the ground as well).

A fully terrestrial arvicolinid, it breeds year-round. The female gestates for approximately 1 mo., gives birth to 2-3 altricial pups, and weans them at around 1 mo. of age. Communicates and perceives along typical *Arborimus* channels: smell and touch, with reduced sight and hearing. Its natural enemies are birds of prey, mustelids, and wild felids. The life span of the rare white-footed vole has not been studied.

WHITE-NOSED COATI (*Nasua narica*): Also called the "coatimundi," it has a shoulder height of 1', grows to a total length of 53", and can weigh up to 25 lb. Its coat coloration ranges from cinnamon, light brown, and gray, to dark brown, black, and grizzled. Like its close cousin the raccoon, this intelligent, lithe, energetic mammal has a face-mask and a ringed-tail. It has small ears; a white spot on each cheek; black feet with furless soles; and white supraocular and subocular arches. Its long pig-like snout has a white band behind its black nose (hence its common name). This flexible appendage can bend 60 degrees in any direction, lending it the nickname the "hog-nosed raccoon."

An adaptable riparian, this nearctic native prefers habitats with good water sources, such as green grassland, savanna, montane forest, canyons, and tropical areas. However, it is also

Palm tree.

known to haunt deserts, dry forests, rocky areas, and sand dunes. Unlike the ubiquitous raccoon, in our region it is found only in the southwestern U.S. (Arizona and New Mexico), south into Mexico and beyond.

An endothermic omnivore, it will journey up to 1 mi. in search of food. Its diet consists of fruit, nuts, cacti, forest floor litter insects and arachnids (spiders, scorpions, beetles, termites), eggs, land crabs, amphibians (frogs), reptiles (snakes, lizards), and rodents (mice). It is also fond of carrion and human refuse.

This social, highly vocal, arboreal procyonid is a proficient tree-climber. Its long tail, which it carries straight up while walking, is semiprehensile and is used mainly for balancing while climbing. With its webbed feet it is also an excellent swimmer. Diurnal, it is most active during the day, at which time it plays, feeds, and grooms. To protect itself from predators, at night it sleeps in a primitively-built nest in the tree canopy. Males are solitary, except during breeding season; females prefer palm trees as nesting locations; females and their male and female young travel in bands of 20-30 members.

White-nosed coati.

It communicates and perceives along standard Procyonidae channels: mainly smell and touch, and sight and hearing to a lesser degree. Utilizes a number of different vocalizations, such as barking, growling, squealing, chattering, chirping, grunting, whining, hissing, screaming, and chuckling. It also telegraphs information to conspecifics chemically by spraying urine and secreting oils from a gland under its tail. A secondary consumer, its natural enemies are snakes (boas), birds of prey (hawks), wild felids (mountain lions, jaguars), and black bears. The white-nosed coati lives up to 8 yrs. in the wild; 15 yrs. in captivity.

WHITE-SIDED JACKRABBIT (*Lepus callotis*): A hare not a true rabbit, this large leporid grows to a total length of 23", its hind foot measures 5¼", its ears are nearly 5" long, and it weighs up to 6½ lb. In summer the short rough fur on its upper body is a grizzled cinnamon-brown (a mixture of white, tan, and black hair), while its lower body and sides are white (the source of its common name).

White-sided jackrabbit.

The outside of its 4 legs are cinnamon colored. From the bicolored tail, which is blackish-white on the dorsum, white on the ventrum, a blackish line runs up onto the hindquarters. The eyes are ringed; the inner ear tips are black; the outer ear tips are white; the inner ear is buff. In winter its upper pelage turns a grizzled gray, while the rest of its fur remains similar to its summer coat (with minor changes).

A nearctic native, it inhabits elevated plateaus, tableland, desert, savanna, and level grassy plains (especially those consisting of tabosa grass) throughout its narrow range: southern New Mexico to central Mexico.

A nocturnal herbivore, granivore, and folivore, it comes out at night to feed on all manner of grasses, sedges, and seeds, but also tree parts, such as leaves, bark, and twigs. By day it usually rests in a form (a shallow depression in the grass), in which it naps and grooms. It breeds from spring to summer. The female gives birth to 1-4 precocial leverets. Unlike many other hares and rabbits, the males and females of this species often form strong monogamous pair-bonds, some which last a lifetime.

Somewhat antisocial for a lagomorph, it communicates with conspecifics primarily through scent; but as with many other members of the Leporidae family, it also uses tail-flashing as well as several vocalizations, such as grunting and screaming. A primary consumer, it has typical hare enemies: reptiles, birds of prey, wild felids, and wild canids. With its lithe body, acrobatic abilities, hardened claws, and long powerful legs, this cursorial mammal is built for speed. It evades predators through camouflage, leaping, running (in a zigzag pattern), and flashing its white flanks, which may temporarily confuse its pursuer. The life span of the white-sided jackrabbit is unknown, but it probably lives an average of 5 yrs. in the wild.

WHITE-TAILED ANTELOPE SQUIRREL (*Ammospermophilus leucurus*): It grows to a total length of 9½", its hind foot measures about 1½", and it attains a weight of up to 5½ oz. This countershaded little ground squirrel has 2 yearly colorations: in summer its upper body is fulvous, in winter it is gray; in both seasons its underbody is white, tan, or buff. There is a predominant single white stripe running laterally along each side of the body from shoulder to thigh; the forehead and outer legs may be washed in rust; the 3½" tail is small and thin, with grizzled black hair on the dorsum, white hair on the ventrum (the trait from which its common name derives). The body is stout and small; the head is small with a slightly concave profile; the eyes are large, dark, and ringed in tan; the ears are small, erect, and rounded; the nose is tapered and snubbed; the limbs and feet are longish; the fingers and toes are clawed. It molts twice annually.

White-tailed antelope squirrel.

This small chipmunk-like sciurid is a nearctic native that inhabits rugged, temperate, arid regions throughout the western portion of North America, from Oregon in the north to Baja California in the south, from California in the west to Colorado in the east. It is thus also found in Arizona, New Mexico, Idaho, Utah, and Nevada. Its habitat includes deserts, shrubland, hills, chaparral, sand dunes, rocky locations, and grassland, preferably with soft gravelly soil, vascular vegetation, and a clean water source.

Builds extensive burrow systems with runways spreading out from their entrances. It sometimes takes over the homes of other subterranean animals, such as the desert rat. In an effort to regulate its body temperature under the blistering noonday sun, this wiry mammal both rests in the shade and practices heat dumping. The latter method it accomplishes by laying spreadeagle on a rock or damp soil—which helps wick away heat, allowing it to cool off. Semi-social in winter, it may den with conspecifics at this time for warmth.

A diurnal, crepuscular, and largely solitary omnivore, its diet consists of seeds, leaves, nuts, grains, fruit, agricultural crops, insects, small reptiles, and carrion. Hoards and caches food, transporting surplus edibles, such as cacti, to its burrow in its large cheek pouches. Its breeding season runs from midwinter to early summer (depending on latitude). The doe gestates for about 1 mo., giving birth to an average of 9-10 altricial pups in a grass- or fur-lined natal nest underground. They are weaned at around 60 days. The buck plays no role in the rearing of his offspring.

Communicates and perceives via traditional Sciuridae channels: sight, smell, hearing, and touch. Its natural enemies are reptiles, birds of prey, mustelids, wild felids, and wild canids. An excellent runner, in times of danger it sprints away with its flashing white-bottomed tail erected over its back, serving as an altruistic warning signal to nearby conspecifics. It is of benefit to us as a seed disperser, which helps the spread and growth of forests and other vegetation. The white-tailed antelope squirrel may live for 1-2 yrs. in the wild.

WHITE-TAILED DEER (*Odocoileus virginianus*): The official state mammal of Arkansas, Georgia, and Nebraska, the official state wildlife animal of Wisconsin, the official game mammal of Michigan, the official state game animal of Oklahoma, and the

White-tailed deer herd.

official state animal of Illinois, New Hampshire, Ohio, Pennsylvania, and South Carolina, it is the most abundant game animal in the eastern section of our region. The muscular "whitetail" stands about 3½' at the shoulder, grows to a total length of 7', and can weigh as much as 500 lb., though 300-400 lb. is more usual. Its dorsal coloration varies, but it is generally reddish-brown during the warm months, grayish-brown during the cold months; its eye rings, upper nose, chin, throat, abdomen, and inside upper leg are white. The tail is brown above with white edging, completely white below (giving it its common name), and there is often a median black stripe on the upper lower half. The male's antlers, which can spread to 3' from tip to tip, grow out in spring, lose their velvet in late summer, and are shed in midwinter. Newborns are adorned with white spots over a cinnamon-brown coat, making for excellent protective camouflage. By about 120 days the spots disappear and the fawn begins to take on adult pelage coloration.

Highly adaptable, this sedentary cursorial cervid inhabits numerous biomes and ecosystems, from tropical swamp and arid desert to farmland and temperate forest. Seeks out areas with good hiding cover (such as dense brush) and clearings that border woodlands (where it can both more easily find food and escape predators). Its range is extensive, covering most of southern Canada and nearly the entire U.S. (except for parts of California, Utah, Nevada, Colorado, Arizona, and New Mexico), running all the way to Central America.

A crepuscular herbivore (that may also be seen during the day), it eats over 500 types of vegetation, which it digests in its 4-chambered stomach. As with many other members of the Cervidae family, its spring-summer diet surrounds green vegetation, such as grasses, leaves, herbs, fruit, and crops; in the fall it eats nuts and corn; in winter it feeds on tree parts,

White-tailed deer (female).

such as the twigs and buds of birch, maple, and poplar trees. Coniferous trees are also consumed. In dry areas it eats cacti; in wet areas it includes aquatic plants in its diet.

Though normally solitary, it forms herds that vary with the season. Through spring, summer, and fall, for instance, bucks congregate in bachelor herds, while does assemble in matriarchal herds. During winter, however, both genders "yard up," forming one large group. The rut takes place in the fall, at which time males antler-clash in

White-tailed deer.

competition over estrus females. The doe gestates for about 7 mos., giving birth to 1-4 precocial fawns in the spring. The young are weaned at around 10 wks., with males becoming fully independent at 1 yr. of age, females at 2 yrs.

Whitetails communicate and perceive using physical cues (body language), aural cues (snorts and bleats), and olfactory cues (scents emitted from numerous glands). It also scrapes its antlers on trees to mark its territory, which strips off bark, leaving bare polished sections on the trunk (known as "buck rubs"). A primary consumer, its natural enemies include bears, coyotes, jaguars, wolves, and mountain lions. As a prey animal it has evolved numerous defenses: it can swim 13 mph, run 40 mph, jump over a 7' fence from a standing position, and leap 20' in a single bound at a full run. It also has a keen sense of smell and hearing. When frightened it raises its tail, exposing the white underside (a "warning flag" to others). The white-tailed deer lives an average of 2-4 yrs. in the wild; more than twice that long under optimum conditions; up to 20 yrs. in captivity.

WHITE-TAILED JACKRABBIT (*Lepus townsendii*): This large lagomorph grows to a total length of 26", its hind foot measures nearly 7", and it can attain a weight of up to 10 lb. Subject to photoperiodicty, its dorsum is a light grizzled gray and the ventrum is whitish during the warm sunny months; in the dark of winter its entire coat becomes whitish to light gray. Its eyes are ringed in white; its long 4½"ears are rimmed in white and tipped in black; and its tail is completely white all year-round—the trait for which it was named.

A nearctic native, it inhabits grassland, fields, prairie, farmland, plains, shrubland, and montane forests throughout its range: central to western Canada, and the central to western U.S. states (California east to Iowa, and south and west to Colorado and New Mexico.) Though this nocturnal leporid is usually solitary, it will congregate in groups during winter storms or times of nutritional abundance. During the day it rests in a form (such as a shallow depression in the grass), where it passes the time cleaning itself and napping.

In the summer this terricolous, cursorial, sedentary herbivore feeds on grasses, clover, forbs, and bushes; in winter it turns to woody vegetation, such as tree twigs, buds, leaves, and bark.

White-tailed jackrabbit.

Courtship and breeding take place from midwinter to midsummer, with the polyestrous doe producing up to 4 litters annually. She gestates for around 5 wks., giving birth to as many as 10 precocial young, who become independent about 1 mo. later. The buck plays no role in the care of his offspring.

It is relatively non-vocal, but squeals when scared or in pain. Its natural enemies are reptiles, birds of prey, wild felids, and wild canids. It is able to evade predators through its camouflaging coat, sharp senses of smell and hearing, by freezing in place, leaping (up to 20'), running (up to 45 mph), zigzagging, and swimming. It is believed that the white-tailed jackrabbit lives 6-7 yrs. in the wild, but its average life span is probably closer to 1-2 yrs.

WHITE-TAILED PRAIRIE DOG (*Cynomys leucurus*): This large stout ground squirrel grows to a total length of almost 15", its hind foot measures nearly 3", and it can attain a weight of close to 4 lb. The cryptic coloration on the upper body is a combination of tan, yellowish, white, pinkish, gray, and brown fur—sometimes with a blackish undercoat; there is an indistinct lateral line; the lower body is a lighter version of the dorsal fur; the 3" tail is short, brushy, and white-tipped (the latter attribute giving it its common name); it

White-tailed prairie dog (juvenile).

has dull brownish-blackish supraocular and subocular markings (though the one below each eye might also be considered a cheek patch); the nose is golden brown; the ears are small. Gender dimorphism is present: the male is larger than the female.

This motile sedentary sciurid is a nearctic native that inhabits cool, elevated, dry, open, temperate grassland, sageland, pastures, valleys, rangeland, shrubland, savanna, chaparral, montane meadows, semidesert, and sand dunes. Its range covers portions of 4 Rocky Mountain states: Montana, Wyoming, Colorado, and Utah. Gregarious, it forms large informal colonies, each comprised of an average of a half dozen individual family clans. Its vast burrow systems are located near or beneath vegetation, with prominent ejecta mounds near the entrances (some up to 3' in height); hundreds of subsidiary burrows spread out over the colony's home range of some 2 million sq. ft. Enters hibernation in the fall, waking periodically to feed; estivates during extremely hot weather conditions; spends most of its life in or close to its burrow, rarely traveling far from the entrance.

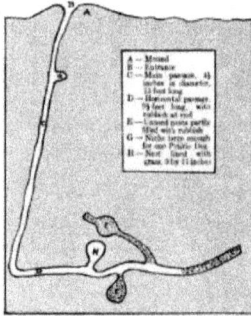

A typical prairie dog burrow.

A diurnal herbivore, its diet consists of typical Sciuridae fare: seeds, grains, sedges, flowers, grasses, weeds, leaves, forbs, herbs, and nuts; does not need to drink as it procures its water needs from the solid food it consumes. Its breeding season begins after emergence from hibernation in the spring. The female gestates for about 5 wks., giving birth to an average of 4-5 altricial pups in a subterranean nest chamber; she nurses, grooms, and protects them until weaning takes place at around 4 wks. of age. The non-paternal male plays no role in the rearing of the young.

Communication and perception run along standard *Cynomys* channels: sight and hearing, with smell and touch being somewhat reduced. This species is well-known for its sharp "bark," a loud, piercing, altruistic alarm call delivered at or near the burrow entrance, warning the rest of the colony of potential danger. Other vocalizations include whistles, yaps, screams, "laughing" calls, and growls. Though not as social as the black-tailed prairie dog, it occasionally engages in such behaviors as allogrooming and "kissing." Commonly stands upright on ejecta mounds to survey the surrounding territory. Its natural enemies include reptiles (rattlesnakes prey on neonates and pups), birds of prey, procyonids, mustelids, wild felids, and wild canids. An important predator of this rodent is the black-footed ferret, which is highly endangered.

Beneficial as a soil aerator, fertilizer, and recycler, as a focus of ecotourism, as a commensal underground house provider (for other animals such as snakes, ferrets, and burrowing owls), and as food for carnivores; detrimental as an agricultural pest, as a ranch pest, and as a carrier of deadly diseases (like sylvatic plague). The life span of the white-tailed prairie dog is 4-5 yrs. in the wild.

WHITE-THROATED WOODRAT (*Neotoma albigula*): This medium-sized, desert-loving rodent grows to a total length of around 16", its hind foot measures nearly 2", and it can attain a weight of up to 11 oz. Its soft countershaded pelage is buff with a gray sheen on the dorsum, white or light gray on the ventrum; one of the so-called "furry-tailed" woodrats, its 7" tail is dichromatic (dark on top, light on the bottom) and covered with sparse hair. The body is thick and ovoid; the head is large and broad with a convex profile; the eyes are large, dark, and protruding; the ears are large, wide, naked, erect, and bat-like; the muzzle is robust and tapered; the nose is snubbed and covered in tactile vibrissae; the limbs are short; the feet are white and clawed. Its genus name, *Neotoma*, is Greek for "new-cut," a reference to its cutting teeth, differentiating it from other similar rodents. Its species name, *albigula*, is Latin for "white throat."

This nocturnal cricetid is a nearctic native that inhabits temperate brushland, low desert, dry

White-throated woodrat.

scrubland, arid plains, and cacti-dominant areas. It is sometimes also found in edgeland, caves, mines, and the rocky foothills of mountains. Its range occupies only the American Southwest, from Nevada, Utah, and Colorado in the north, to Arizona, New Mexico, Texas, and northern Mexico in the south. Territorial, it establishes a home range of about 1,500 sq. ft., which it defends by scent-marking with pheromones. Does not hibernate; active all year.

Its house, typically located at the base of a cactus, is a massive, elaborate mound made of cactus, mesquite, and juniper parts, as well as a wide assortment of sticks, branches, woody stems, leaves, floor litter, bones, dung, deadfall, and, being a packrat, any human debris it comes across. Within its walls lie numerous tunnels as well as separate chambers for eating, sleeping, storage, and nesting. Well-used runways radiate out from multiple entrances. This woodrat's house is continually being upgraded and can grow 7'-9' in width and reach a height of up to 4'.

A terricolous and sedentary omnivore with an herbivorous streak, its diet principally surrounds cacti parts, such as cactus leaves, flowers, fruit, beans, and bark. It also consumes grains, nuts, stalks, seeds, and woody vegetation, as well as some animal matter, including insects, reptiles, and other small mammals. Hoards and caches surplus food.

Depending on geography, its breeding season generally runs year-round. This solitary species is only social during this time. The polyestrous doe can produce at least 2 litters annually. She gestates for about 4-5 wks., giving birth to 2-4 altricial pups in a subterranean grass-lined nest chamber. Weaning takes place at around 3 mos.; dispersal and independence occur at about 6 mos. of age. The buck is non-paternal and plays no role in the raising or care of his offspring.

This common and abundant mammal communicates and perceives via the usual Cricetidae channels: sight, hearing, smell, and touch. Engages in foot-drumming; an expert climber, it can easily scale needle-covered cacti without harm. It benefits its ecosystem by hosting parasites, supplying homes for other animals, and providing food for large carnivores. Injures humans by nesting in our homes, where it can cause great damage. Its natural enemies are reptiles, birds of prey, mustelids, mephitids, procyonids, wild felids, and wild canids. The white-throated woodrat probably lives an average of 2-3 yrs. in the wild.

WILD BOAR (*Sus scrofa*): Also known as the "Eurasian wild pig," "feral hog," "razorback," and "European wild boar" (which in North America includes hybrids and feral hogs), this burly member of the Old World swine family is responsible for producing a well-known subspecies: the far more docile domestic farm pig (*Sus scrofa domesticus*). The wild boar reaches a shoulder height of 3', grows to a total length of 6', and weighs up to 450 lb. (nearly ¼ of a ton). Its pelage coloration ranges from white, tan, tawny, fulvous, and brown, to reddish, gray, black, and grizzled. An adult's fur is sometimes mottled or speckled; piglets are longitudinally stripped for the first month or so.

The adult's coat is made up of long coarse hairs, and it has a mane that covers the neck, shoulders, and spine. Some individuals have cheek ruffs as well as throat manes. Its barrel-shaped body is supported by stout legs. The tail is short and straight, the ears are small, oval, and point backward, and the long snout ends in a tough disk-like appendage that is designed for rooting. The male's upper canines or tusks are used as weapons, and grow to a length of 9". Gender dimorphic, males are typically larger in size and weight than females.

Wild boars.

Highly adaptable (but with a dislike of extreme weather), this crafty suid inhabits a wide variety of biomes and ecosystems, from dry brushland, riverbanks, swamp, ridgeland, coastal areas, semiarid rangeland, orchards, and marshland (both fresh water and brackish water), to flood plains, farmland, savanna, mud flats, forest, and mountainous areas. First introduced to North America by Spanish explorers in the 1500s, it was reintroduced to our region in 1893, when a herd from Germany was brought to New England to serve as a game animal. In the early 1900s still more European hogs were released here, this time in the American South and on the West Coast. Since then this rugged, fast breeding, rapacious mammal has exploded in population and is now one of the most widely distributed terrestrial mammals in the world.

Wild boar (sow with young).

In the U.S. today the wild boar (and its many variations) now ranges over the entire southeast, much of the southwest, and the Pacific Coast north into Oregon, displacing, killing, and eating native wildlife, destroying crops, spreading disease, degrading water, and eroding the soil, causing $2.5 billion in damage a year. Wildlife biologists, conservationists, and hunters alike consider it a threat to biodiversity and to both native and endangered species, for it can wipe out entire ecosystems, leaving biological ruin, financial devastation, and widespread human illness in its wake. At least one conservation group lists the wild boar (feral pig) as one of our planet's worst invasive and most destructive alien species—just one of the reasons why, in our region, it is called "the most hated animal in America."

An opportunistic omnivore, like other members of the Suidae family this hardy porcine will eat nearly anything, from roots, tubers, bulbs, fungi, seeds, grasses, herbs, fruit, grains, nuts, mineral blocks, dung, and agricultural crops, to worms, snails, insects, crayfish, eggs, nestlings, carrion, amphibians (salamanders, frogs), rodents (shrews, voles), reptiles (snakes, lizards, turtles), birds, rabbits, fawns, domestic pets, and the young of farm animals (cows, goats, sheep). Will range up to 50 sq. mi. in search of food. As a result of its enormous destructive impact, in most states hog hunting is permitted year-round.

Both crepuscular and nocturnal, it spends the daylight hours trying to stay cool and avoid predators. Thus it is most likely to be seen between dusk and dawn when it is most active. Sows form matriarchal groups called "sounders," made up of 5-50 mothers, daughters, sisters, cousins, and their (male and female) offspring. Mature boars are typically solitary, usually only joining the sounder when breeding, which (as this mammal has no set mating season) may occur at any time of the year. (Such mixed herds may contain up to 100 individuals.) Breeding reaches its maximum level in early winter. Males compete over females. A prodigious breeder, the sow produces 1-2 litters a year, gestates for around 4 mos., and gives birth to 1-14 piglets (though 4-5 is average) in the spring. The nest, a grassy hollow in the ground, is set up in an isolated thicket away from the sounder. The young

Wild boars.

are weaned at 3 mos., becoming independent about 1 yr. later. While females will defend the piglets in their particular sounder, males play no role in the care of the young.

Besides body language and touching, wild boars communicate and perceive using a variety of other cues, both vocal and chemical. The former include snorts, cries, squeals, shrieks, screams, and grunts; the latter includes scent-marking, a form of communication for which its keen sense of smell and large sensitive snout are ideally suited.

Deadly when provoked, the natural enemies of this invasive species are crocodiles, owls, hawks, foxes, coyotes, bobcats, mountain lions, and bears—which take mainly juveniles due to the formidable tusks of the adults. Mature hogs depend not only on their dental weaponry, but also on their endurance, heightened sense of smell, running speed, and swimming ability to escape predation. As its region-wide population and destruction proceed to escalate, desperate efforts to control it continue. The average life span of the wild boar in nature is about 5 yrs., making the actual range 1-10 yrs. In captivity it may live as long as 30 yrs.

WILD HORSE (*Equus ferus*): An enduring symbol of the American West, and popularly known in our region as the "mustang," it grows to a height of about 5' (15 hands), and can weigh up to 1,000 lb. A descendant of prehistoric North American stock, and more recently the Spanish horse, as might be expected of a creature that is the product of thousands of years of both natural selection and artificial selection, its physical appearance can vary widely. Pelage coloration runs from white, tan, and reddish, to brown, gray, and black (and many shades in between). Coat patterns and markings too are almost endless, ranging from plain to patchy, from spotted to striped. Some of the better known colors and patterns are roan, perlino, bay, grulla, tobiano, chestnut, cremello, pinto, overo, sorrel, buckskin, palomino, Appaloosa, and dun.

The body is smallish but robust; the head is narrow; the eyes are large; the mane is shaggy; the ears are long and upright; the neck is massive; the muzzle is long; the nose is soft with large nostrils; the tail is long; the limbs are muscular; the hooves are extra hard; and its bones are heavier, stronger, and denser than those of its domestic cousins.

Its prehistoric ancestors had 4 or 5 toes, but our modern species has only 1. This makes it (along with the other mammals in this order, the rhinoceros, donkey, zebra, and tapir), a member of the only living group possessing a single toe: the middle one, which forms its rounded hoof. Hence the name of the order to which it belongs: Perissodactyla, meaning "odd-toed" or "uneven digit." Its common name, mustang, derives from the Spanish word *mesteño*: "strayed."

Essentially a prairie animal, this sturdy independent equid mainly inhabits rugged, open, arid plains with sparse but diverse vegetation. Riparian and adaptable, it is also found in or around rocky terrain, marshes, savanna, foothills, harsh desertland, swamps, mixed woodlands, and craggy mountainous regions. Its range extends throughout the states of California, Oregon, Idaho, Utah, Montana, Wyoming, North Dakota, Nevada, New Mexico, and Arizona. (There are Eastern populations as well, living along the Atlantic coast.) Semi-nomadic, where it has the available land it may range over a territory of 500-1,000 sq. mi., moving about with the changing seasons in search of food and water.

Wild horses.

This large grazing herbivore is a dedicated graminivore and a folivore, with a diet made up primarily of grasses, leaves, and general roughage. It also consumes seeds, grains, brush, thistle, nuts, stems, and coarse material such as wood and bark. It gets its water from streams, pools, seeps, watering holes, lakes, marshes, and rivers, and uses mineral licks when found. Will travel some 20 mi. a day foraging.

Wild horse (mustang).

Highly social, it lives in free-roaming, hierarchical herds of some 20 individuals. Led and protected by the dominant stallion (the alpha male), the mustang band is comprised of mares (the stallion's "harem"), fillies, colts (subordinate bachelor males), and foals of both genders. (Note: from birth to 1 yr. of age, all young horses are called "foals"; from 1-2 they are known as "yearlings"; from 2-4 yrs. of age the male is called a "colt," the female is called a "filly"; after 4 yrs. of age the male is known as a "stallion," the female is known as a "mare.") Snorting and aggressively pawing the ground, the alpha stallion challenges rivals and chases off intruders. A mare, or sometimes a group of mares, will occasionally take a leadership-like role in the harem band, helping herd and drive other members when necessary.

The wild horse forms warm social bonds with conspecifics, often engaging in mutual rubbing, nuzzling, touching, caressing, snuggling, allogrooming, and play. Horses from different herds will work together, or even temporarily join bands, to ward off predators. The social structure of the mustang herd can best be described as a well-ordered dominance hierarchy patterned around strict rules. Thus punishment (such as a nip, a headbutt, or a kick) can be swift and harsh if an individual violates them.

With the symmetric and elegant conformation of a fleet-footed cursorial mammal, the wild horse, with its untamable attitude, is in many ways similar to deer and elk. Athletic and powerful, this intelligent endurance runner can reach a top speed of 40 mph and gallop for an hour or more, adaptations for escaping its natural enemies: bears, mountain lions, coyotes, and wolves. Due to the size, speed, strength, and agility of healthy adult mustangs, however,

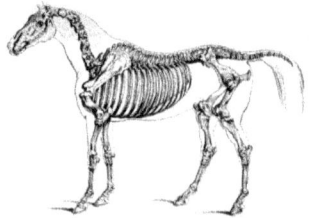

Horse skeleton.

carnivores are usually only able to prey on the young, the infirm, and the old. Diurnal and crepuscular, this unguligrade spends its mornings and evenings grazing and watering in attempt to avoid both the heat and predators. As a prey animal it prefers sleeping standing up, but will occasionally lie on the ground to rest or nap.

Its polygynous breeding season takes place throughout the spring and summer. Stallions fight for breeding rights, with the alpha male violently defending his harem against inferior males. During the rut males employ the flehmen response in order to check the readiness of estrus females. The mare gestates for 11-12 mos., producing 1 precocial foal (rarely 2) at night in an isolated spot. The foal, able to stand, walk, nurse, and run within an hour or two after birth, is fiercely protected by its mother until it becomes a yearling at 1 yr. of age. At between 1 and 2 yrs. it is weaned, then driven from the band by the alpha stallion: fillies join another harem herd, where they will grow up to become one of the "wives" of the dominant male; colts form bachelor herds, where they develop the strength and sparring skills to one day leave and establish their own harem group.

The majestic free-ranging mustang communicates and perceives using channels typical of the Equidae family specifically, and of large primary consumers more generally: sight, smell, touch, and sound. Relies on a wide array of vocalizations, from

Wild horses (mother and foal).

grunting and neighing to whinnying and squealing. Visual cues based on body language are important as well: individuals may paw the ground, shove one another, bob the head, stomp, kick, bite, expose the teeth, and pin the ears back (anger) or lay the ears forward (interest).

This proud perissodactyl is economically valuable to us: not only do we admire the wild horse for its resilience, beauty, and nobility, but it has long and faithfully served humanity as a laborer, pet, and provider of meat and hides. It is little wonder that its cousin, Przewalski's horse (*Equus ferus przewalski*), appears in prehistoric cave paintings tens of thousands of years old. As a seed disperser, parasite host, and food for carnivores, it has biological value as well. The North American mustang lives an average of 17 yrs. in the wild; up to 30 yrs. or more in captivity. (For a discussion on the debate over the native/nonnative status of the mustang, see Appendix D.)

Wolverine on the hunt.

WOLVERINE (*Gulo gulo*): It stands as tall as 17" at the shoulder, it can grow to 44" in total length, its hind foot measures 8", and it can attain a weight of up to 70 lb. This bearlike mammal has a thick brown to blackish oily coat with two lighter cinnamon colored stripes running nearly the full length of its body and out to its bushy 14" tail. A black dorsal "saddle" of dark fur covers its lower back; the limbs and feet are usually dark brown to black in color; the face is sometimes washed with dark or silver hair, giving it a mask-like appearance, while unique blonde or cream patches may appear on the throat and chest.

Its tank-like body, muscular build, and semi-plantigrade stride are designed for traveling in rugged conditions and snowy terrain. It has a large head; small dark eyes with horizontal pupils; a slightly concave muzzle; thick neck; short rounded ears; broad torso; dense fur; stout legs; massive paws (that act as snowshoes); and sharp, semi-retractable claws (that provide traction on rocks and ice). Its powerful jaws house large, robust, bone-crushing teeth, and include 2 specialized upper molars that are set at a 90 degree angle (pointing toward the inside of the mouth)—an adaptation that helps it more easily tear meat off frozen carcasses. The largest terrestrial mustelid, as well as the second largest member of the weasel family, in North America (only the sea otter is bigger), it is gender dimorphic, with the male being larger than the female.

This nearctic and palearctic native inhabits tundra, plains, mountains, forests, and boreal and arctic regions, often above the tree line (the alpine zone). Its original range has been greatly reduced, but this burly powerful cousin of the mink still survives in Alaska, much of Canada, the northern Rocky Mountain states, and in small populations in Washington, California, and Michigan.

Native Americans call it the "skunk bear" for its typical mustelid ability to discharge a feculent musk when alarmed or marking its territory. It can cover distances of up to 20 mi. a day; will go over a mountain rather than around it; can lope for hours without stopping; and it is a skillful swimmer and climber, one that will sometimes leap upon its prey from trees. An animal like this requires a large amount of unrestricted undeveloped space. Indeed, its personal home range may extend to over 1,000 sq. mi. of wilderness, an area it scent-marks and aggressively defends against intruders. Forms grass-lined dens inside rock piles, hollow logs, and upturned stumps.

Arguably the most powerful, aggressive, and ferocious mammal for its size in the world, this solitary wanderer can easily defend itself against

Wolverine.

much larger predators using its speed, power, agility, and pulverizing bite force. Usually ravenous and always opportunistic, this omnivore will hunt, kill, scavenge, and eat almost anything, from fruits, seeds, roots, and insect larvae, to birds, martens, porcupines, marmots, weasels, mink, beaver, and foxes; from eggs, voles, shrews, mice, lemmings, squirrels, hares, and rabbits, to lynxes, sheep, deer, caribou, elk, moose, carrion, and coyote and wolf pups. If available it will consume human garbage, livestock, and pets; breaks into cabins and homes in search of food; it is known for raiding trap lines, devouring both the bait and the trapped animals. It messily hoards and caches excess food under snowpack, scent-marking it as protection from other carnivores.

Though its scientific name, *gulo gulo*, means "glutton glutton," the wolverine is not a true glutton. It is merely an extremely active, non-hibernating animal with a high metabolic rate, living in harsh environments where food is scarce, and which therefore must be consumed rapidly and voraciously whenever and wherever it is found.

Its breeding season takes place in summer. The low reproducing female employs embryonic diapause, or delayed implantation, gestating for nearly 9 mos., then giving birth to as many as 6 all-white altricial kits in rocky debris, dense vegetation, or an excavated snow den. She nurses, grooms, and protects them, occasionally moving them from den to den (to discourage predation); when they are old enough she teaches them to hunt. They are weaned at 3 mos. of age and become independent at around 6 mos. The polygynous male does not directly assist the female in the rearing of their offspring. In fact, he will kill kits that he has not sired.

Wolverine.

Generally nocturnal, but sometimes diurnal, this hardy snow-dependent mammal has finely tuned senses, in particular smell and hearing (though it seems to have somewhat poor vision). Thus, like other members of the Mustelidae family, it communicates primarily via smell and hearing, in its case emitting vocalizations that include hisses, growls, snarls, howls, and grunts. Not only can it gallop through snow over long distances, it is also an excellent climber and swimmer. Active all year long; does not hibernate. Its hydrophobic coat helps insulate it against snow, ice, and cold water.

Though its natural enemies—wolves, mountain lions, and bears—are much larger, it will readily attack them in order to defend itself. Birds of prey will take young wolverines. Hunted and trapped for its pelt, sparsely distributed, occurring mainly in low population densities, and with rapidly disappearing habitat, this secondary consumer is extremely rare and not often seen by humans. As one of Nature's most efficient scavengers, it aids in keeping wild places clean, while as an apex predator it helps maintain the herd health of large game animals by eliminating the weak, the sick, and the old. These factors make it beneficial to both the ecosystem and to us. The wolverine lives to around 5 yrs. in the wild; up to 15 yrs. or longer in captivity.

WOODCHUCK (*Marmota monax*): Also known variously as a "groundhog," "whistle pig," "grass rat," "earth pig," and "marmot," this large chunky member of the squirrel family grows to a total length of 32" and can attain a weight of up to 15 lb. Though the dorsal, lateral, and ventral fur is generally the same color, that color may come in a multitude of shades, from rusty, orangish, and cinnamon, to reddish-brown, gray, and grizzled blackish; the short 9" tail is bushy and usually matches the color of the body. The dark eyes and rounded ears are small; the incisors are white; the limbs are short; the plantigrade paws are hand-like, well-clawed, and dark in color. Gender dimorphism is evident: the male is larger than the female.

This sedentary sciurid is a nearctic native that inhabits temperate, low-lying, sloped, open grassland, pastures, savanna, chaparral, meadows, overgrown fields, shrubland, farmland, orchards, and mixed woodland; often occupies well-drained edge habitat along hedgerows, fence rows, and forest borders, with a clean water source nearby. Found only in our region, with a 1 million sq. mi. range it is the most broadly distributed member of its genus (the marmots), living over the entire width of the middle tier of North America—from Alaska and British Columbia in the west to Nova Scotia and New England in the east, from Canada's Northwest Territories in the north to Mississippi, Alabama, and Georgia in the south (it is not found in the Rocky Mountain states and further west).

Woodchuck.

Fossorial, it excavates massive, multi-chamber burrow systems with numerous 1' wide portals (entrances, exits, spy holes, and escape holes); its tunnels may reach a depth of 6' and a length of over 30'; ejecta mounds and runways are usually present near the main entrance. A true hibernator, throughout the summer it builds up its fat reserves and in the fall retires to its hibernaculum to sleep away the cold months; it may estivate any time during extreme weather. Territorial, solitary, and antisocial, it establishes a home range averaging about 600,000 sq. ft., scent-marks the boundaries, and defends it aggressively against conspecifics and other intruders.

A terricolous, diurnal, crepuscular omnivore, its diet is made up of typical rodent fare: grasses, grains, nuts, stalks, flowers, herbs, leaves, garden vegetables, bark, wood, insects, mollusks, and eggs; it has a predilection for alfalfa, dandelion, plantain, and clover. Eating up to 5 lb. of succulent food a day means that it seldom needs to drink water. Its breeding season begins with the end of hibernation, stretching from late winter to spring. Males compete for estrus females, who gestate for around 1 mo., give birth to an average of 4 altricial pups in an underground grass-lined nest, then wean them approximately 6 wks. later. Neonatal care includes nursing, protection, and grooming. The male and female socialize only briefly during the mating season. Thus, the father plays no role in the rearing of his offspring.

Communication and perception run along standard *Marmota* channels: sight, hearing, smell, and touch. Vocalizations include barking, growling, whistling, squealing, hissing, and teeth-gnashing. Its natural enemies are reptiles, birds of prey, wild felids, wild canids, and ursids. Having its most important sensory organs (eyes, ears, nose) situated at the top of the skull helps it remain undetected by large meat-eaters while surveilling its territory from its den entrance; excellent running, tree-climbing, and swimming abilities provide its primary defenses against predation. Will sun-bathe on warm days. Groundhog Day, which falls on February 2 each year, takes its name from this common species, our region's third largest rodent. "Woodchuck" is an anglicized version of the Algonquin word *wuchak* and the Narragansett word *ockqutchaun*, both meaning "excavator."

Beneficial as a soil aerator, soil fertilizer, soil recycler, seed disperser, commensal homebuilder (for other burrowing animals), host for parasites, food for carnivores, and a subject of scientific research; detrimental as a home, garden, farm, and ranch pest. The largest member of the squirrel family, the woodchuck lives for an average of 2-3 yrs. in the wild; under optimum conditions it may reach 5 yrs. of age; in captivity it can live twice that long.

Woodchucks.

WOODLAND JUMPING MOUSE (*Napaeozapus insignis*): This mid-sized trichromatic rodent, a member of the subfamily *Zapodinae*, grows to a total length of around 10" and can attain a weight of up to 1¼ oz. It has a wide buff-grayish dorsal band running the length of the body; lateral coloration is bright orangish-cinnamon; the ventral pelage is white; the long 6" tail is thin, whip-like, scaly, white-tipped, and bicolored—dark above, light below. (Subspecies are variously colored, depending on latitude and geography.) As its common name indicates, it is a saltatorial species with large specialized hind legs, ankles, and toes. Incisors are grooved and yellowish-orange. Gender dimorphism is present, with the female being larger than the male.

A jumping mouse.

This solitary terricolousis dipodid is a riparian nearctic native that inhabits damp, boreal montane forests; it is also found in cool elevated wetlands, grassland, and mixed tree stands; it requires dense hiding cover or thick forest duff. Strictly a northeastern mouse, its range extends from Manitoba and Labrador in the north to Georgia and South Carolina in the south, from Minnesota and Tennessee in the west to New England and Virginia in the east. It rests, sleeps, and nests in a burrow system—either self-made or appropriated from other fossorial creatures. Builds up stores of body fat in preparation for fall hibernation, spending as much as a ½ year in its hibernaculum.

A crepuscular nocturnal omnivore, its generalist menu consists of standard Dipodidae fare: seeds, grains, fruit, roots, leaves, nuts, tubers, ferns, fungi, worms, insects, insect larvae, and non-insect arthropods; it is especially fond of the fungus *Endogone*, which makes up as much as ⅓ of its diet; it does not need to drink as all of its water needs are met by its consumption of solid food; unlike many other members of its family it does not hoard or cache surplus victuals.

The breeding season of this energetic zapodid begins in the spring, after emergence from hibernation, and ends in the fall; males awaken from hibernation first, followed by the polyestrous females, who produce 1-2 litters annually. Gestation lasts for 3-4 wks., after which the female gives birth to an average of 3-4 altricial pups, weaning them approximately 1 mo. later.

Communication and perception run along standard *Napaeozapus* channels: sight, smell, hearing, and touch; it engages in tail-thumping when alarmed. Its natural enemies are reptiles (snakes), birds of prey (owls), mustelids (skunks, mink), wild felids (bobcats), and wild canids (coyotes, wolves); defense strategies include the ability to run, swim, and jump up to 8'. Beneficial as a seed and spore disperser, a parasite host, and food for carnivores. The woodland jumping mouse lives longer than most of its kind, with an average life span of 2-3 yrs. in the wild; perhaps 1 yr. longer under optimum conditions.

WOODLAND VOLE (*Microtus pinetorum*): Also known as the "pine vole," this tiny common microtine grows to a total length of 5½", its hind foot measures ¾", and it can attain a weight of up to 1½ oz. Its soft dorsal pelage is rusty-brown with a grayish tinge; its ventral pelage is silvery-whitish to tannish; its short 1" tail is rusty-brown; it lacks a lateral line and grooved incisors. The body is cylindrical; the eyes are small, dark, and beady; the ears are small and rounded; long tactile vibrissae cover the muzzle; its front paws are large (for its size)—all evolutionary adaptations to a semi-fossorial lifestyle.

Contrary to its species name, this sedentary cricetid is not usually found in pine (coniferous) woodlands. It is a riparian nearctic native that is most at home in temperate deciduous forests, and sometimes wetlands, fens, grassland, bogs, and agricultural settings; it requires dense hiding cover and friable soil; prefers locations with copious ground litter, fallen trees, stumps, logs, blowdown, deadfall, and rocks. Its range extends over the mid to eastern half of Canada and the U.S., from Nebraska, Oklahoma, and Kansas in the west to Maine, New Jersey, and Virginia in the east, from Ontario, Quebec, and Minnesota in the north to Texas, Louisiana, and Florida in the south. Semicolonial, it excavates its own shallow burrows (using its front paws and large incisors) or takes over the abandoned burrows of other animals, where it remains for most of its life; establishes a home range of about 15,000 sq. ft.

Temperate wetland: woodland vole habitat.

A diurnal, crepuscular, and nocturnal omnivore, its diet consists of normal Cricetidae fare: seeds, grains, nuts, tubers, stalks, roots, fruits, leaves, and insects; hoards and caches surplus food in underground larders. Its breeding season runs from spring to fall, with the iteroparous female able to produce up to 4 litters annually. She gestates for about 3 wks., gives birth to an average of 2 or 3 altricial pups in a globular, well-concealed, grass-lined ground or subterranean nest, and weans them within 3 wks. A social nonterritorial rodent, males and females spend more time together than some other members of the Cricetidae family, and may be temporarily monogamous during the mating season.

Communication and perception run along standard *Microtus* channels: hearing, smell, and touch (sight is reduced). Its natural enemies are reptiles, birds of prey, mustelids, procyonids, wild felids, and wild canids. Beneficial as a seed disperser and as food for carnivores; detrimental as a garden, farm, nursery, orchard, and timber pest. The average life span of the woodland vole in the wild is 6 mos. or less.

WYOMING GROUND SQUIRREL (*Urocitellus elegans*): Once thought to be a subspecies of Richardson's ground squirrel, recent genetic studies now classify it as a distinct species. The two are nearly identical morphologically, however, and share the same basic ecology and life history. Two differences are that the Wyoming ground squirrel is more social, and also its range covers the states of Wyoming, Colorado, Montana, Idaho, and Nevada.

WYOMING POCKET GOPHER (*Thomomys clusius*): Once believed to be conspecific with the northern pocket gopher, it is now accepted as an individual species. Despite this distinction, it continues to possess much of the same ecology and life history as its close cousin. The main difference is their ranges: the northern's extends over much of the American Northwest, while the Wyoming's range is limited to the state of Wyoming.

Northern fur seal.

YELLOW-BELLIED MARMOT (*Marmota flaviventris*): Also known as the "rockchuck" and the "mountain marmot," this common medium-sized rodent grows to a total length of 28" and can attain a weight of up to 11 lb. Its dorsal pelage is rusty-brown with a grizzled wash; the ventral pelage is brownish-yellow, giving it its common name; the 8" tail is short, bushy, and brownish-cinnamon in color. Its head crown is black; there is often a white mark between the eyes, as well as buff-colored subauricular patches. (Overall coloration varies from subspecies to subspecies.) Like many other fossorial mammals, its body is chunky; the eyes, ears, and nose are set near the top of the head; the limbs are muscular; and the clawed paws are designed for digging. Gender dimorphism is evident: the male is much larger than the female.

This terrestrial sedentary sciurid is a nearctic native that inhabits a wide variety of alpine and subalpine biomes and ecosystems, including dry, temperate, elevated grassland, tundra, talus slopes, boulder fields, steppe, meadows, open woodland, savanna, bare rocky situations, semi-desert, canyons, valleys, sand dunes, foothills, and montane forest. A western species, its range extends from British Columbia and Alberta in the north to Colorado and New Mexico in the south, from Washington and California in the west to Nebraska and South Dakota in the east.

Yellow-bellied marmot.

Burrow systems are excavated in well-drained areas and can reach a depth of 3' and a length of 20'; portal openings may reach 10" in diameter; ejecta mounds are usually seen near the main entrance, from which radiate well-used compressed earthen paths; as added protection against predators, den portals are concealed beneath rock piles, inside rock crevices or outcroppings, or on the sides of rocky hummocks.

Highly social, this stocky member of the ground squirrel family lives in colonies made up of hierarchal harem units, each one led by a dominant male who oversees the lives of several breeding females and their combined offspring. Despite this gregariousness, individuals occasionally live on their own. Hibernates each year beginning in the fall after building up its body fat; spends the majority of its life underground, eating, socializing, hibernating, estivating, resting, nesting, and sleeping. Territorial, after establishing an average home range of 1 or 2 acres, it scent-marks the area, aggressively guarding it against conspecifics.

A diurnal omnivore, its diet consists of standard Sciuridae fare: all types of plant matter and some animal matter. This includes seeds, grains, grasses, legumes, forbs, herbs, nuts, fruit, flowers, and insects. It is not known to hoard or cache surplus food. Its breeding season runs throughout the spring, beginning after emergence from its long 8-mo. hibernation. The iteroparous female gestates for about 1 mo., gives birth to an average of 3-4 altricial pups in a subterranean, grass-lined, maternal nest chamber, and weans them within 50 days. Males are agonistic toward other males, as well as their own male offspring, the latter which they drive from the den area after a year or two.

Marmots.

Also known as the "yellow-footed marmot," its communication and perception run along standard *Marmota* channels: sight, smell, touch, and hearing. Vocalizations include teeth-clacking, chirping, growling, alarm calls, and whistling. Rump and cheek glands secret oils that chemically communicate various states, from alarm and territoriality to one's hierarchal position in the colony. It enjoys grooming, playing, sun-bathing, mock-fighting, and resting in its burrow.

Its genus name, *Marmota*, derives from a little-known Swiss language called Romansh, and means "mountain mouse"; its species name, *flaviventris*, is Latin for "yellow belly." Its natural enemies are birds of prey, mustelids, wild canids, and ursids. Beneficial as a seed disperser, commensal homebuilder, and food for carnivores; detrimental as a carrier of deadly diseases, such as Rocky Mountain spotted fever and sylvatic plague. One of the longest living rodents, the yellow-bellied marmot can survive for up to a dozen years in the wild; 15 yrs. under optimum conditions.

YELLOW-CHEEKED CHIPMUNK: (*Neotamias ochrogenys*): Once believed to be a subspecies of the similar Townsend's chipmunk, based on recent genetic studies it has been accorded the distinction of being a separate species. Nonetheless, it shares much of the same morphology, ecology, and life history as its close cousin—though the range of the yellow-cheeked occurs further south (California) than Townsend's (Washington and Oregon).

YELLOW-FACED POCKET GOPHER (*Cratogeomys castanops*): This plump rodent grows to a total length of nearly 13", its hind foot measures 1½", and it can attain a weight of up to almost 12 oz. Coloration on the dorsum is a drab, gunmetal grayish-brown, yellowish-brown, or reddish-brown, often washed with black-tipped hairs; the ventrum is whitish, tannish, or reddish; the 4" tail is brown and hairless. The body appears chunky; the head is massive; the snout is robust; the small dark eyes and tiny ears are located at the top of the head; it has large yellowish-orangish incisors; the vibrissae-covered nose is blunt; the limbs are short and muscular; the paws are large and heavily clawed. The facial area has light, indistinct, yellowish cheek patches, the origin of its common name. It is gender dimorphic, with the male being larger than the female.

Pocket gopher.

This solitary motile geomyid is a nearctic native that inhabits temperate open savanna, prairie, meadows, and grassland; it is also found in cropland, sand dunes, hedgerows, orchards, and arid desert locations; prefers the friable loamy soil found on cultivated farmland. A southwestern species, its range extends over portions of the states of Colorado, Kansas, New Mexico, Oklahoma, and Texas, and south into northeastern Mexico. Solitary and fossorial, it excavates large, multi-chamber burrow systems which it readily defends against conspecifics; the 4" wide portals are located in concealed spots, such as under vegetation (like cacti); it leaves surface ejecta mounds around the den area. Does not hibernate; active year-round.

A strict herbivore, its diet consists of standard Geomyidae fare: tubers, roots, bark, stalks, wood, flowers, forbs, leaves, bulbs, and cultivated crops. Hoards and caches surplus food in underground larders, transporting it in its fur-lined cheek pouches or "pockets"—from which it derives the latter half of its common name. In the southernmost part of its range its polygynous breeding season occurs throughout the year. Northward it tends to have 2 separate seasons: 1 in summer and 1 in winter. The polyestrous female, who can produce 2 litters annually, gives birth to an average of 2 altricial pups, and weans them 1-2 mos. later.

Pocket gopher, showing its cheek pouches or "pockets" (to the right and left of its mouth), and its large incisors.

Communication and perception run along standard *Cratogeomys* channels: smell and touch, with reduced sight and sound. Its natural enemies are reptiles (rattlesnakes), birds of prey (owls, hawks, eagles), mustelids (badgers, weasels), and wild canids (coyotes). Beneficial as a soil aerator and recycler, as a parasite host, and as food for carnivores; detrimental as a home, orchard, and farm pest. The average life span of the yellow-faced pocket gopher in the wild is 6-12 mos.

YELLOW-NOSED COTTON RAT (*Sigmodon ochrognathus*): Similar to a cross between a mouse and a rat, this small vole-like rodent grows to a total length of about 11", its hind foot measures around 1", and it can weigh up to 4 oz. Its countershaded pelage is grizzled black and brown on the upper body, light gray on the lower body; its 4" tail is bicolored, dark on top, light on the bottom. The body is compact and ovoid; the head is small; the eyes are dark, beady, and ringed; the ears are small, rounded, and erect; the muzzle is tapered; the nose is snubbed and covered in tactile vibrissae; the limbs are small; the feet are tiny and delicate. Its Greek genus name, *Sigmodon*, means "s-tooth," a reference to the s-like pattern on the top of its molars; its Latin species name, *ochrognathus*, means "ochre-jaw," an allusion to its yellowish-brownish-orange nose—and also the origin of its common name. It lacks gender dimorphism: males and females are approximately the same size and weight.

This motile endothermic cricetid inhabits a myriad of biomes and ecosystems, from deserts, meadows, and grassy areas (particularly near elevated rocky foothills) to mixed forests, mountain slopes, and

Mountain slope: yellow-nosed cotton rat habitat.

agricultural fields. When possible it prefers loose soil, forest litter, agave-rich habitat, snag, deadfall, blowdown, and well-drained piñon-juniper scrubland with patchy grass. Its range is limited to the American Southwest, and more specifically the states of Arizona, New Mexico, and Texas, and from there south into northern Mexico.

Its house has multiple entrances; though constructed on the ground using dry grass, cactus parts, plant fibers, and dead leaves, it may add subsidiary burrows and an underground nest to the structure. Will appropriate the abandoned burrows of other fossorial animals if necessary. Sign: trails and runways spreading out from entrances and into the surrounding grass.

A diurnal herbivore, its diet is comprised primarily of grasses and seeds; this is supplemented with a wide variety of vegetable matter. Its breeding season runs from spring to fall. The doe may produce more than one litter a year; she gestates for about 5 wks., giving birth to an average of 3-4 pups. It is unlikely that the male contributes to the rearing of his offspring. Its natural enemies are reptiles, birds of prey, wild felids, and wild canids. The yellow-nosed cotton rat probably only lives an average of 6-12 mos. in the wild.

YELLOW-PINE CHIPMUNK (*Neotamias amoenus*): This small brightly colored chipmunk grows to a total length of nearly 10" and can attain a weight of up to almost 3 oz. Its upper body pelage is a vivid rusty-yellow to cinnamon-orange; the lower body pelage is whitish or tan; the brushy 4" tail is grizzled with black, gray, and rust-colored fur on the dorsum, yellowish-tan fur on the ventrum. Its head crown is brownish-gray; the haunches are grayish; the ears are blackish on the anterior, whitish on the posterior. It has 5 distinct dark dorsal stripes interspersed with 4 distinct light dorsal stripes; 3 facial stripes mimic its back stripes. Its yellowish coloration is both cryptic and protective, providing excellent camouflage in its light-streaked woodland environment. Gender dimorphism is evident: the female is larger than the male.

Chipmunk.

This little terricolous sciurid is a riparian nearctic native that inhabits mainly temperate, montane coniferous forests; it is also found in and around tundra, brushland, talus slopes, chaparral, shrubland, open ground, scree, and meadows with abundant snag, bushes, rotting logs, rocks, outcroppings, deadfall, and duff. It is particularly fond of yellow pine, from which it derives its common name. Has a large range that covers much of the American Northwest and the Canadian Southwest, extending from British Columbia and Alberta to Nevada, from Washington, Oregon, and California to Idaho, Montana, Wyoming, and Utah. It may nest above ground (in thickets) or below ground (in burrows under stumps, rocks, and logs). Estivates during hot weather. Unlike most of its kind, it is unable to store body fat. Thus, at the onset of cold temperatures it enters a light state of torpidity, awakening frequently throughout the winter to eat from its food caches. Establishes a home range of several thousand square feet, defending it against intruding conspecifics.

Crepuscular and diurnal, this solitary sedentary omnivore consumes seeds, roots, grains, tubers, grasses, nuts, bulbs, leaves, stems, fruit, flowers, buds, foliage, fungi, worms, insects, birds' eggs, and small mammals. A scatter-hoarder, it caches surplus food in both its burrow and in ground holes, which it digs throughout its territory, transporting it in its cheek pouches.

Its polygynandrous breeding season begins in spring with the emergence of the adult males and females from winter torpor. Bucks instigate a mating chase, pursuing estrus does who gestate for roughly 5 wks., and give birth to about 6 altricial pups in a subterranean natal chamber lined with lichen, leaves, and grass. Weaning takes place at 6 wks.; independence takes place at around 10 wks. The father is non-paternal and plays no role in the rearing of his offspring.

Montane coniferous forest: yellow-pine chipmunk habitat.

Communication and perception occur via standard Sciuridae channels: sight, smell, touch, and hearing. Engages in self-grooming and dust-bathing. Its natural enemies are reptiles, birds of prey, mustelids, wild felids, and wild canids. Beneficial by acting as a parasite host, seed disperser, and as a prey base for carnivores. The yellow-pine chipmunk probably lives an average of 2-3 yrs. in the wild; perhaps longer under optimal conditions.

YUMA MYOTIS (*Myotis yumanensis*): This troglophilic chiropterid is nearly 4" in total length, its forearm measures 1½", its wingspan is a little over 9", and it weighs ¼ oz. Its short dull fur ranges in color from cinnamon-brown to grizzled gray-brown on the dorsum; the ventrum is light brown; the throat is buff to whitish. Its calcar lacks a keel.

A secondary consumer, it inhabits caves, mines, trees, buildings, cliffs, and bridge beams over a wide variety of terrain, from desert to woodland. A migratory species, it ranges across the western coastal portion of our region, from British Columbia south to Mexico, but is also found as far east as Colorado and Oklahoma. The most riparian of all North American bats, it is almost always found near water (creeks, streams, rivers, ponds, lakes, reservoirs).

Bat.

This insectivorous vespertilionid leaves its roost after dark, flying low over water, using echolocation to snatch up aquatic insects with its mouth. It may also scoop up additional prey, such as moths, beetles, and midges, using its pouch-like tail membrane. It will eat up to half its body weight in one night, making it an ideal form of natural pest control. Sitting upright, the female gives birth once a year in late spring, catching the altricial newborn pup in her interfemoral membrane. The life span of the Yuma myotis in the wild is unknown. However, it lives to about 9 yrs. in captivity.

The End

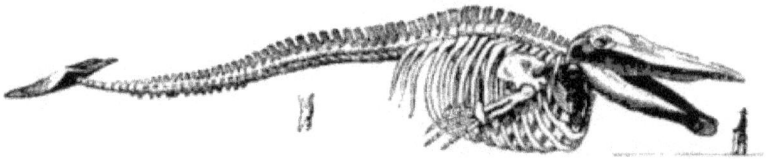

Whale skeleton showing size comparison to a human.

GLOSSARY
Natural History Words & Terms

BY LOCHLAINN SEABROOK

NOTE: THESE DEFINITIONS CONCERN BIOLOGY, ECOLOGY, MAMMALOGY, NATURAL HISTORY, ZOOLOGY, AND THE ANIMAL SCIENCES IN GENERAL, AND THUS MAY NOT INCLUDE ALL POSSIBLE MEANINGS.

Abiotic: Lacking life; the nonliving attributes of an ecosystem; something that does not involve living organisms, is not produced by living organisms, or is not a living organism; that is, something that is not biological (abiological). Examples of abiotic factors are wind, rocks, light, soil, and temperature, all which heavily influence biotic or living organisms.

Aerodynamic: A shape that expedites movement through the air by reducing drag.

Agonistic behavior: From the Greek word *agonizesthai*, meaning to "struggle" or "contend," it refers to aggressive or defensive social contests and conduct between animals, such as fighting, fleeing, and submitting. Agonism often entails threat displays like growling, barking, piloerection (bristling or raising the hair), baring the teeth, and beating the chest. Animals that engage in agonistic behavior are called agonists, that is, "competitors."

Alkali sink: A salty basin in which water evaporation leaves behind high concentrations of salt. This fragile (threatened) plant community is home to a number of North American mammals, including the kangaroo rat, kit fox, Mojave ground squirrel, and coyote.

Allogenic engineering: The act of altering a habitat by changing materials from one state to another. The American beaver, as one North American example, is an allogenic engineer whose activities (felling trees and using them to dam up waterways) can completely transform an area.

Allogrooming: The grooming of one animal by another of the same species. Allogrooming serves to strengthen the social bonds and structures of a family unit or group. A North American mammal that engages in allogrooming is the North American river otter.

Allopatry: In biology, it is when more than one species, particularly related species, live in separate geographic areas; that is, their geographical distribution does not overlap. Two North American mammals that are allopatric are Abert's squirrel and the Kaibab squirrel (a subspecies of the former). Related terms: parapatry, peripatry, sympatry.

Alpine: In biology it is a biome located in mountainous regions above the timberline. It is thus typically a harsh biogeographic zone, in some areas often totally devoid of vegetation and wildlife. More generally alpine refers to anything of, relating to, or resembling a mountain or mountains.

Altricial: The opposite of precocial, it refers to an animal being born or hatched in a helpless or undeveloped state, thereby requiring long-term care and protection by the parents. An example of a North American mammal that gives birth to altricial young is the American black bear.

Ambush predation: A form of hunting and killing in which a predatory animal hides and waits for its prey to pass nearby. A North American mammal that is an ambush predator is the jaguar.

Ambystomatid: A member of the Ambystomatidae family (mole salamanders).

Amensalism: A type of symbiotic biological interaction or relationship between two different species in which one is harmed while the other is not affected. Related terms: commensalism, ectosymbiosis, endosymbiosis, mutualism, parasitism, and symbiosis.

Amphibian: An animal that lives on both land and water. More specifically it is a member of the Amphibia class: cold-blooded vertebrates such as frogs, toads, newts, and salamanders. The larvae of amphibians are generally born in water and breathe through gills; adults are semi-terrestrial, ectothermic, and have scaleless moist skin.

Amphibivorous: Known as amphibivory, it is a carnivorous diet made up partly, primarily, or solely of amphibians (frogs, toads, salamanders, newts). A North American mammal that is an amphibivore is the ornate shrew. (These words were coined by the author.)

Amphipod: A member of the Amphipoda order (marine, freshwater, or terrestrial crustaceans, such as the sand hopper). Marine amphipod species make up part of what is known as zooplankton, an important element in the diet of many North American marine mammals, like the gray whale.

Animal: A member of the Animalia kingdom (animals).

Annelid: A member of the Annelida phylum (metameristic or segmented worms and leeches).

Anthropogenic: Meaning "born of man," it denotes changes in Nature related to, caused by, influenced by, or resulting from humans or their activities. Our anthropogenic impact can be positive (for example, creating wildlife preserves and protecting green spaces) or negative (clearing forests and draining wetlands).

Antitragus: A small cartilaginous tubercle or prominence located on the lower rear part of the external or outer ear just above the intertragic notch and opposite the tragus. It points forward in order to help gather sounds coming from the front of the head. It can be helpful in identifying mammals; the Great Basin pocket mouse, for example, has a lobed antitragus, a trait missing in a number of other mice species.

Antlers: A pair of solid structures that rise from the head of an animal. Unlike horns—which grow from the base, are not branched, are made of keratin and bone, and are permanent—antlers grow from their tips, are usually branched, are made of bone, and are shed each year. Additionally, antlers are not as strong and are usually longer than horns. North American mammals that have antlers are the white-tailed deer and the moose.

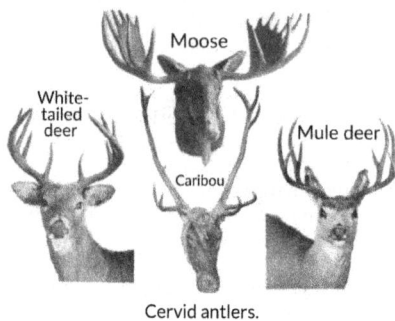

Cervid antlers.

Apex predator: A predator at the top of the food chain, thus it is not preyed upon by any other animal (except, in some cases, its own kind). North American mammals that are apex predators are the polar bear and Sasquatch.

Aplodontid: A member of the Aplodontiidae family (mountain beaver).

Aposematism: The use of striking colors or bright markings which function as visual warning signals to other animals. Predators soon learn to recognize such colorations and avoid the animals who wear them. An example of an aposematic North American mammal is the striped skunk.

Aquadynamic: A shape that expedites movement underwater by reducing drag.

Aquatic: Related to water. An animal that spends the majority or all of its life in water (as opposed to living on land or in trees) is considered aquatic. An example of an aquatic North American mammal is the bottle-nosed dolphin.

Arachnid: A member of the Arachnida class (spiders, scorpions, mites, ticks).

Arachnivorous: Known as arachnivory, it is a carnivorous diet made up partly, primarily, or solely of spiders. A North American mammal that is an arachnivore is Merriam's shrew.

Arboreality: Inhabiting or frequenting trees. An animal that spends the majority or all of its life in trees (as opposed to living on the ground or in water) is considered arboreal. Arboreality is often associated with tree-climbing. An example of an arboreal North American mammal is the opossum. (Not to be confused with scansoriality.)

Ardeid: A member of the Ardeidae family (herons).

Arthropodivorous: Known as arthropodivory, it is a carnivorous diet made up partly, primarily, or solely of members of the Arthropoda phylum, a group of invertebrates that includes spiders, crustaceans, and various types of segmented insects with exoskeletons and jointed appendages. A North American mammal that is an arthropodivore is the little brown bat.

Artiodactyl: A member of the Artiodactyla order (even-toed ungulates or hoofed mammals: deer, antelope, goats, pigs, bison, camels, giraffes, sheep, cattle, hippopotami, etc.)

Arvicolinid: An arvicoline (or arvicolin); a member of the Arvicolinae subfamily (lemmings, muskrats, and voles)—also known as the Microtinae subfamily.

Asocial: In biology, generally an animal that is solitary; that is, it lives alone and not in groups. A North American mammal that is asocial is the margay.

Autotrophy: The condition of being autotrophic; that is, having the capability to manufacture one's own energy (food) from inorganic substances, using light or chemical energy. Examples of autotrophs are algae and green plants, both which form the lowest level of the food chain.

Avivorous: Known as avivory, it is a carnivorous diet made up partly, primarily, or solely of birds. A North American mammal that is an avivore or bird-eater is the eastern fox squirrel.

Balaenid: A member of the Balaenidae family (bowhead whales, right whales). Alternate form is balaenine.

Balaenopterid: A member of the Balaenopteridae family; that is, rorqual whales (minke, humpback, blue, fin, sei, etc.).

Baleen: Also known as "whalebone," it is a tough, flexible, keratinous tissue shaped into bristles. These, in turn, are attached to plates that hang from the upper jaws of baleen whales, allowing them to filter feed.

Benthic: A type of aquatic biome relating to or occurring at the bottom of a body of water (usually an ocean). Can also refer to an animal that lives on or near this region. An example of a benthic North American mammal is the gray whale.

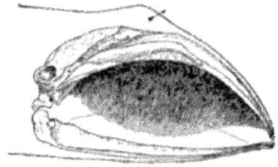

Cross-section view of a baleen whale's head showing position of the whalebone on the upper jaw.

Bilateral symmetry: A body shape, with matching anatomical parts arranged on each side, that is capable of being divided along one plane into two identical or mirrored halves. All mammals (including humans) are bilaterally symmetric because their bodies have a top (dorsal side) and bottom (ventral side), a right and left side (lateral sides), and a front (anterior) and back side (posterior).

Biodegradation: The decomposition of organic substances via the metabolic action of or enzymes produced by living organisms—vital to maintaining a well-balanced ecosystem. A mammal contributes to biodegradation when it eats a dead plant or animal. The microbial organisms in its digestive tract help break down the complex organic molecules of the decomposing plant or animal into smaller simpler molecules, returning them, in a more useful form, to the environment. All mammals, whether herbivorous or carnivorous, participate in the process of biodegradation, making them, in that specific sense, beneficial to humanity.

Biodiversity: The variety of plants and animals living in an environment, including the ecosystem itself, along with its flora and fauna communities as well as its species and genes. The biodiversity of a specific area is determined by what is known as its "species richness," a measure that is calculated by counting the number of each animal and plant species found in it. This figure is used to compare the biodiversity of one environment with another. (Same as "biological diversity.")

Biome: A major ecological community type consisting of distinct plants and animals that are interdependent and share a specific climate and environment; a region that is often characterized by dominant forms of vegetation and a prevailing climate; a large geographical area that is home to organisms adapted to that particular location. Examples of major biomes: grassland, deciduous forest, arctic tundra, meadow, tropical rainforest, jungle, prairie, and desert.

Biotic: Having life; the living attributes of an ecosystem; something that involves living organisms, is produced by living organisms, or is a living organism; that is, something that is biological. Examples of biotic factors are animals, plants, and bacteria.

Biopedturbation: The disturbance of soil by biotic or living organisms (as opposed to, for example, abiotic factors, such as wind, water, earthquakes, etc.).

Bipedalism: Bipedality is a type of locomotion that uses two feet. A mammalian example of a North American biped is Sasquatch; the giant kangaroo rat is occasionally bipedal, as well.

Bisulcate: Marked by, divided by, or having two grooves (in biology used to describe teeth, hooves, etc.).

Blubber: Adipose tissue, more commonly known as "blubber," is a layer of fat that lies between the skin and the muscles, where it acts as an insulator, helping to maintain an animal's core body temperature as well as offer extra protection against sharp-toothed, heavily clawed predators. This dense layer of connective tissue is found mainly in holarctic terrestrial and marine mammals, which in North America would include polar bears, whales, walruses, and seals.

Bolus nest: A ball- or globular-shaped nest made with cut twigs and other forest debris. A north American mammal that constructs bolus nests is Abert's squirrel.

Bosque: Spanish for "woods" or "forest." In biology it refers to a small wooded area; a cluster, grove, or group of trees. Also spelled bosk.

Bovid: A member of the Bovidae family (cow, ox, bison, buffalo, goats, sheep, antelope, gazelles, impala, muskox). Also written bovine, bovids are horned animals, a trait that sets them apart from other hoofed but antlered mammals, such as deer.

Brachiation: A type of arboreal locomotion in which an animal travels by swinging from tree limb to tree limb using its arms. It is used solely by primates, of which there is only one recognized wild North American species in this book: Sasquatch. Since adults are far too heavy to brachiate, this type of locomotion is used only by the young—and indeed, on rare occasions juvenile Sasquatch have been observed doing just this.

Breach: Leaping (partly or completely) out of the water. In marine mammals, such as whales, this is called a "surface behavior."

Broom nest: A nest built in a preexisting dwarf mistletoe patch (actually an infection in a tree limb), also known as "witch's broom." A north American mammal that constructs broom nests is Abert's squirrel.

Brumation: A state of inactivity, lethargy, or torpor exhibited by ectotherms (chiefly amphibians and reptiles) during winter or long periods of cold weather. (Not to be confused with hibernation or estivation.)

Buffer strip: A thin strip of land, often containing vegetation, that may be natural or human-made, and which is used to form a barrier of some kind; for example, against wind, water, snow, human trespassing, vehicles, etc. May also be used to lay out land borders or add privacy to a residence.

Bulbivorous: Known as bulbivory, it is an herbivorous diet made up partly, primarily, or solely of bulbs (a fleshy plant structure, usually formed underground, resembling a bulb). A North American mammal that is a bulbivore or bulb-eater is the camas pocket gopher, whose species name, *bulbivorus*, reflects one of its favorite foods. (This word was coined by the author.)

Caching: In biology, the same as food caching.

Cactivorous: Known as cactivory, it is an herbivorous diet made up partly, primarily, or solely of cactus plants. A North American mammal that is a cactivore or cactus-eater is the desert woodrat.

Calcar: In bats a spur-like projection that extends from the ankle bone and runs along the outside of the interfemoral membrane. It purpose is to add stability and spread to the wing.

Cambium: A formative layer or soft membrane of delicate tissue located between the bark and wood of a tree or plant belonging to the class dicotyledons (also called dicots), such as oak, maple, or sycamore trees. Edible and highly nutritious, it is responsible, for example, for forming a tree's annual rings and for producing the secondary growth of stems and roots.

Canid: A member of the Canidae family (domestic dog, wolf, jackal, fox, coyote, etc.). Also written canine.

Cannibalism: The eating of the flesh of an animal by an individual of the same species.

Caprid: A member of the Caprinae subfamily (chamois, goats, sheep, serows, ibex). Also written caprine.

Caravanning: A behavior practiced by various animals in which the young closely follow behind their mother single file. In some species the offspring hold onto the tail of the individual in front of them, the first one in line holding onto the mother's tail (often presenting a rather comical spectacle). A North American mammal that practices caravanning is the western spotted skunk.

Carcassivorous: Known as carcassivory, it is a carnivorous diet made up partly, primarily, or solely of carrion. A North American mammal that is a carcassivore or carcass-eater, is the collared peccary. (These words were coined by the author.)

Carnivore: A member of the Carnivora order (meat-eaters: wolves, orca, jaguars, bears, seals, etc.)

Carnivorous: Known as carnivory, it is a meat diet made up solely of animal flesh. A North American mammal that is a carnivore or meat-eater is the mountain lion.

Carrion: The decaying flesh of a dead animal.

Castorid: A member of the Castoridae family (beavers).

Cathemeral: In biology, an organism that can be active during the day or night, depending on various factors. Related terms: crepuscular, diurnal, matutinal, nocturnal, vespertine.

Cervid: A member of the Cervidae family (deer, elk, moose). Also written cervine.

Cetacid: A member of the Cetacea order (dolphins, porpoises, whales). Also written cetacean.

Chaetodipid: A member of the *Chaetodipus* genus (rough-haired pocket mice).

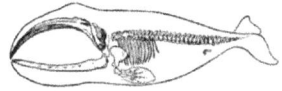

Cross-section of a cetacid or whale, showing its skeleton and body contour.

Chilopod: A member of the Chilopoda class (centipedes).

Chionophilia: Literally "love of snow." In biology it describes organisms that are adapted to winter life and can thus flourish in cold conditions. North American mammals that are chionophiles or snow-lovers are polar bears, arctic foxes, and narwhals.

Chiropterid: A member of the Chiroptera order (bats). Also spelled chiropteran.

Chiropterphily: Bat pollination; or more technically, traits possessed by certain flowers that exist specifically to attract bats. Examples: night-blooming flowers, large tube-shaped blooms, copious amounts of pollen and nectar, and special fruity smells that appeal to bats. A North American mammal that engages in chiroterphily is the lesser long-nosed bat.

Chordate: A member of the Chordata phylum (chordates).

Cingulatid: A member of the Cingulata order (armadillos). (This word was coined by the author.)

Circadian: Characterized by, influenced by, or connected to a 24-hour period. A majority of mammals are considered circadian; that is, their biological activity is governed by earth's 24-hour cycle of light and dark. A North American mammal that is circadian is the northern pocket gopher.

Circumboreal: Related to, distributed in, or occurring primarily throughout the region encompassing the northern parts of North American and Eurasia. An example of a North American mammal that is circumboreal is the wolverine.

Circumpolar: Occurring in or found near the North Pole or South Pole. An example of a North American mammal that is circumboreal is the gray wolf.

Class A sighting: An unobstructed view of an identifiable animal (or its sign)—often up close and by credible witnesses—that is impossible to mistake for anything else. In other words, an unimpeded sighting of a creature (or, for example, its footprint or track) that is manifestly, obviously, and self-evidently what it appears to be.

Class B sighting: A view of an animal that is often obstructed (by, for example, foliage), obscured (by, for example, darkness), or at such a distance that it is impossible to positively identify. In other words, a sighting in which the subject is not clearly observed, and whose identity thus can only be inferred, guessed, or theorized. This classification also includes nonvisual events, such as sounds.

Class C sighting: A secondhand or thirdhand report of a sighting, sometimes anonymous, always ambiguous, and usually open to various interpretations. Due to the possibility of inaccuracies, not considered particularly credible or helpful.

Clear-cutting: The practice of cutting down and removing all trees from an area, typically within a sharply bordered or well-defined area—hence the term "clear-cut."

Coleopteran: A member of the Coleoptera order (beetles).

Colonial: In biology, an animal that lives in a colony with others of its own kind.

Commensalism: A type of symbiotic, biological interaction or relationship between two different species in which one benefits but the other is not affected. Mammalian examples: 1) when a burdock seed sticks to the fur of a wolf (and is carried to another area and deposited); 2) when a barnacle grows on the body of a whale; 3) when an elk walks through an area and birds follow it to feed on the insects it stirs up; 4) when an insect, snake, or mouse shares the large stick house of woodrat. Creatures that benefit from their relationship with another species (without affecting it) are called commensals. Related terms: amensalism, ectosymbiosis, endosymbiosis, mutualism, parasitism, and symbiosis.

Concealing coloration: A camouflage strategy in which an animal of a particular color hides itself in an environment of the same color. Blending into the background by matching colors is an adaptation that is vital in both hunting prey and avoiding being preyed upon. Nearly all North American mammals use concealing coloration in one form or another.

Congener: A thing or animal belonging to the same kind, group, or classification.

Congeneric: A member of the same genus.

Coniferous forest: A woodland comprised of members of the Pinales order (originally known as the Coniferales order). Conifers have soft wood and usually possess needles or scale-like nodes (instead of leaves), and bear true cones (and sometimes fruit). Examples are spruce, fir, larch, yew, juniper, and cedar trees.

Conspecific: A member of the same species.

Copepod: A member of the Copepoda class (generally small or microscopic marine, freshwater, and terrestrial crustaceans, such as the sea louse). Marine copepod species make up part of what is known as zooplankton, an important element in the diet of many North American marine mammals, like the bowhead whale.

Coprophage: An animal that eats dung, either its own or that of other animals.

Coprophagy: The practice of eating dung. In many mammals, such as rabbits and hares, two types of dung are produced: soft, moist, green fecal pellets, which are eaten without chewing, and hard, dry, brown fecal pellets, which are not eaten. The purpose of re-ingesting food a second time is to procure extra nutrients that might not have been extracted by the body the first time it passed through the digestive system. Because coprophagy allows a creature to live on inferior vegetation and in harsh climates and areas where others cannot, this is an important survival tool that can save an animal's life, particularly when and where food is scarce. It is, in other words, a natural form of nutritive recycling. A North American mammal that is a coprophage is the bushy-tailed woodrat.

Countershading: A cryptic coloration strategy in which an animal is dark where light would normally illuminate it, and light where it would normally be in shadow. Examples of North American mammals that utilize countershading are the mule deer, the blue whale, and the western pipistrelle. In all three of these cases, the upper body is darker than the lower body, creating a camouflaging effect by reducing shadows and contours.

Crepuscular: In biology, an organism that is active primarily at dawn and dusk. A North American mammal that is crepuscular is the mule deer. Related terms: cathemeral, diurnal, matutinal, nocturnal, vespertine.

Cricetid: A member of the Cricetidae family (New World mice and rats, as well as voles, hamsters, muskrats, and lemmings).

Crustacean: A member of the Crustacea class (characterized mainly by aquatic arthropods with a segmented exoskeleton, antennae, and modified appendages). This group includes shrimp, lobsters, krill, prawns, wood lice, barnacles, crayfish, and crabs, among many others.

Crustaceanivorous: Known as crustaceanivory, it is a carnivorous diet made up partly, primarily, or solely of crustaceans. North American mammals that are crustaceanivores or crustacean-eaters are the arctic fox, the sea otter, and the hisbid cotton rat.

Crypsis: A type of predator avoidance strategy, it is the ability of an organism to avoid detection by blending in with its environment. Also called cripticity.

Cryptic coloration: An aspect of crypsis, it is when an animal uses its body shape, coloration, markings, a predator avoidance strategy, a predation tactic, etc., to conceal itself within its ecosystem. Stripes, spots, mimicry, and both bright and dull pelage tones, are all examples of cryptic coloration. Most North American mammals rely on cryptic coloration to some extent. Well-known examples are mule deer, American mink, jaguars, ocelots, polar bears, and mountain lions.

Cryptid: A creature whose existence is disputed and is therefore not recognized by mainstream science. Many animals were once considered cryptids, nothing more than the products of myths and legends—that is, until they were "discovered" by mainstream scientists. Examples: the gorilla, the giant squid, and the coelacanth. Related term: cryptozoology.

Cryptozoology: Literally, "the scientific study of secret (or hidden, unknown, mysterious, obscure) animals." The animal subjects of cryptozoologists are called "cryptids." Related term: cryptid.

Cursoriality: Having limbs designed for running. North American mammals that are cursorial are wolves, coyotes, and deer.

Cynomyid: A member of the *Cynomys* genus (prairie dogs). (This word was coined by the author.)

Dactylopatagium: The membranous skin that stretches between the second and fifth digits or fingers on the hand of, for example, a bat.

Dasypodid: A member of the Dasypodidae family (armadillos).

Deciduous forest: A woodland generally comprised of hardwood trees that bear leaves that are shed each year. Also known as "broadleaf trees", in our region examples are maples, birch, aspen, elm, oak, and beech trees.

Delayed implantation: Also known as "embryonic diapause," the females of at least 125 mammal species employ this reproductive strategy to prevent an embryo from developing beyond the blastocyst stage and attaching to the wall of the uterus immediately following fertilization (mating). The purpose is to maintain the embryo in a state of dormancy or suspended growth (diapause) until the arrival of conditions that are favorable to the birth of the young—such as warm weather and an abundant food supply. Thus, while an animal's gestation period can last for 250 days, for example, the unimplanted embryo only attaches to the uterine wall and grows during the final 50 days. North American mammals that utilize delayed implantation are the American black bear, the fisher, the ribbon seal, and the western spotted skunk.

Delphinid: A member of the Delphinidae family (dolphins, pilot whales, killer whales, etc.)

Dendrophilia: Literally "love of trees." In biology it describes organisms that are attracted to, visit, spend time in, or live in trees. North American mammals that are dendrophiles are the Northern yellow bat and the Uinta chipmunk.

Dentary: In higher mammals the mandible (lower jaw). The characteristics of dentaries are often used to help identify species.

Detritivorous: Known as detritivory, it is an omnivorous diet made up partly, primarily, or solely of natural detritus (decomposing or dead plants and animals). An example of a detritivore or waste matter-eater is the earthworm.

Detritus: Another name for forest floor litter.

Diapause: A period of dormancy or suspended development occurring between periods of activity, particularly during unfavorable environmental conditions.

Didelphid: A member of the Didelphidae family (opossums).

Diestrus: The fourth and final stage of the estrous cycle, it is a period of reproductive inactivity between two periods of estrus. Related terms: estrus, metestrus, proestrus.

Digitigrade: An animal that walks on its toes rather than on hooves (unguligrade) or the soles (plantigrade) of its feet. In other words, its heels do not touch the ground during locomotion. Examples of North American mammals that are digitigrades are the coyote and the jaguar.

Dimorphism: This word, from the Greek *di* ("two") and *morph* ("form"), refers to the condition of having two different forms, particularly within the same species. In mammalian zoology, dimorphism is often used to mean "sexual dimorphism" (which, in this book, I more specifically refer to as *gender dimorphism*): the physical differences between the two genders (male and female). The most common application of this term is in reference to body size. However, gender dimorphism can also include male-female differences in bone shape, fur color, fur length, tooth size, and eye color, as well as body ornamentation. Generally speaking, gender dimorphism usually (though not always) indicates some type of a polygynous breeding system, as well as competition between males over females. An example of a North American mammal that exhibits gender dimorphism in body size is the caribou. (Note: humans are also gender dimorphic, and in numerous ways; for example, males tend to be taller and heavier, with more robust bones and larger thicker skulls, than females; they also have, on average, more body hair, as well as bigger lungs, hearts, and muscle tissue.)

Dioecism: Relating to species that have two separate genders; that is, a male and a female. Humans, as well as most other mammals, are dioecious.

Dipodid: A member of the Dipodidae family (jumping mice, jerboas, and birch mice).

Dipteran: A member of the Diptera order (true flies).

Dirt-bath: See dust-bath.

Disruptive coloration: A camouflage strategy in which an animal's (usually bold contrasting) coat markings and colors help break up its outline against its background. Useful in predation and antipredation. A North American mammal that uses disruptive coloration is the jaguar.

Disturbed habitat: When conditions change in such a way as to interrupt the characteristic operation of an ecosystem, it is considered disturbed. This disturbance can result from natural causes (that is, Nature) or artificial causes (that is, humans). Examples of disturbed habitats are forests that have been clear-cut, meadows that have undergone burning, or wetlands that have dried up or were intentionally drained.

Diurnal: In biology, an organism that is active primarily during the day. A North American mammal that is diurnal is the eastern chipmunk. Related terms: cathemeral, crepuscular, matutinal, nocturnal, vespertine.

Dominance hierarchy: A social structure among animals in which a dominant individual, pair, or group rules over the other members, who are considered subordinate, and who are thus required to be submissive to those above them in the "pecking order." Most common among social animals, such as canids and primates.

Dorsal: Of, on, or relating to the upper body or back of an organism. Also written dorsum.

Drey: A squirrel nest, above or below ground.

Drift ice: A type of unattached mobile ice that forms from smaller ice floes, and drifts about lakes and oceans—hence its name. When drift ice is pushed together by wind or currents, it creates a larger mass of ice called pack ice. A North American mammal that uses drift ice is the ringed seal.

Duff: Another name for forest floor litter.

Dust-bath: Many mammals take dust-baths by rolling in dry powdery soil, carefully working the dust into their hides or fur. This is both enjoyable and practical, for it helps rid them of ectoparasites which live on the surface of the body, and can serve as added protection against heat, cold, and insects bites. In the case of territorial mammals, dust-bathing can help scent-mark an area. A North American mammal that engages in dust-bathing is the yellow-pine chipmunk.

Echinodermivorous: Known as echinodermivory, it is a carnivorous diet made up partly, primarily, or solely of echinoderms (starfish, sea lilies, sea cucumbers, etc.). A North American mammal that is a echinodermivore or echinoderm-eater is the sea otter. (These words were coined by the author.)

Echolocation: In biology, a physical process in which an organism emits sound waves that reflect back to it, helping detect the location of invisible or distant objects; known chiefly as a hunting tool among cetaceans and chiropterids, such as North America's Atlantic spotted dolphin and the California leaf-nosed bat.

Ecological flexibility: An organism's ability to adjust to changes in its environment.

Ecology: The scientific study of the relationships between an environment and the organisms that inhabit it.

Ecosystem: An abbreviation of "ecological system," it is a system formed by individual but interdependent living organisms (flora and fauna) that share a specific physical environment; a small, loosely defined community of plants and animals functioning as a unit within a biome; every component—both biotic or animate (plants and animals) and abiotic or inanimate (soil, rocks, water)—of a biological system that interacts as one.

Ecosystem engineering: The formation, change, and upkeep of environments by living creatures. Such animals are called ecosystem engineers. North American mammals that are ecosystem engineers are the California ground squirrel and the black-tailed prairie dog.

Ecosystem succession: Also known as "ecological succession," it refers to the slow and complex species changes an area goes through as one community of plants and animals takes over and supersedes another, and which eventually results in a stable, balanced, and self-sustaining ecosystem. This process of ecosystem succession may take hundreds or even thousands of years.

Ecotone: From the Greek words *eco* (roughly "community") and *tonos* ("tension"). It is the border between two different habitat types; an area where one biome meets another; a piece of land where one distinct habitat transitions into another, such as where grassland changes into forest. A North American mammal that is an ecotone species is the collared pika.

Ecotourism: A type of tourism that focuses on nature appreciation and animal-watching. As our mammals become rarer, ecotourism is becoming more popular.

Ectosymbiosis: A type of symbiotic biological interaction or relationship between two different species in which one organism (the parasite) lives on the outside surface of another organism (the host). An example of North American ectosymbionts are barnacles that live on the skin of whales. Related terms: amensalism, commensalism, endosymbiosis, mutualism, parasitism, and symbiosis.

Ectothermic: In zoology, it relates to animals that are not capable of generating their own internal heat, and which are thus dependent on external sources for warmth. Most invertebrates, fish, amphibians, and reptiles are ectotherms, the so-called "cold-blooded" animals.

Edge species: An organism that is found in, is adapted to, thrives in, or naturally occurs in and around ecotones. An example of a North American mammal that is an edge species is the red fox.

Edificarian: A habitat that is comprised of or includes human structures. A North American mammal that is an edificarian species is the southeastern myotis, which is often found in or near human structures.

Ejecta mound: A pile of excavated soil, usually located near or around the entrance of an underground burrow or tunnel system. Ejecta mounds over a few inches in diameter are almost always the result of the excavation of a fossorial mammal, such as North America's black-tailed prairie dog.

Embryonic diapause: See delayed implantation.

Endosymbiosis: A type of symbiotic biological interaction or relationship between two different species in which one organism (the parasite) lives on the inside of another organism (the host). An example of North American endosymbionts are parasites that live inside the bodies of black rats. Related terms: amensalism, commensalism, ectosymbiosis, mutualism, parasitism, and symbiosis.

Endothermic: In zoology relating to animals that are capable of generating their own internal heat, and which are thus not dependent on external sources for warmth. All mammals are endotherms, the so-called "warm-blooded" animals. (Note that birds and some fish are also endothermic.)

Ephemerid: A member of the Ephemeridae family (mayflies).

Equid: A member of the Equidae family (horses, asses, zebras).

Erethizontid: A member of the Erethizontidae family (New World porcupines).

Ericaceous: Of or related to the heath or heather family of plants, shrubs, and trees.

Eschrichtiid: A member of the Eschrichtiidae family (the gray whale, the only living member).

Estivation: Also spelled aestivation, it is the slowing down or complete cessation of activity during dry or hot periods. It is a survival mechanism, similar to but different from hibernation. Estivation is sometimes further defined as summer sleep (lasting only a few hours, usually during the hottest part of the day), while hibernation is defined as winter sleep (lasting from several days to many months). A North American mammal that estivates is the Mojave ground squirrel. (Note: many insects, mollusks, fish, reptiles, and amphibians also estivate.)

Estrous cycle: The reproductive period in mammals below the higher primates. It is divided into four cycles: 1) proestrus, 2) estrus, 3) metestrus, and 4) diestrus. The "quiet" or non-reproductive period between estrous cycles is called anestrus. (See individual terms.)

Estrus: Also known as "heat" or "rut," it is the second stage of the estrous cycle, marking a recurring period when the female is reproductively receptive to the male and capable of conceiving. Related terms: diestrus, metestrus, proestrus.

Estuarine: Of, related to, or formed in an estuary.

Estuary: The mouth or lower end of a river that connects to the sea. This transition zone, forming a meeting place between freshwater and saltwater, is composed of brackish water and is known as an "estuarine environment." It is an important nexus for a myriad of creatures, from marine crustaceans to riparian mammals. A mammal in our region that inhabits estuaries is the North American river otter.

Estuary.

Eulipotyphlid: A member of the Eulipotyphla order (moles, shrews, hedgehogs, gymnures, solenodons, desmans). (This word was coined by the author.)

Facultative: In biology this refers to something that is not physically essential for an organism's survival. A North American mammal that is, for instance, a facultative hibernator, is the Panamint chipmunk, which may or may not enter hibernation, depending on conditions and geography. A North American mammal that is, for instance, a facultative carnivore is the black bear, which, being omnivorous, can survive on animal flesh, but does not require it to survive.

Fast ice: Also known as "landfast ice," it is a type of ice that forms in shallow water along sea or lake coastlines, "fastening" itself to either the bottom or to land. While its opposites, drift ice and pack ice, are mobile and thus move with water currents and weather, because fast ice is attached, it remains stationary despite surrounding conditions. A North American mammal that relies on landfast ice is the polar bear.

Fawning cover: Among deer, a habitat with the necessary vegetation (usually dense thickets) to conceal the birth of the young from predators.

Felid: A member of the Felidae family (cats; for example, the domestic cat, bobcat, mountain lion, lynx, tiger, jaguar, etc.). Alternate form is feline.

Feral: Technically, an animal once owned by humans that has escaped and become wild; more loosely, an organism that exists in its natural state. Related term: wild.

Fibrillae: A small delicate fiber or filament, such as a hair. A single fiber is called a fibrilla or fibril.

Flehmen: A form of mammalian communication, the flehmen response (often simply known as "flehmen") is a behavior in which an animal curls its upper lip (baring its teeth), lifts its head, and inhales in response to olfactory stimuli (usually from female to male). The flehmen position allows air to more easily transmit scents (such as pheromones) to a chemically-sensitive smelling receptor ("Jacobson's organ") located in the upper part of the mouth. While this chemoreceptor is found in amphibians, reptiles, and mammals, the flehmen lip-curl response itself is a strictly mammalian behavior, and is most familiar to owners of horses, llamas, goats, dogs, and cats.

Florivorous: Known as florivory, it is an herbivorous diet made up partly, primarily, or solely of flowers. A mammalian example of a florivore or flower-eater is the southern flying squirrel.

Folivorous: Known as folivory, it is an herbivorous diet made up partly, primarily, or solely of leaves. A North American mammal that is a folivore or leaf-eater is the chisel-toothed kangaroo rat.

Food caching: Hiding or storing food in concealed locations, either at a single location (known as larder hoarding) or at multiple locations (known as scatter hoarding).

Food chain: A hierarchical arrangement of organisms, each occupying one of five different trophic levels in which a group consumes individuals in the group below it. These five trophic levels are: 1) quaternary consumers; 2) tertiary consumers; 3) secondary consumers; 4) primary consumers; and, at the bottom, 5) producers (also sometimes called primary producers). A food chain allows energy to be transferred in the form of edible nutrients from one organism to another, and is thus part of a larger system called the "food web."

Food web: A network of food chains. Also known as a "consumer-resource system," it shows the relationships or links between food chains and how the animal consumers in those chains are interconnected and integrally related. It can also be a diagram showing the flow of energy and matter as they pass through the five trophic levels of a food chain.

Forb: A broad-leaved flowering herb that is not grass or grasslike. Examples of forbs are sunflowers, lilies, clover, and milkweed.

Forest floor litter: Also known as duff, litterfall, or detritus, it is the top layer of the forest floor, usually consisting of newly fallen dead leaves, twigs, stems, bark, branches, needles, deadfall, and fruit, all in the early stages of decomposition. It is home (and a foraging zone) to a myriad of animals, from bacterium, worms, and insects, to amphibians, reptiles, and small mammals.

Form: A temporary shallow depression made by a hare in grass, soil, or snow in which it rests, grooms itself, or sleeps; forms are used by other animals as well, such as deer, wild boar, and Sasquatch.

Fossoriality: Adapted, designed, or built for digging or burrowing. A North American mammal that is fossorial is the American badger.

Frugivorous: Known as frugivory, it is an herbivorous diet made up partly, primarily, or solely of fruit. A North American mammal that is a frugivore or fruit-eater is the lesser long-nosed bat.

Fungivorous: Known as fungivory, it is an herbivorous-carnivorous diet made up primarily or solely of fungi. A North American mammal that is a fungivore or fungus-eater is the least shrew.

Functional response: The relationship between an animal's rate of food consumption and food availability; the manner in which a predator responds to changes in the density of its prey; the feeding behavior of predators. Related term: numerical response.

Gallery forest: A forest growing along a waterway (creek, stream, river, etc.) in an area otherwise empty of trees. These unique tree corridors, with their usually sharply-defined boundaries, are home to a wide variety of mammals, such as rodents and bats. Gallery forests also serve as important travel-ways for larger mammals, offering shelter, protection, food, and water, as they move from one area to another.

Gastropod: Meaning "stomach foot," it is a land, marine, or freshwater animal that is a member of the Gastropoda class, the largest group in the phylum Mollusca ("mollusks"). This group includes snails, slugs, conchs, abalone, chitons, limpets, and whelks.

Gender dimorphism: See Dimorphism.

Geomyid: A member of the Geomyidae family (pocket gophers).

Geophagia: The practice of eating earthy substances, such as soil, clay, chalk, mud, or ground-up rock, which possess high mineral content. Humans have long participated in geophagy, usually for ritualistic or nutritional reasons. A North American mammal that is a geophage is the nine-banded armadillo.

Gestation: The period of time that young are carried in the uterus between conception and birth; pregnancy.

Gleaning: A foraging behavior used by bats in which they take prey from leaves or the ground.

Gonochorism: The same as dioecism.

Graminivorous: Known as graminivory, it is an herbivorous diet made up partly, primarily, or solely of grass. North American mammals that are graminivores or grass-eaters are the wild horse and the antelope jackrabbit.

Granivorous: Known as granivory, it is an herbivorous diet made up partly, primarily, or solely of grains and seeds. North American mammals that are granivores or grain-eaters (or seed-eaters) are the southern flying squirrel and the black rat.

Habitat: The place where an organism or biological community lives.

Habitat flexibility: An organism's ability to adapt to different habitats.

Hawking: A foraging behavior used by bats in which they capture prey while in flight; also known as "aerial hawking."

Heat dumping: When an animal's body attains its maximum level of heat its survival is in jeopardy. Therefore it must rid itself of excess heat by lowering its body temperature as soon as possible. Different creatures use different methods of thermoregulation, or heat "dumping," such as sweating and panting. One North American mammal, Harris' antelope squirrel, for example, finds a shady spot, then lies flat on its belly with its limbs splayed. This allows it body heat to be wicked away by the cool ground. Some animals lay on shaded rocks to achieve the same effect.

Hematophage: An animal that lives solely on blood.

Herbivorous: Known as herbivory, it is a vegetable diet made up partly, primarily, or solely of plants. A North American that is an herbivore or plant-eater is the plains pocket gopher.

Herpivorous: Known as herpivory, it is a carnivorous diet made up partly, primarily, or solely of reptiles (snakes, turtles, lizards, etc.). A North American mammal that is a herpivore or reptile-eater is the Norway rat. (These words were coined by the author.)

Heteromyid: A member of the Heteromyidae family (kangaroo rats, kangaroo mice, pocket mice).

Heterothermic: In zoology, relating to animals whose body temperature varies with the environment. Heterothermic creatures can regulate their body temperature up or down, making them intermediates between endotherms and ectotherms. In other words, a heterotherm can switch back and forth between endothermy and ectothermy. Other forms of heterothermy: regional heterotherms are able to increase heat in specific parts of their bodies; temporal heterotherms are able to increase heat in their bodies at a specific time. Examples of North American mammals that are heterotherms are brown bears and certain species of rodents and bats.

Hibernaculum: From the Latin word *hibernare* (meaning "to pass the winter"), it is a shelter occupied during winter by a dormant animal. North American mammals that utilize hibernacula are the bat, the marmot, and the brown bear.

Hibernation: Defined differently by different people living in different parts of the world at different times, there is no current consensus on the exact meaning of hibernation. It may be best described in general terms as a state in which an endotherm reduces its activity and lowers its body temperature in order to withstand severe winter weather. This helps it conserve energy, increasing its chances of surviving through the coldest part of the year. Few mammals actually hibernate. Many that are thought to are only entering a state of mammalian dormancy or torpor (a light sleep from which it awakens frequently). Among those North American mammals that engage in true hibernation are the western pipistrelle, the gray myotis, and the little brown bat. (Hibernation should not be confused with estivation or brumation.)

Hiding cover: Vegetation thick enough to conceal up to 90 percent of a creature from a distance of 200 ft. Used primarily (but not solely) by shy or reclusive animals.

Hind foot reversibility: The ability to rotate the hind foot 180 degrees while descending vertically. Found in some scansorial mammals, highly flexible ankles allow a creature to climb down (a tree, for example) headfirst almost as well as it climbs up. It is an adaptation for chasing prey as well as escaping predators. A North American mammal that possesses hind foot reversibility is the margay.

Hoarding: The practice of accumulating food in a single location, such as a burrow or nest. A North American mammal that is a hoarder is the red squirrel.

Holarctic: A vast biogeographic area or kingdom encompassing the nontropical (that is, northern) parts of both the palearctic (Old World) and nearctic (New World) regions. A North American mammal that inhabits this region, and is thus considered holarctic, is the ringed seal.

Holt: The den or home of an animal, in particular an otter.

Homeothermic: In zoology, relates to animals whose body temperature is constant or uniform, and usually warmer than the temperature of the surrounding environment. Thus a homeothermic creature, also known as a "thermoregulator," is able to maintain a stable, relatively high internal body temperature regardless of external temperature changes. Homeothermal or warm-blooded North American mammals include mice, rabbits, wolves, bears, and bison. Also spelled homoiothermic.

Hominid: A member of the Hominidae family (great apes—that is, orangutans, chimpanzees, and gorillas, and their extinct and related ancestors—and modern, unrecognized, related, and extinct humans).

Hominoid: A creature resembling a human or that is related to humans.

A hominid (gorilla).

Hooking: The practice, employed, for example, by some hares, of laying down additional (that is, false) tracks to confuse predators. In snow, hooking also involves repeatedly jumping to one side to break both the visual trail and the scent trail. A North American mammal that uses the hooking technique is the mountain hare.

Horns: Bone-like structures that rise from the head of an animal. Unlike antlers, which grow from their tips, are usually branched, are made of bone, and are shed each year, horns grow from the base, are not branched, are made of a bony core surrounded by keratin, and are permanent. Additionally, horns are stronger and also usually shorter than antlers. North American mammals that have horns are the American bison and bighorn sheep.

Hydric: An environment characterized by, related to, or requiring an abundance of moisture; in other words, a wet habitat. Plants and animals that are adapted to arid conditions are considered hydric.

Hydrophilia: Literally "love of water." In biology it describes molecules that attract water and which are soluble in it. Hydrophilia is thus an integral aspect of the study of the structure and function of cells.

Hydrophobia: Literally "fear of water." In biology it describes molecules that repel water and are insoluble in it. Many (aquatic) birds and mammals possess a hydrophobic coating over their feathers or fur, which serves as a protective layer against cold and damp. An example of a North American mammal that has hydrophobic fur is the wolverine.

Hymenopteran: A member of the Hymenoptera order (wasps, bees, ants).

Hypercarnivorous: Known as hypercarnivory, it is a carnivorous diet made up of at least 70 percent animal flesh. A North American mammal that is a hypercarnivore or high meat-eater is the common bottlenose dolphin.

Hypocarnivorous: Known as hypocarnivory, it is a carnivorous diet made up of less than 30 percent animal flesh. North American mammals that are hypocarnivores or low meat-eaters are the brown bear and the racoon.

Ice pack: Same as pack ice.

Ichnology: A branch of paleontology that involves the scientific study of fossil footprints, which is a type of "trace fossil." As this term implies, trace fossils or ichnofossils are the fossilized traces of biological activity (as opposed to "body fossils," which are the fossilized remains of actual body parts).

Induced ovulation: When ovulation is activated externally (such as via mating) rather than spontaneously (such as via hormones); when a female requires the act of mating to stimulate ovulation. The females of a number of North American mammals are induced ovulators, including some species of weasel, mink, and rabbit.

Insectivorous: Known as insectivory, it is a carnivorous diet made up partly, primarily, or solely of insects. A North American mammal that is an insectivore or insect-eater is the vagrant shrew.

Interfemoral membrane: Also known as the "uropatagium," it is a membrane of smooth skin that is attached to the posterior part of the body, the back legs, and the tail of an animal; it provides stability when the bat is in flight and capturing prey. North American mammals that possess an interfemoral membrane are the big brown bat and the northern flying squirrel.

Intermontane: Located between mountains.

Invertivorous: Known as invertivory, it is a carnivorous diet made up partly, primarily, or solely of invertebrates; that is, creatures without a spine (jellyfish, starfish, crabs, octopus, squid, lobster, worms, snails, insects). North American mammals that are invertivores or invertebrate-eaters are the harp seal and the eastern gray squirrel.

Isopod: A member of the Isopoda order (generally small marine, freshwater, and terrestrial crustaceans, such as the pill bug). Marine isopods make up part of what is known as "zooplankton," an important element in the diet of many North American marine mammals, such as the blue whale.

Iteroparity: A reproductive strategy in which an animal reproduces more than once during its lifetime. A North American female mammal that is iteroparous is the golden mouse, which can produce as many as 12 litters a year. Related term: semelparity.

Karst: A limestone region comprised of irregular landforms and topography, such as caves, sinkholes, caverns, and underground streams. The primary feature of karst landscapes is that they are formed from dissoluble rocks, such as limestone, dolomite, gypsum, and rock salt, that are worn and sculpted by ground water.

Keel: In bats, a small flap of skin that extends out beyond the calcar (an enhanced cartilaginous elongation of the ankle), adding strength and stability to the wing. A number of North American bat species, such as the eastern small-footed bat, possess a keeled calcar, making it useful as an identification tool.

Keystone species: A type of animal that affects its environment in a profound manner;

a manner that is out of proportion with its numbers, and which, if this species were removed, would cause significant deleterious changes in the surrounding ecosystem, such as loss of habitat and biodiversity. A keystone species then plays a vital role in regulating both species balance and the stability of its environment. A North American mammal that is a keystone species is the jaguar.

Kogid: A member of the Kogiidae family (small sperm whales).

Lacustrine: Related to, growing in, living in, formed in, or associated with lakes. Lacustrine animals are those that prefer, inhabit, or live around lakes. An example of a North American mammal that is lacustral is the moose.

Lagomorph: A member of the Lagomorpha order (hares, rabbits, pikas).

Landfast ice: Same as fast ice.

Lanugo: Soft, wooly, embryonic hair that covers the bodies of fetuses and newborns among some mammals. It is eventually shed, signaling the passage from infancy to juvenile status or young adulthood. A North American mammal that is covered in lanugo at birth is the gray seal.

Antelope jackrabbit.

Larder hoarding: The practice of hiding or storing food in a single, usually central, location, such as in an underground chamber, a midden, or a tree cavity. A North American mammal that is a larder hoarder is the red squirrel.

Legumivorous: Known as legumivory, it is an herbivorous diet made up partly, primarily, or solely of legumes (beans). A North American mammal that is a legumivore or bean-eater is the desert kangaroo rat.

Lepidopterans: A member of the Lepidoptera order (butterflies, moths).

Leporid: A member of the Leporidae family (hares, rabbits).

Life history: Biologically speaking, it covers the major stages an organism passes through during its lifetime: birth, infancy, the juvenile period, adulthood, the reproductive period, and death. An animal's life history includes the study of how these various phases influence its physiology and behavior.

Lignivorous: Known as lignivory, it is an herbivorous diet made up partly, primarily, or solely of wood. A North American mammal that is a lignivore or wood-eater is the North American porcupine.

Littoral zone: Related to, situated on, occurring near, or growing on or near the shore of a body of water (for example, a pond, lake, swamp, or estuary), in particular an ocean; the coastline; the shore or shoreline; the seashore. Examples of North American mammals that inhabit the littoral zone are the American mink and the bearded seal.

Macrohabitat: The macroenvironment, or largest habitat, in which an animal lives, and which differs in character from the smaller environments, or microhabitats, that lie within it. Being extensive in size, macrohabitats support a wide and complex variety of flora and fauna compared to their microhabitats, which tend to be much narrower in scope. Examples of macrohabitats: forests, lakes, reefs, deserts, marshes, grassland, rivers, savanna, prairies, steppe, and coastline.

Macrotopography: In ecology, the large or visible surface features of a material, or of the earth or other body.

Malar: Relating to the cheek or the side of the head.

Malpais: Spanish for "bad" (*mal*) "country" (*país*). Rough countryside, traditionally known as "badlands." More scientifically, it is a rugged or barren landscape of volcanic origin that sits on top of basaltic lava fields, often with exposed solidified lava and other signs of past volcanic activity.

Mammal: A member of the Mammalia class (mammals). A mammal is defined as a warm-blooded higher vertebrate that nourishes its offspring with milk secreted by mammary glands, and which possesses skin that is usually more or less covered with hair. We are mammals, making us cousins of every mammal listed in this book.

Mammalivorous: Known as mammalivory, it is a carnivorous diet made up partly, primarily, or solely of mammals. A North American mammalivore or mammal-eater is

the brown bear.

Manus: Latin for "hand" (plural, also manus); in tetrapodal mammals it indicates the front foot or feet (which in humans are called "hands").

Marine: Related to water, or more specifically the sea. An animal that spends the majority or all of its life in the sea (as opposed to living on land or in trees) is considered a marine mammal. An example of a North American marine mammal is the California sea lion.

Marsupial: A member of the Marsupialia order (opossums, kangaroos, wombats, bandicoots, etc.).

Mast: In botany, it specifically refers to nuts that have accumulated on the forest floor. More generally, mast is the fruit or seeds of woody plants, such as shrubs and trees. There is "hard" mast, such as nuts and walnuts, and "soft" mast, like apples and berries. Consumed by both animals and man, mast is an integral aspect of the world ecosystem, whose abundance or shortage can have profound effects on all organisms, from microbes to mammals.

Matutinal: In biology, an organism that is active before dawn. Related terms: cathemeral, crepuscular, diurnal, nocturnal, vespertine.

Megacerine: A member of the Megacerini tribe (giant deer).

Megapterid: A member of the Megaptera genus (humpback whale). Also written megapterin.

Mephitid: A member of the Mephitidae family (skunks, stink badgers).

Mesic: An environment characterized by, related to, or requiring a moderate amount of moisture; in other words, a habitat having a well-balanced supply of water (not too much, not too little). Plants and animals that are adapted to moderate moisture conditions are considered mesic. A North American mammal that inhabits mesic environments is the Arizona shrew.

Giant deer skeleton.

Mesocarnivorous: Known as mesocarnivory, it is a carnivorous diet made up of 50-70 percent animal flesh. North American mammals that are mesocarnivores or middle meat-eaters are the skunk and the fox.

Mesoplodont: A member of the Mesoplodon genus (beaked whales).

Metamerism: Having a body made up of or characterized by metameres or segments. An example of a metameristic animal is the earthworm.

Metestrus: The third stage of the estrous cycle, it is a period of decline in reproductive activity after estrus. Related terms: diestrus, estrus, proestrus.

Microchiroptera: A suborder of the Chiroptera order that includes all bats except fruit-eating bats. A North American mammal that is a microchiropteran is the insectivorous Seminole bat.

Microhabitat: The microenvironment, or smallest habitat, in which an animal lives, and which differs in character from the larger environment, or macrohabitat, in which it is located. Microhabitats support distinct flora and fauna compared to their surroundings. Examples of microhabitats: stumps, leaf litter, rotting logs, roots, rocks, moist soil, underground tunnels, and puddles.

Microtine: A member of the Microtinae subfamily (voles, lemmings, muskrats); also known as the Arvicolinae subfamily.

Microtopography: In ecology, the small or microscopic surface features of a material, or of the earth or other body.

Microtus: A rodent genus that consists of meadow voles.

Midden: A refuse heap; a dunghill; a cess pit; a place, mound, or dump where unwanted debris is deposited. In ecology, it is used to indicate a pile of waste material, such as leaves, bones, sticks, branches, stones, feathers, bark, grass, pollen, urine, and dung, left by or gathered by an animal. A refuse pile that is comprised specifically of food leftovers and uneaten items, such seeds, husks, and pine cone shells, is referred to as a "kitchen midden." Middens perform many functions in Nature: they may serve as territorial markers, concealment barriers, shelters, nest sites, or even homes; or they may be used to transfer chemical information from animal to animal. A midden's decomposing material returns nutrients to the soil. (The middens of prehistoric humans reveal numerous valuable clues to paleoanthropologists about our ancestors' diet and lifestyle. Wildlife biologists and zoologists derive the same type of information from nonhuman mammal middens.) A North American mammal that produces middens is the bushy-tailed woodrat.

Migratory: An organism that travels to a different place, often with the changing seasons, is considered "migratory." A North American mammal that is migratory is the ribbon seal.

Mima mounds: Named after the Mima Prairie in northwest Washington, these odd heaps of earth, also known as "prairie mounds," are found all over the world (mainly on grassland) and may reach 8' in height and 30' in width. Scientists have offered numerous theories as to how these strange, circular, grass-covered domes might have been created—from Indians, birds, UFOs, and pocket gophers, to wind, earthquakes, vegetation, and glacial runoff. Despite these intriguing hypotheses, none explain *all* of the characteristics of Mima mounds, and the truth about their origins remains unknown. A North American mammal that is associated with Mima mounds is the northern pocket gopher.

Mineral lick: Many mammals (such as deer) will lick muddy streambanks and rocks in order to procure minerals (such as sodium and potassium) that they cannot get in their everyday diet. Minerals like calcium and phosphorus, for example, are vital for lactating mothers and horn-growing males.

Mixed forest: A forest with more than two type of trees; more commonly this term refers to a forest with both coniferous (pines) and deciduous (broadleaf) trees.

Molluscivorous: Known as molluscivory, it is a carnivorous diet made up partly, primarily, or solely of mollusks. A North American mammal that is a molluscivore or mollusk-eater is the North American sea otter.

Molossid: A member of the Molossidae family (free-tailed bats).

Molt: In biology, the act of periodically shedding fur, feathers, antlers, skin, etc. A North American mammal that molts is the long-tailed weasel.

Monestrous: Experiencing estrus once a year, or once each breeding season. A North American mammal that is monestrous is the kit fox.

Monodontid: A member of the Monodontidae family (narwhal, beluga).

Monogamy: Being in a relationship with one mating partner at a time. Such an animal and such a relationship are called monogamous. This relationship can last for one breeding season (serial monogamy) or it can be permanent (lifelong monogamy). A North American mammal that practices monogamy is the gray wolf.

Monophagy: A type of specialist diet in which an animal can eat only one type of food. An example of a non-North American mammal that is monophagous or a "single (type of food) eater" is the koala of Australia.

Montane: From *montanus*, Latin for "mountain." Thus, of, pertaining to, growing in, or plants or animals inhabiting mountainous terrain; can also refer to mountains themselves. More scientifically montane is described as a biogeographic zone dominated by coniferous forest and comprised mostly of cool moist slopes located below the timberline. Many North American mammals inhabit montane regions, and some are even named after them, such as the montane shrew and the montane vole.

Mormoopid: A member of the Mormoopidae family (ghost-faced bats, mustached bats, naked backed bats).

Mosaic habitat: An area made up of a variety of habitat types.

Motile: Motility is the ability to move independently; motion produced by one's own energy. All North American mammals are motile.

Mud-bath: See dust-bath.

Murid: A member of the Muridae family (Old world mice and rats, as well as gerbils).

Muroid: A member of the Muroidea superfamily (mice, rats, gerbils, etc.).

Muskeg: From the Native American word *mashkiig*: a "grassy bog." More specifically, it is a mosaic habitat comprised of a thick spongy layer of decaying plant matter deposited on the ground in wet boreal regions. Healthy muskeg ecosystems provide vital habitat for many types of mammals, including shrews, bats, voles, otters, beaver, muskrat, mink, caribou, moose, bobcats, and bears. These, in turn, are supported by a wide range of muskegophilic (my word) plants, insects, amphibians, and birds.

Mustelid: A member of the Mustelidae family (weasels, otters, badgers, etc.).

Mutualism: A type of symbiotic interaction between two different organisms whose biological relationship benefits both; a cooperative and/or mutually beneficial relationship between two dissimilar animals. There are many examples of wild, mammalian, mutualist species who aid humans (through, for example, ecosystem engineering, ecotourism, and hunting), and who are in turn aided by humans, three being pronghorn antelope, American bison, and Barbary sheep. The coyote and the badger form another example of two North American mammal species that are mutualistic. Being carnivores who occupy much the same habitats, they also share the same food preferences. However, instead of competing, they often hunt together: the coyote chases prey (such as a mouse or rat) into the badger's burrow. If the prey sees the badger in time and manages to jump back out, the coyote snatches it from outside. If it does not see the badger in time, the badger snatches it from inside. Related terms: amensalism, commensalism, ectosymbiosis, endosymbiosis, parasitism, and symbiosis.

Mycophagia: An herbivorous diet made up partly, primarily, or solely of fungi and mushroom spores. North American mammals that are mycophages or fungi-eaters are the wild boar and the brown bear.

Myocastorid: A member of the Myocastoridae family (coypus and nutria).

Mysticete: A member of the Mysticeti suborder, that is, baleen whales or filter feeding whales (bowhead whale, blue whale, right whale, gray whale, finback whale, etc.).

Natal philopatry: The tendency of an animal to return to the area of its birthplace, usually to breed. A North American mammal that is natally philopatric is the banner-tailed kangaroo rat. Related term: philopatry.

Natatoriality: Related to, characterized by, or adapted to swimming. North American mammals that are natatorial are the American mink and the common bottlenose dolphin.

Nearctic: A biogeographic region encompassing the arctic and temperate areas of North America, from Alaska and Canada south to Central Mexico; includes all 48 contiguous American states as well as Greenland.

Nectarivorous: Known as nectarivory, it is an herbivorous diet made up partly, primarily, or solely of nectar. A North American mammal that is a nectarivore or nectar-eater is the lesser long-nosed bat.

Neonate: A newborn; an infant less than 30 days old.

Neotropical: Pertains to the biogeographical region covering southern Mexico, the Caribbean (West Indies), and Central and South America. Examples of neotropical or neotropic mammals are the opossum, jaguar, and mountain lion.

Neritic: Of, relating to, or inhabiting the shallow water zone along the coastline of a sea; a shoreline dweller; an inhabitant of the littoral zone. A North American mammal that is neritic is the North American sea otter.

Nocturnal: In biology, an organism that is active primarily at night. A North American mammal that is nocturnal is the white-throated woodrat. Related terms: cathemeral, crepuscular, diurnal, matutinal, vespertine.

Nose-leaf: A leaf-like skin extension on the nose of some bats, which probably aids in sonar detection, camouflage, and its sense of touch.

Nucivorous: Known as nucivory, it is an herbivorous diet made up partly, primarily, or solely of nuts. A North American mammal that is a nucivore or nut-eater is the eastern gray squirrel.

Numerical response: When predators become more abundant as the population of their prey increases (such as through immigration or heightened reproduction); or when predators become less abundant as their prey decreases (such as through extinction, emigration, or diminishing reproduction); in other words, the relationship between predator density and prey density. Related term: functional response.

Obligate: In biology, this refers to something that is biologically essential for an organism's survival. North American mammals that are, for instance, obligate carnivores are the mountain lion, walrus, dolphin, mink, and sea lion, all which must eat meat to survive, and in most cases lack the physiology necessary to digest plant material. Another example: any mammal that would die without drinking water is considered an obligate drinker.

Ochotonid: A member of the Ochotonidae family (pikas).

Odobenid: A member of the Odobenidae family (walruses).

Odontocete: A member of the Odontoceti ("toothed whales") suborder (dolphins, porpoises, sperm whales, etc.).

Oliophagy: A type of semi-specialist diet in which an animal can eat only a few specific types of food, usually foods that are related to one another.

Omnivorous: Known as omnivory, it is a dual diet made up of both animal flesh and vegetation. Most North American mammals fall into this category. An example of an omnivore or "all-eater" is the racoon.

Porpoise skull showing teeth.

Onychomid: A member of the *Onychomys* genus (grasshopper mice). (This word was coined by the author.)

Osteophagy: The practice of eating bones. North American mammals that are osteophages are the brown bear and the red deer.

Otariid: A member of the Otariidae family ("walking" seals, that is, fur seals and sea lions). Due to their small external ears, they are known (somewhat inaccurately) as "eared" seals. Unlike their cousins the phocid seals, they are able to pivot their rear flippers forward, allowing for dexterous movement on both the land and in water.

Orthopteran: A member of the Orthoptera order (crickets, grasshoppers).

Ovid: A member of the *Ovis* genus (bighorn sheep, Dall sheep, Barbary sheep, etc.). A genus within the Bovidae family, it is also spelled ovine.

Oviparity: A reproductive strategy in which an animal produces and bears eggs (as opposed to bearing live young). An example of an oviparous mammal is the duck-billed platypus (not found in North America).

Ovivorous: Known as ovivory, it is a carnivorous diet made up partly, primarily, or solely of eggs. A North American mammal that is an ovivore or egg-eater is the Norway rat.

Pack ice: Also known as "ice pack," it is a type of mobile ice that forms in open seawater or lake water and is not attached to either the sea bottom or to land. Pack ice is made up of smaller pieces of drift ice that have been forced together, forming large masses of free-floating blocks, which are then known as "pack ice." While pack ice moves with water currents and weather, its opposite, fast ice (which is attached to the bottom or to land), remains stationary despite surrounding conditions. A North American mammal that is heavily associated with pack ice is the walrus.

Palearctic: A biogeographic region encompassing the arctic and temperate areas of Europe, Asia (north of the Himalayan Mountains), and Africa (north of the Sahara Desert). A North American mammal that is palearctic is the brown lemming.

Palo verde: Spanish for "green stick," it is a type of small, spiny, bushy plant belonging to the legume family; bears dense, nearly leafless green branches and bark (hence its common name) and is a native of the southwestern U.S. and northeastern Mexico. Provides shelter for numerous North American mammals, such as the desert pocket mouse.

Palustrine: Relating to an inland wetland or marsh that is nontidal (no flowing water) and is dominated or characterized by trees, shrubs, and emergent vegetation such as

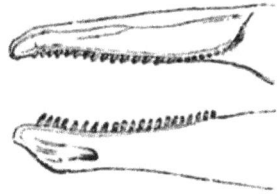

mosses and lichens (whose roots are underwater but whose tops grow above the surface). Palustrine animals are those that prefer or inhabit nontidal wetlands. A North American mammal that is paludal is the hooded skunk.

Palynivorous: Known as palynivory, it is an herbivorous diet made up partly, primarily, or solely of pollen. A North American mammal that is a palynivore or pollen-eater is the lesser long-nosed bat.

Papillae: A vascular connective tissue located in the dermal skin layer; helps nourish body parts like teeth and hair; in some cases, such as on the tongue, it has tactile and tasting properties.

Parapatry: In biology, when more than one species, particularly related species, live in partially overlapping areas. A North American mammal that is parapatric is the least chipmunk, which shares it range with several close cousins. Related terms: allopatry, peripatry, sympatry.

Parasitism: A type of symbiotic, biological interaction or relationship between two different species in which one organism (the parasite) lives in or on another organism (the host).

Parturition: The biological process of giving birth; that is, childbirth.

Patagium: A thin membrane of skin stretching from the body to a limb, forming a wing or wing-like appendage. North American mammals that possess patagia are flying squirrels and bats.

Pectoral: Of, on, or in the chest area.

Pelage: A mammal's fur, wool, or hair.

Peripatry: In biology, when more than one species share adjacent ranges that do not overlap. Related terms: allopatry, parapatry, sympatry.

Perissodactyl: A member of the Perissodactyla order (odd-toed ungulates or hoofed mammals: horses, rhinoceroses, tapirs, zebras, etc.).

Persistence hunting: A form of predation in which a predatory animal chases its fleeing prey at a moderate pace over long distances, eventually wearing it out. An example of a North American mammal that is a persistence hunter is the gray wolf.

Pes: Latin for "foot" (plural, pedes); in tetrapodal mammals it indicates the back foot.

Petraphilia: A lover of stones or rocks; in biology, an organism that is attracted to rocks (of all sizes), lives amidst them, or uses them for some purpose (such as protection, projectiles, nesting, denning). North American mammals that are petraphiles are the Panamint chipmunk, sea otter, bighorn sheep, and Sasquatch.

Petrivorous: Known as petrivory, it is a geophagias diet made up partly, primarily, or solely of rocks and stones. A North American mammal that is a petrivore or stone-eater—or more accurately, is slightly petrivorous—is the northern elephant seal.

Phenology: An especially important branch of science that focuses on the relationships between climate and periodic biological phenomena, such as breeding, migration, and flowering. Can also refer to periodic biological phenomena that correlates with or is influenced by climactic conditions, such as birds building nests in the spring or leaves turning color in the fall. An example of a North American mammal that is influenced phenologically is the southwestern myotis.

Pheromones: A chemical produced by insects, crustaceans, and vertebrates (they are unknown in birds) in order to stimulate one or more responses in other individuals of the same species. Often secreted in substances like urine and sweat, there are numerous types of pheromones, among them: alarm pheromones, food pheromones, parent-young pheromones, territorial pheromones, trail pheromones, aggregation pheromones, and reproductive pheromones (used in courting and mating). Pheromonal communication is also found in humans.

Seal skeleton.

Philopatry: The tendency of an animal to remain in or repeatedly return to a particular area, sometimes year after year. A North American mammal that is philopatric is the

northern fur seal. Related term: natal philopatry.

Phocid: A member of the Phocidae family (true seals or "crawling" seals). Lacking external ear flaps, they are known (somewhat inaccurately) as "earless" seals. Unlike their cousins the otariid seals, they cannot pivot their rear flippers forward, and must awkwardly scoot along on their bellies when on land.

Photoperiod: Daytime; the length of daylight; or more precisely, the number of consecutive hours of light in a 24-hour period (the hours of darkness are not counted) and its effect on plants and animals.

Photoperiodism: Also called photoperiodicity, it is the effect of light (and darkness) on plants and animals, as well as the resulting change in their behavior. Many North American mammals have adapted to this unvarying, dependable natural clock, and for good reason: by relying on it to make assessments regarding food, weather, reproduction, migration, growth, hibernation, and other seasonal activities, it increases their chances of survival. For example, northern populations of long-tailed weasels regulate their fur color by way of photoperiodic response: during the long *sunny* days of summer they shed fur (a thin coat helps dissipate heat) and grow a brown coat on their upper body (excellent camouflage for life in woody environments). During the short *dark* days of winter their fur thickens (a dense coat helps retain heat) and grow a completely white coat (ideal camouflage for life in a snowy environment).

Phyllostomid: A member of the Phyllostomidae family (New World leaf-nosed bats).

Physeterid: A member of the Physeteridae family (sperm whales).

Piedmont: A foothill or an area that lies along the base or foot of a mountain or mountain range; from the French words for "foot" and "mount" (that is, mountain) or "hill." A North American mammal that inhabits piedmont is the giant kangaroo rat.

A phyllostomid or New World leaf-nosed bat.

Piloerection: The bristling or raising of the hair or fur; usually an involuntary reflexive response to panic, fear, agitation, fright, excitement, surprise, shock, arousal, cold, etc. In humans this phenomenon is called "goose bumps"; often accompanies agonistic behavior. A North American mammal that exhibits piloerection is the mountain lion.

Pinniped: A member of the Pinnipedia suborder (seals and walruses).

Piscivorous: Known as piscivory, it is a carnivorous diet made up partly, primarily, or solely of fish. A North American mammal that is a piscivore is the ringed seal.

Plagiopatagium: The part of the patagium that is connected to the last or fifth digits of the hands (or front feet) and the body (including the back legs).

Plankton: Small plant and animal life that drifts, swims, or floats in a body of water. Though tiny, nutrition-rich plankton is extremely abundant and feeds some of the world's largest mammals, such as baleen whales.

Planktivorous: Known as planktivory, it is a carnivorous diet made up partly, primarily, or solely of plankton. A North American mammal that is a planktivore is the baleen whale.

Plantigrade: An animal that walks on its soles rather than on its toes (digitigrade) or on hooves (unguligrade). In other words, its entire foot, including the heel, touches the ground during locomotion. Examples of North American mammals that are plantigrades are the racoon, the brown bear, and Sasquatch. Humans are also plantigrade mammals.

Playa: A shallow temporary lake that evaporates quickly and which is located in a desert basin. It can also refer to the flat dry bed left behind after the lake has evaporated.

Plecopterid: A member of the Plecoptera order (stoneflies).

Phocoenid: A member of the Phocoenidae family (porpoises).

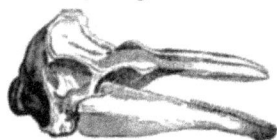

Porpoise skull.

Poikilothermic: Relates to animals whose body temperature varies, because it is regulated by external influences. A poikilotherm then is a creature that cannot maintain its own internal heat, but whose body temperature adjusts to the outside temperature. Most reptiles (though not all) are poikilothermic.

Pollenivorous: Known as pollenivory, it is an herbivorous diet made up partly, primarily, or solely of pollen. A North American mammal that is a pollenivore is the Mexican long-nosed bat.

Pollinator: An organism or agent that pollinates flowers. A North American mammal that is a pollinator is the lesser long-nosed bat.

Polyestrous: Having more than one period of estrus a year. Also spelled polyoestrus.

Polygamy: A mating pattern in which an individual has multiple partners. Such an animal is polygamous. North American mammals that are polygamous are the least chipmunk and the sperm whale.

Polygynandry: A mating system in which both males and females have multiple mating partners. Such an animal is polygynandrous. An example of a polygynandrous North American mammal is the brown bear.

Polynya: An oceanographic term that describes an area of open water surrounded by ice. There are two types: The coastal polynya (which occurs nears shore) and the open-ocean polynya (which occurs at sea, away from shoreline). A North American mammal that inhabits or relies on polynyas is the walrus.

Polyphagy: A type of generalist diet in which an animal can eat a wide variety of foods. An example of a North American mammal that is polyphagous or a "many (types of food) eater" is the coyote.

Pongid: A member of the Pongidae family (orangutan, chimpanzee, gorilla).

Porcid: A member of the Porcidae family (peccary, wild boar, domestic pig). Also spelled porcine.

Postauricular: Behind the ear.

Postocular: Behind the eye.

Precocial: The opposite of altricial, it refers to an animal being born or hatched in a mature or advanced state, able to feed itself and fend for itself almost immediately, thereby requiring little or no parental care. An example of a North American mammal that gives birth to precocial young is the elk.

Predator avoidance: A strategy by which an organism seeks to elude predators. Examples of predator avoidance: staying clear of areas where predators predominate; seeking escape when pursued; use of camouflage; living in dangerous locations; mimicry; the use of striking colors and bright markings; feigning death; living in groups; employing toxins; living underground; living aboveground; switching from diurnal to nocturnal activity.

Predator density: The amount of predators in a specific area at a specific time. Related term: prey density.

Preauricular: In front of the ear.

Preocular: In front of the eye.

Prey density: The amount of prey available to predators in a specific area at a specific time. Related term: predator density.

Primary cavity excavators: Birds (such as woodpeckers, flickers, nuthatches, chickadees) that excavate their own roosting and nesting holes in live, dying, or dead trees. Associated with mammals (such as tree squirrels) that are known as "secondary cavity users."

Primary consumer: Herbivorous organisms that feed on producer organisms (plants) and which serve as prey for secondary consumers (carnivores and omnivores). A North American mammal that is a primary consumer is the mule deer.

Primate: A member of the Primate order (prosimians, monkeys, apes, human beings).

Procyonid: A member of the Procyonidae family (raccoon, ringtail, coati).

Producer: In biology, an autotrophic organism; that is, an organism that makes its own energy and thus serves as the basis for the entire food chain. Producer organisms are chiefly photosynthesizers; that is, aquatic plants (like algae) and terrestrial plants (like grass), which use sunlight to produce sugar, which they use, in turn, to create roots, leaves, wood, and bark. Producer organisms are preyed upon by primary consumers, which are preyed upon by secondary consumers, which are preyed upon by tertiary consumers.

Proestrus: The first stage of the estrous cycle, it is the preparatory period immediately preceding estrus, at which time the female's body readies itself for conception. Related terms: diestrus, estrus, metestrus.

Protective coloration: An organism whose coloring acts like a disguise, helping to camouflage it in its natural environment, possesses protective coloration. A North American mammal that has protective coloration is the cliff chipmunk.

Propatagium: The leading edge of skin of the patagium on, for example, a bat's wing. It is attached to the body and stretches to the thumb or first digit of the hand.

Pursuit predation: A form of hunting and killing in which a predatory animal chases down its fleeing prey. An example of a North American mammal that is a pursuit predator is the common bottlenose dolphin.

Quadrupedalism: Quadrupedality is a type of locomotion that uses four legs. A North American mammal that is quadrupedal is the jaguarundi.

Quaternary consumer: An animal at the top of the food chain; an apex predator, one that eats organisms below it on the food chain (such as tertiary, primary, and secondary consumers), and which is not generally preyed upon by any other animals. North American mammals that are quaternary consumers are the polar bear, orca, gray wolf, and Sasquatch.

Refuge: In biology, a place or area where an organism retreats for protection or shelter. In contrast to refugium, the word refuges (plural) usually refers to places that provide safety for a short period of time only (hours, days, weeks, months).

Refugium: In biology, a relatively stable environment to which organisms retreat during a time of adverse conditions, such as glaciation. In contrast to refuge, the word refugia (plural) usually refers to areas where organisms are able to survive periods of climate change; that is, over a long period of time (millennia).

Reserve: In biology, an area set aside for wildlife; that is, a place that is naturally or legally protected from anthropogenic activity for the benefit of its flora and fauna.

Ricochetalism: Ricochetality is a type of saltatorial locomotion in which an organism travels using short quick hops rather than long leaps—usually employing only the hind limbs. Used mainly when being pursued by predators, ricochetal locomotion allows an animal to change direction quickly, aiding its escape. Two North American mammals that are ricochetal are Heermann's kangaroo rat and the olive-backed pocket mouse.

Rodent: A member of the Rodentia order (rats, mice, squirrels, beavers, chipmunks, prairie dogs, muskrats, voles, gophers, porcupines, marmots, woodchucks, lemmings, etc.).

Rostrum: The beak, snout, or bill of an animal.

Rumination: Also known as "chewing cud," it refers to the act of chewing something slowly and repeatedly over a long period of time.

Sagittal crest: A raised bony ridge situated along the sagittal suture of the skulls of mammals. Its main purpose is to provide an anchor for powerful jaw muscles (aiding in chewing). While modern humans lack a sagittal crest, it was possessed by some types of

Rodents.

extinct hominids (most notably the australopithecines). Male gorillas are well-known for their sagittal crests. This overt raised structure on top of the head has also been observed in Sasquatch, giving it a "pointed skull" appearance—often commented on by eyewitnesses.

Saltatorial: Adapted for leaping. A North American mammal that is saltatorial is the banner-tailed kangaroo rat.

Sand-bath: Many mammals take sand-baths by rolling in dry sand, carefully working the individual grains into their hides or fur. This is both enjoyable and practical, for it helps rid them of ectoparasites which live on the surface of the body. In the case of territorial mammals, sand-bathing can help scent-mark an area. A North American mammal that engages in sand-bathing is the chisel-toothed kangaroo rat.

Sanguivorous: Known as sanquivory, a carnivorous diet made up partly, primarily, or solely of blood. An example of a North American mammal that is a sanguivore or blood-eater is the vampire bat.

Scansoriality: The ability or proclivity to climb, whether it be cliffs, mountains, or trees (the latter which is known as "arboreal locomotion"). A North American mammal that is scansorial is the red squirrel.

Scatter hoarding: The practice of hiding or storing food in a variety of locations, such as underground, or in trees, dens, holes, burrows, and nests. A North American mammal that is a scatter hoarder is the eastern gray squirrel.

Scavenger: An organism that partially or fully feeds on waste, leftovers, refuse, or carrion.

Sciurid: A member of the Sciuridae family (tree squirrels, ground squirrels, flying squirrels, chipmunks, marmots, prairie dogs, woodchucks).

Scrape: Forest litter and dirt scraped into a pile by an animal. Its creator will often urinate or defecate on it as well. A form of animal communication, scrapes are used to mark a creature's home range and territory, and thus are often placed in strategic places, such as along well-used game trails and runs. A North American mammal that makes and uses scrapes is the mountain lion.

Scree: An accumulation of loose stones covering a slope, or lying at the base of a hill.

Seasonal breeding: When an animal mates only during a specific season, it is a seasonal breeder.

Secondary cavity users: Birds and mammals that use or need tree holes made by primary cavity excavators (such as woodpeckers) or that occur naturally. Examples of North American mammals that are secondary cavity users are certain species of mice, voles, rats, tree squirrels, flying squirrels, bats, raccoons, skunks, martens, fishers, porcupines, mink, weasels, bobcats, and bears.

Secondary consumer: Carnivorous or omnivorous organisms that prey on primary consumers (herbivores). A North American mammal that is a secondary consumer is the gray fox.

Sedentary: In biology, when an animal does not wander outside its home range, it is called "sedentary." Terrestrial sedentary mammals, for example, tend to "stick close to home," using the same routes, pathways, and foraging and hunting grounds year after year. An example of a sedentary North American mammal is the American beaver.

Seep: In biology, a seep is a place where liquid (usually water) oozes up from the ground forming a small pool. As sources of food, water, and protection, seeps are important to many North American mammals, such as riparian species like the water shrew.

Seepage slope: An open grassy area that is kept continuously moist from groundwater seepage that accumulates from water flowing down from a gradual slope nearby.

Semelparity: A reproductive strategy in which an animal reproduces only once during its lifetime. Such a creature is considered semelparous. It is rare, or more likely unknown, in North American mammals. An example of at least one semelparous mammal can be found in Australia, where the male gender of a marsupial mouse known as the "brown antechinus" perishes shortly after mating. Related term: iteroparity.

Serial monogamy: Engaging in a succession of monogamous relationships. Such an animal is called a serial monogamist. A North American mammal that practices serial monogamy is the camas pocket gopher.

Shelter form: A temporary shallow depression made by a hare in grass, soil, or snow, which is specifically used for the purposes of protection from weather and predators, as well as resting and grooming.

Sirenid: A member of the Sirenia order (manatees, sea cows, dugongs). Also written sirenian.

Sister species: When one species splits during the process of evolution, the two descendant species are called "sister species." This is because, cladistically, they both share the same ancestor—one not shared by any other species—making them each other's closest relatives.

Snag: Dead trees (such as logs) or other woody debris (branches, sticks, twigs, brushpiles) that have fallen to the ground or into water and are beginning the process of decomposing into soft detritus. Many North American mammals rely on snag for protection and food, while numerous plant species use its nutrient-rich detritus as a nursery. It thus plays an important role in Nature.

Social: In biology, an organism that is basically cooperative in relation to conspecifics is said to be social. Thus, social animals often live in interdependent communities, colonies, pods, "towns," or groups, and display altruistic behaviors. Some mammals are social only for brief periods of the year, such as during mating season, and thus are not technically social. A North American mammal that is social is the black-tailed prairie dog. Related terms: asocial, solitary.

Solitary: In biology, an animal that lives by itself is called solitary. A North American mammal that is solitary is the wolverine.

Soricid: A member of the Soricidae family (shrews).

Sounder: A name for a group of wild pigs.

Speciation: In evolutionary science, the formation of a new and distinct species.

Speleophilia: Literally "love of caves." In biology, it describes animals that are attracted to, visit, spend time in, or live in caves. A North American mammal that is a speleophile is the gray myotis.

Spongivorous: Known as spongivory, it is a carnivorous diet made up partly, primarily, or solely of sponges. Sponge-eating is rare if not unknown among North American mammals.

Spontaneous ovulation: When ovulation is activated spontaneously (such as via hormones) rather than externally (such as via mating); when a female does not require the act of mating to stimulate ovulation. The females of most North American mammal species are spontaneous ovulators, such as, for example, mice, rats, and sheep.

Standing snag: Dying or dead trees that are still standing (on land or in water), and usually in a state of decay. Eventually such trees fall, becoming snag (hollow logs, rotting stumps, blowdown, deadfall, brushpiles). During the next phase of decomposition they transform into woody debris on the forest floor or stream, river, or lake bottom, returning vital elements to the soil. Also known as "wildlife trees," this type of rotting vegetation is used by a myriad of North American mammals (not to mention a host of invertebrates, fish, amphibians, reptiles, and birds) for protection, nesting, shelter, perching, spawning, resting, rearing of young, and food. Standing snag is thus an important part of a healthy biodiverse ecosystem, and should be left alone whenever possible.

Stenotopic: A plant or animal that has a restricted or narrow range of adaptability to changes in environmental conditions; only able to tolerate small fluctuations in an ecosystem. A North American mammal that is stenotopic is the long-tailed shrew.

Stot: Bounding with a stiff-legged gait; it is probably used to confuse predators while fleeing. A number of North American mammals engage in stotting, such as the fallow deer.

Stride length: The distance between the point of initial contact of one foot and the point of initial contact of the opposite foot. Often measured from heel mark to heel mark.

Stump ranch: A fallow or undeveloped farm spread located in the wilderness where animals forage among tree stumps (usually the remains of logging).

Subalpine: In biology, the usually mountainous biogeographic zone below (or just below) the timberline; thus, it indicates anything that relates to, inhabits, or flourishes at lower altitudes where trees are capable of growing. A North American mammal that inhabits subalpine terrain is the alpine chipmunk.

Subnivean: Existing, living, situated, or occurring beneath snow. An example of a North American mammal that is subnivean is the least weasel, which during winter seeks protection from the cold and predators in burrows and runways underneath snow.

Subauricular: Below the ear.

Subocular: Below the eye.

Subspecies: In biology, a taxonomic category ranked immediately below a species. Since it usually indicates a similar but geographically and genetically separate population that can successfully breed throughout its species, the subspecies of many mammal species are difficult to distinguish in the field.

Suid: A member of the Suidae family (hogs and pigs).

Superfetation: A reproductive phenomenon in which a second pregnancy occurs during an initial pregnancy; the occurrence of a second conception during pregnancy, giving rise to embryos of different ages; the presence in the uterus of two fetuses developing from ova fertilized at different times. A North American mammal that experiences superfetation is the snowshoe hare.

Superpod: A group or pod of dolphins or whales of 1,000 or more individuals.

Supra-auricular: Above the ear.

Supraocular: Above the eye.

Symbiosis: Any relationship between two or more biological organisms. Related terms: endosymbiosis, ectosymbiosis, mutualism, amensalism, commensalism, and parasitism.

Sympatry: In biology, when more than one species, particularly related species, live in the same geographic area; that is, their geographical distribution overlaps, yet they still remain separate and distinct species. Two North American mammals that are sympatric are the round-tailed ground squirrel and Harris' antelope squirrel. Related terms: allopatry, parapatry, peripatry.

Synanthropic species: A wild (undomesticated) organism (plant or animal) that benefits by living near humans. Synanthropy is the study of such living things. An example of a North American mammalian synanthrope is the coyote.

Taiga: A damp cool boreal forest in the Northern Hemisphere that is dominated by pine trees. A type of transitional biome with short summers and long winters, it is typically situated between the arctic tundra to the north and the subarctic forests to the south. An example of a North American mammal that inhabits taiga is the taiga vole.

Talpid: A member of the Talpidae family (moles and its relatives).

Talus: Rock debris at the base of a cliff. A talus slope is a rock pile that has accumulated at the base of a cliff or slope, and thus lies at an angle.

Tangle: A dense mass of matted and twisted vegetation.

Tayassuid: A member of the Tayassuidae family (peccaries).

Terrestrial: Related to the earth, land, the ground. An animal that spends the majority or all of its life on land (as opposed to living in water or in trees) is considered terrestrial. An example of a terrestrial North American mammal is the red wolf.

Terricolous: Living on the ground or in the soil.

Territorial: An animal is territorial when it defends its home range against outsiders. A North American mammal that is territorial is the American marten.

Territory: The total geographical area used by an organism. Many mammals "mark off" their territory by scent-marking, then aggressively defend it against intruders. Social mammals tend to be less intolerant of outsiders. All North American mammals establish a territory of one size or another, ranging from a few square yards to thousands of square miles.

Tertiary consumer: An animal that eats both primary and secondary consumers (that is, herbivores as well as carnivores and omnivores). It can be a carnivore or an omnivore itself. A North American mammal that is a tertiary consumer is the Virginia opossum.

Thermal cover: Thickets, brush, and trees that provide protection from the weather; good thermal cover helps keep an animal cool in summer, warm in winter.

Timberline: Also known as the "tree line," it is a transition zone marking the border between the alpine and subalpine zones. Above it trees disappear; thus it marks the latitudinal limit of tree growth. There are many factors that influence the elevation of the timberline, which generally runs unevenly from 10,000'-13,000'—depending on geography. Among them are a region's precipitation amount, length of seasons (particularly summer and winter), type and length of exposure (to, for example, wind, sun, snow), soil quality, and lowest, average, and highest yearly temperatures.

Torpor: A temporary state of inactivity (sometimes described as a "light sleep"), usually accompanied by a drop in body temperature and a decrease in heart rate, respiration, and metabolism. Because torpidity allows an organism to conserve energy, it is useful for animals experiencing harsh conditions, such as drought, food shortages, and extreme heat or cold.

Tragus: A cartilaginous projection or prominence on the outer ear that points backwards, whose purpose is to pick up sounds coming from behind the head. It is one of the key components in helping to identify bats, some species which have greatly enlarged or elaborately shaped tragi.

Trawling: A foraging behavior used by bats in which they capture prey from the water surface of ponds, pools, lakes, etc.

Trichechid: A member of the Trichechidae family (manatees).

Troglobite: An animal that lives only in caves, and which would not be able to survive in any other type of environment. Troglobites are specially adapted to cave life, most having poor eyesight, slow metabolism, and low pigmentation levels. Members of this group include certain types of insects, fish, shrimp, crayfish, and amphibians.

Bat ear showing tragus.

Troglophilia: Literally "love of holes" (that is, caves). It describes an animal capable of living inside or outside of caves. When it does choose to take up residence in a cave, however, it usually only leaves it to search for food. Creatures that are troglophiles include many types of worms, insects, and amphibians. An example of a North American mammal that is a troglophile is the gray fox.

Trogloxene: An animal that only spends a small amount of time in caves, using them primarily for resting, nesting, protection, and hibernation. It has no special adaptations to cave dwelling and lives the majority of its life outside. Members of the trogloxene group are mainly mammals, which in North America include such creatures as bats, skunks, racoons, foxes, bobcats, jaguars, mountain lions, bears, and possibly Sasquatch.

Trophic: Of, related to, or characterized by food, eating, or nutrition.

Tundra: A vast, flat, largely featureless plain, located near the Arctic Circle, in which the subsoil is permanently frozen ("permafrost"). The topsoil, however, thaws in warm weather, allowing the growth of various coarse grasses, mosses, shrubs, and lichens.

Umbrella species: A species of plant or animal—usually occupying a large area—that when protected, helps ensure that co-occurring species of flora and fauna in its ecosystem will potentially be protected as well. A North American mammal that is considered an umbrella species is the brown bear.

Understory: An underlying layer of vegetation—usually pertaining to the plant life (bushes, shrubs, saplings, small trees, etc.) that exists in the shady area between the ground cover and the tree canopy. Also sometimes known as "undergrowth" or "underbrush."

Ungulate: A hoofed, herbivorous, quadrupedal animal. Ungulates are divided into two main branches: even-toed (cattle, deer, pigs) and odd-toed (horses, rhinoceroses).

Unguligrade: An animal that walks on hooves rather than on toes (digitigrade) or the soles (plantigrade) of its feet. Examples of North American mammals that are unguligrades are wild horses, mule deer, and bighorn sheep.

Uropatagium: Same as the interfemoral membrane.

Ursid: A member of the Ursidae family (in our region, black bear, brown bear, polar bear). Also spelled ursine.

Ventral: Of, on, or relating to the belly, underside, or abdomen of an organism. Also

written venter and ventrum.

Vermivorous: Known as vermivory, it is a carnivorous diet made up primarily or solely of worms. A North American mammal that is a vermivore or worm-eater is the hooded skunk.

Vertebrate: A member of the Vertebrata subphylum (vertebrates).

Vespertilionid: A member of the Vespertilionidae family ("evening" bats: mouse-eared bats, pipistrelles, red bats, horay bats, etc.).

Vespertine: In biology, an organism that is active primarily in the early evening. Related terms: cathemeral, crepuscular, diurnal, matutinal, nocturnal.

Vibrissae: In mammalogy, stiff hairs (that is, whiskers) located on or around the snout, and which often function as tactile organs. A single whisker is called a "vibrissa."

Viviparity: A reproductive strategy in which an animal produces and nourishes embryos within its body and gives birth to live young (as opposed to bearing eggs). All currently known North American mammals are viviparous. Examples of viviparous mammals in our region are porcupines, coyotes, and deer.

Wetland: An ecosystem comprised of wet, muddy land and spongy soil that is either permanently or seasonally covered in shallow water; in other words, a bog, swamp, or marsh. A distinguishing characteristic of a wetland is the presence of hydrophytes, aquatic plants that grow partially or completely submerged in water. An example of a mammal that prefers wetland is the marsh rabbit.

Wild: In biology, an organism that is born, lives, and dies in the wild. Related term: feral.

Wildlife trees: Same as standing snag.

Windfall: Something, for example, a tree, that has been blown down by the wind. It is, in other words, a type of snag.

Xeric: An environment characterized by, related to, or requiring only a small amount of moisture; in other words, a dry habitat. Plants and animals that are adapted to arid conditions are considered xeric. A North American mammal that is xeric is the round-tailed ground squirrel.

Zapodid: A member of the Zapodinae subfamily (jumping mice).

Ziphiid: A member of the Ziphiidae family (beaked whales).

APPENDIX A

Important Tips On Interacting With Wild North American Mammals

BY LOCHLAINN SEABROOK

N O MATTER HOW CUTE, CUDDLY, or innocent an animal may appear, *all* wild mammals, being undomesticated, are unpredictable, and thus should be considered potentially dangerous. This is not just true for the obviously large fearless carnivores, like wolves and mountain lions. It is also true for so-called "gentle" vegetarian animals like sheep, deer, elk, and bison. The herbivorous moose, for example, attacks and injures far more people every year than the meat-eating bear. Even the smallest mammals are fully capable of inflicting serious wounds, and even causing death (due to infection, bleeding out, etc).

When observing our mammalian cousins in their natural habitat, be wary and take whatever precautions are necessary to protect yourself and your fellow nature watchers from harm: carry whistles, air horns, bear spray, and weapons (where legal) if necessary, for your life will always be more valuable than a wild animal.

However, instead of assuming that an animal will behave itself *and* respect you (a foolish and deadly assumption if there ever was one), it is much wiser to assume that it will not. In this regard the best advice I can give is to follow the Golden Rule: treat animals as you would want to be treated. You are in their home; they are not in yours. Be courteous toward them; be considerate of their space. When you are looking at a bison from 20' away, realize that this would be the equivalent of a large, strange, uninvited creature standing inside your house staring at you—an overt sign of aggression to most animals. I think we all can agree that, as humans, this would be quite unnerving, if not downright terrifying.

Though wolf attacks on humans in North America are rare, they do occur. Caution is always advised.

The National Park Service has issued guidelines on how to deal with wild *terrestrial* mammals, which I have recorded below. These are important rules that should be followed at all times:

• Keep a safe distance from all wildlife.
• Stay 300' (1 football field) from all bears and wolves.
• Keep 75' from all other wildlife. Females with offspring present can be extremely dangerous.
• Use roadside pull-outs when viewing wildlife.
• Binoculars, spotting scopes, and telephoto lenses help with non-obtrusive wildlife watching.
• Never feed animals, for their safety and yours.
• Do not crowd or block their natural movement. Often you will get a better view if the animals proceed on course.

To these sagacious, time-tested practices let us add the following tips (some which are actual laws) from the NOAA, the National Oceanic and Atmospheric Administration, regarding wild *marine* mammals (my comments are in parentheses):

• Restrict viewing time to 30 min. (to avoid unduly stressing the animals).
• Stay 150' (½ football field) from seals and sea lions (whether they are in the water or onshore); observe dolphins from 150'-300' away; keep at least 300' away from all whales; stay 600' from orcas (killer whales); remain 1,500' (or 5 football fields) from North Atlantic right whales. These rules apply to all vessels and aircraft—including drones, as well as to people using watercraft such as surfboards, kayaks, and jet-skis.
• Never feed marine mammals. (In nearly all cases it is inhumane, dangerous, and illegal.)
• Report sick, injured, entangled, stranded, or dead marine mammals to the proper authorities (usually marine and/or wildlife hotlines, centers, parks, sanctuaries, or refuges—many which are equipped with scientists and professional response and rescue teams).

As with terrestrial mammals, our marine relatives are undomesticated, powerful, intelligent, tough, fast-moving, and unpredictable, and in ways far beyond anything an ordinary person would be familiar with. As a visitor to their world, it is our obligation to behave responsibly and respectfully.

Being in the presence of a wild mammal should be considered a great honor and privilege. This, combined with the knowledge that any interference on our part may cause it undue stress (such as causing a mother to abandon her young), prompts us to act with sensitivity and consideration.

We need not fear Nature. But she demands respect, and this we must give her. Thus, when it comes to animal-watching and study, following these simple instructions, as well as the Golden Rule, will help keep both you and our mammal friends protected, while permitting you to more fully experience and savor the great outdoors.

After over a half-century of working with wild and domestic mammals, my advice is to memorize these tips. They could save a life—maybe your own. Be safe and enjoy!

L.S.

Though a vegetarian, the American bison is one of the more cantankerous and unpredictable mammals you will encounter in the wild, making it potentially extremely dangerous. The largest land animal in our region, the National Park Service recommends that you stay at least 25' from this fleet-footed 1-ton bovid; even more if possible.

APPENDIX B

92 Threatened & Endangered U.S. Mammals

COMPLETE LIST AS OF 2020

SOURCE: U.S. FISH & WILDLIFE SERVICE

(Note: Species are listed by common name; some of these mammals are not included in this book.)

Alabama beach mouse: endangered
Amargosa vole: endangered
Anastastia Island beach mouse: endangered
Bearded seal: threatened
Beluga whale: endangered
Black-footed ferret: endangered
Blue whale: endangered
Bowhead whale: endangered
Buena Vista Lake ornate shrew: endangered
Canada lynx: threatened
Carolina northern flying squirrel: endangered
Choctawhatchee beach mouse: endangered
Columbia Basin pygmy rabbit: endangered
Columbian white-tailed deer: threatened
False killer whale: endangered
Finback whale: endangered
Florida bonneted bat: endangered
Florida panther: endangered
Florida salt marsh vole: endangered
Fresno kangaroo rat: endangered
Giant kangaroo rat: endangered
Gray bat: endangered
Gray wolf: endangered
Grizzly bear: threatened
Guadalupe fur seal: endangered
Gulf Coast jaguarundi: endangered
Hawaiian hoary bat: endangered
Hawaiian monk seal: endangered
Humpback whale: threatened
Indiana bat: endangered
Jaguar: endangered
Key deer: endangered
Key Largo cotton mouse: endangered
Key Largo woodrat: endangered
Killer whale: endangered
Little Mariana fruit bat: endangered
Lower Keys marsh rabbit: endangered
Margay: endangered
Mariana fruit (flying fox) bat: threatened
Mexican long-nosed bat: endangered
Mexican wolf: endangered
Morro bay kangaroo rat: endangered
Mount Graham red squirrel: endangered
New Mexico meadow jumping mouse: endangered
North Atlantic right whale: endangered
Northern Idaho ground squirrel: threatened

Northern long-eared bat: threatened
Northern sea otter: threatened
North Pacific right whale: endangered
Ocelot: endangered
Olympia pocket gopher: threatened
Ozark big-eared bat: endangered
Pacific pocket mouse: endangered
Pacific sheath-tailed bat (*rotensis*): endangered
Pacific sheath-tailed bat (*semicaudata*): endangered
Peninsular bighorn sheep: endangered
Perdido Key beach mouse: endangered
Point Arena mountain beaver: endangered
Polar bear: threatened
Preble's meadow jumping mouse: threatened
Red wolf: endangered
Rice rat: endangered
Ringed seal (*botnica*): threatened
Ringed seal (*hispida*): threatened
Ringed seal (*ladogensis*): endangered
Ringed seal (*ochotensis*): threatened
Riparian brush rabbit: endangered
Riparian woodrat: endangered
Roy Prairie pocket gopher: threatened
Saint Andrew beach mouse: endangered
Salt marsh harvest mouse: endangered
San Bernardino Merriam's kangaroo rat: endangered
San Joaquin kit fox: endangered
Santa Catalina Island fox: threatened
Sei whale: endangered
Sierra Nevada bighorn sheep: endangered
Sinaloan jaguarundi: endangered
Sonoran pronghorn: endangered
Southeastern beach mouse: threatened
Southern sea otter: threatened
Sperm whale: endangered
Spotted seal: threatened
Steller sea lion: endangered
Stephens' kangaroo rat: endangered
Tenino pocket gopher: threatened
Tipton kangaroo rat: endangered
Utah prairie dog: threatened
Virginia big-eared bat: endangered
West Indian manatee: threatened
Wood bison: threatened
Woodland caribou: endangered
Yelm pocket gopher: threatened

Prairie pocket gopher.

American beaver.

APPENDIX C
Common Threats to North American Mammals

COMPILED & ARRANGED BY THE AUTHOR

NOTE: THREATS TO MAMMALS ARE NOT ALWAYS PURELY NEGATIVE. FOR EXAMPLE, MANY OF THE ITEMS LISTED, SUCH AS FIRE, HUNTING, MINING, AND TOURISM, CAN ALSO HAVE POSITIVE EFFECTS ON WILDLIFE.

Acid rain.
Acidification of the oceans.
Agriculture.
Agricultural development.
Agricultural effluents.
Agricultural runoff.
Air pollution.
Alien species.
All-terrain vehicles (ATVs).
Aquaculture.
Avalanches.
Aviation.
Aviation flight paths.
Biodiversity loss.
Bushmeat hunting.
Campfires.
Camping.
Campsites.
Car hits.
Cave entrance sealing.
Civil unrest.
Clear cutting forests.
Climate change.
Coal industry.
Collisions (with vehicles).
Commercial areas.
Commercial development.
Culverts.
Dams.
Deforestation.
Desertification.
Dirt-biking.
Disease (bacteria, parasites, viruses).
Dog-walking.
Domestic pets (dogs, cats).
Domestic waste water.
Droughts.
Earthquakes.
Ecosystem modification.
Energy production.
Environmental degradation.
Erosion.
Fertilizers.
Field mowing.
Fire.
Fire suppression.
Fishing.
Fish farms.
Fishing industry.
Fishing nets.

Flooding (from dams).
Flower collecting.
Folk medicine industry.
Forest fires.
Forest removal.
Forestry effluents.
Fossil fuel burning.
Four-wheeling.
Freshwater aquaculture.
Garbage waste.
Gas drilling.
Gas industry.
Habitat alteration.
Habitat conversion.
Habitat degradation.
Habitat destruction.
Habitat fragmentation.
Habitat isolation.
Habitat loss.
Habitat modification.
Habitat shifting.
Harvesting aquatic resources.
Herbicides.
Hiking.
Hiking trails.
Highway construction.
Housing areas.
Housing development.
Human cave exploration.
Human development.
Human disturbance.
Human encroachment.
Human expansion.
Human generated noise.
Human intrusion.
Human-made byproducts.
Human-made waste.
Human population growth.
Human recreation.
Hunting.
Hurricanes.
Industrial areas.
Industrial development.
Industrial effluents.
Industrial parks.
Industrial sites.
Industrial waste.
International conflicts.
Introduced genetic material.
Invasive and exotic species.

Jet skis.
Land clearing.
Land conversion.
Landslides.
Light pollution.
Livestock farming.
Livestock grazing.
Livestock ranching.
Logging.
Marine aquaculture.
Marine debris (plastic, etc.).
Meat consumption.
Medicine industry.
Meteor strikes.
Military effluents.
Military exercises.
Military operations.
Mine reopenings.
Mining.
Mining waste.
Mountain biking.
Natural resource exploitation.
Noise pollution.
Nonnative species.
Non-timber crops.
Ocean temperatures rising.
Office parks.
Off-road vehicles.
Oil drilling.
Oil industry.
Overgrazing.
Over-harvesting wildlife.
Paper industry.
Parking lots.
Pesticides.
Pet industry.
Pet predation.
Pet trade.
Plant collecting.
Poaching wildlife.
Pollution (water, air, soil, etc.).
Power lines.
Precipitation pattern changes.
Pulp plantations.
Quarrying.
Radioactive pollution.
Railroads.
Railways.
Recreation areas.
Recreational activities.
Recreational climbing.
Recreational development.
Recreational sites.
Recreational trails.
Recreational vehicles.
Renewable energy.
Residential development.
Resource consumption.
River alteration.
River dredging.

Roads.
Road construction.
Road salt.
Sea ice melting.
Sea levels rising.
Septic fields.
Service corridors.
Service lines.
Sewage.
Shipping lanes.
Smuggling wildlife.
Snag disposal.
Soil pollution.
Solid waste.
Storms.
Stream alteration.
Strip malls.
Strip mining.
Suburban expansion.
Tall buildings.
Thermal pollution.
Temperature extremes.
Timber industry.
Tourism areas.
Towers.
Trail-riding.
Transportation corridors.
Trapping wildlife.
Tsunamis.
Urban areas.
Urban development.
Urban expansion.
Urbanization.
Urban waste water.
Utility lines.
Vandalism.
Vegetation zones shifting.
Vehicle strikes (cars, trucks).
Vessel strikes (from boats).
Volcanoes.
War.
Water diversion.
Water impoundments.
Water management.
Water pollution.
Water use.
Wetland destruction.
Wetland drainage.
Wildlife abuse.
Wildlife displacement.
Wildlife trafficking.
Window collisions (home, business).
Wind turbines.
Wood cutting.
Wood harvesting.
Wood plantations.

APPENDIX D
Are Mustangs Native Or Nonnative Horses?

BY LOCHLAINN SEABROOK

DEBATE CONTINUES AS TO WHETHER the mustang is a native or a nonnative species. This is an important question since the answer dictates whether it deserves the same legal protection and management as other native North American wildlife. There are two basic and opposing schools of thought: A) The pro-nonnative group holds that our "wild" horse (*Equus ferus*) is not truly wild in the literal sense of the word—that is, descended from horses that have *always* been wild. This is because, so this group maintains, it is a descendant of the domesticated horse (*Equus ferus caballus*) that was brought to North America by early Spanish explorers in the 15th and 16th Centuries, then rebred and further developed by Native Americans in the 17th and 18th Centuries. Thus it would be more appropriate to call the mustang a "feral" horse; meaning that it is a descendant of a domesticated breed that escaped and then became wild, in the modern era becoming, some say, an invasive pest.

Wild horse.

Furthermore, this group maintains, it must be acknowledged that the mustang is a descendant of a wide variety of horses that were domesticated at different times by different people in different parts of the world. When and where these domestication events occurred, and who the individuals were who initiated them, are not at all clear. This equid's background remains murky and mysterious, and much more study is needed. What we *do* know is that it descends from tamed Eurasian stock; meaning that it is neither truly wild or from North America. Conclusion: the mustang is not a native species.

B) The view of the pro-native group differs considerably, mainly because it includes the mustang's prehistoric lineage. It begins with the argument that the entire horse family, Equidae, got its start in North America. In fact, the Spaniards were merely reintroducing a domesticated modern version of a prehistoric horse, *Equus lambei*, that had lived in North America for several million years, becoming extinct on our

Wild horses on the range.

continent only 7,000-12,000 yrs. ago. Importantly, this makes *Equus lambei* the last horse species to exist in North America before its extinction here.

Long before expiring in our region, however, herds of *Equus lambei* had crossed the Bering land bridge, migrating west into Asia, before spreading further into Europe. It was the domesticated descendant of this wild horse, a tamed descendant scientifically named *Equus caballus*, that Christopher Columbus first brought to North America in

1493, to what is now the Virgin Islands. In 1519, Hernán Cortés arrived on mainland North America (in what is now Mexico), bringing 16 more horses with him. In the ensuing decades and centuries, thousands of additional horses were brought into North America, and by the year 1800, most of what would become the Western states had been populated with horses reintroduced from Europe.

From this domesticated stock, some of which later escaped (or were abandoned, or intentionally released) into the wilderness, *Equus ferus*, or what we call the North American mustang, was born. Hence, after thousands of years of absence, North America's native horse returned to its native soil—though with a different appearance and a different scientific name.

In short, this group claims, after at least 6,000 yrs. of domestication, the mustang should now be considered a native North American species.

This view is buttressed by the fact that, according to some scientists, *Equus lambei* is a genetic duplicate of *Equus caballus*. In essence this means that, despite the appearance of things, North America's wild horses are not merely "escaped livestock," as the pro-nonnative group likes to call them. They are real native equines whose biological ancestors existed here for millions of years prior to the arrival of human beings.

A subspecies of prehistoric wild horse, the tarpan or Eurasian wild horse (*Equus ferus ferus*), went extinct in the late 1800s. If or how it is related to the mustang is not known.

Conclusion: Though they may have disappeared from North America for a time, then returned with European explorers in the 15th and 16th Centuries as domesticated animals, this does not alter the genetic facts. For based on both current paleontological data and molecular biology, our wild horse—no matter how one chooses to define "wild" or what scientific name we give it—is a native species. To put it another way, *Equus ferus*, originated and evolved on our continent, and through no fault of its own, became extinct here, but was reintroduced several thousand years later as a Eurasian descendant with a new scientific name.

From a genetic point of view (despite the passage of millennia and the confusion of various binomials), *Equus lambei*, *Equus ferus*, *Equus caballus*, and *Equus ferus caballus* are all the same species. Thus, in my view, being a "homegrown" species with prehistoric North American ancestry, the mustang is a native and should be treated like all other native North American wildlife—with the same legal protections.

APPENDIX E

Who or What is Making Odd Tree Structures in Our Forests?

A RATIONAL DISCUSSION BY LOCHLAINN SEABROOK

NO ONE WHO HAS EVER come across a bizarre and crudely made tree structure in the middle of an isolated uninhabited wilderness has failed to ponder the question of its origins. It was obviously intentionally made. But by who, or what? Nature, people, Bigfoot? As discussed in this book's entry on Sasquatch, indications point to the latter.

Sasquatch skeptics claim, however, that such structures are accidental products of Nature, the results, for example, of wind, storms, cold, heat, hail, rain, lightning, snowfall, humidity, tornadoes, and hurricanes. Yet this does not explain the often intricate design and overt craftsmanship of these objects; nor does it explain how natural forces could transport massive trees over distances of hundreds of feet, then *lift them up* (defying gravity) and carefully fit, even weave, them in between other trees. Additionally, many of these "weavings" are not only located high off the ground, they are perfectly level.

Proponents of the Sasquatch theory counter that this phenomena demonstrates great strength, intelligence, and creativity, clear evidence that the structures are the result of *intelligent manipulation* by a *large living creature* with *opposable thumbs*. And indeed, a number of eyewitnesses have reported seeing an enormous bipedal anthropoid creature snapping large trees in half, or pulling them up out of the ground.

One scientific hypothesis argues that these structures are created by people. This is certainly possible. But Sasquatch supporters ask what the motivation would be. Who would spend hours, days, even weeks, in the middle of a lonely forest, risking life and limb to make structures that serve no human purpose and, even more oddly, that no one is ever likely to see? What possible use could such structures be to us, and how could a person, or even a group of people, build them in remote areas without roads, heavy machinery, cabins, and truckloads of supplies? There are not even walking paths in most of the areas where these tree structures are found.

Our forests are filled with odd, out-of-place, purposeless tree structures. Who is making them, and why?

There are also the bold facts that human footprints are almost never seen around them, and vehicle tracks leading in and out of the area are nearly always scarce or nonexistent. Furthermore, none of the manipulated trees show signs of artificial cutting from, for instance, hatchets, axes, saws, knives, or chain saws. Instead, most have been snapped off at their bases, while many have been *recently* ripped from the ground and are still alive. There are, in fact, usually no signs of human presence of any kind.

Though Native Americans have known the answer to this mystery for thousands of years, the outside world has ignored it. And so, for skeptics at least, to satisfy the requirements of mainstream science, the ultimate answer to the question of who is constructing strange tree structures in our forests will have to await further study.

Bottle-nose dolphin.

APPENDIX F

On the Accuracy of the Material in This Book

BY LOCHLAINN SEABROOK

BECAUSE MOST WILD MAMMALS ARE by nature intelligent, secretive, shy, unobtrusive, and nocturnal—and in many cases are solitary, sparsely distributed, and possess a heightened (and understandable) distrust of humans—they are often difficult to census, map, and research. Even in those rare cases when they *can* be found, many species so closely resemble one another externally that precise field recognition is out of the question, resulting in a myriad of misidentifications—some that have passed into our science books as "facts." In these types of cases only a detailed, sometimes lengthy and costly laboratory examination of teeth, bones, or DNA can determine the species. An example of this are our nearly two dozen similarly patterned chipmunks, many of which can only be reliably distinguished from one another by examining their skeletons.

The average person would be quite stunned to learn how little is actually understood about many species of North American mammals. Very little is known about many of our lagomorphs (rabbits and hares), for example; and even some of our larger mammals (such as the red wolf) have not been adequately studied, with much of the information we have on them coming only from museum specimens, those caught in traps, and even anecdotal reports. One whale species, for instance, is only known from skeletal remains and has thus never been identified alive in the wild.

Star-nosed mole.

Not surprisingly then, some of the most common words and phrases used in mammalogy literature are: "currently unknown," "little research done," "data is lacking," "poorly studied," "little documented information," "almost nothing is known," "not yet studied," "observations have been infrequent," "severe lack of information," "has not been adequately analyzed," and "requires further study."

What this means for the writer-naturalist is that information for a particular animal can often not only be somewhat speculative, but it can also vary widely from study to study, making it difficult to be scientifically definitive when it comes to the facts. My large collection of encyclopedias, for example, cite widely different weights, pelage coloration, fur markings, distribution range, habits, and populations for the same species. Not unusually, an animal's common name—and even its scientific name—differ from source to source, depending on the time and place they were written, as well as the author and his or her particular scientific viewpoint. Thousands of colloquial, regional, and national nicknames (many of them wholly unscientific and therefore misleading) further complicate this already chaotic taxonomic situation.

Adding to the confusion is the information I have gleaned from my own experiences, studies, observations of, and work with wild and captive mammals over a period of many decades; information that sometimes disagrees with long-accepted reference books and authorities.

While I have done my best to ease around these different views and opinions, my information may (and at times inevitably will) be at variance with details given by others. Since it is impossible to know everything about every mammal in the wild, and because mammals are continually evolving, and because new ones are always being discovered, there will always be some fluidity in the animal sciences. Much of the ecology and life history of North American mammals remains theoretical, not factual. In using this book this reality should be borne in mind.

Wild horse.

APPENDIX G

An Evolutionary Tree According to Early 20th-Century Mainstream Science

FROM 1921

THOUGH LARGELY OBSOLETE, MUCH CAN STILL BE LEARNED FROM THIS DIAGRAM.
MAMMALS, THE SUBJECT OF THIS BOOK, ARE LOCATED IN THE TOPMOST SECTION.

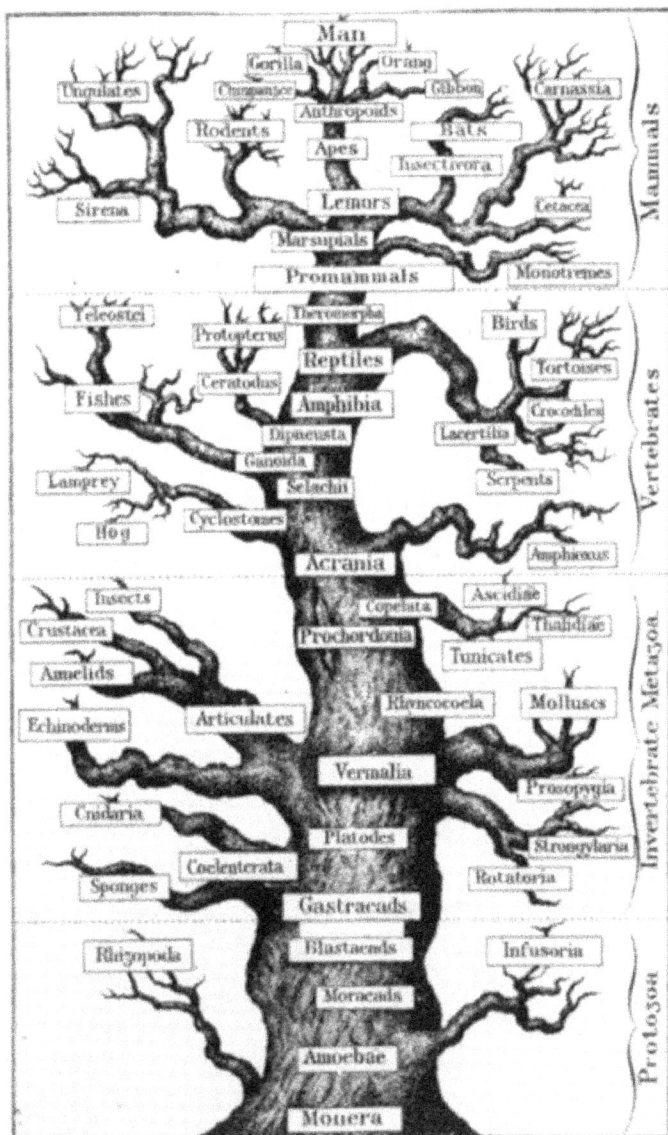

Man

Gorilla   Orang
Ungulates   Chimpanzee   Gibbon   Carnassia
    Anthropoids
Rodents   Bats
    Apes
        Insectivora
Sirena   Lemurs   Cetacea
    Marsupials
    Promammals   Monotremes

Teleostei   Theromorpha   Birds
    Protopterus
    Reptiles   Tortoises
    Ceratodus
Fishes   Amphibia   Crocodiles
    Dipneusta   Lacertilia
    Ganoida
Lamprey   Selachii   Serpents
    Cyclostomes
Hag   Acrania   Amphioxus

Insects   Ascidiae
Crustacea   Copelata   Thalidiae
    Prochordonia
Amelids   Tunicates
Echinoderms   Articulates   Rhyncocoela   Molluscs
    Vermalia
Cnidaria   Prosopygia
    Platodes
Coelenterata   Strongylaria
Sponges   Rotatoria
    Gastraeads

Rhizopoda   Blastaeads   Infusoria
    Moraeads
    Amoebae
    Monera

Mammals
Vertebrates
Invertebrate Metazoa
Protozoa

Canada lynx.

APPENDIX H

Some Useful Governmental Websites

FOR ADDITIONAL INFORMATION ON THE NATURAL HISTORY, PRESERVATION, AND CONSERVATION OF NORTH AMERICAN MAMMALS, PLEASE SEE THE FOLLOWING OFFICIAL RESOURCES.

American Museum of Natural History: www.amnh.org
Canadian Department of Fisheries and Oceans: www.dfo-mpo.gc.ca/index-eng.htm
Canadian Department of Natural Resources: www.nrcan.gc.ca/home
Canadian Department of Parks: www.pc.gc.ca/en/index
Canadian Museum of Nature: www.nature.ca/index.php?q=en/home
Canadian Wildlife Service: www.canada.ca/en/environment-climate-change.html
Mexican Department of Agriculture: www.gob.mx/agricultura
Mexican Department of Natural Resources: www.gob.mx/semarnat
Mexican National Forestry Commission: www.gob.mx/conafor
National Museum of Natural History (U.S.): www.naturalhistory.si.edu
National Oceanic and Atmospheric Administration: www.noaa.gov
United States Bureau of Land Management: www.blm.gov.
United States Bureau of Ocean Energy Management www.boem.gov
United States Bureau of Reclamation: www.usbr.gov
United States Bureau of Safety and Environmental Enforcement: www.bsee.gov
United States Department of Agriculture: www.usda.gov
United States Department of Commerce: www.commerce.gov
United States Department of the Interior: www.doi.gov
United States Fish and Wildlife Service: www.fws.gov
United States Forest Service: www.fs.fed.us
United States Geological Survey: www.usgs.gov
United States National Park Service: www.nps.gov

American black bear.

LIFE LIST

Sightings Record of North American Mammals

COMMON NAME	SCIENTIFIC NAME	DATE	LOCATION	NOTES

COMMON NAME	SCIENTIFIC NAME	DATE	LOCATION	NOTES

BIBLIOGRAPHY

And Suggested Reading

Academic American Encyclopedia. New York: Grolier Academic Reference, 1998.

Adam, Peter James. *Morphological Evolution in Cetacea: Skull Asymmetry and Allometry of Body Size and Prey.* Los Angeles, CA: University of California, 2007.

Adams, Bradley, and Pam Crabtree. *Comparative Osteology: A Laboratory and Field Guide of Common North American Animals.* Academic Press: Oxford, UK: 2012.

Adams, Rick A. *Bats of the Rocky Mountain West: Natural History, Ecology, and Conservation.* Boulder, CO: University Press of Colorado, 2004.

Adams, Rick A., and Scott C. Pedersen (eds.). *Bat Evolution, Ecology, and Conservation.* New York: Springer, 2013.

Adams, William Henry Davenport. *Animal Life Throughout the Globe: An Illustrated Book of Natural History.* London, UK: T. Nelson and Sons, 1876.

——. *The Arctic: A History of Its Discovery, Its Plants, Animals, and Natural Phenomena.* Edinburgh, Scotland: George Tod, 1876.

Agassiz, Elizabeth C., and Alexander Agassiz. *Seaside Studies in Natural History.* Boston, MA: Ticknor and Fields, 1865.

Agassiz, Louis. *An Introduction to the Study of Natural History.* New York: Greeley and McElrath, 1847.

——. *Contributions to the Natural History of the United States of America.* Boston, MA: Little, Brown and Co., 1860.

Agustí, Jordi, and Mauricio Antón. *Mammoths, Sabertooths, and Hominids: 65 Million Years of Evolution in Europe.* New York: Columbia University Press, 2002.

Ahern, Albert M. *Fur Facts: A Book of Knowledge.* St. Louis, MO: C. P. Curran, 1922.

Alden, Peter, and Fred Heath. *National Audubon Society Field Guide to California.* New York: Alfred A. Knopf, 1998.

Allan, Larry. *Florida Animals for Everyday Naturalists.* Gainesville, FL: Seaside Publishing, 2014.

Allen, Glover M. *The Mammals of China and Mongolia: Natural History of Central Asia.* New York: American Museum of Natural History, 1938-1940.

——. *Extinct and Vanishing Mammals of the Western Hemisphere.* American Committee for International Wildlife Protection, 1942.

Allen, Joel Asaph. *History of North American Pinnipeds: A Monograph of the Walruses, Sea-Lions, Sea-Bears and Seals of North America.* Washington, D.C.: U.S. Government Printing Office, 1880.

——. *Bulletin of the American Museum of Natural History.* New York: American Museum of Natural History, 1906.

Allen, Sarah G., and Joe Mortenson. *Field Guide to Marine Mammals of the Pacific Coast: Baja, California, Oregon, Washington, British Columbia.* Berkeley, CA: University of California Press, 2011.

Allen, Thomas B. (ed.). *Wild Animals of North America.* Washington, D.C.: National Geographic Society, 1995.

Alpers, Antony. *Dolphins: The Myth and the Mammal.* New York: Houghton Mifflin, 1961.

Altringham, John D. *Bats: Biology and Behaviour.* Oxford, UK: Oxford University Press, 1996.

Alves, Paulo Célio, Nuno Ferrand, and Klaus Hackländer (eds.). *Lagomorph Biology: Evolution, Ecology, and Conservation.* New York: Springer, 2008.

Ammerman, Loren K., Christine L. Hice, and David J. Schmidly. *Bats of Texas.* College Station,

TX: Texas A&M University Press, 2012.

Amos, William H. *The Life of the Seashore*. New York: McGraw-Hill, 1966.

Amsel, Sheri. *Vermont Nature Guide: A Field Guide to Birds, Mammals, Trees, Insects, Wildflowers, Amphibians, Reptiles, and Where to Find Them*. Brattleboro, VT: Echo Point Books, 2013.

Anderson, Sydney, and J. Knox Jones. *Orders and Families of Recent Mammals of the World*. New York: John Wiley and Sons, 1984.

Anthony, Harold Elmer (ed.). *Mammals of America*. New York: University Society, 1917.

——. *Field Book of North American Mammals: Descriptions of Every Mammal Known North of the Rio Grande, Together with Brief Accounts of Habits, Geographical Ranges, Etc*. New York: Putnam, 1928.

Arctic Research of the United States. Alexandria, VA: National Science Foundation, 2002.

Ardrey, Robert. *African Genesis: A Personal Investigation Into the Animal Origins and Nature of Man*. New York: Dell, 1961.

Armstrong, David M. *Rocky Mountain Mammals: A Handbook of Mammals of Rocky Mountain National Park and Vicinity*. 1975. Boulder, CO: University Press of Colorado, 2008 ed.

Armstrong, David M., James P. Fitzgerald, and Carron E. Meaney. *Mammals of Colorado*. Boulder, CO: University Press of Colorado, 2011.

Asdell, Sydney Arthur. *Patterns of Mammalian Reproduction*. Ithaca, NY: Comstock Publishing, 1964.

Attenborough, David. *Life on Earth: A Natural History*. New York: Little, Brown and Co., 1981.

——. *The Living Planet: A Portrait of the Earth*. New York: Little, Brown and Co., 1984.

——. *The Life of Mammals*. London, UK: BBC Books, 2002.

Aubry, Keith B., William J. Zielinski, Martin G. Raphael, Gilbert Proulx, and Steven W. Buskirk (eds.). *Biology and Conservation of Martens, Sables, and Fishers: A New Synthesis*. Ithaca, NY: Cornell University Press, 2012.

Audubon, John James, and John Bachman. *The Quadrupeds of North America*. 3 vols. London, UK: Wiley and Putnam, 1847.

Bailey, Vernon. *Harmful and Beneficial Animals of the Arid Interior, With Special Reference to the Carson and Humboldt Valleys, Nevada*. Washington, D.C.: U.S. Government Printing Office, 1908.

——. *The Mammals and Life Zones of Oregon*. Washington, D.C.: United States Department of Agriculture, 1936.

Baker, Mary L. *Whales, Dolphins, and Porpoises of the World*. New York: Doubleday, 1987.

Baker, Rollin H. *Michigan Mammals*. East Lansing, MI: Michigan State University Press, 1991.

Ballard, Jack. *Grizzly Bears*. Guilford, CT: Morris Book Publishing, 2012.

Banfield, Alexander William Francis. *The Mammals of Canada*. Toronto, Ontario, CAN: University of Toronto Press, 1975.

Bangs, Outram. *The Land Mammals of Peninsular Florida and the Coast Region of Georgia*. Boston, MA: Boston Society of Natural History, 1898.

Barash, David P. *Marmots: Social Behavior and Ecology*. Stanford, CA: Stanford University Press, 1989.

Barbour, Roger W. *Bats of America*. Lexington, KY: University of Kentucky Press, 1997.

Barkalow, Frederick Schenck, and Monica Shorten. *The World of the Gray Squirrel*. Philadelphia: J. B. Lippencott Co., 197.

Barnard, Edward S., and Sharon Fass Yates (eds.). *Reader's Digest North American Wildlife: Mammals, Reptiles, and Amphibians*. Pleasantville, NY: Reader's Digest, 1998.

Barnett, Samuel Anthony. *The Rat: A Study in Behavior*. New York: Transaction Publishers, 1963.

Barrett, Gary W., and John D. Peles (eds.). *Landscape Ecology of Small Mammals*. New York: Springer, 1999.

Beale, Thomas. *The Natural History of the Sperm Whale*. London, UK: John Van Voorst, 1838.

Beddington, J. R., R. J. H. Beverton, and D. M. Lavigne (eds.). *Marine Mammals and Fisheries*. London, UK: George Allen and Unwin, 1985.

Beer, Amy-Jane, and Pat Morris. *Encyclopedia of North American Mammals: An Essential Guide to Mammals of North America*. San Diego, CA: Thunder Bay Press, 2004.

Beeton, Samuel Orchart. *Beeton's Dictionary of Natural History: A Compendious Encyclopedia of the Animal Kingdom*. London, UK: Ward, Lock, and Tyler, 1871.

Belwood, Jacqueline Janine. *Rare and Endangered Biota of Florida: Mammals*. Gainesville, FL: University Press of Florida, 1992.

Benyus, Janine M. *The Field Guide to Wildlife Habitats of the Eastern United States*. New York: Fireside, 1989.

Beolens, Bo, Michael Watkins, and Michael Grayson. *The Eponym Dictionary of Mammals*. Baltimore, MD: Johns Hopkins University Press, 2009.

Berry, William D., and Elizabeth Berry. *Mammals of the San Francisco Bay Region*. Berkeley, CA: University of California Press, 1959.

Berta, Annalisa. *The Rise of Marine Mammals: 50 Million Years of Evolution*. Baltimore, MD: Johns Hopkins University Press, 2017.

Berta, Annalisa, James L. Sumich, and Kit M. Kovacs. *Marine Mammals: Evolutionary Biology*. 1999. London, UK: Academic Press, 2015 ed.

Best, Troy L., and Julian L. Dusi. *Mammals of Alabama*. Tuscaloosa, AL: University of Alabama Press, 2014.

Bindernagel, John A. *North America's Great Ape: The Sasquatch - A Wildlife Biologist Looks at the Continent's Most Misunderstood Large Mammal*. Courtenay, B.C., CAN: Beachcomber Books, 1998.

——. *The Discovery of the Sasquatch: Reconciling Culture, History, and Science in the Discovery Process*. Courtenay, B.C., CAN: Beachcomber Books, 2010.

Blackman, W. Haden. *The Field Guide to North American Monsters*. New York: Three Rivers Press, 1998.

Blix, Arnoldus, Schytte. *Arctic Animals: And Their Adaptations to Life on the Edge*. Bergen, Norway: Fagbokforlaget, 2005.

Bockstoce, John R. *Whales, Ice, and Men: The History of Whaling in the Western Arctic*. Seattle, WA: University of Washington Press, 1995.

Boitani, Luigi, and Stefania Bartoli. *Simon and Schuster's Guide to Mammals*. New York: Fireside, 1982.

Bolgiano, Chris, and Jerry Roberts (eds.). *The Eastern Cougar: Historic Accounts, Scientific Investigations, New Evidence*. Mechanicsburg, PA: Stackpole Books, 2005.

Bonner, William Nigel. *The Natural History of Seals*. New York: Facts on File, 1990.

——. *Whales of the World*. New York: Facts on File, 2003.

Booth, Earnest Sheldon. *How to Know the Mammals*. Dubuque, IA: William Brown and Co., 1950.

Boreman, Thomas. *A Description of 300 Animals*. London, UK: self-published, 1769.

Borror, Donald Joyce. *Dictionary of Word Roots and Combining Forms: Compiled from the Greek, Latin, and other Languages, With Special Reference to Biological Terms and Scientific Names*. 2 vols. Mountain View, CA: Mayfield, 1960.

Borror, Donald Joyce, and Richard E. White. *A Field Guide to the Insects: America North of Mexico*. Boston, MA: Houghton Mifflin, 1970.

Bourlière, François. *The Natural History of Mammals*. New York: Knopf, 1954.

Bowers, Nora, Rick Bowers, and Kenn Kaufman. *Mammals of North America*. New York: Houghton Mifflin, 2004.

Bown, Thomas M., and Kenneth D, Rose (eds.). *Dawn of the Age of Mammals in the Northern Part of the Rocky Mountain Interior, North America*. Boulder, CO: Geological Society of America, 1990.

Box, Hilary O., and Kathleen R. Gibson (eds.). *Mammalian Social Learning: Comparative and Ecological Perspectives*. Cambridge, UK: Cambridge University Press, 1999.

Boyce, Mark S. (ed.). *Evolution of Life Histories of Mammals: Theory and Pattern*. New Haven, CT: Yale University Press, 1988.

Boyd, Lee, and Katherine A. Houpt (eds.). *Przewalski's Horse: The History and Biology of an Endangered Species*. Albany, NY: State University of New York Press, 1994.

Braun, Dieter. *Wild Animals of North America*. London, UK: Flying Eye Books, 2015.

Brehm, Alfred Edmund. *Brehm's Life of Animals: A Complete Natural History for Popular Home Instruction and for the Use of Schools*. Chicago, IL: A. N. Marquis and Co., 1896.

Brown, Gary. *The Bear Almanac: A Comprehensive Guide to the Bears of the World*. 1993. Guilford, CT: Globe Pequot Press, 2009 ed.

Brown, Larry N. *A Guide to the Mammals of the Southeastern United States*. Knoxville, TN: University of Tennessee Press, 1997.

Buckner, Eldon L., and Jack Reneau (eds.). *Boone and Crockett Club's Records of North American Big Game*. (12th ed.) Missoula, MT: Boone and Crockett Club, 2005.

Bulletin of the American Museum of Natural History (vol. 2, 1887-1890). New York: American Museum of Natural History, 1890.

Burde, John H., and George A. Feldhamer. *Mammals of the National Parks*. Baltimore, MD: Johns Hopkins University Press, 2005.

Burnie, David. *The Concise Animal Encyclopedia*. Boston, MA: Kingfisher, 2000.

Burnie, David, and Don E. Wilson (eds.). *Animal: The Definitive Visual Guide*. London, UK: DK Publishing, 2011.

Burt, William Henry. *A Field Guide to the Mammals: North America North of Mexico*. 1952. New

York: Houghton Mifflin, 1976 ed.
Burton, Maurice, and Robert Burton. *International Wildlife Encyclopedia*. Tarrytown, NY: Marshall Cavendish, 2002.
Buseth, Marit Emilie, and Richard Saunders. *Rabbit Behaviour, Health and Care*. Oxfordshire, UK: CABI, 2015.
Buskirk, Steven W. *Wild Mammals of Wyoming and Yellowstone National Park*. Oakland, CA: University of California Press, 2016.
Butterworth, Andy (ed.). *Marine Mammal Welfare: Human Induced Change in the Marine Environment and its Impacts on Marine Mammal Welfare*. Cham, Switzerland: Springer Nature, 2017.
Byrne, Peter. *The Search for Bigfoot: Monster, Myth, or Man?* New York: Pocket Books, 1976.
Cahalane, Victor H. *Mammals of North America*. New York: Macmillan, 1964.
Caire, William, Bryan P. Glass, Michael A. Mares, and Jack D. Tyler. *Mammals of Oklahoma*. Norman, OK: University of Oklahoma Press, 1989.
Calhoun, John B., and James U. Casby. *Calculation of Home Range and Density of Small Mammals*. Washington, D.C.: U.S. Government Printing Office, 1958.
Caro, Tim. *Antipredator Defenses in Birds and Mammals*. Chicago, IL: University of Chicago Press, 2005.
Carrington, Richard. *The Mammals*. New York: Time-Life, 1965.
Carroll, Robert L. *Vertebrate Paleontology and Evolution*. New York: W. H. Freeman, 1990.
Carson, Mary Kay. *The Bat Scientists*. Boston, MA: Houghton Mifflin Harcourt, 2010.
Carson, Rachel. *The Sea Around Us*. New York: Oxford University Press, 1951.
——. *Silent Spring*. Boston, MA: Houghton Mifflin Co., 1962.
——. *The Sense of Wonder*. New York: Harper and Row, 1965.
Carwardine, Mark, Erich Hoyt, R. Ewan Fordyce, and Peter Gill. *Whales, Dolphins and Porpoises*. New York: Time-Life, 1998.
Castellini, Michael A., and Jo-Ann Melish (eds.). *Marine Mammal Physiology: Requisites for Ocean Living*. Boca Raton, FL: CRC Press, 2016.
Castelló, José R. *Bovids of the World: Antelopes, Gazelles, Cattle, Goats, Sheep, and Relatives*. Princeton, NJ: Princeton University Press, 2016.
——. *Canids of the World: Wolves, Wild Dogs, Foxes, Jackals, Coyotes, and Their Relatives*. Princeton, NJ: Princeton University Press, 2018.
Ceballos, Gerardo (ed.). *Mammals of Mexico*. Baltimore, MD: Johns Hopkins University Press, 2014.
Chambers's Encyclopaedia: A Dictionary of Universal Knowledge for the People. Edinburgh, Scotland: Chambers Publishing Company, 1860.
Chambers, Robert. *Vestiges of the Natural History of Creation*. New York: William H. Colyer, 1846.
Chapman, Joseph A., and George A. Feldhammer (eds.). *Wild Mammals of North America: Biology, Management, and Economics*. Baltimore, MD: Johns Hopkins University Press, 1982.
Chapman, Joseph A., and John E. C. Flux (eds.). *Rabbits, Hares, and Pikas: Status Survey and Conservation Action Plan*. Gland, Switzerland: IUCN, 1990.
Chester, Sharon. *The Arctic Guide: Wildlife of the Far North*. Princeton, NJ: Princeton University Press, 2016.
Chinsamy-Turan, Anusuya. *Forerunner of Mammals: Radiation, Histology, Biology*. Bloomington, IN: Indiana University Press, 2012.
Choate, Jerry R., J. Knox Jones Jr., and Clyde Jones. *Handbook of Mammals of the South-Central States*. Baton Rouge, LA: Louisiana State University Press, 1994.
Christiansen, Per. *The Encyclopedia of Animals*. London, UK: Amber Books, 2006.
Churchfield, Sara. *The Natural History of Shrews*. Ithaca, NY: Comstock Publishing, 1990.
Clark, Tim W., and Mark R. Stromberg. *Mammals in Wyoming*. Lawrence, KS: University Press of Kansas, 1987.
Cleave, Andrew. *Seals and Sea Lions: A Portrait of the Animal World*. Cliffside Park, NJ: New Line Books, 1998.
Clutter, Mary E. (ed.). *Dormancy and Developmental Arrest: Experimental Analysis in Plants and Animals*. New Haven, CT: Academic Press, 1978.
Clutton-Brock, Juliet. *A Natural History of Domesticated Animals*. Cambridge, UK: Cambridge University Press, 1999.
Cockrum, E. Lendell. *Laboratory Manual for Mammalogy*. New York: Ronald Press Co., 1962.
——. *Mammals of the Southwest*. Tucson, AZ: University of Arizona Press, 1982.
Cohen, Joshua G., Michael A. Kost, Bradford S. Slaughter, and Dennis A. Albert. *A Field Guide to the Natural Communities of Michigan*. East Lansing, MI: Michigan State University Press, 2015.

Collier's Encyclopedia. New York: Macmillan, 1984.
Collins, Henry Hill (ed.). *Harper and Row's Complete Field Guide to North American Wildlife* (Eastern ed.). 1959. New York: Harper and Row, 1981 ed.
Collins, Scott L., and Linda L. Wallace (eds.). *Fire in North American Tallgrass Prairies*. Norman, OK: University of Oklahoma Press, 1990.
Compton's Pictured Encyclopedia. Chicago, IL: F. E. Compton and Co., 1929.
Corbet, Gordon Barclay. *A World List of Mammalian Species*. London, UK: Natural History Museum Publications, 1991.
Coues, Elliot. *Fur-bearing Animals: A Monograph of North American Mustelidae*. Washington, D.C.: U.S. Government Printing Office, 1877.
Cousteau, Jacques. *The Whale: Mighty Monarch of the Sea*. New York: Doubleday, 1972.
——. *Jacques Cousteau: The Ocean World*. 1979. New York: Harry N. Abrams, 1993 ed.
——. *Whales*. New York: Harry N. Abrams, 1988.
Cozzi, Bruno, Stefan Huggenberger, and Helmut Oelschläger. *Anatomy of Dolphins: Insights Into Body Structure and Function*. Amsterdam, The Netherlands: Elsevier, 2017.
Craighead, John Johnson. *The Grizzly Bears of Yellowstone*. Washington, D.C.: Island Press, 1995.
Craighead, Lance. *Bears of the World*. Minneapolis, MN: Voyageur Press, 2003.
Cuvier, Georges. *The Animal Kingdom: Arranged in Conformity to its Organization*. New York: G. and C. and H. Carvill, 1831.
Cuvier, Georges, et al. *A System of Natural History; Containing Scientific and Popular Descriptions of Various Animals*. Brattleboro, VT: Peck and Wood, 1834.
Dagg, Anne Innis. *Mammals of Ontario*. Waterloo, Ontario, CAN: Otter Press, 1974.
Dalquest, Walter W., and Norman V. Horner. *Mammals of North Central Texas*. Wichita Falls, TX: Midwestern University Press, 1984.
Dalrymple, Byron W. *North American Big-Game Animals*. New York: Outdoor Life Books, 1985.
Davis, Buddy, and Kay Davis. *Magnificent Mammals: Marvels of Creation*. Green Forest, AR: Master Books, 2006.
Davis, David Edward, and Frank B. Golley. *Principles in Mammalogy*. New York: Reinhold, 1963.
Davis, Lance E., Robert E. Gallman, and Karin Gleiter. *In Pursuit of Leviathan: Technology, Institutions, Productivity, and Profits in American Whaling, 1816-1906*. Chicago, IL: University of Chicago Press, 1997.
Davis, Randall W. *Marine Mammals: Adaptations for an Aquatic Life*. London, UK: Springer, 2019.
Davis, Russell, Ronnie Sidner. *Mammals of Woodland and Forest Habitats in the Rincon Mountains of Saguaro National Monument, Arizona*. Washington, D.C.: United States Department of the Interior, 1992.
DeBlase, Anthony F. *A Manual of Mammalogy: With Keys to Families of the World*. Dubuque, IA: William C. Brown, 1980.
DeGraff, Richard M., and Mariko Yamasaki. *New England Wildlife: Habitat, Natural History, and Distribution*. Lebanon, NH: University Press of New England, 2001.
De La Rosa, Carlos L., and Claudia C. Nocke. *A Guide to the Carnivores of Central America*. Austin, TX: University of Texas Press, 2000.
Denis, Armand. *Cats of the World*. New York: Houghton Mifflin, 1964.
Dent, David. *Insect Pest Management*. Cambridge, MA: CAB International, 2000.
Dewhurst, Henry William. *The Natural History of the Order Cetacea and the Oceanic Inhabitants of the Arctic Regions*. London, UK: self-published, 1834.
Dines, Lisa. *The American Mustang Guidebook: History, Behavior, and State-By-State Directions on Where to Best View America's Wild Horses*. Minocqua, WI: Willow Creek Press, 2001.
Dinets, Vladimir. *Peterson Field Guide to Finding Mammals in North America*. New York: Houghton Mifflin, 2015.
Dobbyn, Jon Sandy. *Atlas of the Mammals of Ontario*. Toronto, Canada: Federation of Ontario Naturalists, 1994.
Drew, Liam. *I Mammal: The Story of What Makes Us Mammals*. London, UK: Bloomsbury, 2017.
Drury, Herbert Rentschler. *Natural History, Conservation, and Ecology of the Musk Ox*. Ann Arbor, MI: University of Michigan, 1955.
Duff, Andrew, and Ann Lawson. *Mammals of the World: A Checklist*. New Haven, CT: Yale University Press, 2004.
Durrell, Lee. *State of the Ark*. London, UK: Gaia Books, 1986.
Dutson, Judith. *Horse Breeds of North America: The Pocket Guide to 96 Essential Breeds*. North Adams, MA: Storey Publishing, 2006.
Eder, Tamara. *Whales and Other Marine Mammals of British Columbia and Alaska*. Edmonton, Alberta, CAN: Lone Pine, 2002.

———. *Squirrels of North America.* Edmonton, Alberta, CAN: Lone Pine, 2009.

Edwards, Elwyn Hartley. *The Ultimate Horse Book.* London, UK: DK Publishing, 1991.

———. *Horses (Smithsonian Handbooks).* London, UK: DK Publishing, 1993.

———. *The Horse Encyclopedia.* London, UK: DK Publishing, 2016.

Eisenberg, John F. *The Mammalian Radiations: An Analysis of Trends in Evolution, Adaptation, and Behavior.* Chicago, IL: University of Chicago Press, 1983.

Eisenberg, John F., and Kent H. Redford. *Mammals of the Neotropics.* 3 vols. Chicago, IL: University of Chicago Press, 1999.

Elbroch, Mark. *Mammal Tracks and Sign: A Guide to North American Species.* Mechanicsburg, PA: Stackpole Books, 2003.

———. *Animal Skulls: A Guide to North American Species.* Mechanicsburg, PA: Stackpole, 2006.

Elbroch, Mark, and Kurt Rinehart. *Behavior of North American Mammals.* New York: Houghton Mifflin, 2011.

Elliot, Daniel Giraud. *The Land and Sea Mammals of Middle America and the West Indies.* Chicago, IL: Field Columbian Museum, 1904.

———. *A Check List of Mammals of the North American Continent, the West Indies, and the Neighboring Seas.* Chicago, IL: Field Columbian Museum, 1905.

———. *A Catalogue of the Collection of Mammals in the Field Columbian Museum.* Chicago, IL: Field Columbian Museum, 1907.

Elliot, Henry W. *Report on the Seal Islands of Alaska.* Washington, D.C.: U.S. Government Printing Office, 1884.

Elliot, Lang. *A Guide to Wildlife Sounds: The Sounds of 100 Mammals, Birds, Reptiles, Amphibians, and Insects.* Mechanicsburg, PA: Stackpole, 2005.

Ellis, Richard. *The Book of Whales.* New York: Alfred A. Knopf, 1985.

Elton, Charles. *Voles, Mice and Lemmings: Problems in Population Dynamics.* Oxford, UK: Clarendon Press, 1942.

Emmons, Louise. *Neotropical Rainforest Mammals: A Field Guide.* Chicago, IL: University of Chicago Press, 1997.

English Cyclopaedia. London, UK: Bradbury, Evans, and Co., 1867.

English, Douglas. *Beasties Courageous: Studies of Animal Life and Character.* London, UK: S. H. Bousfield and Co., 1905.

———. *One Hundred Photographs from Life of the Shrew-Mouse, Dormouse, House-Mouse, Field-Mouse, and Harvest-Mouse.* London, UK: Cassell and Co., 1909.

Ensminger, Marion Eugene. *Horses and Horsemanship.* Danville, IL: Interstate Printers and Publishers, 1969.

Entwistle, Abigail, and Nigel Dunstone (eds.). *Priorities for the Conservation of Mammalian Diversity: Has the Panda Had its Day?* Cambridge, UK: Cambridge University Press, 2000.

Erdoes, Richard, and Alfonso Ortiz (eds.). *American Indian Myths and Legends.* New York: Pantheon, 1984.

Errington, Paul L. *Muskrats and Marsh Management.* Lincoln, NE: University of Nebraska Press, 1978.

Estes, James, Douglas P. DeMaster, Daniel F. Doak, Terrie M. Williams, and Robert L. Brownell Jr. (eds.). *Whales, Whaling, and Ocean Ecosystems.* Berkeley, CA: University of California Press, 2006.

Evans, P. G. H. *The Natural History of Whales and Dolphins.* London, UK: Christopher Helm, 1987.

Evans, Phyllis Roberts. *The Sea World Book of Seals and Sea Lions.* Boston, MA: Harcourt, 1986.

Evans, Will F. *Hunting Grizzlys, Black Bear and Lions "Big-Time" on the Old Ranches.* Silver City, NM: High Lonesome Books, 2001.

Evers, David C. *A Guide to Michigan's Endangered Wildlife.* Ann Arbor, MI: University of Michigan Press, 1992.

Everyman's Encyclopaedia. London, UK: J. M. Dent and Sons, 1958.

Ewer, R. F. *The Carnivores.* Ithaca, NY: Comstock Publishing, 1986.

Fahrenbach, W. H. "Sasquatch: Size, Scaling, and Statistics," *Cryptozoology,* vol. 13, 1998.

Farrand, John. *Familiar Mammals of North America (Audubon Society Pocket Guide).* New York: Alfred A. Knopf, 1988.

Fazio, Patricia Mabee. "The Fight to Save a Memory: Creation of the Pryor Mountain Wild Horse Range (1968) and Evolving Federal Wild Horse Protection Through 1971." Doctoral dissertation, Texas A&M University, College Station, TX, 1995.

Feldhammer, George A., Lee C. Drickamer, Stephen H. Vessey, Joseph F. Merritt, and Carey Krajewski. *Mammalogy: Adaptation, Diversity, Ecology.* Baltimore, MD: Johns Hopkins University Press, 2015.

Feldhamer, George A., Bruce C. Thompson, and Joseph A. Chapman (eds.). *Wild Mammals of North America: Biology, Management, and Conservation*. 1982. Baltimore, MD: Johns Hopkins University Press, 2003 ed.

Fenton, Melville Brock. *Communication in the Chiroptera*. Bloomington, IN: Indiana University Press, 1985.

Fenton, Melville Brock, and Nancy B. Simmons. *Bats: A World of Science and Mystery*. Chicago, IL: University of Chicago Press, 2014.

Ferguson, Steven H., Lisa L. Loseto, and Mark L. Mallory (eds.). *A Little Less Arctic: Top Predators in the World's Largest Northern Inland Sea, Hudson Bay*. London, UK: Springer, 2010.

Findley, James S. *The Natural History of New Mexican Mammals*. Albuquerque, NM: University of New Mexico Press, 1987.

Findley, James S., and Terry L. Yates. *The Biology of the Soricidae*. Albuquerque, NM: University of New Mexico Press, 1991.

Flores, Dan. *Coyote America: A Natural and Supernatural History*. New York: Basic Books, 2016.

——. *American Serengeti: The Last Big Animals of the Great Plains*. Lawrence, KS: University Press of Kansas, 2016.

Flower, William Henry. *The Horse: A Study in Natural History*. London, UK: Kegan Paul, Trench, Trübner, and Co., 1891.

Foresman, Kerry R. *Wild Mammals of Montana*. Missoula, MT: Mountain Press, 2012.

Forsyth, Adrian. *Mammals of the Canadian Wild*. East Camden, Ontario, CAN: Camden House, 1985.

——. *Mammals of North America: Temperate and Arctic Regions*. 1999. Ontario, CAN: Firefly Books, 2006 ed.

Fowler, Charles W., and Tim D. Smith (eds.). *Dynamics of Large Mammal Populations*. Caldwell, NJ: Blackburn Press, 2004.

Fox, Michael W. (ed.). *The Wild Canids: Their Systematics, Behavioral Ecology and Evolution*. 1975. Wenatchee, WA: Dogwise Publishing, 2009 ed.

Franzen, Jens Lorenz. *The Rise of Horses: 55 Million Years of Evolution*. Baltimore, MD: Johns Hopkins University Press, 2010.

Franzmann, Albert W., and Charles C. Schwartz (eds.). *Ecology and Management of the North American Moose*. Boulder, CO: University Press of Colorado, 2007.

Freeman, Dan. *The Love of Monkeys and Apes*. London, UK: Octopus Books, 1977.

Freeman, Milton M. R., and Lee Foote (eds.). *Inuit, Polar Bears, and Sustainable Use: Local, National and International Perspectives*. Edmonton, Alberta, CAN: CCI Press, 2009.

Freeman, Richard. *Adventures in Cryptozoology: Hunting for Yetis, Mongolian Deathworms and Other Not-So-Mythical Monsters*. Coral Gables, FL: Mango Publishing, 2019.

Fryxell, John M., Anthony R. E. Sinclair, and Graeme Caughley. *Wildlife Ecology, Conservation, and Management*. 1994. Oxford, UK: John Wiley and Sons, 2014 ed.

Funk and Wagnalls New Standard Encyclopedia of Universal Knowledge. New York: Funk and Wagnalls, 1931.

Gales, Nick, Mark Hindell, and Roger Kirkwood (eds.). *Marine Mammals: Fisheries, Tourism, and Management Issues*. Collingwood, Victoria, Australia: CSIRO, 2003.

Galloway, Patricia (ed.). *The Hernando de Soto Expedition: History, Historiography, and "Discovery" in the Southeast*. Lincoln, NE: University of Nebraska Press, 1997.

Gannon, Michael R., Allen Kurta, Armando Rodríguez-Durán, and Michael R. Willig. *Bats of Puerto Rico: An Island Focus and a Caribbean Perspective*. Lubbock, TX: Texas Tech University Press, 2005.

Garrott, Robert A., P. J. White, and Fred G. R. Watson (eds.). *The Ecology of Large Mammals in Central Yellowstone: Sixteen Years of Integrated Studies*. San Diego, CA: Academic Press, 2009.

Gaskin, David Edward. *The Ecology of Whales and Dolphins*. Portsmouth, NH: Heinemann, 1982.

Gauvin, Marshall J. *The Illustrated Story of Evolution*. New York: Peter Eckler, 1921.

Geist, Valerius. *Wild Animals of North America*. Washington, D.C.: National Geographic Society, 1979.

——. *Deer of the World: Their Evolution, Behaviour, and Ecology*. Mechanicsburg, PA: Stackpole Books, 1998.

Geraci, Joseph R., and Valerie J. Lounsbury. *Marine Mammals Ashore: A Field Guide for Strandings*. Baltimore, MD: National Aquarium, 2005.

Gerhard, Ken. *The Essential Guide to Bigfoot*. San Antonio, TX: Crypto Excursions, 2019.

Gibbons, Diane K. *Mammal Tracks and Sign of the Northeast*. Lebanon, NH: University Press of New England, 2003.

Gibson, David M., and Robert A. Harris. *Metabolic Regulation in Mammals*. London, UK: Taylor and Francis, 2002.

Gittleman, John L. *Carnivore Behavior, Ecology, and Evolution*. Berlin, Germany: Spring Science and Business Media, 2013.

Godfrey, Linda S. *American Monsters: A History of Monster Lore, Legends, and Sightings in America*. New York: Tarcher Perigee, 2014.

Godin, Alfred J. *Wild Mammals of New England* (Field Guide Edition). Yarmouth, ME: Delorme, 1983.

Godman, John D. *American Natural History*. Philadelphia, PA: Stoddart and Atherton, 1831.

Golley, Frank Benjamin. *Mammals of Georgia: A Study of Their Distribution and Functional Role in the Ecosystem*. Whitefish, MT: Literary Licensing, 2012.

Golley, Frank Benjamin, K. Petrusewicz, and L. Ryszkowski (eds.). *Small Mammals: Their Productivity and Population Dynamics*. Cambridge, UK: Cambridge University Press, 1975.

Goodrich, S. G. *Illustrated Natural History of the Animal Kingdom*. New York: Derby and Jackson, 1859.

Gordon, Iain J., and Herbert H. T. Prins (eds.). *The Ecology of Browsing and Grazing*. Berlin, Germany: Springer, 2008.

Gore, Thomas, Paula Gore, and James M. Giffin. *Horse Owner's Veterinary Handbook*. Hoboken, NJ: Howell Book House, 2008.

Gorman, Martyn L., and R. David Stone. *The Natural History of Moles*. Ithaca, NY: Comstock, 1990.

Gortor, Uko. *Mexico Field Guide: Marine Mammals (Baja California, Sea of Cortez, Pacific Coast)*. Conway, WA: Rainforest Publications, 2002.

Goswami, Anjali, and Anthony Friscia (eds.). *Carnivoran Evolution: New Views on Phylogeny, Form and Function*. Cambridge, UK: Cambridge University Press, 2010.

Gotch, Arthur Frederick. *Mammals - Their Latin Names Explained: A Guide to Animal Classification*. London, UK: Blandford Press, 1979.

Gottshang, Jack L. *A Guide to the Mammals of Ohio*. Columbus, OH: Ohio State University Press, 1983.

Graham, Gary L. *Bats of the World*. New York: St. Martin's Press, 2002.

Grambo, Rebecca. *Wolf: Legend, Enemy, Icon*. Ontario, CAN: Firefly Books, 2015.

Green, John. *Sasquatch: The Apes Among Us*. 1981. Surrey, B.C., CAN: Hancock House, 2017 ed.

Griffin, Donald R. *Listening in the Dark: The Acoustic Orientation of Bats and Men*. Ithaca, NY: Cornell University Press, 1986.

Grinnell, George Bird (ed.). *American Big Game in its Haunts: The Book of the Boone and Crockett Club*. New York: Harper and Brothers, 1904.

Grinnell, Hilda Wood. *A Synopsis of the Bats of California*. Berkeley, CA: University of California Press, 1918.

Grinnell, Joseph. *An Account of the Mammals and Birds of the Lower Colorado Valley: With Especial Reference to the Distributional Problems Presented*. Berkeley, CA: University of California, 1914.

——. *A Geographical Study of the Kangaroo Rats of California*. Berkeley, CA: University of California Press, 1922.

Grinnell, Joseph, and Joseph S. Dixon. *Natural History of the Ground Squirrels of California*. Sacramento, CA: California State Printing Office, 1919.

Grinnell, Joseph, and Tracy Irwin Storer. *Animal Life in the Yosemite: An Account of the Mammals, Birds, Reptiles, and Amphibians in a Cross-Section of the Sierra Nevada*. Berkeley, CA: University of California, 1924.

Grinnell, Joseph, Joseph S. Dixon, and Jean M. Linsdale. *Fur-Bearing Mammals of California: Their Natural History, Systematic Status, and Relations to Man*. 2 vols. Berkeley, CA: University of California Press, 1937.

Groves, Colin, and Peter Grubb. *Ungulate Taxonomy*. Baltimore, MD: Johns Hopkins University Press, 2011.

Grzimek, Bernhard. *Grzimek's Encyclopedia: Mammals*. New York: McGraw-Hill, 1989.

Gubernick, David J., and Peter H. Klopfer. *Parental Care in Mammals*. New York: Plenum Press, 1981.

Guest, Kristen, and Monica Mattfield (eds.). *Horse Breeds and Human Society: Purity, Identify and the Making of the Modern Horse*. London, UK: Routledge, 2020.

Guggisberg, Charles Albert Walter. *Wild Cats of the World*. New York: Taplinger Publishing, 1975.

Gunderson, Harvey L. *Mammalogy*. New York: McGraw-Hill, 1976.

Gunderson, Harvey L., and James R. Beer. *The Mammals of Minnesota*. Minneapolis, MN:

University of Minnesota Press, 1953.

Gurnell, J. *The Natural History of Squirrels.* London, UK: Christopher Helm, 1987.

Gutteridge, Anne C. *Barnes and Noble Thesaurus of Biology: The Principles of Biology Explained and Illustrated.* New York: Barnes and Noble Books, 1983.

Haley, Delphine. *Marine Mammals of Eastern North Pacific and Arctic Waters.* Seattle, WA: Pacific Search Press, 1978.

Halfpenny, James. *A Field Guide to Mammal Tracking in North America.* Boulder, CO: Johnson Books, 1986.

Hall, Eugene Raymond. *Mammals of Nevada.* Berkeley, CA: University of California Press, 1946.

——. *American Weasels.* Lawrence, KS: University of Kansas, 1951.

——. *The Mammals of North America.* 2 vols. New York: John Wiley and Sons, 1981.

Halpin, Marjorie M., and Michael M. Ames (eds.). *Manlike Monsters on Trial: Early Records and Modern Evidence.* Vancouver, CAN: University of British Columbia Press, 1980.

Hamilton, Robert. *The Natural History of the Amphibious Carnivora, Including the Walrus and Seals, Also of the Herbivorous Cetacea.* Edinburgh, Scotland: self-published, 1839.

Hamilton, W., Jr. *American Mammals: Their Lives, Habitats, and Economic Relations.* New York: McGraw-Hill, 1939.

Hampton, Bruce. *The Great American Wolf.* New York: Henry Holt and Co., 1997.

Handley, Charles O. Jr. *Virginia's Endangered Species.* Blacksburg, VA: McDonald and Woodward, 1991.

Handley, Charles O. Jr., and Clyde P. Patton. *Wild Mammals of Virginia.* VA: Virginia Commission of Game and Inland Fisheries, 1947.

Hanney, Peter W. *Rodents: Their Lives and Habits.* New York: Taplinger Publishing, 1975.

Harmsworth's Universal Encyclopedia. London, UK: The Educational Book Co., 1921.

Harris, Larry D. *The Fragmented Forest: Island Biogeography and the Preservation of Biotic Diversity.* Chicago, IL: University of Chicago Press, 1984.

Harrison, Richard, and Michael Bryden (eds.). *Whales, Dolphins, and Porpoises.* New York: Facts on File, 1988.

Hartman, Carl G. *Possums.* Austin, TX: University of Texas Press, 1952.

Harvey, Michael J., J. Scott Altenbach, and Troy L. Best. *Bats of the United States and Canada.* Baltimore, MD: Johns Hopkins University Press, 2011.

Hawksworth, David L., and Alan t. Bull (eds.). *Vertebrate Conservation and Biodiversity.* Dordrecht, The Netherlands: Springer, 2007.

Hayssen, Virginia, and Teri Orr. *Reproduction in Mammals: The Female Perspective.* Baltimore, MD: Johns Hopkins University Press, 2017.

Hayssen, Virginia, Ari Van Tienhoven, and Ans van Tienhoven. *Asdell's Patterns of Mammalian Reproduction: A Compendium of Species-Specific Data.* 1946. Ithaca, NY: Comstock, 1993 ed.

Hazard, Evan B. *The Mammals of Minnesota.* Minneapolis, MN: University of Minnesota Press, 1982.

Heffelfinger, Jim. *Deer of the Southwest: A Complete Guide to the Natural History, Biology, and Management of Southwestern Mule Deer and White-Tailed Deer.* College Station, TX: Texas A&M University Press, 2006.

Heide-Jørgensen, Mads Peter (ed.). *Ringed Seals in the North Atlantic.* Nuuk, Greenland: North Atlantic Marine Mammal Commission, 1998.

Heintzelman, Donald S. *A World Guide to Whales, Dolphins, and Porpoises.* Tulsa, OK: Winchester Press, 1981.

Hendricks, Bonnie L. *International Encyclopedia of Horse Breeds.* Norman, OK: University of Oklahoma Press, 2007.

Herman, Louis. *Cetacean Behavior: Mechanisms and Functions.* New York: John Wiley and Sons, 1988.

Heuvelmans, Bernard. *On the Track of Unknown Animals.* 1955. London, UK: Kegan Paul International, 1995 ed.

——. *The Natural History of Hidden Animals.* London, UK: Kegan Paul International, 2007.

Hewer, Humphrey Robert. *British Seals.* London, UK: William Collins, 1974.

Hickling, Grace. *Grey Seals and the Farne Islands.* London, UK: Routledge and Kegan Paul, 1962.

Hill, John E., and James D. Smith. *Bats: A Natural History.* Austin, TX: University of Texas Press, 1984.

Hill, Richard W., Gordon A. Wyse, and Margaret Anderson. *Animal Physiology.* Oxford, UK: Oxford University Press, 2016 ed.

Hillier, S. H., D. W. H. Walton, and D. A. Wells (eds.). *Calcareous Grasslands: Ecology and Management.* Bluntisham, UK: Bluntisham Books, 1990.

Hoare, Ben. *Animal Migration: Remarkable Journeys in the Wild.* Berkeley, CA: University of

California Press, 2009.
Hoffman, Robert S., and Donald L. Pattie. *A Guide to Montana Mammals: Identification, Habitat, Distribution, and Abundance.* Missoula, MT: University of Montana, 1968.
Hoffmeister, Donald Frederick. *Mammals of Arizona.* Tucson, AZ: University of Arizona Press, 1986.
——. *Mammals of Illinois.* 1989. Urbana, IL: University of Illinois Press, 2002 ed.
Holder, Charles Frederick. *Half Hours With the Mammals.* New York: American Book Co., 1907.
Hollis, Durwood. *Hunting North American Big Game.* Iola, WI: Krause Publications, 2002.
Holmes, Martha, and Michael Gunton. *Life: Extraordinary Animals, Extreme Behaviour.* London, UK: BBC Books, 2009.
Honacki, James H., Kenneth E. Kinman, and James W. Koeppl (eds.). *Mammal Species of the World: A Taxonomic and Geographic Reference.* Washington, D.C.: Natural Science Collections Alliance, 1982.
Hood, Donald W., and John A. Calder (eds.). *The Eastern Bering Sea Shelf: Oceanography and Resources.* Washington, D.C.: Office of Marine Pollution Assessment of the National Oceanic and Atmospheric Administration, 1981.
Hoogland, John L. *The Black-Tailed Prairie Dog: Social Life of a Burrowing Mammal.* Chicago, IL: University of Chicago Press, 1995.
Hornaday, William Temple. *Notes on the Mountain Sheep of North America, with a Description of a New Species.* New York: New York Zoological Society, 1901.
——. *The American Natural History: A Foundation of Useful Knowledge of the Higher Animals of North America.* New York: Charles Scribner's Sons, 1904.
——. *Camp-fires on Desert and Lava.* New York: Charles Scribner's Sons, 1908.
Hornocker, Maurice, and Sharon Negri (eds.) *Cougar: Conservation and Ecology.* Chicago, IL: University of Chicago Press, 2009.
Houston, Harry. *British Mammals: An Attempt to Describe and Illustrate the Mammalian Fauna of the British Isles From the Commencement of the Pleistocene Period Down to the Present Day.* London, UK: Hutchinson and Co., 1903.
Howell, Alfred Brazier. *Anatomy of the Woodrat: Comparative Anatomy of the Subgenera of the American Wood Rat (Genus Neotoma).* Baltimore, MD: Williams and Wilkins Co., 1926.
Howell, Arthur H. *Revision of the North American Ground Squirrels, With a Classification of the North American Sciuridae.* Washington, D.C.: U.S. Government Printing Office, 1938.
Howell, Catherine Herbert. *National Geographic Guide to the Mammals of North America.* Washington, D.C.: National Geographic Society, 2016.
Howell, F. Clark. *Early Man.* New York: Time-Life Books, 1965.
Howitt, Mary. *Sketches of Natural History.* London, UK: Effingham Wilson, 1834.
Hulbert, Richard C. (ed.). *The Fossil Vertebrates of Florida.* Gainesville, FL: University Press of Florida, 2001.
Hulme, F. Edward. *Natural History: Lore and Legend.* London, UK: Bernard Quaritch, 1895.
Hummel, Monte, and Justina C. Ray. *Caribou and the North: A Shared Future.* Toronto, CAN: Dundurn Press, 2008.
Hunter, John. *Essays and Observations on Natural History, Anatomy, Physiology, Psychology, and Geology.* London, UK: John Van Voorst, 1861.
Hunter, Luke. *Carnivores of the World.* 2011. Princeton, NJ: Princeton University Press, 2018 ed.
Hurn, Samantha (ed.). *Anthropology and Cryptozoology: Exploring Encounters with Mysterious Creatures.* London, UK: Routledge, 2017.
Huryn, Alex, and John Hobbie. *Land of Extremes: A Natural History of the Arctic North Slope of Alaska.* Fairbanks, AK: University of Alaska Press, 2012.
Hutson, Anthony M., Simon P. Mickleburgh, and Paul A. Racey. *Microchiropteran Bats: Global Status Survey and Conservation Action Plan.* Gland, Switzerland: IUCN, 2001.
Hutyra, Franz, and Josef Marek. *Special Pathology and Therapeutics of the Diseases of Domestic Animals.* Chicago, IL: Alexander Eger, 1912.
Huxley, Thomas Henry. *Man's Place in Nature, and Other Anthropological Essays.* London, UK: Macmillan and Co., 1894.
Ingles, Lloyd Glenn. *Mammals of California.* Stanford, CA: Stanford University Press, 1946.
——. *Mammals of the Pacific States: California, Oregon, Washington.* Stanford, CA: Stanford University Press, 1965.
Jackson, Hartley H. T. *A Taxonomic Review of American Long-tailed Shrews (Genera Sorex and Microsorex).* Washington, D.C.: U.S. Government Printing Office, 1928.
——. *Mammals of Wisconsin.* Madison, WI: University of Wisconsin Press, 1961.
Jackson, Stephen. *Gliding Mammals of the World.* Collingwood, Victoria, Australia: CSIRO,

2012.

Jackson, Tom. *The Illustrated Encyclopedia of Animals, Birds and Fish of North America: A Natural History and Identification Guide to the Captivating Indigenous Wildlife of the United States of America and Canada.* London, UK: Lorenz Books, 2012.

Jaeger, Edmund C. *A Source-Book of Biological Names and Terms.* Springfield, IL: Charles C. Thomas, 1950.

Jaeger, Ellsworth. *Tracks and Trailcraft.* New York: Macmillan, 1948.

Jameson, Everett Williams, Jr., and Hans J. Peeters. *Mammals of California.* Berkeley, CA: University of California Press, 2004.

Janis, Christine M., Kathleen M. Scott, and Louis L. Jacobs (eds.). *Evolution of Tertiary Mammals of North America.* Cambridge, UK: Cambridge University Press, 1998.

Jardine, William (ed.). *The Naturalist's Library.* London, UK: Henry G. Bohn, 1866.

Jefferson, Thomas A., Marc A. Webber, and Robert L. Pitman. *Marine Mammals of the World: A Comprehensive Guide to Their Identification.* 2008. London, UK: Academic Press, 2015 ed.

Jenyns, Leonard. *Observations in Natural History.* London, UK: John Van Voorst, 1846.

Johanson, Donald, and James Shreeve. *Lucy's Child: The Discovery of a Human Ancestor.* New York, Avon, 1989.

Jones, J. Know Jr., David M. Armstrong, and Jerry R. Choate. *Guide to Mammals of the Plains States.* Lincoln, NE: University of Nebraska Press, 1985.

Jones, J. Knox Jr., David M. Armstrong, Robert S. Hoffman, and Clyde Jones. *Mammals of the Northern Great Plains.* Lincoln, NE: University of Nebraska Press, 1983.

Jones, J. Knox Jr., and Elmer C. Birney. *Handbook of Mammals of the North-Central States.* Minneapolis, MN: University of Minnesota Press, 1988.

Jones, J. Knox Jr., and Richard W. Manning. *Illustrated Key to Skulls of Genera of North American Land Mammals.* Lubbock, TX: Texas Tech University Press, 1992.

Jones, Mary Lou, Steven L. Swartz, and Stephen Leatherwood (eds.). *The Gray Whale: Eschrichtius robustus.* Orlando, FL: Academic Press, 1984.

Jordan, David Starr. *The Fur Seals and Fur-seal Islands of the North Pacific Ocean.* Washington, D.C.: U.S. Government Printing Office, 1899.

Journal of Mammalogy. Baltimore, MD: American Society of Mammalogists, 1919-present.

Katona, Steven K., Valerie Rough, and David Richardson. *A Field Guide to the Whales, Porpoises, and Seals of the Gulf of Maine and Eastern Canada: Cape Cod to Newfoundland.* New York: Charles Scribner's Sons, 1983.

Kays, Roland W., and Don E. Wilson. *Mammals of North America.* 1971. Princeton, NJ: Princeton University Press, 2002 ed.

Keast, Allen, Frank C. Erk, and Bentley Glass (eds.). *Evolution, Mammals, and Southern Continents.* Chicago, IL: University of Chicago Press, 1972.

Kemp, T. S. *The Origin and Evolution of Mammals.* Oxford, UK: Oxford University Press, 2005.

Kerr, Robert. *A General History and Collection of Voyages and Travels.* Edinburgh, Scotland: self-published, 1815.

Kielan-Jaworowska, Zofia, Richard L. Cifelli, and Zhe-Xi Luo. *Mammals From the Age of Dinosaurs: Origins, Evolution, and Structure.* New York: Columbia University Press, 2004.

Kieran, John. *A Natural History of New York City.* New York: Fordham University Press, 1959.

King, Carolyn M., and Roger A. Powell. *The Natural History of Weasels and Stoats: Ecology, Behavior, and Management.* Oxford, UK: Oxford University Press, 2007.

King, John A. (ed.). *Biology of Peromyscus (Rodentia).* Stillwater, OK: American Society of Mammalogists, 1968.

King, Judith E. *Seals of the World.* Oxford, UK: Oxford University Press, 1983.

Kingdon, Jonathan. *The Kingdon Field Guide to African Mammals.* 1997. London, UK: Bloomsbury 2012 ed.

Kingsley, John Sterling. *The Riverside Natural History: Vol. 5, Mammals.* London, UK: Kegan Paul, Trench and Co., 1888.

Kinze, Carl Christian. *Marine Mammals of the North Atlantic.* 1994. Princeton, NJ: Princeton University Press, 2001 ed.

Kirkland, Gordon L., and James N. Layne (eds.). *Advances in the Study of Peromyscus (Rodentia).* Lubbock, TX: Texas Tech University Press, 1989.

Kitchener, Andrew. *The Natural History of the Wild Cats.* Ithaca, NY: Comstock Publishing, 1998.

Klinowska, Margaret. *Dolphins, Porpoises and Whales of the World: The IUCN Red Data Book.* Gland, Switzerland: International Union for Conservation of Nature, 1991.

Knight, Charles. *Natural History.* London, UK: Bradbury, Evans, and Co., 1866.

Knight, Richard L. *Forest Fragmentation in the Southern Rocky Mountains.* Boulder, CO: University

Press of Colorado, 1999.

Knowledge and Illustrated Scientific News. London, UK: Knowledge Office, 1906.

Köhler, Wolfgang. *The Mentality of Apes.* London, UK: Kegan Paul, Trench, Trübner, and Co., 1925.

Kramer, Raymond J. *Hawaiian Land Mammals.* Rutland, VT: C. E. Tuttle, 1971.

Krantz, Grover S. *Big Footprints: A Scientific Inquiry into the Reality of Sasquatch.* Boulder, CO: Johnson Books, 1992.

———. *Bigfoot Sasquatch Evidence: The Anthropologist Speaks Out.* Boulder, CO: Johnson Books, 1992.

Krantz, Grover S., and Roderick Sprague (eds.). *The Scientist Looks at the Sasquatch.* Moscow, ID: University of Idaho Press, 1977.

Krantz, Grover S., and Vladimir Markotić. *The Sasquatch and Other Unknown Hominoids.* Calgary, CAN: Western Publishing, 1984.

Kruuk, Hans. *Wild Otters: Predation and Populations.* 1995. Oxford, UK: Oxford University Press, 2001 ed.

Kunz, Thomas H. (ed.). *Ecology of Bats.* New York: Plenum Press, 1982.

Kunz, Thomas H, and Paul A. Racey (eds.). *Bat Biology and Conservation.* Washington, D.C.: Smithsonian Institution Scholarly Press, 1998.

Kunz, Thomas H., and Stuart Parsons (eds.). *Ecological and Behavioral Methods for the Study of Bats.* Baltimore, MD: Johns Hopkins University Press, 2009.

Kurta, Allen. *Mammals of the Great Lakes Region.* Ann Arbor, MI: University of Michigan Press, 2017.

Kurta, Allen, and Jim Kennedy (eds.). *The Indiana Bat: Biology and Management of an Endangered Species.* Austin, TX: Bat Conservation International, 2002.

Kurtén, Björn, and Elaine Anderson. *Pleistocene Mammals of North America.* New York: Columbia University Press, 1980.

Lacey, Eileen A., James L. Patton, and Guy M. Cameron (eds.). *Life Underground: The Biology of Subterranean Rodents.* Chicago, IL: University of Chicago press, 2000.

Lacki, Michael J., John P. Hayes, and Allen Kurta (eds.). *Bats in Forests: Conservation and Management.* Baltimore, MD: Johns Hopkins University Press, 2007.

Larrison, Earl Junior. *Mammals of the Northwest: Washington, Oregon, Idaho, and British Columbia.* Seattle, WA: Seattle Audubon Society, 1976.

Larrison, Earl Junior, and Donald R. Johnson. *Mammals of Idaho.* Moscow, ID: University of Idaho Press, 1981.

Larsen, Thor. *The World of the Polar Bear.* London, UK: Hamlyn Publishing, 1978.

Lawlor, Timothy E. *Handbook to the Orders and Families of Living Mammals.* Eureka, CA: Mad River Press, 1979.

Lawrence, W. *Lectures on Physiology, Zoology, and the Natural History of Man.* London, UK: self-published, 1819.

Leach, Michael. *The Rabbit.* London, UK: Shire Publications, 1999.

Leakey, Richard E. *Origins: What New Discoveries Reveal About the Emergence of Our Species and Its Possible Future.* New York: E. P. Dutton, 1977.

———. *Origins Reconsidered: In Search of What Makes Us Human.* New York: Doubleday, 1992.

Leatherwood, Stephen, and Randall R. Reeves. *The Sierra Club Handbook of Whales and Dolphins.* San Francisco, CA: Sierra Club Books, 1983.

Leatherwood, Stephen, David K. Caldwell, and Howard E. Winn. *Whales, Dolphins, and Porpoises of the Western North Atlantic: A Guide to Their Identification.* (NOAA Technical Report.) Washington, D.C.: U.S. Government Printing Office, 1987.

Lechleitner, R. R. *Wild Mammals of Colorado: Their Appearance, Habits, Distribution, and Abundance.* Boulder, CO: Pruett, Publishing Co., 1969.

LeGro, Shannon, and G. Michael Hopf. *Beyond the Fray: Bigfoot.* San Diego, CA: Beyond the Fray Publishing, 2019.

Leopold, Aldo Starker. *Wildlife of Mexico: The Game Birds and Mammals.* Berkeley, CA: University of California Press, 1959.

Levy, Buddy. *Conquistador: Hernán Cortés, King Montezuma, and the Last Stand of the Aztecs.* New York: Bantam, 2009.

Lidicker, William Z. (ed.). *Landscape Approaches in Mammalian Ecology and Conservation.* Minneapolis, MN: University of Minnesota Press, 1995.

Lindsay, Everett H., Volker Fahlbusch, and Pierre Mein (eds.). *European Neogene Mammal Chronology.* New York: Plenum Press, 1989.

Linsdale, Jean Myron. *The California Ground Squirrel: A Record of Observations Made on the Hastings Natural History Reservation.* Berkeley, CA: University of California Press, 1946.

———. *The Dusky-footed Wood Rat.* Berkeley, CA: University of California Press, 1951.
Linzey, Donald W. *The Mammals of Virginia.* Knoxville, TN: University of Tennessee Press, 2001.
———. *A Natural History Guide to Great Smoky Mountains National Park.* Knoxville, TN: University of Tennessee Press, 2008.
———. *Vertebrate Biology.* Baltimore, MD: Johns Hopkins University Press, 2012.
Lloyd, H. G. *The Red Fox.* London, UK: B. T. Batsford, 1980.
Long, Charles Alan. *The Mammals of Wyoming.* Lawrence, KS: University of Kansas Publications, 1965.
———. *The Wild Mammals of Wisconsin.* Sofia, Bulgaria: Pensoft, 2008.
Long, John L. *Introduced Mammals of the World: Their History, Distribution and Influence.* Oxford, UK: CABI Publishing, 2003.
Loughry, W. J., and Colleen M. McDonough. *The Nine-Banded Armadillo: A Natural History.* Norman, OK: University of Oklahoma Press, 2013.
Lowery, George Hines. *The Mammals of Louisiana and its Adjacent Waters.* Baton Rouge, LA: Louisiana State University Press, 1974.
Lydekker, Richard. *The Great and Small Game of Europe, Western and Northern Asia and America: Their Distribution, Habits, and Structure.* London, UK: Rowland Ward, 1901.
———. *Wild Life of the World: A Descriptive Survey of the Geographical Distribution of Animals.* London, UK: Frederick Warne and Co., 1916.
Lydekker, Richard, et al. *Natural History.* New York: D. Appleton and Co., 1897.
———. *Guide to the Galleries of Mammals in the Department of Zoology of the British Museum (Natural History).* London, UK: British Museum of Natural History, 1906.
Lynch, Patrick J. *A Field Guide to Long Island Sound: Coastal Habitats, Plant Life, Fish, Seabirds, Marine Mammals, and Other Wildlife.* New Haven, CT: Yale University Press, 2017.
MacClintock, Dorcas. *Squirrels of North America.* New York: Van Nostrand Reinhold, 1970.
Macdonald, David W. (ed.). *The Encyclopedia of Mammals.* 1995. Oxford, UK: Oxford University Press, 2006 ed.
———. *The Princeton Encyclopedia of Mammals.* Princeton, NJ: Princeton University Press, 2009.
Macdonald, David W., and Katrina Service (eds.). *Key Topics in Conservation Biology.* Malden, MA: Blackwell Publishing, 2007.
MacDonald, S. O., and Joseph A. Cook. *Recent Mammals of Alaska.* Fairbanks, AK: University of Alaska Press, 2009.
MacGillivray, William. *A History of British Quadrupeds.* Edinburgh, Scotland: self-published, 1838.
MacMahon, James A. *The Audubon Society Nature Guides: Deserts.* New York: Alfred A. Knopf, 1985.
Mammal Anatomy: An Illustrated Guide. Tarrytown, NY: Marshall Cavendish Corp., 2010.
Mangin, Arthur. *The Desert World.* London, UK: T. Nelson and Sons, 1872.
Mann, Janet, Richard C. Connor, Peter L. Tyack, and Hal Whitehead. *Cetacean Societies: Field Studies of Dolphins and Whales.* Chicago, IL: University of Chicago Press, 2000.
Marais, Eugène. *The Soul of the Ape.* New York: Atheneum, 1969.
Mares, Michael A., and David J. Schimdly (eds.). *Latin American Mammalogy: History, Biodiversity, and Conservation.* Norman, OK: University of Oklahoma Press, 1991.
Mares, Michael A., Ricardo A. Ojeda, and Rubén M. Barquez. *Guide to the Mammals of Salta Province, Argentina.* Norman, OK: University of Oklahoma Press, 1989.
Marks, Cynthia S., and George E. Marks. *Bats of Florida.* Gainesville, FL: University of Florida Press, 2006.
Martin, Alexander C., Herbert S. Zim, and Arnold L. Nelson. *American Wildlife and Plants: A Guide to Wildlife Food Habits.* New York: McGraw-Hill, 1951.
Martin, Robert E., Ronald H. Pine, and Anthony F. DeBlase. *A Manual of Mammalogy: With Keys to Families of the World.* 1974. Long Grove, IL: Waveland Press, 2011 ed.
Martyn, Thomas. *Elements of Natural History.* Cambridge, UK: J. Archdeacon, printer to the university, 1775.
Maser, Chris. *Mammals of the Pacific Northwest: From the Coast to the High Cascades.* Corvallis, OR: Oregon State University Press, 1998.
Mathews, F. Schuyler. *Familiar Life in Field and Forest: The Animals, Birds, Frogs, and Salamanders.* New York: D. Appleton and Co., 1898.
Matthews, Leonard Harrison. *Sea Elephant: The Life and Death of the Elephant Seal.* London, UK: MacGibbon and Kee, 1952.
———. *The Life of Mammals.* 2 vols. New York: Universe Books, 1970.
———. *The Natural History of the Whale.* New York: Columbia University Press, 1978.

Maunder, Samuel. *The Treasury of Natural History: Or a Popular Dictionary of Animated Nature.* London,. UK: Longman, Brown, Green, and Longmans, 1852.

Mayer, William V., and Richard G. Van Gelder (eds.). *Physiological Mammalogy.* New York: Academic Press, 1963.

Mayr, Ernst. *Systematics and the Origin of Species from the Viewpoint of a Zoologist.* Cambridge, MA: Harvard University Press, 1999.

McBane, Susan. *The Illustrated Encyclopedia of Horse Breeds.* New York: Wellfleet Press, 2008.

McCabe, Thomas Tonkin, and Barbara Dewing Blanchard. *Three Species of Peromyscus.* Santa Barbara, CA: Rood and Associates, 1950.

McConnaughey, Bayard H., and Evelyn McConnaughey. *Pacific Coast.* New York: Alfred A. Knopf, 1985.

McIntyre, Joan. *Mind in the Waters: A Book to Celebrate the Consciousness of Whales and Dolphins.* New York: Charles Scribner's Sons, 1974.

McIntyre, Rick. *A Society of Wolves: National Parks and the Battle Over the Wolf.* Stillwater, MN: Voyageur Press, 1994.

McKenna, Malcolm C., and Susan K. Bell. *Classification of Mammals: Above the Species Level.* New York: Columbia University Press, 1997.

Mearns, Edgar Alexander. *Mammals of the Mexican Boundary of the United States.* Washington, D.C.: U.S. Government Printing Office, 1907.

Mech, L. David. *The Wolf: The Ecology and Behavior of an Endangered Species.* Garden City, NY: Natural History Press, 1970.

Mech, L. David, and Luigi Boitani (eds.). *Wolves: Behavior, Ecology, and Conservation.* Chicago, IL: University of Chicago Press, 2003.

Meffe, Gary K., and C. Ronald Carroll. *Principles of Conservation Biology.* Sunderland, MA: Sinauer Associates, 1997.

Meldrum, Jeff. *Sasquatch: Legend Meets Science.* New York: Tom Doherty Associates, 2006.

——. *Sasquatch, Yeti and Other Wildmen of the World: A Field Guide to Relict Hominoids.* Blue Lake, CA: Paradise Cay Publications, 2016.

Melletti, Mario, and Erik Meijaard (eds.). *Ecology, Conservation and Management of Wild Pigs and Peccaries.* Cambridge, UK: Cambridge University Press, 2018.

Memoirs of the Wernerian Natural History Society. Edinburgh, Scotland: self-published, 1811.

Mercer, Henry Chapman. *The Hill-Caves of Yucatan: A Search for Man's Antiquity in the Caverns of Central America.* Philadelphia, PA: J. B. Lippincott Co., 1896.

Merriam, Clinton Hart. *The Mammals of the Adirondack Region: Northeastern New York.* New York: Henry Holt and Co., 1886.

Merritt, Joseph F. *Guide to the Mammals of Pennsylvania.* Pittsburgh, PA: University of Pittsburgh Press, 1987.

——. *The Biology of Small Mammals.* Baltimore, MD: Johns Hopkins University Press, 2010.

Meyer, Karl. *Wild Animals of North America.* North Adams, MA: Storey Publishing, 2006.

Meyerhof, Wolfgang, and Sigrun Korsching (eds.). *Chemosensory Systems in Mammals, Fishes, and Insects.* Berlin, Germany: Springer, 2009.

Milanich, Jerald T. *Florida Indians and the Invasion from Europe.* Gainesville, FL: University of Florida Press, 1995.

Milanich, Jerald T., and Susan Milbrath (eds.). *First Encounters: Spanish Explorations in the Caribbean and the United States, 1492-1570.* Gainesville, FL: University of Florida Press, 1989.

Miller, Char, and Clay S. Jenkinson (eds.). *Theodore Roosevelt: Naturalist in the Arena.* Lincoln, NE: University of Nebraska Press, 2020.

Miller, Gerrit S., Jr. *Proceedings of the Boston Society of Natural History.* Boston, MA: Boston Society of Natural History, 1896.

——. *Revision of the North American Bats of the Family Vespertilionide.* Washington, D.C.: U.S. Government Printing Office, 1897.

——. *Key to the Land Mammals of Northeastern North America.* Albany, NY: University of the State of New York, 1900.

——. *List of North American Land Mammals in the United States National Museum, 1911.* Washington, D.C.: U.S. Government Printing Office, 1912.

——. *List of North American Recent Mammals, 1923.* Washington, D.C.: U.S. Government Printing Office, 1924.

Miller, William John. *An Introduction to Physical Geology: With Special Reference to North America.* New York: D. Van Nostrand Co., 1924.

Mills, Enos A. *The Grizzly.* Boston, MA: Houghton Mifflin, 1919.

Minasian, Stanley M., Kenneth C. Balcomb, and Larry Foster. *The World's Whales: The Complete*

Illustrated Guide. Washington, D.C.: Smithsonian Institution Scholarly Press, 1984.

Mirarchi, Ralph E. *Alabama Wildlife.* 4 vols. Tuscaloosa, AL: University of Alabama Press, 2004.

Mish Frederick (ed.). *Webster's Ninth New Collegiate Dictionary.* 1828. Springfield, MA: Merriam-Webster, 1984.

Mohr, Charles E. *The World of the Bat.* Philadelphia, PA: J. B. Lippincott, 1976.

Monks, Gregory G. *The Exploitation and Cultural Importance of Sea Mammals.* Oxford, UK: Oxbow Books, 2004.

Morgan, Ben. *Guide to Mammals: A Wild Journey With These Extraordinary Beasts.* London, UK: DK Publishing, 2003.

Morgan, Robert W. *Bigfoot Observer's Field Manual: A Practical and Easy-to-follow Step-by-step Guide to Your Very Own Face-to-face Encounter With a Legend.* Enumclaw, WA: Pine Winds Press, 2008.

Morrell, G. Herbert. *Comparative Anatomy, and Guide to Discussion, Part 1: Mammalia (Anatomy and Dissection).* London, UK: Longman and Co., 1872.

Moskowitz, David. *Wildlife of the Pacific Northwest: Tracking and Identifying Mammals, Birds, Reptiles, Amphibians, and Invertebrates.* Portland, OR: Timber Press, 2010.

Muir, Gillie, and Pat Morris. *How to Find and Identify Mammals.* London, UK: The Mammal Society, 2013.

Muir, John (ed.). *Picturesque California: The Rocky Mountains and the Pacific Slope.* New York: J. Dewing Publishing, 1888.

——. *The Mountains of California.* New York: The Century Co., 1894.

——. *Our National Parks.* Boston, MA: Houghton Mifflin Co., 1901.

——. *My First Summer in the Sierra.* Boston, MA: Houghton Mifflin Co., 1911.

——. *The Yosemite.* New York: The Century Co., 1912.

——. *The Boyhood of a Naturalist.* Boston, MA: Houghton Mifflin Co., 1913.

——. *Travels in Alaska.* Boston, MA: Houghton Mifflin Co., 1915.

——. *The Story of my Boyhood and Youth.* Boston, MA: Houghton Mifflin Co., 1916.

——. *A Thousand-Mile Walk to the Gulf.* Boston, MA: Houghton Mifflin Co., 1916.

Murie, Jan O., and Gail R. Michener. *The Biology of Ground-Dwelling Squirrels: Annual Cycles, Behavioral Ecology, and Sociality.* Lincoln, NE: University of Nebraska Press, 1984.

Murie, Olaus J. *A Field Guide to Animal Tracks.* Boston, MA: Houghton Mifflin, 1954.

Murray, John A. *Wildlife in Peril: The Endangered Mammals of Colorado : River Otter, Black-Footed Ferret, Wolverine, Lynx, Grizzly Bear, Gray Wolf.* New York: Rinehart and Co., 1987.

Nachtigall, Paul E., and Patrick W. B. Moore (eds.). *Animal Sonar: Processes and Performance.* New York: Plenum Press, 1988.

Nagle, John, J. B. Ruhl, and Kalyani Robbins. *The Law of Biodiversity and Ecosystem Management.* St. Paul, MN: Foundation Press, 2012.

Nagorsen, David W. *Opossums, Shrews and Moles of British Columbia.* 3 vols. Vancouver, CAN: UBC Press, 1996.

Nagorsen, David W., and R. Mark Brigham. *Bats of British Columbia.* 2 vols. Vancouver, British Columbia, CAN: University of British Columbia Press, 1993.

Napier, John Russell. *Bigfoot: The Yeti and Sasquatch in Myth and Reality.* Boston, MA: E. P. Dutton, 1973.

National Antarctic Expedition, 1901-1904. *Natural History.* London, UK: The British Museum of Natural History, 1908.

Naughton, Donna. *The Natural History of Canadian Mammals: Hoofed Mammals.* Toronto, Ontario, CAN: University of Toronto Press, 2012.

Neal, Ernest G. *The Natural History of Badgers.* New York: Facts on File, 1986.

Nelson, Edward William. *The Eskimo About Bering Strait.* Washington, D.C.: U.S. Government Printing Office, 1900.

——. *The Rabbits of North America.* Washington, D.C.: U.S. Government Printing Office, 1909.

——. *Wild Animals of North America.* Washington, D.C.: National Geographic Society, 1930.

Neuweiler, Gerhard. *The Biology of Bats.* New York: Oxford University Press, 2000.

Newton, Michael. *Encyclopedia of Cryptozoology.* Jefferson, NC: McFarland and Co., 2005.

Nieuwland, Julius A. (ed.). *The American Midland Naturalist: Devoted to the Natural History, Primarily That of the Prairie States.* Notre Dame, IN: University of Notre Dame, 1909.

Norris, K. S., and K. Pryor (eds.). *Dolphin Societies: Discoveries and Puzzles.* Los Angeles, CA: University of California Press, 1991.

Nowak, Ronald M. *Walker's Mammals of the World.* 2 vols. 1964. Baltimore, MD: Johns Hopkins University Press, 1999 ed.

——. *Walker's Bats of the World.* Baltimore, MD: Johns Hopkins University Press, 1994.

O'Brien, Stephen J., Joan C. Menninger, and William G. Nash (eds.). *Atlas of Mammalian Chromosomes*. Hoboken, NJ: John Wiley and Sons, 2006.
Oelke, Hardy. *Born Survivors on the Eve of Extinction: Can Iberia's Wild Horse Survive Among America's Mustangs?* Wipperfürth, Germany: Ute Kierdorf Verlag, 1997.
Ohlin, Axel. *Some Remarks on the Bottlenose-whale (Hyperoodon)*. Lund, Sweden: self-published, 1893.
Osgood, Wilfred H. *Biological Investigations in Alaska and Yukon Territory*. Washington, D.C.: U.S. Government Printing Office, 1909.
Owen, Richard. *On the Anatomy of Vertebrates, Vol. 3: Mammals*. London, UK: Longmans, Green, and Co., 1868.
Page, Roderic D. M. (ed.). *Tangled Trees: Phylogeny, Cospeciation, and Coevolution*. Chicago, IL: University of Chicago Press, 2003.
Paine, Stefani. *The World of the Sea Otter*. San Francisco, CA: Sierra Club Books, 1995.
——. *The World of the Arctic Whales: Belugas, Bowheads, and Narwhals*. San Francisco, CA: Sierra Club Books, 1997.
Palmer, Ephraim Lawrence. *Fieldbook of Natural History*. New York: Whittlesey House, 1949.
Parker, Gerry. *Eastern Coyote: The Story of its Success*. Halifax, CAN: Nimbus, 1995.
Parsons, E. C. M., et al. *An Introduction to Marine Mammal Biology and Conservation*. Burlington, MA: Jones and Bartlett Learning, 2013.
Patterson, Bruce D., and Leonora P. Costa (eds.). *Bones, Clones, and Biomes: The History and Geography of Recent Neotropical Mammals*. Chicago, IL: University of Chicago Press, 2012.
Pattie, Don, Chris Fisher, and Tamara Hartson. *Mammals of the Rocky Mountains*. Edmonton, Alberta, CAN: Lone Pine, 2000.
Payne, Roger. *Among Whales*. New York: Scribner, 1995.
Perrin, William F., Bernd Würsig, and J. G. M. Thewissen (eds.). *Encyclopedia of Marine Mammals*. San Diego, CA: Academic Press, 2002.
Peterson, Randolph L. *The Mammals of Eastern Canada*. Oxford, UK: Oxford University Press, 1966.
Peterson, Russell Francis. *Silently, by Night*. New York: McGraw-Hill, 1964.
Pfeiffer, Carl J. *Molecular and Cell Biology of Marine Mammals*. Malabar, FL: Krieger Publishing, 2002.
Phipps, Constantine John. *A Voyage Towards the North Pole Undertaken by His Majesty's Command, 1773*. London, UK: J. Nourse, 1774.
Pickeral, Tamsin. *The Majesty of the Horse: An Illustrated History*. Hauppauge, NY: B.E.S. Publishing, 2011.
Pickering, Charles. *United States Exploring Expeditions During the Years 1838, 1839, 1840, 1841, 1842*. Boston, MA: Gould and Lincoln, 1863.
Potts, Steve. *The Armadillo*. Mankato, MN: Capstone Books, 1998.
Powell, Roger A. *The Fisher: Life History, Ecology, and Behavior*. Minneapolis, MN: University of Minnesota Press, 1993.
Proctor, Noble S., and Patrick J. Lynch. *A Field Guide to North Atlantic Wildlife: Marine Mammals, Seabirds, Fish, and Other Sea Life*. New Haven, CT: Yale University Press, 2005.
Prothero, Donald R. *After the Dinosaurs: The Age of Mammals*. Bloomington, IN: Indiana University Press, 2006.
——. *The Princeton Field Guide to Prehistoric Mammals*. Princeton, NJ: Princeton University Press, 2017.
Prothero, Donald R., and Robert M. Schoch. *The Evolution of Perissodactyls*. Oxford, UK: Oxford University Press, 1989.
Prothero, Donald R., and Robert M. Schoch. *Horns, Tusks, and Flippers: The Evolution of Hoofed Mammals*. Baltimore, MD: Johns Hopkins University Press, 2002.
Prothero, Donald R., and Scott E. Foss (eds.). *The Evolution of Artiodactyls*. Baltimore, MD: Johns Hopkins University Press, 2007.
Putman, Rory J. (ed.). *Mammals as Pests*. London, UK: Chapman and Hall, 1989.
——. *The Natural History of Deer*. Ithaca, NY: Comstock Publishing, 1998.
Pyle, Robert Michael. *Where Bigfoot Walks: Crossing the Dark Divide*. Boston, MA: Mariner, 1995.
Ransford, Sandy. *The Kingfisher Illustrated Horse and Pony Encyclopedia*. Boston, MA: Kingfisher, 2004.
Ransom, Jason I., and Petra Kaczensky (eds.). *Wild Equids: Ecology, Management, and Conservation*. Baltimore, MD: Johns Hopkins University Press, 2016.
Redfern, Nick. *The Bigfoot Book: The Encyclopedia of Sasquatch, Yeti and Cryptid Primates*. Canton, MI: Visible Ink Press, 2016.
Redford, Kent H. *Advances in Neotropical Mammalogy*. Gainesville, FL: Sandhill Crane Press,

1989.
Redford, Kent H., and John F. Eisenberg. *Mammals of the Neotropics*. 2 vols. Chicago, IL: University of Chicago Press, 1992.
Reed, Andrew. *Porpoises*. Stillwater, MN: Voyageur Press, 1999.
Rees, Paul A. *An Introduction to Zoo Biology and Management*. Oxford, UK: Wiley-Blackwell, 2011.
Reeves, Randall R. *The Sierra Club Handbook of Whales and Dolphins*. San Francisco, CA: Sierra Club Books, 1983.
Reeves, Randall R., Brent S. Stewart, Phillip J. Clapham, and James A. Powell. *National Audubon Society Guide to Marine Mammals of the World*. New York: Alfred A. Knopf, 2002.
Reeves, Randall R., Brent S. Stewart, and Stephen Leatherwood. *The Sierra Club Handbook of Seals and Sirenians*. San Francisco, CA: Sierra Club Books, 1992.
Reeves, Randall R., Brent S. Stewart, Phillip J. Clapham, and James A. Powell. *Sea Mammals of the World: A Complete Guide to Whales, Dolphins, Seals, Sea Lions and Sea Cows*. London, UK: A and C Black, 2002.
Reid, Fiona A. *A Field Guide to Mammals of North America North of Mexico*. New York: Houghton Mifflin, 2006.
——. *A Field Guide to the Mammals of Central America and Southeast Mexico*. Oxford, UK: Oxford University Press, 2009.
Rennick, Penny. *Seals, Sea Lions and Sea Otters*. Anchorage, AK: Alaska Geographic Society, 2000.
Renouf, Deane (ed.). *Behaviour of Pinnipeds*. Dordrecht, The Netherlands: Springer, 1991.
Reynolds, Sidney H. *British Pleistocene Mammalia*. London, UK: Paleontographical Society, 1906.
Rice, Dale W. *Marine Mammals of the World: Systematics and Distribution*. Yarmouth Port, MA: Society for Marine Mammalogy, 1998.
Richardson, W. John, Charles R. Greene Jr., Charles I. Malme, and Denis H. Thomson. *Marine Mammals and Noise*. San Diego, CA: Academic Press, 1995.
Ridgway, Sam H. *Mammals of the Sea: Biology and Medicine*. Springfield, IL: Charles Thomas, 1972.
Ridgway, Sam H., and Richard Harrison (eds.). *Handbook of Marine Mammals: River Dolphins and the Larger Toothed Whales*. Cambridge, MA: Academic Press, 1989.
Riedman, Marianne. *The Pinnipeds: Seals, Sea Lions, and Walruses*. Berkeley, CA: University of California Press , 1990.
Rinella, Steven. *American Buffalo: In Search of a Lost Icon*. New York: Spiegel and Grau, 2008.
Roest, Aryan I. *Key-Guide to Mammal Skulls and Lower Jaws*. Eureka, CA: Mad River Press, 1986.
Roosevelt, Theodore. *The Wilderness Hunter*. New York: G. P. Putnam's Sons, 1893.
Rose, Kenneth D. *The Beginning of the Age of Mammals*. Baltimore, MD: Johns Hopkins University Press, 2006.
Rose, Kenneth D., and J. David Archibald (eds.). *The Rise of Placental Mammals: Origins and Relationships of the Major Extant Clades*. Baltimore, MD: Johns Hopkins University Press, 2005.
Rousseau, Élise. *Horses of the World*. Princeton, NJ: Princeton University Press, 2017.
Rue, Leonard Lee, III. *Pictorial Guide to the Mammals of North America*. New York: Thomas Y. Crowell and Co., 1967.
——. *Sportsman's Guide to Game Animals: A Field Book of North American Species*. New York: Outdoor Life Books, 1968.
——. *Furbearing Animals of North America*. New York: Crown, 1981.
——. *The Deer of North America*. Guilford, CT: Globe Pequot Press, 2004.
Ryan, James M. *Mammalogy Techniques: Lab Manual*. Baltimore, MD: Johns Hopkins University Press, 2018.
Sadleir, Richard M. F. S. *The Ecology of Reproduction in Wild and Domestic Mammals*. 1969. New York: Springer, 2012 ed.
Salvin, Osbert (ed.). *Barton's Fragments of the Natural History of Pennsylvania*. London, UK: self-published, 1883.
Samuel, David. *Understanding Whitetails*. Beverly, MA: Creative Publishing International, 1996.
Sanderson, Ivan T. *Abominable Snowmen: Legend Come to Life*. Philadelphia, PA: Chilton Book Co., 1966.
Savage, R. J. G., and M. R. Long. *Mammal Evolution: A Brief Guide*. London, UK: Natural History Museum Publications, 1986.
Sawyer, G. J., Viktor Deak, et al. *The Last Human: A Guide to Twenty-two Species of Extinct Humans*. New Haven, CT: Yale University Press, 2007.
Scammon, Charles Melville. *The Marine Mammals of the North-western Coast of North America*. San

Francisco, CA: John H. Carmany and Co., 1874.

Schaller, George B. *Mountain Monarchs: Wild Sheep and Goats of the Himalaya.* Chicago, IL: University of Chicago Press, 1989.

Scharff, Robert Francis. *Distribution and Origin of Life in America.* New York: Macmillan Co., 1912.

Scheffer, Theo. H. *The Pocket Gopher.* Manhattan, KS: Kansas State Agricultural College, 1910.

Scheffer, Victor B. *Seals, Sea Lions and Walruses: A Review of the Pinnipedia.* Redwood City, CA: Stanford University Press, 1958.

Schimdly, David J. *The Mammals of Trans-Pecos Texas: Including Big Bend National Park and Guadalupe Mountains National Park.* College Station, TX: Texas A&M University Press, 1977.

——. *Texas Mammals East of the Balcones Fault Zone.* College Station, TX: Texas A&M University Press, 1983.

——. *The Bats of Texas.* College Station, TX: Texas A&M University Press, 1991.

——. *The Mammals of Texas.* Austin, TX: University of Texas Press, 1994.

Schimdly, David J., and Robert D. Bradley. *The Mammals of Texas.* 1994. Austin, TX: University of Texas Press, 2016 ed.

Schober, Wilfried, and Eckard Grimmberger. *The Bats of Europe and North America: Knowing Them, Identifying Them, Protecting Them.* Neptune City, NJ: T.F.H., 1997.

Schwartz, Charles W., and Elizabeth R. Schwartz. *The Wild Mammals of Missouri.* Columbia, MO: University of Missouri Press, 1964.

Sclater, William Lutley, and Philip Lutley Sclater. *The Geography of Mammals.* London, UK: Kegan Paul, Trench, Trübner, and Co., 1899.

Seabrook, Lochlainn. *The Concise Book of Owls: A Guide to Nature's Most Mysterious Birds.* Spring Hill, TN: Sea Raven Press, 2019.

——. *The Concise Book of Tigers: A Guide to Nature's Most Remarkable Cats.* Spring Hill, TN: Sea Raven Press, 2020.

Seal, Ulysses S., E. Tom Thorne, Michael A. Bogan, and Stanley H. Anderson (eds.). *Conservation Biology and the Black-Footed Ferret.* New Haven, CT: Yale University Press, 1989.

Sealander, John A. *A Guide to Arkansas Mammals.* New Orleans, LA: River Road Press, 1979.

Sealander, John A., and Gary A. Heidt. *Arkansas Mammals: Their Natural History, Classification, and Distribution.* Fayetteville, AR: University of Arkansas Press, 1990.

Seidensticker, John. *Great Cats: Majestic Creatures of the Wild.* Sydney, Australia: Murdoch Books, 1991.

Seton, Ernest Thompson. *Wild Animals I Have Known.* New York: Grosset and Dunlap, 1898.

——. *The Biography of a Grizzly.* New York: The Century Co., 1900.

——. *The Trail of the Sandhill Stag.* New York: Charles Scribner's Sons, 1902.

——. *Woodmyth and Fable.* New York: The Century Co., 1905.

——. *Animal Heroes.* New York: Grosset and Dunlap, 1905.

——. *Life-Histories of Northern Animals.* 2 vols. New York: Charles Scribner's Sons, 1909.

——. *The Arctic Prairies.* New York: Charles Scribner's Sons, 1912.

——. *The Book of Woodcraft and Indian Lore.* Garden City, NY: Doubleday, Page and Co., 1912.

——. *Wild Animal Ways.* Garden City, NY: Doubleday, Page and Co., 1917.

——. *Lives of Game Animals.* 4 vols. New York: Doubleday, Page and Company, 1925.

Shackleton, David M. (ed.). *Wild Sheep and Goats and Their Relatives: Status Survey and Conservation Action Plan for Caprinae.* Gland, Switzerland: IUCN, 1997.

——. *Hoofed Mammals of British Columbia.* Vancouver, British Columbia, CAN: University of British Columbia Press, 1999.

Shaw, Harley. *Soul Among Lions: The Cougar as a Peaceful Adversary.* Tucson, AZ: University of Arizona Press, 2000.

Sheldon, Jennifer W. *Wild Dogs: The Natural History of the Nondomestic Canidae.* San Diego, CA: Academic Press, 1992.

Shirihai, Hadoram, and Brett Jarrett. *Whales, Dolphins, and Other Marine Mammals of the World.* Princeton, NJ: Princeton University Press, 2006.

Shufeldt, Robert Wilson. *Chapters on the Natural History of the United States.* New York: Studer Brothers, 1900.

Shupe, Scott. *Kentucky Wildlife Encyclopedia: An Illustrated Guide to Birds, Fish, Mammals, Reptiles, and Amphibians.* New York: Skyhorse Publishing, 2018.

Simberloff, Daniel, Don C. Schmitz, and Tom C. Brown (eds.). *Strangers in Paradise: Impact and Management of Nonindigenous Species in Florida.* Washington, D.C.: Island Press, 1997.

Simpson, George Gaylord. *The Meaning of Evolution: A Study of the History of Life and of its*

Significance for Man. 1949. New Haven, CT: Yale University Press, 1967 ed.
Slijper, E. *Whales.* Ithaca, NY: Cornell University Press, 1979.
Smith, Andrew T., Charlotte H. Johnston, Paulo C. Alves, and Klaus Hackländer (eds.). *Lagomorphs: Pikas, Rabbits, and Hares of the World.* Baltimore, MD: Johns Hopkins University Press, 2018.
Smith, Charles Hamilton. *The Natural History of Horses: The Equidae or Genus Equus of Authors.* Edinburgh, Scotland: W. H. Lizars, 1841.
Smith, Hugh C. *Alberta Mammals: An Atlas and Guide.* Edmonton, Alberta, CAN: Provincial Museum of Alberta, 1993.
Southern, Henry Neville. *The Handbook of British Mammals.* London, UK: Blackwell Science, 1965.
Sowerby, Arthur De Carle. *A Naturalist's Holiday By the Sea.* London, UK: George Routledge and Sons, 1923.
Spinage, Clive A. *The Natural History of Antelopes.* Kent, UK: Croom Helm, 1986.
Stark, John. *Elements of Natural History, Adapted to the Present State of Science.* Edinburgh, Scotland: Adam Black and John Stark, 1828.
Steele, Michael A., and John L. Koprowski. *North American Tree Squirrels.* Washington, D.C.: Smithsonian Books, 2003.
Steffanson, Vilhjalmur. *The Friendly Arctic: The Story of Five Years in Polar Regions.* New York: Macmillan, 1922.
Stensworth, Nils C., and Rolf Anker Ims. *The Biology of Lemmings.* London, UK: Academic Press, 1993.
Stephens, Frank. *California Mammals.* San Diego, CA: The West Coast Publishing Co., 1906.
Stephens, James Francis. *General Zoology.* London, UK: self-published, 1826.
Stidworthy, John. *Mammals: The Large Plant-Eaters.* New York: Facts on File, 1989.
Stone, Witmer, and William Everett Cram. *American Animals: A Popular Guide to the Mammals of North America North of Mexico.* New York: Doubleday, Page, and Co., 1902.
Storer, Tracy I., and Lloyd P. Tevis, Jr. *California Grizzly.* Berkeley, CA: University of California Press, 1996.
Storer, Tracy I., Robert L. Usinger, and David Lukas. *Sierra Nevada Natural History.* Berkeley, CA: University of California Press, 2004.
Strain, Kathy Moskowitz. *Giants, Cannibals, and Monsters: Bigfoot in Native Culture.* Surrey, B.C., CAN: Hancock House Publishers, 2008.
Streubel, Donald P. *Small Mammals of the Yellowstone Ecosystem.* New York: Roberts Rinehart, 1995.
Strier, Karen B. *Primate Behavioral Ecology.* 2000. Oxford, UK: Routledge, 2011 ed.
Strong, Asa B. (ed.). *Illustrated Natural History of the Three Kingdoms.* New York: Green and Spencer, 1849.
Stuart, Chris and Tilde. *Field Guide to the Larger Mammals of Africa.* 1997. Cape Town, South Africa: Struik Nature, 2006 ed.
Sumner, L., and J. S. Dixon. *Birds and Mammals of the Sierra Nevada with Records From Sequoia and Kings Canyon National Parks.* Berkeley, CA: University of California Press, 1953.
Sunquist, Mel, and Fiona Sunquist. *Wild Cats of the World.* Chicago, IL: University of Chicago Press, 2002.
Szalay, Frederick S. *Evolutionary History of the Marsupials and an Analysis of Osteological Characters.* Cambridge, UK: Cambridge University Press, 1994.
Szalay, Frederick S., Michael J. Novacek, and Malcolm C. McKenna (eds.). *Mammal Phylogeny: Mesozoic Differentiation, Multituberculates, Monotremes, Early Therians, and Marsupials.* New York: Springer-Verlag, 1993.
Taylor, Marianne. *Bats: An Illustrated Guide to All Species.* Washington, D.C.: Smithsonian Institution Scholarly Press, 2019
Tekiela, Stan. *Mammals of Michigan Field Guide.* Cambridge, MN: Adventure Publications, 2005.
——. *Mammals of Texas Field Guide.* Cambridge, MN: Adventure Publications, 2009.
Terres, J. K. *The World of the Coyote.* Philadelphia, PA: Lippincott, 1964.
Terwilliger, Karen. *Virginia's Endangered Species.* Knoxville, TN: University of Tennessee Press, 2001.
The Anthropological Review. (Multiple issues). London, UK: Trübner and Co., 1863.
The Columbia Encyclopedia. New York: Columbia University Press, 1964.
The English Encyclopaedia. London, UK: G. Kearsley, 1802.
The Hutchinson Encyclopedia. Oxford, UK: Helicon Publishing, 2001.
The New American Encyclopedia. New York: D. Appleton and Co., 1862.
The New Larousse Encyclopedia of Animal Life. 1967. New York: Bonanza Books, 1980 ed.

The Oracle Encyclopedia. London, UK: George Newnes, 1896.

The Pine Trees of the Rocky Mountain Region. Washington, D.C.: U.S. Government Printing Office, 1917.

The World Book Encyclopedia. Chicago, IL: Field Enterprises, 1976.

Thomas, Hugh. *Conquest: Cortes, Montezuma, and the Fall of Old Mexico.* New York: Simon and Schuster, 1995.

Thomas, J. W. (ed.) *Wildlife Habitats in Managed Forests: The Blue Mountains of Oregon and Washington.* Washington, D.C.: U.S. Government Printing Office, Department of Agriculture, 1979.

Thomas, Jeanette A., Cynthia F. Moss, and Marianne Vater (eds.). *Echolocation in Bats and Dolphins.* Chicago, IL: University of Chicago Press, 2004.

Thomas, Peggy. *Marine Mammal Preservation.* Brookfield, CT: Twenty-First Century Books, 2000.

Thompson, Harry V., and Carolyn M. King (eds.). *The European Rabbit: The History and Biology of a Successful Colonizer.* Oxford, UK: Oxford University Press, 1994.

Thorington, Richard W. Jr., John L. Koprowski, Michael A. Steele, and James F. Whatton. *Squirrels of the World.* Baltimore, MD: Johns Hopkins University Press, 2012.

Tinker, Spencer Wilkie. *Whales of the World.* Leiden, The Netherlands: E. J. Brill, 1988.

Townsend, Colin R., Michael Begon, and John L. Harper. *Essentials of Ecology* (2nd ed.). 2000. Malden, MA: Blackwell Science, 2003 ed.

Trani, Margaret K., W. Mark Ford, and Brian Chapman (eds.). *The Land Manager's Guide to Mammals of the South.* Durham, NC: The Nature Conservancy, 2007.

Trimmer, Mary. *A Natural History of the Most Remarkable Quadrupeds, Birds, Fishes, Serpents, Reptiles, and Insects.* Chiswick, UK: self-published, 1825.

True, Frederick W. *A Review of the Family Delphinidae.* Washington, D.C.: U.S. Government Printing Office, 1889.

——. *An Account of the Beaked Whales of the Family Ziphidae in the Collection of the United States National Museum, With Remarks on Some Specimens in Other American Museums.* Washington, D.C.: U.S. Government Printing Office, 1910.

——. *Description of Mesoplodon Mirum, a Beaked Whale Recently Discovered on the Coast of North Carolina.* Washington, D.C.: U.S. Government Printing Office, 1913.

Turbak, Gary. *America's Great Cats.* Flagstaff, AZ: Northland Publishing, 1986.

——. *Survivors in the Shadows: Threatened and Endangered Mammals of the American West.* Flagstaff, AZ: Northland Publishing, 1993.

Turner, Dennis C., and Patrick Bateson (eds.). *The Domestic Cat: The Biology of its Behaviour.* 1988. Cambridge, UK: Cambridge University Press, 2014 ed.

Turner, William. *The Marine Mammals in the Anatomical Museum of the University of Edinburgh.* London, UK: Macmillan and Co., 1912.

Tuttle, Merlin D. *Texas Bats.* Austin, TX: Bat Conservation International, 2002.

——. *America's Neighborhood Bats.* Austin, TX: University of Texas Press, 2005.

——. *The Secret Lives of Bats: My Adventures with the World's Most Misunderstood Mammals.* New York: Houghton Mifflin Harcourt, 2015.

Twiss, John R., and Randall R. Reeves. *Conservation and Management of Marine Mammals.* Washington, D.C.: Smithsonian Institution Scholarly Press, 1999.

Ulrich, Tom J. *Mammals of the Northern Rockies.* Missoula, MT: Mountain Press, 1986.

Ungar, Peter S. *Mammal Teeth: Origin, Evolution, and Diversity.* Baltimore, MD: Johns Hopkins University Press, 2010.

United States Army Corps of Engineers. *Washington Environmental Atlas.* Seattle, WA: U.S. Government Printing Office, 1975.

——. *Species Profile: Indiana Bat (Myotis sodalis) on Military Installations in the Southeastern United States.* Washington, D.C.: U.S. Government Printing Office, 1998.

United States Department of Agriculture. *North American Fauna.* Washington, D.C.: U.S. Government Printing Office, 1895.

——. *Potash Salts and Other Salines in the Great Basin Region.* Washington, D.C.: U.S. Government Printing Office, 1914.

——. *A Taxonomic Review of the American Long-tailed Shrews.* Washington, D.C.: U.S. Government Printing Office, 1928.

——. *Revision of the American Chipmunks.* Washington, D.C.: U.S. Government Printing Office, 1929.

——. *Mammals of New Mexico.* Washington, D.C.: U.S. Government Printing Office, 1931.

——. *Alaska-Yukon Caribou.* Washington, D.C.: U.S. Government Printing Office, 1935.

——. *Pocket Mice of Washington and Oregon in Relation to Agriculture.* Washington, D.C.: U.S.

Government Printing Office, 1938.

——. *Hart Mountain Antelope Refuge: A National Wildlife Refuge in Oregon*. Washington, D.C.: U.S. Government Printing Office, 1939.

——. *Appraising Farmland: Soil Surveys Can Help You*. Washington, D.C.: U.S. Government Printing Office, 1974.

——. *Pocket Gophers in Forest Ecosystems*. Washington, D.C.: U.S. Government Printing Office, 1983.

——. *Small Mammal Use of a Desert Riparian Island and Its Adjacent Scrub Habitat*. Washington, D.C.: U.S. Government Printing Office, 1987.

——. *Biology of Bats in Douglas-Fir Forests*. Washington, D.C.: U.S. Government Printing Office, 1993.

——. *Small Mammals of the Bitterroot National Forest: A Literature Review and Annotated Bibliography*. Washington, D.C.: U.S. Government Printing Office, 1999.

——. *Roost Tree Selection by Maternal Colonies of Northern Long-eared Myotis in an Intensely Managed Forest*. Washington, D.C.: U.S. Government Printing Office, 2002.

——. *Conservation Assessments for Five Forest Bat Species in the Eastern United States*. Washington, D.C.: U.S. Government Printing Office, 2006.

United States Department of Commerce. *The Status of Endangered Whales: An Overview*. Washington, D.C.: U.S. Government Printing Office, 1984.

——. *Marine Mammal Protection Act of 1972: Annual Report 1983-1984*. Washington, D.C.: U.S. Government Printing Office, 1984.

——. *U.S. Atlantic and Gulf of Mexico Marine Mammal Stock Assessments, 1996*. Washington, D.C.: U.S. Government Printing Office, 1997.

——. *Marine Fisheries Review*. Washington, D.C.: U.S. Government Printing Office, 1998.

United States Department of the Interior. *The Pocket Gophers (Genus Thomomys) of Arizona*. Washington, D.C.: U.S. Government Printing Office, 1947.

——. *Habitat Management Series for Endangered Species*. Washington, D.C.: U.S. Government Printing Office, 1973.

——. *Habitat Suitability Index Models: Swamp Rabbit*. Washington, D.C.: U.S. Government Printing Office, 1985.

——. *Restoring America's Wildlife, 1937-1987*. Washington, D.C.: U.S. Government Printing Office, 1987.

——. *Checklist of Vertebrates of the United States, the U.S. Territories, and Canada*. Washington, D.C.: U.S. Government Printing Office, 1987.

——. *Prevention and Control of Animal Damage to Hydraulic Structures*. Denver, CO: U.S. Government Printing Office, 1991.

——. *Wetlands Stewardship*. Washington, D.C.: U.S. Government Printing Office, 1992.

——. *Wildlife Review* (Issues 246-247). Washington, D.C.: U.S. Government Printing Office, 1995.

——. *Recovery Plan for the Amargosa Vole (Microtus Californicus Scirpensis)*. Washington, D.C.: U.S. Government Printing Office, 1997.

United States Fish and Wildlife Service. *Operation of the National Wildlife Refuge System*. Washington, D.C.: U.S. Government Printing Office, 1975.

——. *Endangered and Threatened Vertebrate Animals and Vascular Plants of Illinois*. Washington, D.C.: U.S. Government Printing Office, 1981.

——. *Habitat Suitability Index Models: Eastern Cottontail*. Washington, D.C.: U.S. Government Printing Office, 1984.

——. *Salt Marsh Harvest Mouse and California Clapper Rail Recovery Plan*. Portland, OR: U.S. Fish and Wildlife Service, 1984.

——. *The Ecology of Peat Bogs of the Glaciated Northeastern United States: A Community Profile*. Washington, D.C.: U.S. Government Printing Office, 1987.

——. *The Ecology of the Lower Colorado River From Davis Dam to the Mexico-United States International Boundary: A Community Profile*. Washington, D.C.: U.S. Government Printing Office, 1988.

——. *Draft Recovery Plan for the Pacific Pocket Mouse (Perognathus longimembris pacificus)*. Washington, D.C.: U.S. Government Printing Office, 1997.

United States Forest Service. *Desert Rodent Abundance in Southern Arizona in Relation to Rainfall*. Washington, D.C.: U.S. Government Printing Office, 1977.

——. *Mammals of the Superior National Forest*. Washington, D.C.: U.S. Government Printing Office, 1981.

United States Marine Mammal Commission. *Annual Report to Congress, 2003*. Washington, D.C.: U.S. Government Printing Office, 2004.

United States National Museum. *The American Bats of the Genera Myotis and Pizonyx*. Washington,

D.C.: U.S. Government Printing Office, 1928.

Valdez, Raul. *The Wild Sheep of the World*. Mesilla, NM: Wild Sheep and Goat International, 1982.

Valdez, Raul, and Paul R. Krausman (eds.). *Mountain Sheep of North America*. Tucson, AZ: University of Arizona Press, 1999.

Van Gelder, Richard George. *Mammals of the National Parks*. Baltimore, MD: Johns Hopkins University Press, 1982.

Van Zyll de Jong, C. G. *Handbook of Canadian Mammals*. 2 vols. Ottawa, Ontario, CAN: Canada Communication Publication Group, 1995.

Vasey, George. *A Monograph of the Genus Bos: The Natural History of Bulls, Bison, and Buffaloes*. London, UK: John Russell Smith, 1851.

Vaughan, Terry A., James M. Ryan, and Nicholas J. Czaplewski. *Mammalogy*. Sudbury, MA: Jones and Bartlett, 2011.

Vernberg, Winona B., and F. John Vernberg. *Environmental Physiology of Marine Animals*. Berlin, Germany: Springer-Verlag, 1972.

Verts, B. J. *The Biology of the Striped Skunk*. Urbana, IL: University of Illinois Press, 1967.

Verts, B. J., and Leslie N. Carraway. *Land Mammals of Oregon*. Berkeley, CA: University of California Press, 1998.

Vizcaino, Sergio F., and W. J. Loughry (eds.). *The Biology of Xenarthra*. Gainesville, FL: University of Florida Press, 2008.

Voigt, Christian C., and Tigga Kingston (eds.). *Bats of the Anthropocene: Conservation of Bats in a Changing World*. New York: Springer International Publishing, 2016.

Vorhies, Charles T., and Walter P. Taylor. *Life History of the Kangaroo Rat*. Washington, D.C.: U.S. Government Printing Office, 1922.

Wainright, Mark. *The Mammals of Costa Rica: A Natural History and Field Guide*. 2002. Ithaca, NY: Cornell University Press, 2007 ed.

Walker, Ernest Pillsbury. *Mammals of the World*. 2 vols. 1968. Baltimore, MD: Johns Hopkins University Press, 1975 ed.

Wallace, David Rains. *The Klamath Knot: Explorations of Myth and Evolution*. San Francisco, CA: Sierra Club Books, 1983.

Wallmo, Olof C. *Mule and Black-Tailed Deer of North America*. Lincoln, NE: University of Nebraska Press, 1981.

Warren, Edward Royal. *The Mammals of Colorado: An Account of the Several Species Found Within the Boundaries of the State, Together With a Record of Their Habits and of Their Distribution*. Norman, OK: University of Oklahoma Press, 1942.

Waterton, Charles. *Essays on Natural History*. London, UK: Frederick Warne and Co., 1870.

Watson, Lyall. *Sea Guide to Whales of the World*. London, UK: Hutchinson, 1981.

Webster, William David, James F. Parnell, and Walter C. Biggs Jr. *Mammals of the Carolinas, Virginia, and Maryland*. Chapel Hill, NC: University of North Carolina Press, 2004.

Weidenreich, Franz. *Apes, Giants, and Man*. Chicago, IL: University of Chicago Press, 1946.

Weigl, Richard. *Longevity of Animals in Captivity: From the Living Collections of the World*. Stuttgart, Germany: Kleine Senckenberg-Reihe, 2005.

Wells-Gosling, Nancy. *Flying Squirrels: Gliders in the Dark*. Washington, D.C.: Smithsonian Institution Press, 1985.

Wells, William Bittle (ed.). *The Pacific Monthly* (July-December 1908). Portland, OR: The Pacific Monthly Company, 1908.

Wheeler, Jill C. *Ghost-faced Bats*. Edina, MN: ABDO Publishing, 2006.

Whitaker, John O., Jr. *The Audubon Society Field Guide to North American Animals*. New York: Alfred A. Knopf, 1980.

Whitaker, John O., and Russell E. Mumford (eds.). *Mammals of Indiana*. Bloomington, IN: Indiana University Press, 2008.

Whitaker, John O., Jr., and William J. Hamilton Jr. *Mammals of the Eastern United States*. 1943. Ithaca, NY: Comstock Publishing, 1998 ed.

Whitehead, G. Kenneth. *The Whitehead Encyclopedia of Deer*. Minneapolis, MN: Voyageur Press, 1993.

Whitehead, Hal. *Sperm Whales: Social Evolution in the Ocean*. Chicago, IL: University of Chicago Press, 2003.

Whitney, William Dwight. *The Century Dictionary: An Encyclopedic Lexicon of the English Language*. 1889. New York: The Century Co., 1911 ed.

Whyte, Adam Gowans. *The Wonder World We Live In*. New York: Alfred A. Knopf, 1921.

Williams, John H. *A Deer Watcher's Field Guide: Whitetails of the Midwest*. Troy, MI: Momentum Books, 1996.

Williams, Wendy. *The Horse: The Epic History of Our Noble Companion.* New York: Scientific American/Farrar, Straus, and Giroux, 2015.
Wilson, Don E., and DeeAnn M. Reeder (eds.). *Mammal Species of the World: A Taxonomic and Geographic Reference.* 2 vols. Baltimore MD: Johns Hopkins University Press, 2005.
Wilson, Don E., and F. Russell Cole. *Common Names of Mammals of the World.* Washington, D.C.: Smithsonian Institution Scholarly Press, 2000.
Wilson, Don E., and Sue Ruff (eds.). *The Smithsonian Book of North American Mammals.* Washington, D.C.: Smithsonian Institution Scholarly Press, 1999.
Wilson, Edward O. *Sociobiology: The New Synthesis.* Cambridge, MA: Belknap Press, 1975.
Wimsatt, William A. (ed.). *Biology of Bats.* 3 vols. New York: Academic Press, 1970.
Wishner, Lawrence Arndt. *Eastern Chipmunks: Secrets of Their Solitary Lives.* Washington, D.C.: Smithsonian Institution Press, 1982.
Wolfe, James L. *Mississippi Land Mammals: Distribution, Identification, Ecological Notes.* Jackson, MS: Mississippi Museum of Natural Science, 1971.
Wood, John George. *The Illustrated Natural History.* London, UK: George Routledge and Sons, 1853.
——. *Natural History Picture Book for Children.* London, UK: Routledge, Warne, and Routledge, 1861.
——. *The Natural History of Man; Being an Account of the Manners and Customs of the Uncivilized Races of Men.* London, UK: George Routledge and Sons, 1868.
Woodburn, Michael O. *Late Cretaceous and Cenozoic Mammals of North America: Biostratigraphy and Geochronology.* New York: Columbia University Press, 2004.
Woodin, Sarah J., and Mick Marquiss (eds.). *Ecology of Arctic Environments.* (Number 13.) London, UK: Blackwell Science, 1997.
Woods, Shirley. *The Squirrels of Canada.* Ontario, CAN: Canada Communication Group, 1999.
World Checklist of Threatened Mammals. Peterborough, UK: Joint Nature Conservation Committee, 1993.
Wright, John (ed.). *A Natural History of the Globe, of Man, of Beasts, Birds, Fishes, Reptiles, Insects, and Plants.* 5 vols. Philadelphia, PA: Thomas DeSilver Jr., 1831.
Wright, Mabel Osgood. *Four-footed Americans and Their Kin.* New York: Macmillan Co., 1898.
Wright, William H. *The Grizzly Bear.* New York: Scribner, 1909.
Wrobel, Murray (ed.). *Elsevier's Dictionary of Mammals.* Amsterdam, The Netherlands, Elsevier, 2007.
Wursig, Bernd, J. G. M Thewissen, and Kit M. Kovacs (eds.). *Encyclopedia of Marine Mammals.* London, UK: Academic Press, 2018.
Wynne, Kate. *Guide to Marine Mammals of Alaska.* Juneau, AK: University of Alaska, 2007.
Yahner, Richard H. *Fascinating Mammals: Conservation and Ecology in the Mid-Eastern States.* Pittsburgh, PA: University of Pittsburgh Press, 2001.
Young, Stanley Paul, and E. A. Goldman. *The Wolves of North America.* Washington, D.C.: American Wildlife Institute, 1944.
——. *The Puma: Mysterious American Cat.* Washington, D.C.: American Wildlife Institute, 1946.
Young, Stanley Paul, and Hartley H. T. Jackson. *The Clever Coyote.* Lincoln, NE: University of Nebraska Press, 1978.
Zachos, Frank E. *Species Concepts in Biology: Historical Development, Theoretical Foundations and Practical Relevance.* Basel, Switzerland: Springer, 2016.
Zachos, Frank E., and Robert J. Asher (eds.). *Mammalian Evolution, Diversity and Systematics.* Berlin, Germany: De Gruyter, 2018.
Zeveloff, Samuel I. *Mammals of the Intermountain West.* Salt Lake City, UT: University of Utah Press, 1988.
Zubaid, Akbar, Gary F. McCracken, and Thomas H. Kunz (eds.). *Functional and Evolutionary Ecology of Bats.* Oxford, UK: Oxford University Press, 2006.

Steller sea lion.

INDEX

American bison.

MEET THE AUTHOR

LOCHLAINN SEABROOK is a lifelong writer as well as a naturalist and an award-winning author and editor of nearly 100 books ranging in topic from nature and science to history and religion. He wrote his first natural history composition at age nine, an essay on the fisher (*Martes pennanti*).

His love of animals, natural history, and the great outdoors has taken him to nearly every corner of the U.S., from working as a hunting guide in the Rocky Mountains to ranch hand on the Great Plains, from farrier assistant in the Northeast to stableman in the Southwest, from wrangler in the Midwest to fish farmer in the Deep South. One of his most memorable experiences was helping deliver a foal at midnight on a freezing spring evening as a snowstorm raged outside the barn.

An eagle scout, a Kentucky Colonel, a nature photographer and videographer, an avid outdoorsman, and a 17th-generation Southerner of Appalachian heritage, the Rocky Mountain resident also worked as a zookeeper at a private wildlife park and sanctuary, where he cared for the following North American animals: birds of prey (owl, hawk, eagle, vulture); reptiles (alligator, snake, turtle); small mammals (mouse, rat, chipmunk); medium-sized mammals (rabbit, skunk, prairie dog, otter); large mammals (bear, deer, horse); canids (fox, coyote); medium-sized felids (bobcat, lynx, margay); large felids (mountain lion, jaguar).

In addition to his monumental volume, *North America's Amazing Mammals: An Encyclopedia for the Whole Family*, Colonel Seabrook is the author of the popular works, *The Concise Book of Owls: A Guide to Nature's Most Mysterious Birds* (endorsed by the World Bird Sanctuary, St. Louis, Missouri), and *The Concise Book of Tigers: A Guide to Nature's Most Remarkable Cats*.

For more info visit

LOCHLAINNSEABROOK.COM

PRESERVING OUR NATURAL WEALTH

All the people of a country have a direct interest in
conservation. For some, as for the commercial fishermen
and trappers, the interest is financial. For others,
successful conservation means preserving a favorite
recreation—hunting, fishing, the study and observation
of wildlife, or nature photography. For others,
contemplation of the color, motion, and beauty of form
in living nature yields esthetic enjoyment of as high an
order as music or painting. But for all the people, the
preservation of wildlife and of wildlife habitat means also
the preservation of the basic resources of the earth,
which men, as well as animals, must have in order to
live. Wildlife, water, forests, grasslands—all are parts of
man's essential environment; the conservation and
effective use of one is impossible except as the others also
are conserved. — RACHEL LOUISE CARSON (1907-1964)

If you enjoyed this book you will be interested in Colonel Seabrook's related titles:

☛ THE CONCISE BOOK OF OWLS: A GUIDE TO NATURE'S MOST MYSTERIOUS BIRDS
☛ THE CONCISE BOOK OF TIGERS: A GUIDE TO NATURE'S MOST REMARKABLE CATS

Available from Sea Raven Press and wherever fine books are sold.

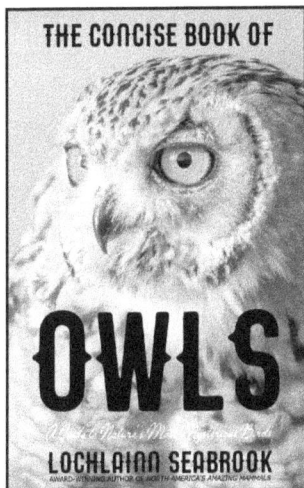

(This book is endorsed by the World Bird Sanctuary, St. Louis, Missouri)

All of our book covers are available as 11" X 17" posters, suitable for framing.

www.ingramcontent.com/pod-product-compliance
Lightning Source LLC
Chambersburg PA
CBHW030728280326
41926CB00086B/559